Mathematical Morphology: 40 Years On

T0135199

Computational Imaging and Vision

Managing Editor

MAX A. VIERGEVER
Utrecht University, Utrecht, The Netherlands

Volume 30

Mathematical Morphology: 40 Years On

Proceedings of the 7th International Symposium on Mathematical Morphology, April 18–20, 2005

Edited by

Christian Ronse

LSIIT, UMR 7005 CNRS-ULP, Illkirch,
France

Laurent Najman

A2SI-ESIEE/IGM, UMR 8049 CNRS-UMLV, Noisy-le-Grand,
France

and

Etienne Decencière

CMM, École des Mines de Paris, Fontainebleau,
France

 Springer

A C.I.P. Catalogue record for this book is available from the Library of Congress.

ISBN-13 978-90-481-6866-8 (PB)
ISBN-13 978-1-4020-3443-5 (e-book)

Published by Springer,
P.O. Box 17, 3300 AA Dordrecht, The Netherlands.

Printed on acid-free paper

*This book is dedicated to the
memory of
Georges Matheron, who
gave the study of
morphology its initial
impetus.*

Contents

Contents

Foreword

Mathematical Morphology (MM) was born in 1964 through the collaboration of Georges Matheron and Jean Serra, who established its basic concepts and tools, coined the name in 1966, and set up in 1968 the "Centre de Morphologie Mathématique" on the Fontainebleau site of the Paris School of Mines.

MM gained a wide recognition after the publication of the three books "Random Sets and Integral Geometry" by G. Matheron (1975), "Image Analysis and Mathematical Morphology" by J. Serra (1982), and "Image Analysis and Mathematical Morphology, Vol. 2: Theoretical Advances" edited by J. Serra (1988). It has now spread worldwide, with active research teams in several countries. This led to the organization of a specific international forum for presenting the most recent advances in the field: the International Symposium on Mathematical Morphology (ISMM). Its first six venues were held in Barcelona (1993), Fontainebleau (1994), Atlanta (1996), Amsterdam (1998), Palo Alto (2000) and Sydney (2002).

In May 2003, on the occasion of a MM day workshop of the GDR ISIS (a French national action linking many laboratories and researchers in image processing) of the CNRS, held in Paris, it was proposed to organize the 7th ISMM in 2005 in Paris, and to make it a celebration of the 40 years of MM. We were pleased by the warm welcome that this proposal met among our colleagues from all countries, so we went ahead and planned the meeting to take place in April 2005 in Paris, just after DGCI'05 (the 12th International Conference on Discrete Geometry for Computer Imagery) in Poitiers. We were lucky to be able to make use of the rooms in the historical buildings of the Paris School of Mines.

We received 62 submissions, from which 41 were accepted for oral presentation, after being reviewed by at least two independent referees. Accepted papers originate from 14 countries: Australia, Austria, Brazil, France, Germany, Greece, Israel, Italy, Japan, Mexico, Netherlands, Portugal, Spain and United Kingdom. Several came from "new" researchers or teams, that is, who had not presented papers at previous ISMM's. We are also honoured by the presence of 3 eminent guest speakers: Jean Serra himself, Jean-Marc Chassery, leader of the GDR ISIS of the CNRS, and member of the steering committee of DGCI, and Olivier Faugeras, leading researcher on computer vision at INRIA, and member of the French Academy of Sciences and of the French Academy of Technologies.

The high quality of the papers, the variety of their topics, and the geographical diversity of their authors, show that MM is a lively research field.

This ISMM is held in honour of Jean Serra, on the occasion of his 65th birthday in March 2005, and these Proceedings are dedicated to the memory of Georges Matheron, who passed away in 2000 at the age of seventy.

Christian Ronse, Laurent Najman and Etienne Decencière

Preface

This 7th ISMM has brought a wide variety of papers, whose topics generally circle around "mainstream morphology", namely considering images as geometrical objects, which are analysed by their interactions with predefined geometrical shapes. However, this circle is ever growing, and is extending towards the frontiers of morphology with neighbouring specialities; we are pleased to welcome a few papers at the interface with other aspects of imaging and computer science. Theory, methodology and practical applications are all represented. It has been difficult to sort the contributions into well-defined categories, and any classification, like the one used to make this book, is somewhat artificial.

Operator design

The design of morphological operators remains an important topic of research. Barrera and Hashimoto show how binary decision diagrams can be used to obtain the decomposition of a local translation-invariant operator in terms of erosions and dilations. Lerallut et al. describe the construction of morphological operators with adaptive translation-varying structuring elements. Beucher extends to grey-level images some set operators like the ultimate erosion or the skeleton by openings. Appleton and Talbot present an efficient algorithm for computing the opening or closing by "approximately straight" segments. Laveau and Bernard combine morphological filters with motion compensation, in order to process video sequences. Then Vidal et al. present a morphological interpolation method for binary images, based on the concept of median set.

Connected filters and reconstruction

About ten years ago emerged the idea that morphological operators can act at the level of flat zones instead of individual pixels; the basic tool for this purpose is the geodesical reconstruction. This led to the concept of connected filters. Ouzounis and Wilkinson propose a tree structure adapted to the implementation of attribute filters in the case where flat zones are based on clustering or contraction connectivity. Terol-Villalobos and Mendiola-Santibañez introduce a modification of geodesic reconstruction, where the latter can be stopped on the basis of size criteria; this is illustrated with applications in image filtering and segmentation. Wilkinson generalizes connected filters thanks to a space-attribute connectivity. Urbach et al. give a variant of attribute filtering, where the attribute is not a scalar, but a vector; this leads to granulometries. Géraud presents a form of Tarjan's union-find algorithm dedicated to connected operators. Buckley and Lagerstrom give an algorithm for reconstructing the labelled branches of an object from its skeleton and the associated quench function. Then Braga-Neto introduces a multi-scale connectivity on grey-level images, based on the connectivity of threshold sets for a varying threshold. Finally Keshet investigates morphological operators on trees of shapes (corresponding to grains and pores), in order to obtain morphological image operators acting in the same way on bright and dark image areas.

Segmentation

At the basis of the morphological approach to segmentation stands the watershed transformation. Since then much work has been done towards building hierarchical groupings of segmentation classes, in particular to reduce oversegmentation. A very interesting paper by Serra investigates colour spaces suitable for segmentation, and shows in particular how one can segment a colour image by combining a luminance-based segmentation in unsaturated areas with a chrominance-based one in saturated areas. Marcotegui and Beucher propose a new kind of hierachical waterfall algorithm, and show how to implement it on the basis of the minimum spanning tree of the neighbourhood graph. Najman et al. investigate the divide set produced by watershed algorithms, and show that they give a separation in the mosaic if and only if the watershed is obtained via a topological thinning. Bertrand proposes a new definition for the dynamics and studies its links with minimum spanning trees. Pratikakis et al. apply hierarchical watershed segmentation for content-based image retrieval. Then Beare gives an algorithm implementing the locally constrained watershed transform.

Geometry and topology

By its concern with shapes, morphology has always been linked with geometrical and topological approaches in image processing. Chassery and Coeurjolly propose new algorithms for the "convex skull", also called "potato peeling problem", that is: construct a largest convex polygon inside a non-convex one. Banon investigates metrics supporting some version of the concept of straight line segment, and the relations between such metrics and morphological operators. Hesselink et al. give an algorithm for constructing a digital skeleton through an Euclidean distance transform. Stelldinger shows how to discretize shapes whose contours have bounded curvature, in such a way as to preserve topology. Decencière and Bilodeau propose an adaptive discretization scheme, suited to the preservation of significant geometrical and topological features of the figure. Meyer studies lexicographic-type metrics that can be used in segmentation, by allowing to partition an image from a set of markers. Finally Bloch et al. apply morphology to the formalization of the spatial relation "X is between Y and Z", in a fuzzy framework.

PDEs and evolutionary models

Partial differential equations (PDEs) govern scale spaces corresponding to the evolution of a filter when the size of its kernel increases. Welk and Weickert provide a theory for such equations for shock filters on one-dimensional signals. Maragos gives a variational interpretation of PDEs for dilation, erosion and levelling.

From the beginning, morphology has sought to relate its operations to probabilistic models of shapes and images. Caciu et al. solve a constrained shape optimisation problem by a combination of genetic algorithms and simulated annealing methods. Spodarev and Schmidt use local connectivity numbers to estimate the Minkowski functionals of realizations of a random closed set.

Texture, colour and multivalued images

Morphology has classically been applied on binary and grey-level images, but it can also be extended to multivalued images, in particular colour ones, and to the analysis of textures. Asano et al. introduce a new characterization of texture based on intersize correlation of grain occurrences. Fletcher and Evans give supervised and unsupervised texture segmentation algorithms, based on the area pattern spectrum. Hanbury et al. investigate the use of the granulometry and the variogram for characterizing texture in colour and greyscale images, and introduce a method for minimising the effect of varying illumination conditions. Angulo proposes a unified framework for morphological processing of colour images in a luminance/saturation/hue representation. Brunner and Soille study the segmentation of multi-channel images by an iterative method based on seeded region growing and quasi-flat zones. Burgeth et al. extend basic morphological operators to tensor-valued images by the use of the Loewner ordering, illustrated on DT-MRI images.

Applications in imaging sciences

The success of an image processing theory is sooner of later verified through its practical applications. Passat et al. segment brain arteries in MRA brain images by an interaction between watershed segmentations on MRI and MRA images. Naegel et al. reconstruct the liver portal vein in CT scan images by grey-level hit-or-miss operators and topological reconstruction. Tek et al. give an algorithm for segmenting red blood cells in microscopic blood images, by using an area granulometry, a minimum area watershed and a circle Radon transformation. Domingo et al. show how to calculate the mean shape of footprint images by use of morphological means. Ramos and Pina analyse images of Portuguese granites by a classification in feature space based on a genetic algorithm and a nearest neighbour rule. Finally Faucon et al. investigate watershed based methods for the segmentation of horizons in seismic images.

This collection is a sample of current research topics in mathematical morphology. It shows the vitality, diversity and maturity of this field, which attracts an ever growing number of researchers and practitioners.

CHRISTIAN RONSE, LAURENT NAJMAN AND ETIENNE DECENCIÈRE

Acknowledgments

The organization of this 7th ISMM and the publication of these proceedings were made possible thanks to the help of many people. We are grateful to Michel Schmitt, deputy director for research at the Paris School of Mines, for putting at our disposal the resources and rooms of the school. We thank also Dominique Deville and Danielle Gozlan for administrative support, Christophe Dietrich for web-system administration, Bruno Restif, Emmanuelle Lafouge and Jocelyne Roger for financial management, as well as Isabelle Bloch and Petr Dokládal for their help in the organization.

Finally we must acknowledge the work of the more than fifty reviewers, who anonymously evaluated and commented the papers. We list them below, and thank them all.

List of reviewers

Vincent Agnus	Jesús Angulo
Akira Asano	Junior Barrera
Gilles Bertrand	Serge Beucher
Michel Bilodeau	Isabelle Bloch
Gunilla Borgefors	Ulisses M. Braga-Neto
Thomas Brox	Michael Buckley
Bernhard Burgeth	Lilian Buzer
Costin Alin Caciu	Michel Couprie
José Crespo	Olivier Cuisenaire
Alain Daurat	Etienne Decencière
Petr Dokládal	Gabriel Fricout
Luis Garrido	John Goutsias
Frédéric Guichard	Marcin Iwanowski
Paul T. Jackway	Dominique Jeulin
Renato Keshet	Ron Kimmel
Sébastien Lefèvre	Petros Maragos
Beatriz Marcotegui	Fernand Meyer
Jean-Michel Morel	Laurent Najman
Haris Papasaika	Ioannis Pitas
Antony T. Popov	Gerhard Ritter
Jos Roerdink	Christian Ronse
Philipe Salembier	Michel Schmitt
Jean Serra	Pierre Soille
Mohamed Tajine	Hugues Talbot
Ivan R. Terol-Villalobos	Jean-Philippe Thiran
Rein van den Boomgaard	Marc Van Droogenbroeck
Luc Vincent	Thomas Walter
Joachim Weickert	Martin Welk

Christian Ronse, Laurent Najman and Etienne Decencière

I

MORPHOLOGICAL OPERATORS

BINARY DECISION DIAGRAMS
AS A NEW PARADIGM
FOR MORPHOLOGICAL MACHINES

Junior Barrera[1] and Ronaldo Fumio Hashimoto [1]

[1]*Departamento de Ciencia da Computacao*
Instituto de Matematica e Estatistica - USP
Rua do Matao, 1010
05508-090 Cidade Universitaria - Sao Paulo - SP - Brasil
jb@ime.usp.br and ronaldo@ime.usp.br

Abstract   Mathematical Morphology (MM) is a general framework for studying mappings
between complete lattices. In particular, mappings between binary images that
are translation invariant and locally defined within a window are of special interest in MM. They are called W-operators. A key aspect of MM is the representation of W-operators in terms of dilations, erosions, intersection, union,
complementation and composition. When W-operators are expressed in this
form, they are called morphological operators. An implementation of this decomposition structure is called morphological machine (MMach). A remarkable
property of this decomposition structure is that it can be represented efficiently
by graphs called Binary Decision Diagrams (BDDs). In this paper, we propose a
new architecture for MMachs that is based on BDDs. We also show that reduced
and ordered BDDs (ROBDDs) are non-ambiguous schemes for representing W-
operators and we present a method to compute them. This procedure can be
applied for the automatic proof of equivalence between morphological operators, since the W-operator they represent are equal if and only if they have the
same ROBDD.

Keywords: Binary Decision Diagram, Morphological Machine, Morphological Language,
Morphological Operator.

1.   Introduction

Mathematical Morphology (MM) is a theory that studies images and signals based on transformations of their shapes. These transformations can be
viewed as mappings between complete lattices [8, 13]. In particular, mappings
between binary images are of special interest in MM and they are called *set
operators*. A central paradigm in MM is the representation of set operators in

C. Ronse et al. (eds.), Mathematical Morphology: 40 Years On, 3–12.

terms of dilations, erosions, union, intersection, complementation and composition. This decomposition structure can be described by a formal language called *morphological language* [2, 3]. Sentences of the morphological language are called *morphological operators* or *morphological expressions*. An implementation of the morphological language is called *morphological machine* (MMach). The first known MMach was the Texture Analyzer, created in the late sixties in Fontainebleau by Serra and Klein. Nowadays, a large number of these machines are available.

The motivation of this work comes from the search for non-ambiguous and compact representation for a large class of operators. It is also desired that this representation leads to efficient algorithms for morphological image processing and that it satisfactorily solves the issue of determining whether two representations are equivalent. In this context, we propose a new architecture for MMachs implemented as software for sequential machines. This new architecture is based on the representation of Boolean functions by Binary Decision Diagrams (BDDs) and was first used in MM in a special algorithm to compute the thinning operator [12].

An important class of set operators is that of W-operators, i.e., set operators that share the properties of translation invariance and local definition within a window. W-operators are extensively used in morphological image processing and this family of operators is the focus of this paper. This paper extends the use of BDD as a representation scheme for the whole class of W-operators.

The class of BDDs studied here (reduced and ordered BDDs) provides a trivial algorithm for determining whether morphological operators are equivalent. Algorithms to convert sentences of the morphological language to this new form of representation are also presented in this work.

2.   Binary Mathematical Morphology

In this section, we recall some basic concepts of binary MM. Let E be a nonempty set and let $\mathcal{P}(E)$ denote the power set of E. Let $\subseteq$ denote the usual set inclusion relation. The pair $(\mathcal{P}(E), \subseteq)$ is a *complete Boolean lattice* [4]. A *set operator* is any mapping defined from $\mathcal{P}(E)$ into itself. The set $\mathbf{\Psi}$ of all set operators inherits the complete lattice structure of $(\mathcal{P}(E), \subseteq)$ by setting $\psi_1 \leq \psi_2 \Leftrightarrow \psi_1(X) \subseteq \psi_2(X), \forall \psi_1, \psi_2 \in \mathbf{\Psi}, \forall X \in \mathcal{P}(E)$. Let $X, Y \in \mathcal{P}(E)$. The operations $X \cup Y$, $X \cap Y$ and $X \backslash Y$, X^c are the usual set operations of union, intersection, difference and complementation, respectively.

Let $(E, +)$ be an *Abelian group* with zero element $o \in E$, called *origin*. Let $h \in E$ and $X, B \subseteq E$. The set X_h, defined by $X_h = \{x + h : x \in X\}$, is the *translation* of X by h. The set $X^t = \{-x : x \in X\}$ is the *transpose* of X. The set operations $X \oplus B = \cup_{b \in B} X_b$ and $X \ominus B = \cap_{b \in B} X_{-b}$ are the *Minkowski addition* and *Minkowski subtraction*, respectively.

A set operator ψ is called *translation invariant* (t.i.) if and only if, $\forall h \in E, \forall X \in \mathcal{P}(E), \psi(X_h) = \psi(X)_h$.

Let W be a finite subset of E. A set operator ψ is called *locally defined* (l.d.) *within a window* W if and only if, $\forall h \in E, \forall X \in \mathcal{P}(E), h \in \psi(X) \Leftrightarrow h \in \psi(X \cap W_h)$.

Let $\boldsymbol{\Psi}_W$ denote the collection of all t.i. operators that are also l.d. within a window W. The elements of $\boldsymbol{\Psi}_W$ are called *W-operators*. The pair $(\boldsymbol{\Psi}_W, \leq)$ constitutes a sub-lattice of the lattice $(\boldsymbol{\Psi}, \leq)$ [3]. Furthermore, $(\boldsymbol{\Psi}_W, \leq)$ is isomorphic to the complete Boolean lattice $(\mathcal{P}(\mathcal{P}(E)), \subseteq)$, since the mapping $\mathcal{K}_W : \boldsymbol{\Psi}_W \to \mathcal{P}(\mathcal{P}(E))$ defined by $\mathcal{K}_W(\psi) = \{X \in \mathcal{P}(W) : o \in \psi(X)\}$, is bijective and preserves the partial order [3]. The set $\mathcal{K}_W(\psi)$ is the *Kernel* of the set operator ψ.

3. Morphological Language

In this section, we recall the main concepts of MM from the viewpoint of a formal language. Let $\psi_1, \psi_2 \in \boldsymbol{\Psi}$. The *supremum* $\psi_1 \vee \psi_2$ and *infimum* $\psi_1 \wedge \psi_2$ operations are defined by $(\psi_1 \vee \psi_2)(X) = \psi_1(X) \cup \psi_2(X)$ and $(\psi_1 \wedge \psi_2)(X) = \psi_1(X) \cap \psi_2(X), \forall X \in \mathcal{P}(E)$. They can be generalized as $(\vee_{i \in I} \psi_i)(X) = \cup_{i \in I} \psi(X)$ and $(\wedge_{i \in I} \psi_i)(X) = \cap_{i \in I} \psi(X)$, where I is a finite set of indices. The composition operator $\psi_2 \psi_1$ is given by $(\psi_2 \psi_1)(X) = \psi_2(\psi_1(X)), \forall X \in \mathcal{P}(E)$.

The set operators $\imath$ and ν defined by $\imath(X) = X$ and $\nu(X) = X^c, \forall X \in \mathcal{P}(E)$, are called, respectively, the *identity* and *negation* operators. These operators are l.d. within the window $\{o\}$. The *dual* of an operator $\psi \in \boldsymbol{\Psi}$, denoted ψ^*, is defined by $\psi^*(X) = (\psi(X^c))^c, \forall X \in \mathcal{P}(E)$. Note that $\psi^* = \nu \psi \nu$.

For any $h \in E$, the set operator τ_h defined by $\tau_h(X) = X_h, \forall X \in \mathcal{P}(E)$, is called *translation operator* by h. This operator is l.d. within the window $\{-h\}$. For a t.i. set operator ψ, $\tau_h \psi = \psi \tau_h$.

Let $B \in \mathcal{P}(E)$ be finite. The t.i. set operators δ_B and ε_B defined by $\delta_B(X) = X \oplus B$ and $\varepsilon(X) = X \ominus B, \forall X \in \mathcal{P}(E)$, are the *dilation* and *erosion* by the *structuring element* B. These set operators are l.d. within the window B^t and B, respectively. One can also write $\delta_B = \vee_{b \in B} \tau_b$ and $\varepsilon_B = \wedge_{b \in B} \tau_{-b}$.

The following proposition, proved by Barrera and Salas [3], gives some properties of W-operators.

PROPOSITION 1 *If ψ, ψ_1 and ψ_2 are set operators l.d. within windows W, W_1 and W_2, respectively, then they have the following properties:* (1) ψ *is l.d. within any window* $W' \supseteq W$; (2) $\psi_1 \wedge \psi_2$ *and* $\psi_1 \vee \psi_2$ *are l.d. within* $W_1 \cup W_2$; (3) $\psi_2 \psi_1$ *is l.d. within* $W_1 \oplus W_2$.

The following corollary gives some properties of W-operators that are particular case of Property 3 of Proposition 1.

COROLLARY 2 *If ψ is a set operator l.d. within a window W, then we have the following properties:* (1) $\forall h \in E$, $\tau_h \psi$ *is l.d. within* W_{-h}; (2) $\forall B \subseteq W$, $\delta_B \psi$ *and* $\varepsilon_B \psi$ *are l.d. within* $W \oplus B^t$ *and* $W \oplus B$, *respectively.* (3) *$\imath\psi$, $\psi\imath$, $\nu\psi$, $\psi\nu$ and ψ^* are l.d. within W.*

Morphological operators (i.e., sentences of the morphological language), are built as strings of elementary operators $(\varepsilon_B, \delta_B, \imath, \nu)$ bound by the operations $\vee, \wedge$ and composition. As an example, let $A, B \in \mathcal{P}(W)$ such that $A \subseteq B$. The collection $[A, B] = \{X \in \mathcal{P}(W) : A \subseteq X \subseteq B\}$ is called an *interval*. The t.i. *sup-generating* operator $\lambda_{A,B}^W$, defined by $\lambda_{A,B}^W(X) = \{x \in E : (X_{-x}) \cap W \in [A, B]\}, \forall X \in \mathcal{P}(E)$, can be described as $\lambda_{A,B}^W = \varepsilon_A \wedge \nu\delta_{\overline{B}^t}$, where $\overline{B} = W \backslash B$. Note that, $\lambda_{A,B}^W$ is l.d. within $A \cup \overline{B}$, and, hence, within W.

Let X be a collection of intervals. An element $[A, B] \in X$ is said to be *maximal* if and only if (iff) no other element of X properly contains it, that is, $[A, B]$ is maximal iff $\forall [A', B'] \in X$, $[A, B] \subseteq [A', B'] \Rightarrow [A, B] = [A', B']$.

The set $\mathcal{B}_W(\psi)$ of all maximal intervals contained in the kernel $\mathcal{K}_W(\psi)$ of a W-operator ψ is called *basis* of ψ [2]. Any W-operator ψ can be written by their canonical sup-decompositions, that is, $\psi = \vee\{\lambda_{A,B}^W : [A, B] \subseteq \mathcal{K}_W(\psi)\}$ or $\psi = \vee\{\lambda_{A,B}^W : [A, B] \in \mathcal{B}_W(\psi)\}$.

As a conclusion of this section, morphological language is complete, in the sense that, any W-operator can be represented by canonical forms, which are valid sentences of morphological language [3]. It is also expressive, since many useful operators can be described with short sentences.

4.   Binary Decision Diagrams

A Boolean function (on n variables) is a function $f : \{0, 1\}^n \rightarrow \{0, 1\}$. This section presents a representation method of Boolean functions called Binary Decision Diagrams (BDDs). In Section 6, we extend the use of BDD as a representation scheme for the whole class of W-operators.

A variety of representation methods for Boolean functions have been developed. In the classical ones, such as truth tables and canonical sum-of-products form, the representation of a Boolean function with n variables has size $O(2^n)$. Other approaches, such as the set of prime implicants (or equivalently, the set of maximal intervals) or a subset of irrendundant ones, yield compact representation for many Boolean functions. Here we describe a graph-based representation method that is very useful for a large class of Boolean functions.

Boolean Functions Represented by BDDs

The representation of a Boolean function by a decision-based structure was introduced by Lee [9] and further popularized by Akers [1] under the name of BDD. Algorithms for BDD manipulation are described in [6]. Efficient implementation directions are found in [5]. Applications of BDDs in Digital Image Processing have been recently developed [10–12].

A BDD of a Boolean function $f : \{0,1\}^n \rightarrow \{0,1\}$ is a rooted, directed acyclic graph (DAG) with two types of nodes: terminal and nonterminal. Each terminal node v has as attribute a value $value(v) \in \{0,1\}$. The nonterminal nodes v have two children nodes, $low(v)$ and $high(v)$. Each nonterminal v is labeled with an input variable index $index(v) \in \{1,2,\ldots,n\}$.

For a given assignment to the input variable vector $\mathbf{x} = (x_1,\ldots,x_n)$, the value of the function is determined by transversing the graph from the root to a terminal node: at each nonterminal node v with $index(v) = i$, if $x_i = 0$, then the arc to $low(v)$ is followed. Otherwise (i.e., $x_i = 1$), the arc to $high(v)$ is followed. The value of the function is given by the value of the terminal node.

A node v in a BDD represents a Boolean function f_v such that: (i) if v is a terminal node with $value(v) = 0$, then $f_v = 0$; (ii) if v is a terminal node with $value(v) = 1$, then $f_v = 1$; (iii) if v is a nonterminal node and $index(v) = i$, then $f_v = \overline{x}_i \cdot f_{low(v)} + x_i \cdot f_{high(v)}$. The mathematical background of the BDD construction is the well known *Shannon expansion* of Boolean functions: $f = x_i \cdot f|_{x_i} + \overline{x}_i \cdot f|_{\overline{x}_i}$. The *restrictions* $f|_{x_i} = f(x_1,\ldots,x_{i-1},1,x_{i+1},\ldots,x_n)$ and $f|_{\overline{x}_i} = f(x_1,\ldots,x_{i-1},0,x_{i+1},\ldots,x_n)$ are the *cofactors* of f with respect to the literal x_i and $\overline{x}_i$.

Reduced and Ordered BDDs

An *ordered* BDD (OBDD) is a BDD such that any path in the graph from the root to a terminal node visits the variables in ascending order of their indices, i.e., $index(v) < index(low(v))$ whenever v and $low(v)$ are nonterminal nodes, and $index(v) < index(high(v))$ whenever v and $high(v)$ are nonterminal nodes. Since a variable appears at most once in each path, a function is evaluated in time $O(n)$ in an OBDD.

If an OBDD contains no pair of nodes $\{u,v\}$ such that the graph rooted by u and v are isomorphic, and if it contains no node v such that $low(v) = high(v)$, it is called a *reduced* OBDD (ROBDD). An OBDD of N vertices can be transformed into an equivalent ROBDD in time $O(N \cdot \log N)$ by the REDUCE algorithm presented in [6]. The following theorem, proved in [6], states the canonicity of the ROBDD representation.

THEOREM 3 *For any Boolean function f, there is a unique ROBDD representing f. Furthermore, any other OBDD representing f contains more vertices.*

The Lattice of ROBDDs and W-operators

Let $\mathcal{F}_n$ denote the set of all functions $f : \{0,1\}^n \rightarrow \{0,1\}$ and let Φ_W denote the set of all functions $\varphi : \mathcal{P}(W) \rightarrow \{0,1\}$, such that $W = \{w_1,\ldots,w_n\}$.

We establish a one-to-one correspondence between elements of $\mathcal{F}_n$ and $\mathbf{\Phi}_W$ by making $f(x_1, \ldots, x_n) = \varphi(\{w_i \in W : x_i = 1\})$. Let $\leq$ denote the usual order in $\{0, 1\}$. This order induces a parcial order in $\mathcal{F}_n$, that is, $f_1, f_2 \in \mathcal{F}_n$, $f_1 \leq f_2 \Leftrightarrow f_1(x) \leq f_2(x), \forall x \in \{0, 1\}^n$.

The partially ordered sets $(\mathcal{F}_n, \leq)$ and $(\mathbf{\Phi}_W, \leq)$ are complete Boolean lattices isomorphic to $(\mathcal{P}(\mathcal{P}(W)), \subseteq)$. This is observed from the bijective mapping $K : \mathbf{\Phi}_W \to \mathcal{P}(\mathcal{P}(W))$ given by $K(\varphi) = \{X : \varphi(X) = 1\}$.

Let $\mathbf{G}_W$ denote the set of all ROBDDs representing functions in $\mathcal{F}_n$. From Theorem 3, ROBDDs are unique and non-ambiguous representations of Boolean functions. Thus, there is a bijective mapping between $\mathbf{G}_W$ and $\mathcal{F}_n$, and, hence, between $\mathbf{G}_W$ and $\mathbf{\Phi}_W$. The ROBDD of a function $\varphi \in \mathbf{\Phi}_W$ is denoted by $G(\varphi)$ and is simply called "the graph of φ". The pair $(\mathbf{G}_W, \sqsubseteq)$ is a complete Boolean lattice isomorphic to the lattice $(\mathbf{\Phi}_W, \leq)$, where the partial order $\sqsubseteq$ between two graphs $G(\varphi_1)$ and $G(\varphi_2)$ in $\mathbf{G}_W$ is defined by $G(\varphi_1) \sqsubseteq G(\varphi_2) \Leftrightarrow \varphi_1 \leq \varphi_2$.

5.   Algorithms for Operations on BDDs

In this section, we present some algorithms on BDDs. First, we show some algorithms for logical operations between graphs and, then, we give an algorithm for translations of BDDs.

Algorithms for Logical Operations between BDDs

Let $\varphi, \varphi_1, \varphi_2 \in \mathbf{\Phi}_W$ and let $G, G_1, G_2 \in \mathbf{G}_W$ be their corresponding graphs. The graph complement, infimum and supremum, respectively denoted by $\overline{\cdot}$, $\sqcap$ and $\sqcup$, are computed by the algorithm APPYOPERATION presented in [6]. This algorithm takes as input two graphs with N_1 and N_2 nodes, respectively, and a logical operation AND, OR, or XOR as parameters, and return the resulting graph in time $O(N_1 \cdot N_2)$. The algorithm is based on the following property of Boolean functions: $f_1 \otimes f_2 = \overline{x}_i \cdot (f_1|_{\overline{x}_i} \otimes f_2|_{\overline{x}_i}) + x_i \cdot (f_1|_{x_i} \otimes f_2|_{x_i})$, where $\otimes$ denotes one of the sixteen logical function of two variables [7]. Thus,

$$\overline{G} = G(\overline{\varphi}) = \text{APPLYOPERATION}(G, 1, \text{XOR})$$
$$G_1 \sqcap G_2 = G(\varphi_1 \cdot \varphi_2) = \text{APPLYOPERATION}(G_1, G_2, \text{AND})$$
$$G_1 \sqcup G_2 = G(\varphi_1 + \varphi_2) = \text{APPLYOPERATION}(G_1, G_2, \text{OR})$$

where 1 is the trivial leaf "1". The graph of the complement of $\varphi \in \mathbf{\Phi}_W$ can be alternatively computed by simply exchanging the values of all terminal nodes. The dual graph of $G \in \mathbf{G}_W$, denoted by G^*, is defined as $G^*(\varphi(X)) = G(\varphi^*(X)) = G(\overline{\varphi(X^c)})$. It can be computed from G by swapping $low(v)$ with $high(v)$ in each node of G and also by swapping the terminal nodes.

Algorithm for Translation of BDDs

Let $h \in E$ and $\varphi \in \Phi_W$. The *translation* of φ by h, denoted by φ_h is defined by $\varphi_h(X) = \varphi(X_h), \forall X \in \mathcal{P}(W_{-h})$. Thus $\varphi_h \in \Phi_{W_{-h}}$ and the Boolean expression of φ_h can be obtained from the Boolean expression of φ by a changing of variables. The *translation* of a graph $G(\varphi) \in \mathbf{G}_W$ by h, denoted by $G(\varphi) + h$, is defined as $G(\varphi) + h = G(\varphi_h)$. Note that $G(\varphi) + h \in \mathbf{G}_{W_{-h}}$. For a trivial implementation of $G + h$, the window $W \cup W_{-h}$ must be consistently ordered, otherwise a reordering of the BDD may be mandatory.

6. Translating Morphological Operators into BDDs

In this section, we present a way to incrementally compute the ROBDD of a W-operator described by a sentence of the morphological language.

We presented (Section 2) that $(\mathbf{\Psi}_W, \leq)$ is isomorphic to $(\mathcal{P}(\mathcal{P}(E)), \subseteq)$, and also (Section 4) that $(\mathbf{\Phi}_W, \leq)$ is isomorphic to $(\mathbf{G}_W, \sqsubseteq)$ and to $(\mathcal{P}(\mathcal{P}(E)), \subseteq)$. This demonstrates, by transitivity of isomorphisms, that there is a one-to-one correspondence between elements of $\mathbf{G}_W$ and $\mathbf{\Phi}_W$. This shows that a graph non-ambiguously and uniquely represents a W-operator.

We denote by $\mathcal{G}_W(\psi)$ the corresponding graph of the W-operator ψ. The calligraphic $\mathcal{G}(\cdot)$ is to stress that the argument is an operator, while $G(\cdot)$ has as argument a function. The subscript W is used in both notations to emphasize the window within which the graph is defined.

The following proposition establishes the algorithms that perform the basic operations on graphs of morphological operators. The proof of this proposition is based on the isomorphisms mentioned in the beginning of this section.

PROPOSITION 4 *Let $h \in E$ and $B \subseteq E$. Let ψ, ψ_1 and ψ_2 be W-operators, respectively, within the windows W, W_1 and W_2. If $\mathcal{G}_W(\psi)$, $\mathcal{G}_{W_1}(\psi_1)$ and $\mathcal{G}_{W_2}(\psi_2)$ are the corresponding graphs to ψ, ψ_1 and ψ_2, respectively, then*

a. $\mathcal{G}_W(\nu\psi) = \overline{\mathcal{G}}_W(\psi)$;

b. $\mathcal{G}_{W_1 \cup W_2}(\psi_1 \wedge \psi_2) = \mathcal{G}_{W_1 \cup W_2}(\psi_1) \sqcap \mathcal{G}_{W_1 \cup W_2}(\psi_2)$;

c. $\mathcal{G}_{W_1 \cup W_2}(\psi_1 \vee \psi_2) = \mathcal{G}_{W_1 \cup W_2}(\psi_1) \sqcup \mathcal{G}_{W_1 \cup W_2}(\psi_2)$;

d. $\mathcal{G}_W(\psi^*) = \mathcal{G}_W^*(\psi)$;

e. $\mathcal{G}_{W_{-h}}(\tau_h\psi) = \mathcal{G}_W(\psi) + h$;

f. $\mathcal{G}_{W \oplus B^t}(\delta_B\psi) = \sqcup_{b \in B}(\mathcal{G}_{W \oplus B^t}(\psi) + b)$;

g. $\mathcal{G}_{W \oplus B}(\varepsilon_B\psi) = \sqcap_{b \in B^t}(\mathcal{G}_{W \oplus B}(\psi) + b)$.

From Proposition 4 and since $\lambda_{A,B}^W = \varepsilon_A \wedge \nu\delta_{\overline{B}^t}$, we have the following corollary.

COROLLARY 5 *Let ψ be a W-operator and $A, B, W' \in \mathcal{P}(E)$ such that $A \subseteq B \subseteq W'$. If $\overline{B} = W \backslash B$ and $\mathcal{G}_W(\psi)$ is the graph corresponding to ψ, then $\mathcal{G}_{W \oplus W'}(\lambda_{A,B}^{W'}\psi) = \mathcal{G}_{W \oplus W'}(\varepsilon_A\psi) \sqcap \overline{\mathcal{G}}_{W \oplus W'}(\delta_{\overline{B}^t}\psi)$.*

DEFINITION 6 *Given a set of indices I, we say that a collection of intervals $\{[A_i, B_i] : i \in I\}$ is an* exact cover *of a W-operator ψ if $\psi = \vee_{i \in I} \lambda^W_{A_i, B_i}$.*

For example, the basis of a W-operator ψ is an exact cover of ψ. And so is the set $\{[A, A] : A \in \mathcal{K}_W(\psi)\}$.

PROPOSITION 7 *Let ψ_1 and ψ_2 be W-operators, respectively, within the windows W_1 and W_2. If $\mathcal{G}_{W_1}(\psi_1)$ and $\mathcal{G}_{W_2}(\psi_2)$ are the corresponding graphs to ψ_1 and ψ_2, respectively, and $\{[A_i, B_i] : i \in I\}$ is an exact cover of ψ_2, then $\mathcal{G}_{W_1 \oplus W_2}(\psi_2\psi_1) = \sqcup_{i \in I}\mathcal{G}_{W_1 \oplus W_2}(\lambda^{W_2}_{A_i, B_i}\psi_1)$.*

It is possible to compute the graph of any W-operator, whenever its description in the morphological language is known. For the incremental computation of the graph of a W-operator ψ described by a morphological expression, we start with the graph of the identity operator and successively apply the algorithms that corresponding to Propositions 4 and 7, according to the parsing of the sentence that describes ψ. Each step is initiated by the modification of the window by applying Proposition 1 and, eventually, Corollary 2.

In the following, we give an example illustrating how to compute the ROBDD representation from a W-operator ψ l.d. within $W = \{w_1, w_2, w_3\}$, where $w_1 = (-1, 0)$, $w_2 = o = (0, 0)$ and $w_3 = (1, 0)$, that detects the vertical borders. There are several ways of representing this operator such as: (i) the standard decomposition $\psi = \lambda^W_{\{w_2\},\{w_1,w_2\}} \vee \lambda^W_{\{w_2\},\{w_2,w_3\}}$; (ii) the morphological operator by means of erosion and dilations $\psi = (\varepsilon_{\{w_2\}} \wedge \nu\delta_{\{w_1\}}) \vee (\varepsilon_{\{w_2\}} \wedge \nu\delta_{\{w_3\}})$; and (iii) the last graph (ROBDD) $\mathcal{G}_W(\psi)$ of Fig. 1. Using the representation by means of erosion and dilations, we have

$$\begin{aligned}\mathcal{G}_W(\psi) &= \mathcal{G}((\varepsilon_{\{w_2\}} \wedge \nu\delta_{\{w_1\}}) \vee (\varepsilon_{\{w_2\}} \wedge \nu\delta_{\{w_3\}})) \\ &= (\mathcal{G}(\varepsilon_{\{w_2\}}) \sqcap \mathcal{G}(\nu\delta_{\{w_1\}})) \sqcup (\mathcal{G}(\varepsilon_{\{w_2\}}) \sqcap \mathcal{G}(\nu\delta_{\{w_3\}})).\end{aligned}$$

Thus, we first have to compute $G_1 = \mathcal{G}(\varepsilon_{\{w_2\}})$, $G_2 = \mathcal{G}(\delta_{\{w_1\}})$, $G_3 = \mathcal{G}(\delta_{\{w_3\}})$, $G_4 = \mathcal{G}(\nu\delta_{\{w_1\}})$ and $G_5 = \mathcal{G}(\nu\delta_{\{w_3\}})$; and then $G = (G_1 \sqcap G_4) \sqcup (G_1 \sqcap G_5)$. Let G_0 be the BDD for the identity operator, that is, $G_0 = \mathcal{G}(\imath)$. Applying the corresponding algorithm, we have

$G_1 = \mathcal{G}(\varepsilon_{\{w_2\}}\imath) = G_0 + (-w_2) = G_0$, by Proposition 4(g).
$G_2 = \mathcal{G}(\delta_{\{w_1\}}\imath) = G_0 + w_1$, by Proposition 4(f).
$G_3 = \mathcal{G}(\delta_{\{w_3\}}\imath) = G_0 + w_3$, by Proposition 4(f).
$G_4 = \mathcal{G}(\nu\delta_{\{w_1\}}\imath) = \overline{G}_2$ and $G_5 = \mathcal{G}(\nu\delta_{\{w_3\}}\imath) = \overline{G}_3$, by Proposition 4(a).
$G_6 = (G_1 \sqcap G_4)$ and $G_7 = (G_1 \sqcap G_5)$, by Proposition 4(b).
$G = G_6 \sqcup G_7$, by Proposition 4(c).

7.   Automatic Proof of Equivalence

An important task in MM is to determine whether two morphological representations correspond to the same operator. Determining the equivalence of

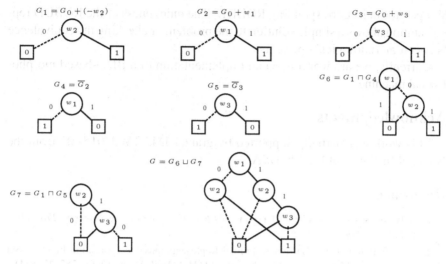

Figure 1. Incrementally computed ROBDD of the vertical border detector.

two morphological operators ψ_1 and ψ_2 involves manipulation of their expressions using the well known set-theoretic properties, and it is often a difficult task. On the other hand, if we know the graphs G_1 and G_2 of two W-operators, then the determination of their equivalence is trivial: since the graph representation of a W-operator is unique, we simply compare if both graphs are equal. Alternatively, one could verify it by calling APPLYOPERATION(G_1, G_2, XOR), and test if the resulting graph if the trivial leaf "0".

8.   Computational Complexity Time

The analysis of several implementations of morphological operators (such as median filters, four-homotopic thinnings, supremum of openings), presented in [10], shows that the proposed architecture is the most efficient alternative to implement MMachs with pixels represented by bytes. Furthermore, in general, it is beneficial to compact parallel operators (i.e., built by union or intersection of several operators). In complex hybrid morphological operators, it may be beneficial to compact just pieces of the operator in BDDs.

9.   Conclusion

In this work we showed that the ROBDD is a good alternative for the representation of W-operators. Its efficient application time makes it a good candidate as the core representation scheme of nonlinear signal and image processors.

The main contribution of this work is the development of a well defined procedure to convert any expression of the morphological language for a given

W-operator in its corresponding ROBDD. The uniqueness of the ROBDD representation allows a simple solution to the problem of checking the equivalence between morphological operators.

Currently, we are working on an implementation of a BDD-based morphological machine.

Acknowledgments

This work was partially supported by grant 1 D43 TW07015- 01 from the National Institutes of Health, USA.

References

[1] S. B. Akers. Binary Decision Diagrams. *IEEE Transactions on Computers*, C-27(6):509–516, June 1978.

[2] G. J. F. Banon and J. Barrera. Minimal Representations for Translation-Invariant Set Mappings by Mathematical Morphology. *SIAM J. Appl. Math.*, 51(6):1782–1798, December 1991.

[3] J. Barrera and G. P. Salas. Set Operations on Closed Intervals and Their Applications to the Automatic Programming of Morphological Machines. *Electronic Imaging*, 5(3):335–352, July 1996.

[4] G. Birkhoff. *Lattice Theory*. American Mathematical Society Colloquium Publications, Rhode Island, 1967.

[5] K. S. Brace, R. L. Rudell, and R. E. Bryant. Efficient Implementation of a BDD Package. In *Proceedings of the ACM/IEEE Design Automation Conference (DAC)*, pages 40–45. ACM/IEEE, 1990.

[6] R. E. Bryant. Graph-Based Algorithms for Boolean Function Manipulation. *IEEE Transactions on Computers*, C-35(8):677–691, August 1986.

[7] G. de Micheli. *Synthesis and Optimization of Digital Circuits*. McGraw-Hill Higher Education, 1994.

[8] H. J. A. M. Heijmans. *Morphological Image Operators*. Academic Press, Boston, 1994.

[9] C.Y. Lee. Representation of Switching Circuits by Binary-Decision Programs. *Bell Systems Technical Journal*, 38:985–999, July 1959.

[10] H. M. F. Madeira, J. Barrera, R. Hirata Jr., and N. S. T. Hirata. A New Paradigm for the Architecture of Morphological Machines: Binary Decision Diagrams. In *SIBGRAPI'99 - XII Brazilian Symposium of Computer Graphic and Image Processing*, pages 283–292. IEEE Computer Society, November 1999.

[11] H. M. F. Madeira and J. Barrera. Incremental Evaluation of BDD-Represented Set Operators. In *SIBGRAPI 2000 - XIII Brazilian Symposium of Computer Graphic and Image Processing*, pages 308–315. IEEE Computer Society, 2000.

[12] L. Robert and G. Malandain. Fast Binary Image Processing Using Binary Decision Diagrams. *Computer Vision and Image Understanding*, 72(1):1–9, October 1998.

[13] J. Serra. *Image Analysis and Mathematical Morphology. Volume 2: Theoretical Advances*. Academic Press, 1988.

IMAGE FILTERING USING MORPHOLOGICAL AMOEBAS

Romain Lerallut, Étienne Decencière and Fernand Meyer
Centre de Morphologie Mathématique, École des Mines de Paris
35 rue Saint-Honoré, 77305 Fontainebleau, France
lerallut@cmm.ensmp.fr

Abstract   This paper presents morphological operators with non-fixed shape kernels, or amoebas, which take into account the image contour variations to adapt their shape. Experiments on grayscale and color images demonstrate that these novel filters outperform classical morphological operations with a fixed, space-invariant structuring element for noise reduction applications.

Keywords: Anisotropic filters, noise reduction, morphological filters, color filters

1.   Introduction

Noise is possibly the most annoying problem in the field of image processing. There are two ways to work around it: either design particularly robust algorithms that can work in noisy environments, or try to eliminate the noise in a first step while losing as little relevant information as possible and consequently use a normally robust algorithm.

There are of course many algorithms that aim at reducing the amount of noise in images. Most are quite effective but also often remove thin elements such as canals or peninsulas. Even worse, they can displace the contours and thus create additional problems in a segmentation application.

In mathematical morphology we often couple one of these noise-reduction filters to a reconstruction filter that attempts to reconstruct only relevant information, such as contours, and not noise. However, a faithful reconstruction can be problematic when the contour itself is corrupted by noise. This can cause great problems in some applications which rely heavily on clean contour surfaces, such as 3D visualization, so a novel approach was proposed.

C. Ronse et al. (eds.), Mathematical Morphology: 40 Years On, 13–22.
©2005 *Springer. Printed in the Netherlands.*

2.   Amoebas: dynamic structuring elements

Principle

Classic filter kernel.   Formally at least, classic filters work on a fixed-size sliding window, be they morphological operators (erosion, dilation) or convolution filters, such as the diffusion by a Gaussian. If the shape of that window does not adapt itself to the content of the image (see figure 1), the results deteriorate. For instance, an isotropic Gaussian diffusion smooths the contours when its kernel steps over a strong gradient area.

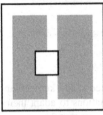 

Figure 1 Closing of an image by a large structuring element. The structuring element does not adapt its shape and merges two distinct objects.

Amoeba filter kernel.   Having made this observation, Perona and Malik [1] (and others after them) have developed anisotropic filters that inhibit diffusion through strong gradients. We were inspired by these examples to define morphological filters whose kernels adapt to the content of the image in order to keep a certain homogeneousness inside each structuring element (see figure 2). The coupling performed between the geometric distance between pixels and the distance between their values has similarities with the work of Tomasi and Manduchi described in [5].

The interest of this approach, compared to the analytical one pioneered by Perona and Malik is that it does not depart greatly from what we use in mathematical morphology, and therefore most of our algorithms can be made to use amoebas with little additional work. Most of the underlying theoretical groundwork for the morphological approach has been described by Jean Serra in his study [2] of structuring functions, although until now it has seen little practical use.

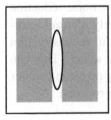 

Figure 2 Closing of an image by an amoeba. The amoeba does not cross the contour and as such preserves even the small canals.

The shape of the amoeba must be computed for each pixel around which it is centered. Figure 3 shows the shape of an amoeba depending on the position of its center. Note that in flat areas such as the center of the disc, or the

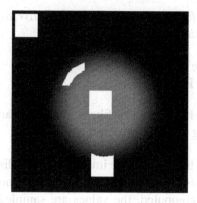

Figure 3 Shape of an amoeba at various positions on an image.

background, the amoeba is maximally stretched, while it is reluctant to cross contour lines.

When an amoeba has been defined, most morphological operators and many other types of filters can be used on it: median, mean, rank filters, erosion, dilation, opening, closing, even more complex algorithms such as reconstruction filters, levelings, floodings, etc.

Construction

Amoeba distance. In general, a filtering kernel of radius r is formally defined on a square (or a hexagon) of that radius, that is to say on the ball of radius r relative to the norm associated to the chosen connectivity. We will keep this definition changing only the norm, using one that takes into account the gradient of the image.

DEFINITION 1 *Let d_{pixel} be a distance defined on the values of the image, for example a difference of gray-value, or a color distance.*

Let $\sigma = (x = x_0, x_1, \dots, x_n = y)$ a path between points x and y. Let λ be a real positive number. The length of the path σ is defined as

$$L(\sigma) = \sum_{i=0}^{n} 1 + \lambda.d_{pixel}(x_i, x_{i+1})$$

The "amoeba distance" with parameter λ is thus defined as:

$$\begin{cases} d_\lambda(x, x) &= 0 \\ d_\lambda(x, y) &= \min_\sigma L(\sigma) \end{cases}$$

It it important to realize that d_{pixel} has no geometrical aspect, it is a distance computed only on the values of the pixels of the image. Furthermore, if n is the number of pixels of a path σ, then $L(\sigma) \geq n$ (since $\lambda \geq 0$), which bounds the maximal extension of the amoeba.

This distance also offers an interesting inclusion property:

PROPERTY 1 *At a radius r given the family of the balls $\mathcal{B}_{\lambda,r}$ relative to the distance d_λ is decreasing (for the inclusion),*

$$0 \leq \lambda_1 \leq \lambda_2 \;\Rightarrow\; \forall(x,y), d_{\lambda_1}(x,y) \leq d_{\lambda_2}(x,y)$$
$$\Rightarrow\; \forall r \in \mathbf{R}^+, \mathcal{B}_{\lambda_1,r} \supset \mathcal{B}_{\lambda_2,r}$$

Which may be useful when building hierarchies of filters, such as a family of alternate sequential filters with strong gradient-preserving properties.

The pilot image.   We have found that the noise in the image can often distort the shape of the amoeba. As such, we often compute the *shape* of the amoeba on another image. Once the shape is computed, the values are sampled on the *original* image and processed by the filter (mean, median, max, min, ...). Usually, the other image is the result of a strong noise removal filtering of the original image that dampens the noise while preserving as much as possible the larger contours. A large Gaussian works fairly well, and can be applied very quickly with advanced algorithms, however we will see below that iterating amoeba filters yields even better results.

3.   Amoebas in practice

Adjunction

Erosions and dilations can easily be defined on amoebas. However it is necessary to use *adjoint* erosions and dilations when using them to define openings and closings:

$$\delta(X) = \bigcup_{x \in X} B_{\lambda,r}(x)$$
$$\epsilon(X) = \{x / B_{\lambda,r}(x) \subset X\}$$

These two operations are at the same time adjoint and relatively easy to compute, contrary to the symmetrical ones that use the transposition, which is not easy to compute for amoebas. See [2] for a discussion of the various forms of adjunction and transposition of structuring functions.

Algorithms

The algorithms used for the erosion and dilation are quite similar to those used with regular structuring elements, with the exception of the step of computing the shape of the amoeba.

Erosion (gray-level):

for each pixel x:

compute the shape of the amoeba centered on x

compute the minimum M of the pixels in the amoeba

set the pixel of the output image at position x to value M

Dilation (gray-level):
for each pixel x:
compute the shape of the amoeba centered on x
for each pixel y of the amoeba:
value(y)=max(value(y),value(x))

The opening using these algorithms can be seen as the gray-level extension of the classic binary algorithm of first taking the centers of the circles that fit inside the shape (erosion), and then returning the union of all those circles (dilation).

Complexity

The theoretical complexity of a simple amoeba-based filter (erosion, dilation, mean, median) can be asymptotically approximated by:

$$T(n, k, op) = O\left[n * \left(op(k^d) + amoeba(k, d)\right)\right]$$

Where n is the number of pixels in the image, d is the dimensionality of the image (usually 2 or 3), k is the maximum radius of the amoeba, $op(k^d)$ is the cost of the operation and $amoeba(k, d)$ is the cost of computing the shape of the amoeba for a given pixel.

The shape of the amoebas is computed by a common region-growing implementation using a priority queue. Depending on the priority queue used, the complexity of this operation is in slightly more than $O(k^d)$ (see [3] and [4] for advanced queueing data structures).

Therefore, for erosion, dilation or mean as operators, we have a complexity of a little more than $O(n * k^d)$ which is the complexity of a filter on a fixed-shape kernel. It has indeed been verified in practice that, while being quite slower than with fixed-shape kernels (especially optimized ones), filters using amoebas tend to follow rather well the predicted complexity, and do not explode (tests have been performed on 3D images, size 512x512x100, with amoebas with sizes up to 21x21x21).

4. Results

Alternate sequential filters

The images of figure 4 compare the differences between alternate sequential filters built on classic fixed shape kernels and ASFs on amoebas in the filtering of the image of a retina.

Median and mean

In the context of image enhancement, we have found that a simple mean or median coupled with an amoeba forms a very powerful noise-reduction filter.

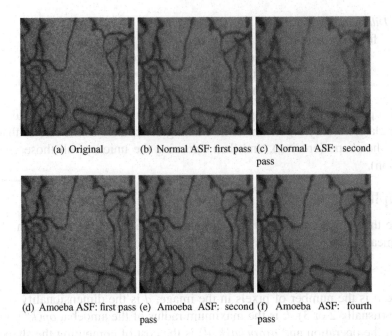

(a) Original (b) Normal ASF: first pass (c) Normal ASF: second pass

(d) Amoeba ASF: first pass (e) Amoeba ASF: second pass (f) Amoeba ASF: fourth pass

Figure 4. Alternate sequential filters on classic kernels and on amoebas. The amoeba preserves extremely well the blood vessels while strongly flattening the other areas.

The images in figure 5 show median and the mean computed on amoebas compared to those built on regular square kernels. The pilot image that drives the shape of the amoeba is the result of a standard Gaussian filter of size 3 on the original image, and the distance d_{pixel} is the absolute difference of gray-levels.

For the filters using amoebas, the median filter preserves well the contour, but the mean filter gives a more "aesthetically pleasing" image. In either case, the results are clearly superior to filterings by fixed-shape kernels, as seen in the figure 5.

Mean and median for color images

In the case of color images, the mean is replaced by the mean on each color component of the RGB color space. For the "median", the point closest to the barycenter is chosen. Other distances or colorspaces can be used, such as increasing the importance of the chrominance information with respect to

(a) Original   (b) Usual median   (c) Amoeba median   (d) Amoeba mean

Figure 5. Results of a "classic" median filtering and two amoeba-based filterings: a median and a mean on Edouard Manet's painting "Le fifre".

luminance, or the other way around, depending on the application, the type of noise and the quality of the color information.

Iteration

The quality of the filtering strongly depends on the image that determines the shape of the amoeba. The previous examples have used the original image filtered by a Gaussian, but this does not always yield good results (also see [6]).

It is frequent indeed that a small detail of the image be excessively smoothed in the pilot image, and thus disappears completely in the result image. On the other hand, noisy pixels may be left untouched if the pilot image does not eliminate them. A possible solution is to somewhat iterate the process, using the first output image not as an input for filtering, as it would commonly be done, but as a new *pilot* image instead.

   (a) Original   (b) Usual median   (c) Amoeba median   (d) Amoeba mean

Figure 6.   Color images: results of a "classic" median filtering, and two amoeba-based filter-ings: a median and a mean. As a simple extension of the grayscale approach, each channel of the pilot image has been independently smoothed by a Gaussian of size 3.

There are two steps at each iteration: the first one follows the scheme de-scribed earlier, using the Gaussian-filtered original image as a pilot, with ag-gressive parameters, and outputs a well-smoothed image in flat areas while preserving as much as possible the most important contours. The second step takes the original image as input and the filtered image as a pilot, with less de-structive parameters, and preserves even more the finer details, while removing a lot of the noise.

In practice, we have found that performing those two steps only once is enough to reduce the noise dramatically (see figure 7), although further itera-tions may be required, depending on the image and the noise.

This method is also very useful for color images, since the amoeba-based pilot image provides better color coupling through the use of an appropriate color distance than simply merging the results of a Gaussian filtering of each channel independently.

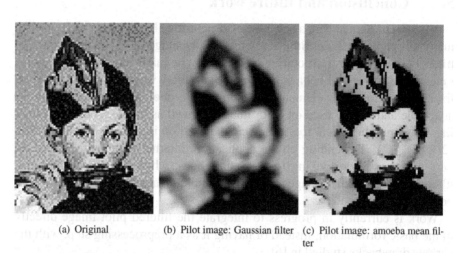

(a) Original   (b) Pilot image: Gaussian filter   (c) Pilot image: amoeba mean filter

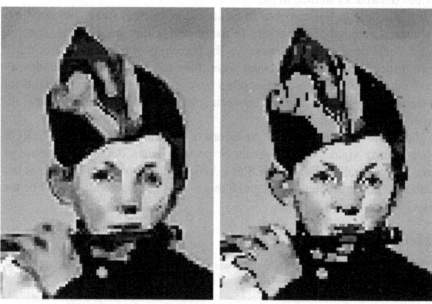

(d) Result image: amoeba mean with Gaussian pilot   (e) Result image: amoeba mean with amoeba pilot

Figure 7. Comparison between two pilot images: a Gaussian one, and one based on a strong amoeba-based filtering. With the amoeba pilot image the hand is better preserved, and the eyebrows do not begin to merge with the eyes, contrary to the Gaussian-based pilot image. Having both less noise and stronger contours in the pilot image also enables the use of smaller values on the lambda parameter so that the amoeba will stretch more in the flatter zones, and thus have a stronger smoothing effect in those zones, while preserving the position and dynamics of the contours

5.   Conclusion and future work

We have presented here a new type of structuring element that can be used in many morphological algorithms. By taking advantage of outside information, filters built upon those structuring elements can be made more robust on noisy images and in general behave in a "more sensible" way than those based on fixed-shape structuring elements. In addition, morphological amoebas are very adaptable and can be used on color images as well as monospectral ones and, like most morphological tools, they can be used on images of any dimension (2D, 3D, ...). Depending on the application, alternate sequential filters are very effective when looking for very flat zones, whereas median and mean filters output smoother images that may be more pleasing to the eye but could be harder to segment.

Work is currently in progress to integrate the filtered pilot image directly in the basic formulation, instead of having it as a preprocessing step, with the various drawbacks studied in [6].

It is possible to use amoebas to create reconstruction filters and floodings that take advantage of the ability to parameterize the shape of the amoebas based on the image content. However, the behaviors of the amoebas are much more difficult to take into account when they are used in such complex algorithms. In particular, amoebas often have a radius larger than one, so for instance the identification made between conditional dilation and geodesic dilation is no longer valid.

The results show that simple extensions of the scalar algorithms to the RGB space already yield excellent results, especially when iterating. The use of more "perceptual" distances (HLS or LAB) would probably prevent most unwanted blending of features, although this is as yet conjectural and will be the basis of further work.

References

[1] Perona, P. and Malik, J., "Scale-space and edge detection using anisotropic diffusion", *IEEE Transactions on Pattern Analysis and Machine Intelligence*, vol. 12, no. 7, July 1990

[2] Serra, J. *et al*, "Mathematical morphology for boolean lattices", *Image analysis and mathematical morphology - Volume 2: Theoretical advances*, Chapter 2, pp 37-46, Academic Press, 1988

[3] Cherkassky, B. V. and Goldberg, A V., "Heap-on-top priority queues", TR 96-042, NEC Research Institute, Princeton, NJ, 1996

[4] Brodnik, A. *et al.*, "Worst case constant time priority queue", Symposium on Discrete Algorithms, 2001, pp 523-528

[5] Tomasi, C. and Manduchi, R.,"Bilateral Filtering for Gray and Color Images", Proceedings of IEEE International Conference on Computer Vision, Bombay, India, 1998

[6] Catté, F. *et al.*, "Image selective smoothing and edge detection by nonlinear diffusion, SIAM J. Numerical Analysis, Vol. 29 No. 1, pp 182-193, February 1992

NUMERICAL RESIDUES

Serge Beucher

Centre de Morphologie Mathématique, Ecole des Mines de Paris
35, rue Saint Honoré,77300 Fontainebleau, France
Serge.Beucher@ensmp.fr

Abstract

Binary morphological transformations based on the residues (ultimate erosion, skeleton by openings, etc.) are extended to functions by means of the transformation definition and of its associated function based on the analysis of the residue evolution in every point of the image. This definition allows to build not only the transformed image itself but also its associated function, indicating the value of the residue index for which this evolution is the most important. These definitions are totally compatible with the existing definitions for sets. Moreover, they have the advantage of supplying effective tools for shape analysis on one hand and, on the other hand , of allowing the definition of new residual transforms together with their associated functions. Two of these numerical residues will be introduced, called respectively ultimate opening and quasi-distance and, through some applications, the interest and efficiency of these operators will be illustrated.

1. Introduction

In binary morphology there are some operators based on the detection of residues of parametric transformations. Among these operators, the ultimate erosion or the skeleton by maximal balls can be quoted. They can more or less easily be extended to greytone images. These extensions are however of little use because it is difficult to exploit them. This paper explains the reasons of this difficulty and proposes a means to obtain interesting information from these transformations. It also introduces new residual transformations and illustrates their use in applications.

2. Binary residues: reminder of their definition

Only operators corresponding to the residues of two primitive transforms will be addressed here. A residual operator θ on a set X is defined by means

23

of two families of transformations (the primitives) depending on a parameter i, $(i \in I)$, ϕ_i and ζ_i, with $\phi_i \geq \zeta_i$. The residue of size i is the set: $r_i = \phi_i / \zeta_i$, the transformation θ is then defined as: $\theta = \cup_{i \in I} r_i$

Usually, ϕ_i is an erosion ϵ_i. According to the choice of ζ_i, we get the different following operators:

- The ultimate erosion [2]; the operator ζ_i is then the elementary opening by reconstruction of the erosion ϵ_i: $\zeta_i = \gamma_{rec}(\epsilon_i)$

- The skeleton by maximal balls; in that case the operator ζ_i is the elementary opening of the erosion of size i: $\zeta_i = \gamma(\epsilon_i)$

Generally a function q, called residual or associated function is linked to these transformations. The support of q is the transformed $\theta(X)$ itself. This function takes in every point x, the value of index i of residue r_i containing point x (or more exactly the value i+1, so that this function is different from zero for r_0). Indeed, in the binary case, if the primitives are correctly chosen, to every point x corresponds a unique residue. One has then:

$$q(x) = i + 1 : \; x \in r_i$$

For the ultimate erosion, this function corresponds to the size of the ultimate components. For the skeleton, it is called *quench function* and corresponds to the size of the maximal ball centered in x.

3.   Extension to greytone images

It is common to read or to hear that these operators can be extended without any problem to the numerical case (greytone images). It is just as remarkable to notice that there is practically no interesting application of these operators in greytone image analysis. Two factors explain this established fact:

- Extension is maybe not "as evident" as it appears to be, for the transformation θ itself but also and especially for the associated function q.
- The amount of information is often too excessive and little relevant, a fact which does not ease the use of these numerical transformations.

Definition of the operator θ in the numerical case

A "simple" definition of θ can be written:

$$\theta = \sup_{i \in I} (\psi_i - \zeta_i)$$

by using the numerical equivalents of the set union and difference operators.

However by doing so, a first problem appears. The subtraction of functions is not really equivalent to the set difference. In the binary case, we had, for a sensible choice of the primitive:

$$\forall i, j \; i \neq j : r_i \cap r_j = \varnothing$$

In the numerical case, it is not true any more. The residues r_i and r_j may have a common support, which entails that $\inf(r_i, r_j) \neq 0$.

It follows that, in the numerical case, the definition of function q associated to transformation θ is not as evident as in the binary case where every r_i has a different support.

Definition of a simplified q function

Let us define a simplified q function by observing the construction of the transformed function θ and the evolution of this construction in every point x of the domain of definition of the initial function f. To do so, let us come back to the design of the transformations in the binary case by replacing set X by its indicator function k_X and by observing how the indicator function $k_{r_i}(x)$ of the residues at point x evolves for each transformation step.

In the case of an indicator function, this evolution is obvious: all $k_{r_j}(x)$ are equal to zero except the indicator function of the residue r_i containing x. It can be written:

$$q(x) = i + 1 \; : \; k_{r_i}(x) \neq 0$$

if we replace the indicator of X by a two-level function (with b<a),

$$f(x) \quad = \quad a \text{ if } x \in X$$
$$f(x) \quad = \quad b \text{ if not}$$

the phenomenon does not change (Figure 1).

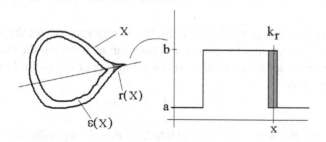

Figure 1.   Residue and its indicator function

Let us take now the case of a general function f. In that case, there are several values of index i for which the difference $r_i(x) = \psi_i(x) - \zeta_i(x)$ at point x is different from zero.

Therefore, we define a residual function q with a value at point x equal to index i for which $r_i(x)$ is positive and maximal.

$$q = \arg\max(r_i) + 1 = \arg\max\left(\psi_i - \zeta_i\right) + 1$$
$$\{q(x) = i + 1 : r_i(x) > 0 \text{ and maximal}\}$$

If this maximum appears for several values of i, only the highest value will be retained:

$$q(x) = \{\max(i) + 1 : r_i((x) > 0 \text{ and maximal}\}$$

The ultimate erosion obtained by applying these definitions is illustrated below (Figure 2).

Initial image	Ultimate erosion	Associated function

Figure 2.   Ultimate erosion for a greytone image

Notice that, when the original image is more or less a two-level one, the result is not very different from the one obtained by using the binary versions of these operators on a thresholded image. The advantage of the approach is to avoid this thresholding step (which in this particular case could be problematic).

Obviously, it is not possible any more to entirely reconstruct the initial image from its skeleton and the associated function as it was in the binary case. One can however define a partial reconstruction $\varrho(f)$ of the initial function f in the following way:

$$\rho(f) = \sup_{x \in E} \left(\theta(x) \oplus B_{q(x)}\right)$$

At every point x a cylinder is implanted, its base being a disc with a radius equal to the value of the associated function in this point and its height being given by the value of the residue at the same point.

4.   New operators

All previous residues are residues of differences of erosions and openings. One can however define many other operators in binary as well as in numerical

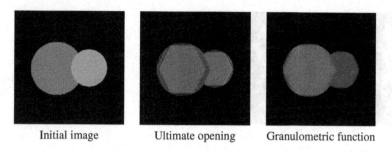

Initial image   Ultimate opening   Granulometric function

Figure 3.   Ultimate opening and granulometric function

cases from different primitive transformations ψ_i and ζ_i . Indeed it is enough that they depend on a parameter i and that they verify the relation $\psi_i \geq \zeta_i$ to be "eligible". However, many of these transformations seem to be of low interest because the obtained results are either too simple, or available by simpler means. Nevertheless, some operators are really interesting. Some, indeed, provide self-evident residues but are far from being uninteresting when the associated function is considered. Others, while presenting low interest in binary, become very useful for greytone images. To illustrate this, let us introduce two new residual operators named respectively *ultimate opening* and *quasi-distance*.

Ultimate opening

Let us consider the residual operator ν where the primitives ψ_i and ζ_i are respectively an opening by balls of size i and an opening by balls of size i+1:

$$
\begin{aligned}
\psi_i &= \gamma_i \\
\zeta_i &= \gamma_{i+1} \\
\nu &= \sup_{i \in I} \left(\gamma_i - \gamma_{i+1} \right)
\end{aligned}
$$

The operator ν does not present any interest in the binary domain. Indeed, in that case, it is easy to show that it is equal to the identity. In the numerical domain, it replaces the initial image by an union of the most significant cylinders included in the sub-graph of the initial function.

The associated function s, even in the binary case, presents a bigger interest. In every point x, s(x) is equal to the size of the biggest disk covering this point x (binary case) or to the radius of the biggest significant cylinder of the partial reconstruction covering (numerical case)(Figure 3). Function s is called *granulometric function*.

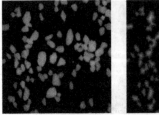
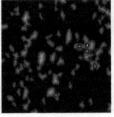

Figure   4 Quasi-distance (right) of the initial image (left)

Quasi-distance

In the previous definition, openings can be replaced by erosions. A new residual operator τ is then defined, which interest lies also in its associated function.

$$\psi_i = \epsilon_i$$
$$\zeta_i = \epsilon_{i+1}$$
$$\tau = \sup_{i \in I} (\epsilon_i - \epsilon_{i+1})$$

In the binary case, this operator does not present any interest because it is equal to the identity and its associated function d is nothing else than the distance function.

In the numerical case, the physical interpretation of the residue itself is not very explicit. The associated function d is more interesting: it is very close to a distance function calculated on the significant flat or almost flat zones of the initial function. By significant, one means a zone corresponding to an important variation of the erosion.

Figure 4 shows this transformation applied to an almost two-level image. Even on this relatively simple image, one notices in certain places the appearance of rather high values of this quasi-distance. These values come from the erosion of relatively flat zones which appear when zones above have been eroded and have disappeared (Figure 5). They correspond to "hung up" distance functions. When the initial function is arranged in terraces (flat zones which are not extrema), its quasi-distance is not symmetric on the flat zones which do not correspond to maxima.

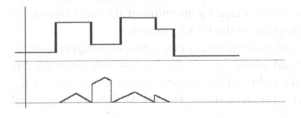

Figure 5 Multi-level function and its quasi distance

Different strategies can be used to correct this phenomenon. One promising technique consists in looking for the zones where the quasi-distance is not 1-lipschitzian and to correct these zones by an iterative approach.

Corrected quasi-distance

A classical distance function d is 1-lipschitzian. It means that, given two points x and y, the following relation holds:

$$|d\left(x\right) - d\left(y\right)| \leq \|x - y\|$$

In particular, when x and y are two adjacent digital points, their distance is at the most equal to 1. It is obviously not the case for the quasi-distance due to the "hung up" distances and to the non symmetric distances on some plateaus. It is however possible to force this quasi-distance to be 1-lipschitzian by means of an iterative procedure of "descent of hung-up distances ". It consists in subtracting from the function d distances larger than 1 between a point and its neighbours (Figure 6).

- For any point x where $[d - \epsilon\left(d\right)]\left(x\right) > 1$, do $d\left(x\right) = \epsilon\left(d\right)\left(x\right) + 1$
- The procedure is iterated until idempotence.

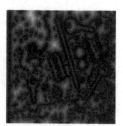

Figure 6. Quasi-distance before and after correction.

Applications

In the same way as the residues were used in set segmentation, their numerical versions as well as the new residues described above constitute remarkable tools of granulometric description and of markers generation for segmentation. To illustrate the potentialities of these transformations, let us present two applications in greytone segmentation.

Size distribution and segmentation of blocks

An application of granulometric functions consists in defining real size distributions of objects in an image without the necessity of extracting them beforehand. Furthermore, this size distribution is certainly much closer to the real size distribution of the analyzed objects than the size distribution obtained by successive openings of the image.

Figure 7a represents a heap of rocks. A granulometric function can be built, associated to the ultimate opening (Figure 7b, openings are isotropic). As the value of every pixel corresponds to the size of the biggest opening which contains this pixel, the histogram of the granulometric function (Figure 8) produces a size distribution curve very close to the real size distribution of blocks (at least in 2D).

(a)   (b)   (c)

Figure 7.   Blocks of rocks (CGES/ENSMP): a) original image; b) Granulometric function c) Blocks marking .

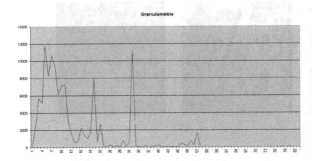

Figure 8   Size distribution of blocks (histogram of the granulometric image)

Granulometric functions can also be used to mark blocks. Then, markers can be counted or used to perform segmentations of the image by watersheds. The generation of these markers is made by performing on every threshold of the granulometric function an erosion of a size proportional to the threshold value. Figure 7c illustrates this algorithm.

Image segmentation

The second application will use quasi-distances. This application is only a sketch of the possibilities offered by this type of tool.

One saw previously that the quasi-distance allows to build a distance function for the relatively flat and relevant zones of a greytone image. This property is used here to exhibit the markers of these regions. Then these markers can

Figure 9 Use of quasi-distances in image segmentation.

be used to control the watershed transformation of the quasi-distance in order to segment the homogeneous regions of the image. The various steps of the algorithm are the following (Figure 9):

- Computation of the quasi-distance of the initial image f.
- Image inversion and computation of the quasi-distance of f^c.
- Supremum of the two quasi-distances.
- A threshold of this new function at a given level i allows the extraction of the homogeneous regions of the image of size larger than I.
- Computation of the watershed transform of the supremum controlled by the previous markers.

The calculation of the quasi-distance of the inverted image is compulsory to exhibit the dark regions which can correspond to minima of the image. One saw previously that, in that case, quasi-distances are either equal to zero or "hung up". The calculation of this quasi-distance after inversion allows to take into account the real sizes of these structures. Notice also that this segmentation does not use the image gradient.

5. Conclusions

The definition of numerical transformations based on residues, not only provides the extensions of efficient tools in binary and numerical morphology but, furthermore, allows to introduce new operators whose potentialities are enormous. The importance of the doublet constituted by the transformation and by its associated function has also been emphasized, this last one being sometimes more interesting in numerical morphology than in binary one.

The extension of these notions and especially the definition of a simplified associated function were possible by changing our point of view: rather than to focus on the neighborhood relationships between the image points, we point out the modifications which occur vertically in a point as we "unwind" the transformation, the most significant changes and especially the moment when they occur constituting the core of information provided by these operators.

The applications presented as illustrations of the potentialities of the granulometric functions and of the quasi-distances still deserve additional developments. However, the efficiency of these operators can already be verified and a large number of tracks of future applications can be considered.

The granulometric function is a powerful segmentation and filtering tool. By associating every point of the image to the size of the highest cylinder included in the sub-graph, it allows ipso facto to adapt the size of the filters which are applied in each of its thresholds. It is also possible to eliminate too deeply covered components or, on the contrary, to extract non covered blocks. This capability is interesting in numerous applications where objects appear in heap and where random sets models ("dead leaves" models notably) are used. The topology of every threshold of the granulometric function and in particular the presence of holes is very important. These holes indicate generally the presence of superimposed structures. This constitutes an important tool for describing stacked structures.

The quasi-distance is the missing link between sets distance functions and a tool allowing the direct extraction of dimensional information on the homogeneous regions in greytone images. The efficiency of the distance function to generate segmentation markers is well known in the binary case. Quasi-distance allows to extend this capability to greytone images. In fact, this operator performs many tasks at the same time: it is a filter which equalizes the homogeneous regions of the image; it quantifies the size of these homogeneous regions and finally, it enhances the most contrasted regions in the image, in a similar way a waterfalls algorithm acts. It is not so surprising that segmentations obtained with this operator are very close to those provided by the hierarchical segmentation by waterfalls. However, while the waterfalls algorithm proceeds by grouping regions, the use of quasi-distance leads directly to a similar result. One can say that, whereas the approach by waterfalls is a "bottom-up" approach, quasi-distance supplies at once a "top-down" hierarchical organization [2].

6.   References

[1] BEUCHER Serge: Watershed, hierarchical segmentation and waterfall algorithm. Proc. Mathematical Morphology and its Applications to Image Processing, Fontainebleau, Sept. 1994, Jean Serra and Pierre Soille (Eds.), Kluwer Ac. Publ., Nld, 1994, pp. 69-76.

[2] BEUCHER Serge & LANTUEJOUL Christian: On the use of the geodesic metric in image analysis. Journal of Microscopy, Vol. 121, Part 1, January 1981, pp. 39-49.

EFFICIENT PATH OPENINGS AND CLOSINGS

Ben Appleton[1] and Hugues Talbot[2]

[1]*School of Information Technology and Electrical Engineering,*
The University of Queensland
Brisbane, Australia
appleton@itee.uq.edu.au

[2]*CSIRO Mathematical and Information Sciences*
Locked Bag 17, North Ryde NSW 2113 Australia
Hugues.Talbot@csiro.au

To Henk.

Abstract   Path openings and closings are algebraic morphological operators using families of thin and oriented structuring elements that are not necessarily perfectly straight. These operators are naturally translation invariant and can be used in filtering applications instead of operators based on the more standard families of straight line structuring elements. They give similar results to area or attribute-based operators but with more flexibility in the constraints.

Trivial implementations of this idea using actual suprema or infima of morphological operators with paths as structuring elements would imply exponential complexity. Fortunately a linear complexity algorithm exists in the literature, which has similar running times as an efficient implementation of algebraic operators using straight lines as structuring elements.

However even this implementation is sometimes not fast enough, leading practitioners to favour some attribute-based operators instead, which in some applications is not optimal.

In this paper we propose an implementation of path-based morphological operators which is shown experimentally to exhibit logarithmic complexity and comparable computing times with those of attribute-based operators.

Keywords: Algebraic morphological operators, attributes, complexity.

Introduction

Many problems in image analysis involve oriented, thin, line-like objects, for example fibres [22, 20], hair [19, 15], blood vessels [8], grid lines on stamped metal pieces [21] and others.

For bright and elongated structures, a common approach for detecting these features is to use to use an infimum of openings using lines as structuring el-

33

C. Ronse et al. (eds.), Mathematical Morphology: 40 Years On, 33–42.
©2005 *Springer. Printed in the Netherlands.*

ements oriented in many directions [10]. The result is an isotropic operator if the line structuring element lengths are adjusted to be independent of orientation [11]. Recursive implementations of openings at arbitrary angles have been proposed and yield linear-time algorithms [17] with respect to the length of the structuring elements. If desired features are very thin, a translation-invariant algorithm should be used [18].

Area and attributes openings [1, 12, 23] are also often used for the analysis of thin structures. An area opening of parameter λ is equivalent to the supremum of all the openings by connected structuring elements of area λ. Obviously this includes all the straight line structuring elements of this length. Practitioners often note that using only straight line structuring elements removes too much of the desired features, while using area operators does not allow them to distinguish between long and narrow features on the one hand, and short compact ones on the other. It is sometimes, but not always possible to combine these operators, or to use morphological reconstruction to obtain the desired outcome.

Recently efficient morphological operators equivalent to using families of narrow, elongated but not necessarily perfectly straight structuring elements were proposed in [3] and [6], together with an algorithm for computing the operator with linear complexity with regards to the length of the structuring elements. These path operators constitute a useful medium between operators using only straight lines and those using area or other attributes.

In the remainder we propose a significantly faster algorithm for implementing path operators, with observed logarithmic complexity with respect to the length of the structuring elements.

1.   Path-based morphological opening

The theory of path openings is explained in detail in [6] and in a shorter fashion in [5]. We only summarize the main points here.

Adjacency and paths

Let E be a discrete 2-D image domain, a subset of $\mathbb{Z}^2$. Then $\mathcal{B} = \mathcal{P}(E) = 2^E$ is the set of binary images and $\mathcal{G} = \mathcal{T}^E$ the space of grey-scale functions, where $\mathcal{T}$ is the set of grey values. We assume E is endowed with an adjacency relation $x \mapsto y$ meaning that there is a directed edge going from x to y. Using the adjacency relation we can define the dilation $\delta(\{x\}) = \{y \in E, x \mapsto y\}$. The L-tuple $\boldsymbol{a} = (a_1, a_2, ..., a_L)$ is called a δ-*path of length* L if $a_{k+1} \in \delta(\{a_k\})$ for $k = 1, 2, ..., L - 1$. Given a path $\boldsymbol{a}$ in E, we denote by $\sigma(\boldsymbol{a})$ the set of its elements, i.e: $\sigma(a_1, a_2, ..., a_L) = \{a_1, a_2, ..., a_L\}$. We denote the set of all δ-paths of length L by Π_L, and the set of δ-paths of length L contained in a subset X of E is denoted by $\Pi_L(X)$.

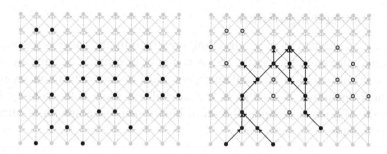

Figure 1. A set $X \subseteq E$ (black points on the left) and its opening $\alpha_6(X)$ (black points on the right). Unfilled points on the right have been discarded. The underlying adjacency graph is in light grey.

Path openings

We define the operator $\alpha_L(X)$ as the union of all δ-paths of length L contained in X:

$$\alpha_L(X) = \bigcup \{\sigma(\boldsymbol{a}), \boldsymbol{a} \in \Pi_L(X)\} \qquad (1)$$

It is easy to establish that α_L has all the properties of an opening. Figure 1 offers an illustration. For an adjacency graph similar to that of this figure, and for an unbounded image, there are 3^{L-1} distinct paths of length L starting from any point. The path opening α_L is in fact the supremum of the morphological opening using these paths as structuring elements, which would suggest an inefficient way to compute the operator. Fortunately [6] proposes a useful recursive decomposition which allows the operator α_L to be computed in linear time with respect to L (not presented here).

Grey-level operator and practical considerations

The binary operator defined above extends to the grey-level domain in the usual way by replacing the union with a supremum. The recursive decomposition in [6] also extends to the grey-level domain.

The current definition of a path opening with an adjacency graph such as in Fig. 1 is not sufficiently useful in a context where features are distributed isotropically : only paths generally oriented North – South will be preserved by the opening. We need to take a supremum with openings using adjacency graphs oriented East – West, North-East – South-West and North-West – South-East. More complex adjacency graphs can also be devised for more constrained path operators, however in the remainder we assume this basic scheme.

2.   Ordered algorithm

The grayscale path opening algorithm presented in this paper is based on a few simple observations. Firstly, the principle of threshold decomposition allows the construction of grayscale morphological operators from binary morphological operators. Secondly, in the case of grayscale path openings it is possible to efficiently compute the set of binary path openings for all thresholds in sequence.

Threshold decomposition

Here we equivalently redefine binary images as functions of the form $b : E \to \{\textbf{false}, \textbf{true}\}$ rather than subsets of the image domain E. Then, given a grayscale image $g \in G$, a threshold operator $T_t : \mathcal{G} \to \mathcal{B}$ with threshold t, and a binary opening $\gamma_B : \mathcal{B} \to \mathcal{B}$, there exists a grayscale opening $\gamma_G : \mathcal{G} \to \mathcal{G}$ such that for all thresholds t we have $T_t \circ \gamma_G = \gamma_B \circ T_t$, where $\circ$ is the composition operator.

This grayscale opening $\gamma_G(g)$ may be constructed explicitly by 'stacking' the results of the binary opening applied to each threshold of the original image. This stacking assigns to each pixel p the highest threshold t for which the binary opening $\gamma_B \circ T_t(g)$ remains **true**.

Updating binary path openings

The second observation is that the binary images produced in this construction tend to vary little between sequential thresholds. In the case of path openings we will show how to efficiently update the result of the binary opening $\gamma_B \circ T_t(g)$ from the result of the binary opening at the previous threshold $\gamma_B \circ T_{t-1}(g)$.

For brevity we here describe only the case of North – South paths. In this case, for a binary image b we store at each pixel p two values: the length $\lambda^-[p]$ (not including p itself) of the longest path travelling upward from pixel p, and the length $\lambda^+[p]$ of the longest path travelling downward from pixel p. Then the length of the longest path passing through pixel p (where $b[p] = \textbf{true}$) is $\lambda[p] = \lambda^-[p] + \lambda^+[p] + 1$. If $b[p] = \textbf{false}$ then we define $\lambda[p] = 0$. The recursive computation of λ^- and λ^+ is described in more detail in [6]. Here, in short, we may state that in the North – South case, if we denote $p = (p^1, p^2)$:

$$\lambda^-[p] = 1 + \max(\lambda^-[(p^1-1,p^2+1)], \lambda^-[(p^1,p^2+1)], \lambda^-[(p^1+1,p^2+1)]) \quad (2)$$
$$\lambda^+[p] = 1 + \max(\lambda^+[(p^1-1,p^2-1)], \lambda^+[(p^1,p^2-1)], \lambda^+[(p^1+1,p^2-1)]) \quad (3)$$

where $b[p] = \textbf{true}$ and 0 otherwise. Note that this allows us to easily compute the *opening transform* for a given threshold, i.e. the operator which at each pixel associates the length of the longest path going through that pixel.

In order to update the binary opening $\gamma_B \circ T_t(g)$ given the result from the previous threshold $\gamma_B \circ T_{t-1}(g)$, we must compute the new binary opening transform λ and hence λ^- and λ^+. Rather than recomputing these from the image $b = T_t(g)$, we may compute the changes to λ^- and λ^+ due solely to the pixels which made the transition from **true** to **false** between $T_{t-1}(g)$ and $T_t(g)$. This is performed in the following steps:

- Initialisation:
    - Set all pixels with $g[p] = t$ to *active* and enqueue
- For each row from top to bottom:
    - For all *active* pixels p in this row:
        * Recompute $\lambda^-[p]$ according to Equation 2
        * If $\lambda^-[p]$ changed, set as *active* and enqueue the dependent pixels $(p^1 - 1, p^2 + 1), (p^1, p^2 + 1), (p^1 + 1, p^2 + 1)$
- For each row from bottom to top:
    - For all *active* pixels p in this row:
        * Recompute $\lambda^+[p]$ according to Equation 3
        * If $\lambda^+[p]$ changed, set as *active* and enqueue the dependent pixels $(p^1 - 1, p^2 - 1), (p^1, p^2 - 1), (p^1 + 1, p^2 - 1)$

The queueing system for *active* pixels consists of a first-in-first-out (FIFO) queue for each row as well as a queue of rows which contain active pixels. This queueing system is necessary to comply with the dependencies in Equations 2 and 3 and also avoids inefficiently scanning the entire image.

Recursive ordered path opening

Here we present an algorithm to compute a grayscale path opening. L denotes the desired path length. Note that, as we are interested in the specific path length L, path lengths λ^-, λ^+ greater than $L - 1$ may be treated as equal to $L - 1$ in Algorithm 2. This limits the propagation of changes to the binary opening transform and hence improves the efficiency of the grayscale path opening.

- Initialisation:
    - Sort the pixels by their intensities
    - Set $b[p] = $ **true** for all pixels p
    - Compute λ^+, λ^- from b.
- For each threshold t in $\mathcal{T}$:
    - Using Algorithm 2, update λ^-, λ^+ for the new threshold
    - For all *active* pixels p whose path length $\lambda[p]$ consequently dropped below L, set $\gamma_G(g)[p] = t$

Sorting the pixels by their intensities is a necessary preprocessing step in order to efficiently locate the pixels whose threshold changes in the step $t-1 \rightarrow$

t. For integer data a linear-time sorting algorithm such as the Radix sort is recommended [9]. Alternatively a suitable priority queue data structure [2] can be used.

A simple heuristic has been found to further improve the efficiency of this algorithm in practice. When the maximal path length of a pixel p drops below L, it cannot contribute to a path of length L or greater at any further threshold. Therefore we may remove this pixel from further consideration by setting $b[p] = \textbf{false}$. We refer to this as the *length heuristic* in the remainder of this paper. We believe that the average running time of this algorithm is $O(N \log L)$ on images containing N pixels. However the formal derivation of this average running time would require the selection of an appropriate stochastic image model and is not pursued in this paper.

3.   Grayscale opening transform

The algorithm presented in Section 2 may be extended in a simple manner to compute the grayscale path opening transform, i.e. the operator which at each pixel associates the length of the longest path going through that pixel.

In the course of Algorithm 2, the path opening transforms for all binary thresholds were computed in sequence. Instead of discarding these intermediate results we may store them in compressed form allowing them to be queried at a later point. At each threshold, those *active* pixels whose maximal path length $\lambda[p]$ has decreased store a point $(t, \lambda[p])$ in a linked list. This linked list is monotonically increasing in t and monotonically decreasing in $\lambda[p]$. Once computed, we may query this structure with any desired path length to extract the associated grayscale path opening.

- Initialisation: As per Algorithm 2
- For each threshold t in $\mathcal{T}$:
  - Using Algorithm 2, update λ^-, λ^+ for the new threshold.
  - For all *active* pixels p whose path length $\lambda[p]$ decreased, append the node $(t, \lambda[p])$ to the linked list at pixel p.

This algorithm requires the same order of computation as Algorithm 2, that is $O(N \log L)$. The number of linked list nodes generated in Algorithm 3 must be less than the number of operations in Algorithm 2, and therefore the average memory required by Algorithm 3 is $O(N \log L)$.

4.   Results

Figure 2 illustrates the usefulness of path operators. On this toy example we wish to eliminate the compact round object and retain the line-like features. An area opening is not sufficient in this case because the compact round noise is too big and one feature is eliminated before the noise as the parameter increases. Conversely the supremum of openings by lines suppresses features that are not perfectly straight. On the other hand the path opening delivers the expected result. Note that path openings do not afford control over the thickness of detected paths, both the thin wavy line and the thicker straight line are detected. Path of various thickness be separated using standard top-hats for example.

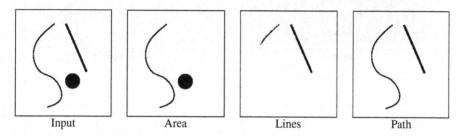

| Input | Area | Lines | Path |

Figure 2. Toy example: On the input we wish to retain the line-like features while eliminating compact noise. Only the path opening works in this case.

A more realistic example is shown in Fig. 3. We wish to detect the small thin fibres in this electron micrograph present at the bottom of this image. The large fibres are detected by a different method [20] which is of no interest here. The thin fibres are present on a noisy background which requires some filtering. A supremum of openings by lines is too crude here (result not shown). An area opening does not eliminate enough of the noise, but a path opening works as expected.

Discussion

Results obtained by path openings depend greatly on the adjacency graph. It is for example possible to define narrower or wider cones for paths as discussed in [6]. Using more narrower cones one can define paths that are increasingly similar to lines, and obtain results similar to regular openings by line structuring elements. Using fewer wider cones one can obtain results more similar to area openings.

In the results above we used 90 degree angle cones which are easy to implement and seem to strike a good balance between area openings and line-based openings.

Timings

Table 1 shows the running times of the proposed algorithm compared with various alternatives. We observe that the proposed ordered path opening implementation has a running time approximately logarithmic (plus a constant) with respect to L, while both the recursive path opening and the supremum of openings by lines have approximately linear running times. The individual

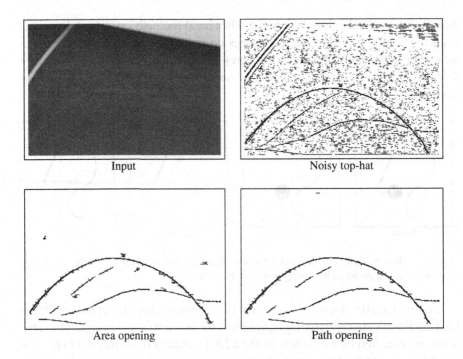

Figure 3. Electron micrograph of glass fibres: to detect the small thin fibres in the bottom of the image, a white top-hat is useful but noisy. When this top-hat image is filtered by an area opening some compact noise remain while a path opening yields a better result.

openings by lines in the latter algorithm are all running in constant time irrespective of L, but for larger L more orientations need to be explored. Note also that the presented algorithm for the supremum of opening by lines is not the translation-invariant implementation, which would be slower still. The area opening algorithm seems to converge to a constant-time algorithm with low constant. The area parameter was simply L, although $k \times L$ with k small and constant (e.g: 3) could have been chosen without significantly affecting the result.

Memory demands for these algorithms are all low except the recursive path opening implementation which requires an amount of memory proportional to LN, with N the number of pixels in the image.

We observe that the area opening is the fastest algorithm, but that the presented path opening algorithm comes second and significantly faster than the other two algorithms for most useful values of L.

Table 1. Comparison of algorithm running times. From left to right the columns are the or-dererd path opening presented in this paper, the recursive path opening of Heijmans et. al, the supremum of openings by lines and the area opening. Timings are in seconds, image was $560 \times 510 \times 8$-bit. Processor was a single Pentium IV 1.7GHz.

L	Ordered PO	Recursive PO	Supremum Lines	Area
1	0.56	0.08	0.14	0.13
5	0.69	0.54	0.65	0.17
10	0.73	1.16	1.38	0.17
50	0.90	14.24	3.29	0.21
100	0.93	30.74	11.43	0.22

5. Conclusion and future work

We have presented a new, ordered implementation of the path opening and closing operators. This operator is identical to the supremum (resp. infimum) of openings (resp. closing) by a family of structuring elements described as oriented paths. The family of paths is of exponential size with respect to their length L, but we have proposed a new implementation with experimental logarithmic complexity with respect to L, improving on a known recursive implementation except for very small L. The proposed method is also significantly faster than the usual operator by unions/intersection of lines s tructuring elements used for the study of oriented structures.

Area operators are still faster than the proposed algorithm, by a nearly constant factor of 4 to 5. However, the new algorithm is fast enough for many applications and can be used in cases where using an area or attribute operator is not appropriate, e.g. in the presence of sufficiently large compact noise.

The path operators are intuitive, translation-invariant methods useful for the analysis of thin, elongated but not necessarily perfectly straight structures.

Future work will include incomplete path openings, i.e. paths which are not necessarily connected, and work on a constant-time algorithm.

References

[1] E. Breen and R. Jones. Attribute openings, thinnings and granulometries. *Computer Vision, Graphics, and Image Processing: Image Understanding*, 64(3):377–389, 1996.

[2] E. J. Breen and D. Monro. An evaluation of priority queues for mathematical morphology. In J. Serra and P. Soille, editors, ISMM'94, Fontainebleau, pages 249–256. Kluwer.

[3] M. Buckley and H. Talbot. Flexible linear openings and closings. In L. Vincent and D. Bloomberg, editors, ISMM'2000, Palo Alto. Kluwer.

[4] H. Heijmans. *Morphological image operators*. Advances in Electronics and Electron Physics Series. Academic Press, Boston, 1994.

[5] H. Heijmans, M. Buckley, and H. Talbot. Path-based morphological openings. In *ICIP'04*, pp. 3085–3088, Singapore, October 2004.

[6] H. Heijmans, M. Buckley, and H. Talbot. Path openings and closings. *Journal of Mathematical Imaging and Vision*, 2004. In press.

[7] H.J.A.M. Heijmans and C. Ronse. The algebraic basis of mathematical morphology, part I: dilations and erosions. *Computer Vision, Graphics, and Image Processing*, 50:245–295, 1990.

[8] F. Jaumier. Application de la morphologie mathématique à la détection de formes sur les clichés angiofluorographiques de la rétine. Master's thesis, July 1988.

[9] D. Knuth. *The art of computer programming. Volume 3: Sorting and searching*. Addison-Wesley, 1973.

[10] M.B. Kurdy and D. Jeulin. Directional mathematical morphology operations. In *Proceedings of the 5th European Congress for Stereology*, volume 8/2, pages 473–480, Freiburg im Breisgau, Germany, 1989.

[11] C. L. Luengo Hendriks and van Vliet L. J. A rotation-invariant morphology for shape analysis of anisotropic objects and structures. In C. Arcelli, L.P. Cordella, and G. Sanniti di Baja, editors, *Visual Form 2001: 4th International Workshop on Visual Form*, page 378, Capri, Italy, 2001. Springer LNCS.

[12] A. Meijster and H.F. Wilkinson. A comparison of algorithms for connected set openings and closings. *IEEE Transactions on Pattern Analysis and Machine Intelligence*, 24(4):484–494, April 2002.

[13] C. Ronse and H.J.A.M. Heijmans. The algebraic basis of mathematical morphology, part ii: openings and closings. *Computer Vision, Graphics, and Image Processing: Image Understanding*, 54:74–97, 1991.

[14] J. Serra. *Image analysis and mathematical morphology*. Academic Press, 1982.

[15] V.N. Skladnev, A. Gutenev, S. Menzies, L.M. Bischof, H.G.F Talbot, E.J. Breen, and M.J. Buckley. Diagnostic feature extraction in dermatological examination. Patent WO 02/094098 A1, Polartechnics Limited, 2001.

[16] P. Soille. *Morphological Image Analysis, principles and applications*. Springer-Verlag, 2nd edition, 2003. ISBN 3-540-42988-3.

[17] P. Soille, E. Breen, and R. Jones. Recursive implementation of erosions and dilations along discrete lines at arbitrary angles. *IEEE Transactions on Pattern Analysis and Machine Intelligence*, 18(5):562–567, May 1996.

[18] P. Soille and H. Talbot. Directional morphological filtering. *IEEE Transactions on Pattern Analysis and Machine Intelligence*, 23(11):1313–1329, 2001.

[19] Lee T., Ng V., Gallagher R., Coldman A., and McLean D. Dullrazor: A software approach to hair removal from images. *Computers in Biology and Medicine*, 27:533–543, 1997.

[20] H. Talbot, T. Lee, D. Jeulin, D. Hanton, and L. W. Hobbs. Image analysis of insulation mineral fibres. *Journal of Microscopy*, 200(3):251–268, Dec 2000.

[21] A. Tuzikov, P. Soille, D. Jeulin, H. Bruneel, and M. Vermeulen. Extraction of grid lines on stamped metal pieces using mathematical morphology. In *Proc. 11th ICPR*, pages 425–428, The Hague, September 1992.

[22] G. van Antwerpen, P.W. Verbeek, and F.C.A Groen. Automatic counting of asbestos fibres. In I.T. Young et al., editor, *Signal Processing III: theories and applications*, pages 891–896, 1986.

[23] L. Vincent. Grayscale area openings and closings, their efficient implementation and applications. ISMM'93, pages 22–27, Barcelona, Spain.

STRUCTURING ELEMENTS FOLLOWING THE OPTICAL FLOW

Combining Morphology and Motion

Nicolas Laveau[1] and Christophe Bernard[2]

[1]*Centre de Morphologie Mathématique*
35, rue Saint-Honoré, 77305 Fontainebleau, France
laveau@cmm.ensmp.fr

[2]*Let It Wave*
XTEC, École Polytechnique, 91128 Palaiseau, France
bernard@letitwave.fr

Abstract   This paper deals with the combination of classical morphological tools and motion compensation techniques. Morphological operators have proven to be efficient for filtering and segmenting still images. For video sequences however, using motion information to modify the morphological processing is necessary. In previous work, iterative frame by frame segmentation using motion information has been developed in various forms. In this paper, motion is used at a very low level, by locally modifying the shape of the structuring element in a video sequence considered as a 3D data block. Motion adapted morphological tools are described and their use is demonstrated on video sequences. Moreover, the features of the motion model best suited to our purpose are also discussed.

Keywords: Mathematical Morphology, Motion compensation, Optical Flow, Structuring Element

Introduction

Mathematical morphology provides very efficient tools for tasks like filtering (alternate sequential filters, levelings, etc) and for segmenting still images. These tools have been also used for 3-dimensional datasets like medical volume imaging where they perform equally well.

Segmentation tools have also been designed to segment video sequences, and to extract video objects ([4], [3]). These segmentation tools mostly dissociate computation performed along the time axis and along the space axes:

C. Ronse et al. (eds.), Mathematical Morphology: 40 Years On, 43–52.
©2005 *Springer. Printed in the Netherlands.*

segmentation is done frame by frame, and temporal correlation in the video sequence is exploited by propagating segmentation results along time. For instance, markers are computed from a segmentation of a frame t, and are used to segment frame $t + 1$ ([7]). Alternatively, frames are segmented separately beforehand, and the resulting segmentation graphs are matched in a second step ([1], [5]). In these examples, the temporal correlation between the frames is weakly enforced, and as a result the boundaries of the segmented video objects tend to flicker. Another approach is 3D filtering. This approach is more promising because it enforces a stronger temporal correlation between frames.

However either method is limited by time aliasing: when large displacements occur between frames, pixels belonging to the same object in time are not connected together and segmentation can fail. For similar reasons, video sequence filtering is also bound to be less efficient.

Our way around this limitation consists in introducing motion information provided by a motion estimation method, and incorporating this motion information to modifiy locally the grid connectivity of the 3D video volume, so that pixels that belong to the same object stay connected together along the time axis.

In a first part, we will detail the 3D morphological filtering, and why they are not sufficiently efficient. In the second part, new structuring elements following the optical flow will be defined that answer the issue of temporal aliasing, and example and results will be given in a third part. Then we will discuss the issue of the influence of the motion model on the quality of our new structuring elements.

1.   3D morphology on video sequences

3D morphology as already applied for volumic data can be applied to video sequences. In a standard setting, the grid is square and the neighbourhood at (t, x, y) is $(t, x, y) + \{(0, 0, 0), (\pm 1, 0, 0), (0, \pm 1, 0), (0, 0, \pm 1)\}$ and is octaedric. The corresponding structuring element is the set $S = \{(\delta t, \delta x, \delta y) : |\delta x| + |\delta y| + |\delta t| \leq 1\}$.

However, the contents of a video sequence does not behave the same way along spatial axes and along the time axis. While the video sequence is at most only slightly aliased along spatial axes, it can be very strongly aliased along the time axis.

As a result, a moving object may not be considered with respect to the resulting connectivity as a single connected component. This is illustrated in fig. 2. On both left and side parts, the grayed squares represent a moving object. On the left figure, an unslanted structuring element is represented. According to the resulting connectivity, the moving object is not a single connected component. If the structuring element is slanted in order to take into account visual motion (right figure), the resulting connectivity then recognizes the object as a single connected component along time.

Likewise, the ignorance of motion information can make a morphological filtering of a video sequence perform badly. This illustrated in fig. 1 where a single erosion of the video sequence with the above standard structuring element results in foreman having 3 nostrils.

Figure 1. Erosion of the foreman sequence taken as a 3D-block by an octaedric structuring element (frame 2).

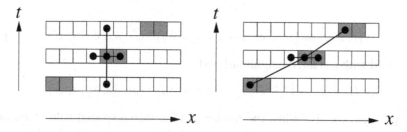

Figure 2. Time aliasing and the effect of different structuring elements. On the left, a standard structuring element, on the right, a structuring element slanted along optical flow.

It is thus necessary to incorporate the motion information into the definition of the morphological operators in order to incorporate temporal redundancy without introducing time aliasing problems and artifacts. Some kind of temporal morphological operators have been proposed (ex : [1]), but they often lack the generality they have in the spatial domain, and they often assume that the spatial segmentation has already been done. In this paper, we introduce motion information in a very early phase, in order to modify locally the structuring elements and make any morphological operator "motion-aware".

2.   3D morphology slanted along optical flow

To construct a new class of morphological operators that take into account
the motion along time, it is sufficient to (1) define the shape of the basic discrete structuring elements, and then (2) to define with this the basic operators
(erosion and dilation). Once this is done, we can unroll a wide set of morphological tools, including levelings and watershed.

We assume that at each time t, and for each pixel (x, y), we know a forward and backward motion vector $v^+(t, x, y)$ and $v^-(t, x, y)$ that have been
measured by some optical flow technique. These vectors are supposed to have
integer coordinates. The interpretation of these motion vectors is that the pixel
at (x, y) at time t moves at time $t + 1$ to $(x, y) + v^+(t, x, y)$ and was at time
$t - 1$ at $(x, y) + v^-(t, x, y)$. How these vectors are computed is explained with
more detail in Sect. 4. From this, we define forward and backward motion
operators:

$$M^+(t, x, y) \quad = \quad (t + 1, x + v_x^+(t, x, y), y + v_y^+(t, x, y)) \qquad (1)$$

$$M^-(t, x, y) \quad = \quad (t - 1, x + v_x^-(t, x, y), y + v_y^-(t, x, y)) \qquad (2)$$

where v_x^+ and v_y^+ are the coordinates of vector v^+, and likewise for v^-. We
define M^p as being the identity when $p = 0$, $(M^+)^p$ if $p > 0$ and $(M^-)^{(-p)}$
else.

With this, a standard neighbourhood at location (t, x, y) which is assumed
to be:

$$N(t, x, y) = \{(t + \delta t, x + \delta x, y + \delta y) : (\delta t, \delta x, \delta y) \in S\} \qquad (3)$$

is replaced with a slanted neighbourhood:

$$N'(t, x, y) = \{M^{\delta t}(t, x + \delta x, y + \delta y) : (\delta t, \delta x, \delta y) \in S\} \qquad (4)$$

In practice, we also allow the motion field to take a special value "discon-
nected", meaning that a pixel disappears at the next frame if v^+ takes such
a value, or just appeared at frame t if v^- takes such a value. We can then for
instance decide that whenever $v^+(t, x, y)$ takes such a value, $M^+(t, x, y)$ is defined to be (t, x, y) instead of some $(t + 1, x', y')$ (and likewise for M^-). This
can result in the slanted structuring element having less pixels than a standard
structuring element.

Some extra care is then required to define dilations and erosions of a video
sequence f. The classical implementations

$$\varepsilon(f)(t, x, y) = \min_{S'(t,x,y)} f \qquad \text{and} \qquad \delta(f)(t, x, y) = \max_{\check{S}'(t,x,y)} f$$

do not guarantee that the resulting operators are dual: the structuring element
are not necessarily translation-invariant. Instead, one of both is defined as

above (say the erosion), and the other (the dilation) is defined as the explicit dual of the first. The resulting erosion and dilation are computed in practice as explicited below:

Erosion algorithm – minimum collect

```
for each (t, x, y) {
    Iout(t, x, y) := +∞
    for each (t', x', y') ∈ S'(t, x, y) {
        Iout(t, x, y) := min(Iout(t, x, y), Iin(t', x', y'))
    }
}
```

Dilation algorithm – maximum send

```
for each (t, x, y) {
    Iout(t, x, y) := −∞
}
for each (t, x, y) {
    for each (t', x', y') ∈ S'(t, x, y) {
        Iout(t', x', y') := max(Iout(t', x', y'), Iin(t, x, y))
    }
}
```

Once these operators have been defined, all other morphological operators can be deduced by applying again standard definitions in terms of dilations and erosions.

3. Experiments

We have experimented the new structuring element design with some simple morphological tools, by modifying a classical octaedric structuring element. They were applied on blocks made by the ten first frames of the foreman sequence, and of the mobile and calendar sequence, in CIF resolution.

Comparison between erosion of size 1 on the Foreman sequence (Fig. 3 and Fig. 4) show clearly the interest of the motion compensation in morphological filtering, at least from a visual quality point of view. While in the classic case, the erosion truly damages the pictures, in the motion-compensated case, the quality is comparable with a 2D erosion, except that we have taken advantage of the temporal structure of the sequence.

The same is true for openings and closings. Opening of size 3 on the Mobile and Calendar sequence (Fig. 5), when applied as usual, would remove the white points on the ball, that would not have been removed by a 2D opening, and which are not removed by a motion-compensated opening (Fig. 6). On the other hand, closing of size 3 (Fig. 7 and Fig. 8) shows a trail left by the white points of the ball, which also degrades the visual quality of the pictures.

Slightly more advanced test have been done by applying the same segmentation protocol to the Foreman sequence (Fig. 9 and Fig. 10). Results show that the object boundaries are more stable in time using our new scheme, with less variations of object size. Advantage over segmenting separately the different pictures is that we also get the temporal connectivity.

Figure 3.   Foreman (frames 3-4): Erosion of size 1 using the classic octaedric structuring element.

Figure 4.   Foreman (frames 3-4): Erosion of size 1 using the motion-compensated octaedric structuring element.

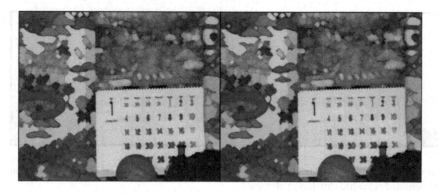

Figure 5.   Mobile and Calendar (frames 3-4): Opening of size 3 using the classic octaedric structuring element.

Figure 6.   Mobile and Calendar (frames 3-4): Opening of size 3 using the motion-compensated octaedric structuring element.

Figure 7.   Mobile and Calendar (frames 3-4, detail): Closing of size 3 using the classic octaedric structuring element.

Figure 8.   Mobile and Calendar (frames 3-4, detail): Closing of size 3 using the motion-compensated octaedric structuring element.

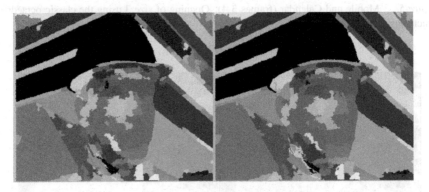

Figure 9.   Foreman (frames 3-4): Segmentation using the classic octaedric structuring element.

Figure 10.   Foreman (frames 3-4): Segmentation using the motion-compensated octaedric structuring element.

4.   Influence of the motion model

Obviously, the quality of the structuring elements defined is dependent on the motion model, and on the quality of the motion analysis. One important

feature of the motion measuring system is that it should not introduce discontinuities in the flow that do not actually correspond to a true motion discontinuity
in the video sequence. Discontinuous flow fields introduce visible artifacts in
morphological filtering, by introducing discontinuities in the filtered sequence.
This thus disqualifies block-matching.

Region-based motion estimation sounds appealing, but since regions are to
be segmented with motion-based morphological tools, this becomes a chicken-
and-egg problem. We thus decided to use a motion model represented by the
warping of a lattice, such as TMC ([6],[2]) or CGI ([8]). With such models,
it is possible to obtain accurate flow fields, and the resulting flow field is always continuous. These flow models are typically bijective and easy to invert
(this might be untrue on the picture boundary). The flow vector coordinates
obtained with methods relying on such models are usually non-integer. They
are thus rounded at each pixel to the closest integer. The estimation is done by
a multi-scale resolution, minimizing the error prediction between filtered and
downsampled versions of the pictures. Between two resolutions, the motion
field is refined by spliting each tile of the lattice into four smaller tiles (Fig.
11).

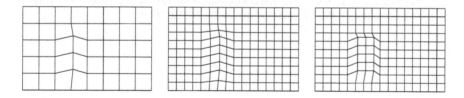

Figure 11. Refining of the lattice warping. Left : lattice obtained at scale n. Middle : splitting
of the tiles. Right : refinement

5. Conclusion

In this paper, we have defined a whole new category of structuring elements,
which are built using the optical flow to find the temporal neighbours. The results of some experiments with basic morphological operations are promising.
However, several issues are to be further investigated. The motion model described in Sect. 4 does not allow to describe discontinuous motion, as would be
necessary around object occlusions. Also, when using motion estimation, it is
natural to think about subpixel accuracy. Can it be implemented? does it provide additional accuracy in segmentation or filtering? if yes, does it mean that

some of these findings might be used successfully for enhancing still image segmentation tools?

References

[1] V. Agnus. *Segmentation Spatio-Temporelle de Séquences d'Images par des Opérateurs de Morphologie Mathématique*. PhD thesis, Université Louis Pasteur de Strasbourg, 2001.

[2] H. Brusewitz. Motion compensation with triangles. In *Proc. 3rd Int. Conf. on 64 kbit Coding of Moving Video*, Rotterdam, the Netherlands, Sept. 1990.

[3] L. Garrido. Extensive operators in partition lattices for image sequence analysis. *Signal Processing*, 66(2):157–180, 1998.

[4] B. Marcotegui, P. Correia, F Marqués, R. Mech, R. Rosa, M. Wollborn, and F. Zanoguera. A video object generation tool allowing friendly user interaction. pages 391–395.

[5] Ferran Marqués and B. Marcotegui. Tracking areas of interest for content-based functionalities in segmentation-based video coding.

[6] Y. Nakaya and H. Harashima. An iterative motion estimation method using triangular patches for motion compensation. In *Proc. SPIE Visual Comm. Image Processing*, volume 1606, pages 546–557, Nov. 1991.

[7] Philippe Salembier, Ferran Marqués, Montse Pardàs, Josep Ramon Morros, Isabelle Corset, Sylvie Jeannin, Lionel Bouchard, Fernand G. Meyer, and Beatriz Marcotegui. Segmentation-based video coding system allowing the manipulation of objects. *IEEE Transactions on Circuits and Systems for Video Technology*, 7(1):60–74, 1997.

[8] G. Sullivan and R. L. Baker. Motion compensation for video compression using control grid interpolation. In *Proc. ICASSP'91*, pages 2713–2716, 1991.

RECURSIVE INTERPOLATION TECHNIQUE FOR BINARY IMAGES BASED ON MORPHOLOGICAL MEDIAN SETS

Javier Vidal[1,2], Jose Crespo [1] and Víctor Maojo [1]

[1]*Artificial Intelligence Laboratory*
Facultad de Informática
Universidad Politécnica de Madrid
28660 Boadilla del Monte (Madrid), SPAIN

jvidal@infomed.dia.fi.upm.es, {jcrespo, vmaojo}@fi.upm.es

[2]*Departamento Ingeniería Informática y Ciencias de la Computación*
Facultad de Ingeniería
Universidad de Concepcion, CHILE

jvidal@udec.cl

Abstract   Interpolation is an important step in many applications of image processing. This paper presents a morphological interpolation technique for binary images based on the median set concept. A characteristic of our method is that it treats recursively the connected components of input slices. This technique uses the minimal skeleton by pruning (MSP) as reference points for translating connected components; this fact guarantees the non-empty intersection between them.

Keywords: Mathematical Morphology, Image Processing, Image Analysis, Interpolation, Median Set.

Introduction

In many applications of imaging, data are composed of different slices. In particular, this situation occurs when we process volumetric images and video data. In the first case, slices arise when the spatial dimension is sampled, whereas in the second case slices correspond to different instants of time. Frequently, the distance between adjacent elements within adjacent slices is much larger (until 10 times) than the distance between adjacent image elements in a slice. Thus, it is often useful to be able to interpolate data between adjacent slices, and many interpolation techniques have been developed [5] for this purpose.

C. Ronse et al. (eds.), Mathematical Morphology: 40 Years On, 53–62.

The objective of interpolation techniques is normally to produce a set of intermediary slices between two known ones. Particularly, there exists a recent category of interpolation techniques, called shape-based interpolation [13], which attempt to incorporate knowledge about the image structures to the interpolation process. In mathematical morphology [7][8][12][2], interpolation is considered as a particular case of shape-based interpolation techniques [13] [1][6][9][10][3][4].

This paper presents a morphological interpolation technique based on median sets [11][1][3][4]. Our algorithm is characterized by the recursive treatment of the *connected components* (CCs) of input slices. Besides, due to the fact that the CCs to be interpolated must overlap, our technique uses minimal skeleton by pruning (MSP) as reference points for translating them and force the overlapping of the CCs. MSP points are also used for matching purposes.

This paper is organized as follows. Section 1 provides some definitions and concepts about median sets. In Sec. 2, we present our technique, including a description of its main steps, and Sec. 3 discusses and compares some experimental results.

1.   Theoretical Background

In this section, we provide some definitions and the basic concepts about median sets.

Binary images, Slices and Connected Components.   As mentioned above, our technique interpolate *binary images*, which are functions $f : \mathbf{D} \to \{0, 1\}$, where $\mathbf{D} \subset \mathbf{Z}^2$, 0 defines a background point, and 1 defines a foreground point.

The term *slice* refers to each bi-dimensional image used as input or generated as output by the interpolation method. Each slice can contain 0 or more disjoint *connected components* of image pixels. We will see that our interpolation technique processes both the CCs of the foreground ("grains") and the CCs of the background ("holes"). In this work, 8-connectivity is assumed.

Median Set.   The notion of *median set* is an extension of the *influence zone* (IZ) concept, which is defined in the following. Let us consider two sets X and Y, where $X \subseteq Y$. The influence zone of X with respect to Y^C (which is also called the influence zone of X inside of Y) is:

$$\mathrm{IZ}_Y(X) = \{x : d'(x, X) \leq d'(x, Y^C)\} \qquad (1)$$

where d' is the distance between a point and a set. The distance $d'(p, A)$ is equal to $\min\{d(p, a) : a \in A\}$, where d is the Euclidean distance between two points.

In the case of partially intersected CCs (i.e., $X \cap Y \neq \emptyset$, $X \not\subseteq Y$ and $Y \not\subseteq X$), it is possible to extend the notion of median set as the influence zone of $(X \cap Y)$ within $(X \cup Y)$:

$$M(X, Y) = IZ_{(X \cup Y)}(X \cap Y) \qquad (2)$$

Using morphological dilations and erosions, the median set can be defined as:

$$M(X, Y) = \bigcup_{\lambda \geq 0} \{\delta_\lambda(X \cap Y) \cap \varepsilon_\lambda(X \cup Y)\} \qquad (3)$$

where δ_λ and ε_λ represent, respectively, the dilation and the erosion using a disk of radius λ.

Interpolation sequence.   Using the median set, the interpolation sequence is obtained recursively. If we denote X and Y as the input sets, n as the level of recursion (assume n is a power of 2) and K_i as the output sets, the sequence of interpolated sets can be defined as:

$$K_0 = X; K_n = Y; K_{\frac{n}{2}} = M(K_0, K_n); K_{\frac{n}{4}} = M(K_0, K_{\frac{n}{2}}); \ldots \qquad (4)$$

2.   Description of the Technique

In this section, we will present our binary image interpolation technique based on the notion of median sets. This method is motivated by the following idea. Suppose A_1 and B_1 are two sets of input slice 1, such as $B_1 \subset A_1$; and suppose C_2 and D_2 are two sets of input slice 2, such as $D_2 \subset C_2$. If there is a correspondence between these two pairs (i.e., we want to interpolate A_1 with C_2, and B_1 with D_2), then the following condition should be satisfied:

$$\text{Inter}[A_1 \setminus B_1, C_2 \setminus D_2] = \text{Inter}[A_1, C_2] \setminus \text{Inter}[B_1, D_2] \qquad (5)$$

where "Inter" denotes our interpolation technique, and "\" symbolizes the set difference.

We consider Eq.(5) as fundamental in order to consider inclusion relationships between interpolated structures. Our technique can be viewed as a generalization of that expression, where the CCs of the slices are treated recursively.

The general algorithm pseudo-code is shown in Fig. 1. The algorithm is divided in three main steps: (1) the separation of "outer" CCs within the input slices, (2) the matching between the "filled" CCs of input slices, and (3) the interpolation step.

As can be observed in Fig. 1, our algorithm uses "filled" CCs. If G is a CC of an image, the filled CC of G, denoted by FG, will be the union of G and the holes surrounded by G.

At the first step of the recursion, the "outer" CCs are extracted. These are the filled grains that are adjacent to the background parts that touch the image borders. The results of the separation step of the algorithm are two vectors of

```
INTERPOLATOR (Current_S₁: Slice ,Current_S₂: Slice): Slice {
    // (1) Separation of outer CCs
    OS₁ = Extract_Outer_CC(Current_S₁);                    // Vector of CCs
    OS₂ = Extract_Outer_CC(Current_S₂);                    // Vector of CCs
    // (2) Matching
    P = MATCHING (OS₁,OS₂);
    // P is a vector of pairs of matched CCs and their MSPs
    S = ∅;                              // initialize binary image result (set notation)
    // (3) Interpolation
    WHILE (P ≠ ∅) {
        (FS₁, MSP₁, FS₂, MSP₂) = Extract_Pair_CCs (P);
        // (FS₁, FS₂) is a pair of matched filled CCs
        // Computation of median set MS
        MS = MSF (FS₁, MSP₁, FS₂, MSP₂);
        // MSF : median set function using MSPs
        // Holes, if exists, are stored as CCs
        HS₁ = (FS₁ ∩ Current_S₁)ᶜ \ (FS₁)ᶜ;
        HS₂ = (FS₂ ∩ Current_S₂)ᶜ \ (FS₂)ᶜ;
        // Resulting image is updated; inner structures
        // (holes and grains), if exists, are treated recursively
        IF ((HS₁ = ∅) and (HS₂ = ∅))
            S = S ∪ MS;
        ELSE
            S = S ∪ (MS \ INTERPOLATOR(HS₁, HS₂));         // Recursive call
    }
    RETURN (S);
}
```

Figure 1.   General interpolation algorithm.

images, OS_1 and OS_2, where each element of each vector contains one outer CC of its corresponding slice.

The final statement of the while loop in the pseudo-code is, in fact, the extension of expression (5) to consider recursively all the CCs of the input images.

In successive iterations of recursions, the operations are confined to the mask defined by the input parameters *Current_S₁* and *Current_S₂*, and the "outer" CCs are computed within these binary images (masks).

Matching CCs

This step allows to determine which pairs of outer CCs (one from input slice 1 and the other one from input slice 2) are related to each other. The CC matching is mainly achieved by using the next criterion: an outer CC i of slice 1 (OS_1^i) matches an outer CC j of slice 2 (OS_2^j) if:

$$\delta_{\lambda_i}(OS_1^i) \cap OS_2^j \neq \emptyset \text{ or } OS_1^i \cap \delta_{\lambda_j}(OS_2^j) \neq \emptyset \qquad (6)$$

where λ_i and λ_j represent the *radius* of OS_1^i and OS_2^j, respectively. The λ parameter is computed as follows: if G is an outer CC and x is its MSP point (MSP will be commented below), the radius of G is calculated as:

$$\lambda = \bigwedge_\alpha \{\alpha : G \subseteq \delta_\alpha(x)\} \qquad (7)$$

It could be possible that an outer CC is not related to any outer CC of the other slice. We will see later that the condition (6) must be supplemented.

We reduce each CC to a point using the Minimal Skeleton by Pruning (MSP) [12], i.e., the original skeleton is reduced by pruning until a final point is reached (in the case of a CC with no holes). This minimal reference point of a set is better for our purposes than others (such as, for example the centroid), because the MSP will always be part of the CC (necessary in general for the computation of median set).

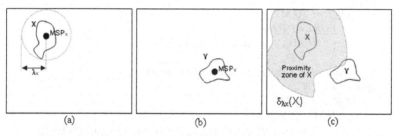

Figure 2. First matching criterion: (a) a slice with one CC X, indicating its MSP and its radius λ_X; (b) a second slice with CC Y; and (c) the proximity zone of X. Since the proximity zone of X touches Y, then X matches Y.

The algorithm of the matching process is showed in Fig.3, where three steps can be distinguished. The computations of the so-called proximity tests (corresponding to Eq.(6)) and distances between CCs are performed first, as described below. The other steps of the matching will be commented afterwards.

Figure 2 illustrates a simple example of the proximity test between two CCs, X and Y. We define the *proximity zone* of X as the dilation of X by a disk-shaped structuring element of radius λ_X, where λ_X is the radius of X as defined in Eq.(7). If the proximity zone of X intersects Y or the proximity zone of Y intersects X, then X matches Y (and Y matches X, since expression (6) is symmetrical).

We have to supplement the previous criterion in cases where there are multiple possible matches. In Fig. 4(a), for example, both CCs at the upper left of slice 1 (OS_1^1 and OS_1^2) match both CCs at the upper left of slice 2 (OS_2^1 and OS_2^2), and viceversa. The proximity zones of OS_1^1 and OS_1^2 of slice 1 touch both OS_2^1 and OS_2^2 of slice 2, and the proximity zones of OS_2^1 and OS_2^2 in slide 2 touch both OS_1^1 and OS_1^2 of slice 1. Figure 4(b) shows this situation, where the proximity zones appear in dark gray. Therefore, there exist multiple

MATCHING (OS$_1$: *VSlice,OS$_2$*: *VSlice*): *PSlice* {
    $D_{N \times M}$: *Real*;   *// N and M are the number of CCs of slices 1 and 2*
    P: PSlice;
    <u>FOR</u> *i*:= *1* <u>TO</u> *N*
        <u>FOR</u> *j*:= *1* <u>TO</u> *M*
            // (1) The proximity test is applied
            <u>IF</u>$((\delta_{\lambda_1}(OS_1^i) \cap OS_2^j) \neq \emptyset \; or \; (OS_1^i \cap \delta_{\lambda_2}(OS_2^j)) \neq \emptyset)$
                // There is a possible match; the distance between MSPs is stored
                $D_{(i,j)} = d(MSP(OS_1^i), MSP(OS_2^j))$;
            <u>ELSE</u> $D_{(i,j)} = \infty$;
    // (2) Selection of matched CCs with minimal distance between them
    <u>IF</u> $(N \geq M)$
        <u>FOR</u> *i*:=*1* <u>TO</u> *N*
            <u>FOR</u> *j*:=*1* <u>TO</u> *M*
                <u>IF</u> $(D_{(i,j)} \neq \min_{k \in [1,M]}(D_{(i,k)}))$
                    $D_{(i,j)} = \infty$;
    <u>ELSE</u>
        <u>FOR</u> *j*:=*1* <u>TO</u> *M*
            <u>FOR</u> *i*:=*1* <u>TO</u> *N*
                <u>IF</u> $(D_{(i,j)} \neq \min_{k \in [1,N]}(D_{(k,j)}))$
                    $D_{(i,j)} = \infty$;
    // The pairs of CCs that match are stored as result
    <u>FOR</u> *i*:=*1* <u>TO</u> *N*
        <u>FOR</u> *j*:=*1* <u>TO</u> *M*
            <u>IF</u> $(D_{(i,j)} \neq \infty)$
                $P = Store\_Pair\_CCs \; (OS_1^i, MSP(OS_1^i), OS_2^j, MSP(OS_2^j))$;
    // (3) Treatment for isolated CCs
    <u>FOR</u> *i*:=*1* <u>TO</u> *N*
        <u>IF</u> $(D_{(i,j)} = \infty)(\forall j = \overline{1,M})$
            $P = Store\_Pair\_CCs \; (OS_1^i, MSP(OS_1^i), MSP|(OS_1^i), MSP(OS_1^i))$;
    <u>FOR</u> *j*:=*1* <u>TO</u> *M*
        <u>IF</u> $(D_{(i,j)} = \infty)(\forall i = \overline{1,N})$
            $P = Store\_Pair\_CCs(MSP(OS_2^j), OS_2^j, MSP(OS_2^j), MSP(OS_2^j))$;
    <u>RETURN</u> *(P)*;
}

Figure 3.   Matching algorithm.

possible matches in this example. We will use a second criterion, which is explained next, in order to deal with these multiplicities. In this example, what we desire is that OS_1^1 matches OS_2^1 (and viceversa), and that OS_1^2 matches OS_2^2 (and viceversa).

Note that step (2) is necessary; if not, the interpolation of OS_1^1 and OS_1^2 with OS_2^1 and OS_2^2 in the example of Fig. 4 would be a merged CC in the interpolated image. Such a result would be normally undesirable.

We employ the Euclidean distance between the MSP points of both CCs to select the "best" match. If the proximity test between two CCs is not satisfied, we consider that their distance is ∞. As we can see in the matching algorithm

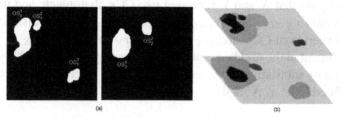

Figure 4.   Multiple matching and isolated CCs.

pseudo-code in Fig. 3, these distances are stored in matrix D. Table 1 shows, for the example of Fig. 4(a), the distances between the CCs of slice 1 and the CCs of slice 2 (all combinations). From this table we can determine the best matches between CCs: OS_1^1 matches OS_2^1 (because their distance, 13.53, is smaller than the distance between OS_1^1 and OS_2^2, which is 66.7875). Similarly, OS_1^2 matches OS_1^2. The loops in step (2) of the pseudo-code in Fig. 3 compute this selection of the best matches by processing the matrix of distances D.

==	OS_2^1	OS_2^2
OS_1^1	13.5300	66.7875
OS_1^2	47.6476	33.7567
OS_1^3	∞	∞

Table 1.   Distance table between CCs of Fig. 4.

In addition, the case of isolated CCs is also consider in the matching process. In the example of Fig. 4, the CC at the bottom-right of slice 1 (OS_1^3) is isolated, since it does not match any CC in slice 2. We can see this fact in the last row of Table 1, where all values are ∞. The case of isolated CCs is considered in step (3) of the pseudo-code in Fig. 3. In our technique, isolated CCs are matched with "artificial" points in the other slice. Each artificial point corresponds to the MSP point of the corresponding isolated CC.

The overall result of the matching process is a list of pairs (OS_1^i, OS_2^j), where OS_1^i matches OS_2^j, and their corresponding MSP points. As commented previously, in some cases OS_1^i or OS_2^j (one of them) could be an "artificial" point (where no CC matching occurs).

Median Set Computation of Matched CC Pairs

After the matching process we perform the computation of the median sets of the resulting matched pairs of CCs (S_1^i, S_2^j). The CCs of each pair must have a non-empty intersection, condition which is achieved by using their MSP points for translating them.

Let us suppose that X and Y represent two CCs that match, and that their MSP points are (x_X, y_X) and (x_Y, y_Y), respectively. Then the distance components between both CCs are: $d_X = |x_X - x_Y|$ and $d_Y = |y_X - y_Y|$. X and Y are aligned in the middle point between them by translating X by vector $(\frac{d_X}{2}, \frac{d_Y}{2})$ and by translating Y by vector $(-\frac{d_X}{2}, -\frac{d_Y}{2})$ (see Fig. 5). Let us call these translated CCs as X' and Y'.

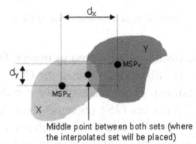

Middle point between both sets (where
the interpolated set will be placed)

Figure 5.   MSP points are used to calculate the point where the intersected set will be placed

Then, the median set between X' and Y' is computed by using the algorithm proposed in [3], which implements Eq.(3). First, three auxiliary sets are initialized: $Z_0 = W_0 = (X' \cap Y')$ and $M_0 = (X' \cup Y')$. Then, the following three equations are defined:

$$\begin{aligned} Z_i &= \delta_B(Z_{i-1}) \\ W_i &= \varepsilon_B(W_{i-1}) \\ M_i &= (Z_i \cap W_i) \cup M_{i-1} \end{aligned} \qquad (8)$$

where δ_B and ε_B are, respectively, a dilation and an erosion with the elementary disk-shaped structuring element B of radius one. The median set M is obtained by iterating until idempotence the previous equations, i.e., until $M_i = M_{i-1}$.

So far, the shapes of the pairs of outer CCs (as extracted in the first step of our technique) have been interpolated. We need now to process and, ultimately, to interpolate the structures that are *inside* the outer CCs. We achieve this by applying recursively the three steps of our technique to the structures that are located inside the outer CCs that have already been interpolated. These inner structures are denoted as HS_1 and HS_2 in the pseudo-code in Fig. 1. The recursive application is the purpose of the recursive call at the end of the pseudo-code.

3.   Experimental Results and Discussion

This section discusses some results of our technique in different situations.

Figure 6 displays a simple case that corresponds to a pair of matched CCs with non-empty intersection. The shape in slice 1 is a circle, and the shape in slice 2 is a square. Even though the input CCs intersection is non-empty, translations are needed to perform adequately the interpolation of shapes.

Figure 6.   Interpolation result 1.

The situation shown in Fig. 7 is more complex. We can see that there are several CCs in both input slices. The CC matching problem that arises has already been discussed in Section 2. The pairs of CCs at the upper-left parts of the input slices are successfully matched (after resolving possible multiple matches), and the isolated CC at the bottom-right part of the input slice 2 is matched with an "artificial" point created in input slice 1.

Figure 7.   Interpolation result 2.

Figure 8 compares the results of our technique with other previous methods using input slices composed of a grain with a hole. Figures 8(a) and (b) display the input slices. Fig. 8(c) shows the intersection between the CCs of the input slices, and Fig. 8(d) visualizes their median set computed using Eq (3). Then, Figs. 8(e) and (f) display the translated CCs, whose intersection is shown in Fig. 8(g). Fig. 8(h) displays the interpolated slice computed according to [4]. Finally, Fig. 8(i) shows the interpolation result computed using our technique. It can be said that the result in Fig. 8(i) compares favorably with the other results. Further analysis and comparison with other methods can be found in [14].

Acknowledgments

This work has been supported in part by the University of Concepción (Chile), by the MECESUP programme of the Education Ministry of Chile and

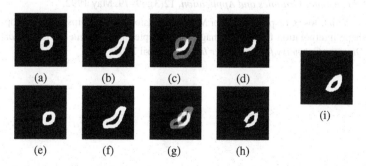

Figure 8.   Comparison between our technique and previous methods.

by the Artificial Intelligence Laboratory of School of Computer Science of
"Universidad Politécnica de Madrid".

References

[1] Serge Beucher. Interpolations d'ensembles, de partitions et de fonctions. Technical Report N-18/94/MM, Centre de Morphologie Mathématique, May 1994.

[2] Edward R. Dougherty and Roberto A. Lotufo. *Hands-on Morphological Image Processing*. SPIE Press, Bellingham, WA, 2003.

[3] Marcin Iwanowski. *Application of Mathematical Morphology to Image Interpolation*. PhD thesis, School of Mines of Paris - Warsaw University of Technology, 2000.

[4] Marcin Iwanowski and Jean Serra. The morphologycal-affine object deformation. In Luc Vincent John Goutsias and Dan S. Bloomberg, editors, *International Symposium on Mathematical Morphology, Palo Alto, CA*, page 445. Kluwer Academics Publishers, 2000.

[5] Erik Meijering. A chronology of interpolation: From ancient astronomy to modern signal and image processing. *Proceedings of the IEEE*, 90(3):319–342, March 2002.

[6] Fernand Meyer. Interpolations. Technical Report N-16/94/MM, Centre de Morphologie Mathématique, May 1994.

[7] Jean Serra. *Mathematical Morphology. Volume I*. London: Academic Press, 1982.

[8] Jean Serra, editor. *Mathematical Morphology. Volume II: Theoretical advances*. London: Academic Press, 1988.

[9] Jean Serra. Interpolations et distances of Hausdorff. Technical Report N-15/94/MM, Centre de Morphologie Mathématique, May 1994.

[10] Jean Serra. Hausdorff distances and interpolations. In Henk J.A.M. Heijmans and Jos B.T.M. Roerdink, editors, *Mathematical Morphology and its Applications to Images and Signal Processing, Dordrecht, The Netherlands*. Kluwer Academics Publishers, 1998.

[11] Pierre Soille. Spatial distributions from contour lines: an efficient methodology based on distance transformations. *Journal of Visual Communication and Image Representation*, 2(2):138–150, June 1991.

[12] Pierre Soille. *Morphological Image Analysis: Principles and Applications*. Springer-Verlag, 2nd edition, 2003.

[13] Gabor T.Herman, Jingsheng Zheng, and Carolyn A. Bucholtz. Shape-based interpolation. *IEEE Computer Graphics and Application*, 12(3):69–79, May 1992.

[14] Javier Vidal, Jose Crespo, and Victor Maojo. Inclusion relationships and homotopy issues in shape interpolation for binary images. In accepted in $12^t h$ *International Conference on Discrete Geometry for Computer Imagery*, April 2005.

II

CONNECTED FILTERS AND RECONSTRUCTION

SECOND-ORDER CONNECTED ATTRIBUTE FILTERS USING MAX-TREES

Georgios K. Ouzounis and Michael H. F. Wilkinson

Institute of Mathematics and Computing Science
University of Groningen
(georgios, michael)@cs.rug.nl

Abstract   The work presented in this paper introduces a novel method for second-order connected attribute filtering using Max-Trees. The proposed scheme is generated in a recursive manner from two images, the original and a modified copy by an either extensive or an anti-extensive operator. The tree structure is shaped by the component hierarchy of the modified image while the node attributes are based on the connected components of the original image. Attribute filtering of second-order connected sets proceeds as in conventional Max-Trees with no further computational overhead.

Keywords: second-order connectivity, Max-Tree, attribute filters, clustering, partitioning

Introduction

The concept of second-order connectivity [6, 8] is a generalization of conventional connectivity summarizing two perceptual conditions known as *clustering* and *partitioning*. In brief, when clustering objects close enough to each other in morphological terms, are considered as a single entity, while when partitioning isolated object regions interconnected by thin elongated segments are handled as independent objects. The theoretic framework developed to formalize this [8, 1] defines the two cases by means of connected openings that consider the intersection of the original image with the generalized connectivity map. Extensions to a multi-scale approach employing a hierarchical representation of connectivity have also been made. Two examples are connectivity pyramids [2] and *Connectivity Tree* [9], which quantify how strongly or loosely objects or object regions are connected.

Algorithmic realizations of this framework originally suggested the use of generalized binary and gray-scale reconstruction operators [1] for recovering the object clusters or partitions. This introduced a family of filters based on topological object relations with width as the attribute criterion. Efficient al-

C. Ronse et al. (eds.), Mathematical Morphology: 40 Years On, 65–74.

gorithms for the more general class of gray-scale attribute filters using second-order connectivity have not yet been proposed. In this paper we will present a method based on Max-Trees [7].

Our method builds a hierarchical representation based on gray scale image pairs comprising the original image and a modified copy by an increasing and either extensive or anti-extensive operator. The algorithm, referred to as *Dual Input Max-Tree* is inspired by [7, 10] and demonstrates an efficient way of computation of generalized area openings . The results extend easily to other attribute filters.

A presentation of our method is given in this paper which is organized as follows: The first section gives a brief overview of the concept of connectivity and attribute filters. A short description of second-order connectivities follows in the second section where the two cases of clustering and partitioning are described in a connected opening form. A review of the Max-Tree algorithm is given in the third section complemented by a description of our implementation while results and conclusions are discussed in the fourth section.

1.   Connectivity and Connected Filters

This section briefly outlines the concept of connectivity from the classical morphological prospective. For the purpose of this analysis we assume a universal (non-empty) set E and we denote by $\mathcal{P}(E)$ the collection of all subsets of E. A set X representing a binary image such that $X \subseteq E$ is said to be connected if it cannot be partitioned into two non-empty closed or opened sets. Expressing this using the notion of *connectivity classes*, Serra [8] derived the following definition:

DEFINITION 1 *A family $\mathcal{C} \subseteq \mathcal{P}(E)$ with E an arbitrary non-empty set, is called a connectivity class if it satisfies:*

1 $\emptyset \in \mathcal{C}$ and $\{x\} \in \mathcal{C}$ for $x \in E$,

2 if $C_i \in \mathcal{C}$ with $i = 1, ...I$ and $\bigcap_{i=1}^{N} C_i \neq \emptyset$, then $\bigcup_{i \in I} C_i \in \mathcal{C}$

where $\{x\}$ denotes a singleton.

The class $\mathcal{C}$ in this case defines the connectivity on E and any subset of $\mathcal{C}$ is called a *connected set* or a *connected component*.

Given the connected sets $C_x \in \mathcal{C}$ containing $x \in X$, the connected opening connected, opening Γ_x can be expressed as the union of all C_x:

$$\Gamma_x(X) = \bigcup \{C_x \in \mathcal{C} | x \in C_x \text{ and } C_x \subseteq X\} \qquad (1)$$

With all sets C_x containing at least one point of X in their intersection, i.e. x, their union $\Gamma_x(X)$ is also connected. Furthermore $\forall x \notin X, \Gamma_x(X) = \emptyset$.

Attribute Filters

Binary attribute openings attribute filter, opening [3] are a subclass of connected filters incorporating an increasing criterion T. The increasingness of T implies that if a set A satisfies T then any set B such that $B \supseteq A$ satisfies T as well. Using T to accept or reject a connected set C involves a trivial opening Γ_T which returns C if T is satisfied and $\emptyset$ otherwise. Furthermore, $\Gamma_T(\emptyset) = \emptyset$. The binary attribute opening is defined as follows:

DEFINITION 2 *The binary attribute opening* Γ^T *of a set* X *with increasing criterion* T *is given by:*

$$\Gamma^T(X) = \bigcup_{x \in X} \Gamma_T(\Gamma_x(X)) \tag{2}$$

The binary attribute opening is equivalent to performing a trivial opening on all connected components in the image. Note that if T is non-increasing we have an attribute thinning rather that an attribute opening.

2.   Second-Order Connectivity

The concept of second-order connectivity is briefly reviewed in this section by visiting two characteristic cases. The clustering and partitioning operators presented next, exploit the topological properties of image objects by modifying the underlying connectivity while preserving the original shape.

Clustering Based Connected Openings

The first case concerns groups of image objects that can be perceived as clusters of connected components if their relative distances are below a given threshold. In morphological terms this is verified by means of an increasing and extensive operator ψ_c which modifies the connectivity accordingly. Merged objects in the resulting connectivity class $\mathcal{C}^{\psi_c}$ define the morphology of the clusters.

DEFINITION 3 *Let* ψ_c *be an increasing and extensive operator that modifies the original connectivity from* $\mathcal{C}$ *to* $\mathcal{C}^{\psi_c}$. *The clustering based connected opening* $\Gamma_x^{\psi_c}$ *associated with the generalized connectivity class* $\mathcal{C}^{\psi_c}$ *is given by:*

$$\Gamma_x^{\psi_c}(X) = \begin{cases} \Gamma_x(\psi_c(X)) \bigcap X & \textit{if } x \in X \\ \emptyset & \textit{if } x \notin X \end{cases} \tag{3}$$

Thus $\Gamma_x^{\psi_c}$ extracts the connected components according to Γ_x in $\psi_c(X)$, rather than X, and then restricts the results to members of X [8, 1].

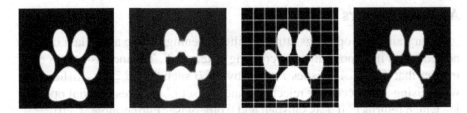

Figure 1. First pair: original image and the clustered connectivity map $\psi_c(X)$, Second pair: the original image in front of a grid (background) and the partitioned connectivity map $\psi_p(X)$.

Partitioning Based Connected Openings

Partitioning operators split wide object regions connected by narrow bridging segments which are often present due to image noise, background texture or out of focus details. The corresponding generalized binary connected opening extracts the intersection of the original image with the partitioned connectivity map $\psi_p(X)$, with ψ_p increasing and anti-extensive. To maintain the integrity of the original shape, all object level regions discarded by ψ_p are preserved as singletons in $\mathcal{C}$:

DEFINITION 4 *Let ψ_p be an increasing and anti-extensive operator that modifies the original connectivity from $\mathcal{C}$ to $\mathcal{C}^{\psi_p}$. The partitioning based connected opening $\Gamma_x^{\psi_p}$ associated with the generalized map $\mathcal{C}^{\psi_p}$ is given by:*

$$\Gamma_x^{\psi_p}(X) = \begin{cases} \Gamma_x(\psi_p(X)) \bigcap X & \text{if } x \in \psi_p(X) \\ \{x\} & \text{if } x \in X \setminus \psi_p(X) \\ \emptyset & \text{if } x \notin X \end{cases} \qquad (4)$$

The problem that $X \setminus \psi_p(X)$ is fragmented into singletons is discussed in [11].

Second-Order Attribute Filters

Attribute filters as mentioned earlier apply a trivial opening Γ_T on the output of a binary connected opening Γ_x. Replacing Γ_x with a second-order connected opening Γ_x^ψ with ψ a generalizing operator (clustering or partitioning), gives rise to the concept of second-order attribute filters which in the binary case can be expressed as:

DEFINITION 5 *The binary second-order attribute opening of a generalized set $\psi(X)$ with increasing criterion T is given by:*

$$\Gamma_\psi^T(X) = \bigcup_{x \in X} \Gamma_T(\Gamma_x^\psi(X)) \qquad (5)$$

The increasingness of these operators makes it possible to extend them directly to gray scale by threshold decomposition [4] of f, the mapping from the image domain $\mathbf{M}$ to $\mathbb{R}$. Assuming that f can be decomposed to a set of binary images $T_h(f)$ resulting from thresholding f at all levels h, given by:

$$T_h(f) = \{x \in \mathbf{M} | f(x) \geq h\} \tag{6}$$

then superimposing them by taking their supremum leads to :

DEFINITION 6 *For a mapping* $f : \mathbf{M} \rightarrow \mathbb{R}$, *the gray scale second-order attribute opening* $\gamma_\psi^T(f)$ *is given by:*

$$(\gamma_\psi^T(f))(x) = \sup\{h | x \in \Gamma_\psi^T(T_h(f))\} \tag{7}$$

Thus, the second-order attribute opening of a gray scale image assigns each point of the original image the highest threshold at which it still belongs to a connected foreground component according to the second-order connectivity class $\mathcal{C}^\psi$. Other types of gray scale generalizations can be found in [7, 10].

3. The Max-Tree Algorithm

The Max-Tree is a hierarchical image representation algorithm introduced by Salembier [7] in the context of anti-extensive attribute filtering. The tree structure reflects the connected component hierarchy obtained by threshold decomposition of the given image with nodes and leaves corresponding to *peak components* and *regional maxima* respectively. A peak component P_h at level h is a connected component of the thresholded image $T_h(f)$ while a regional maximum M_h at level h is a level component no members of which have neighbors of intensity larger that h. A Max-Tree node C_h^k (k is the node index) corresponding to a certain peak component contains only those pixels in P_h^k which have gray-level h. In addition each node except for the root, points towards its parent $C_{h'}^{k'}$ with $h' < h$. The root node is defined at the minimum level $h_{\min}$ and represents the set of pixels belonging to the background.

Node attributes are parameters stored in the tree structure and are computed during the construction of the tree. In the case of the increasing attribute of node area the connected component k at level h inherits the area of all the peak components $P_{h'}^k$ connected to C_h^k at levels $h' > h$. Computing an area opening reduces to removing all nodes with area smaller than the attribute criterion λ from the tree. Note that the node filtering is a separate stage from the computation of attributes and connected component analysis [7] therefore consumes only a short fraction of the total computation time. Extensions to other types of attributes are trivial [3, 5, 7, 10].

Construction Phase

Max-Trees are constructed in a recursive manner from data retrieved from a set of hierarchical queues. The queues are allocated at initialization in the

```
/* flood(h, thisAttribute) : Flooding function at level h      */
attribute = thisAttribute        /* accounts for child attributes */
while (not HQueue-empty(h))              /* First step: propagation */
{ p = HQueue-first(h)                    /* retrieve priority pixel */
  STATUS[p] = NumberOfNodes[h]           /* STATUS = the node index */
  for (every neighbor q of p)            /* process the neighbors   */
  {   if (STATUS[q] == "NotAnalyzed")
      {   HQueue-add(ORI[q],q)           /* add in the queue        */
          STATUS[q] = "InTheQueue"
          NodeAtLevel[ORI[q]] = TRUE     /* confirm node existance  */
          if (ORI[q] > ORI[p])           /* check for child nodes   */
          {   m = ORI[q]
              child_attribute = 0
              do{                        /* recursive child flood   */
                  m = flood(m,child_attribute)
              } while (m != h)
              attribute += child_attribute   }}}}
NumberOfNodes = NumberOfNodes[h] + 1  /* update the node index    */
m = h-1                               /* 2nd step: defines father*/
while ((m >= 0) and (NodeAtLevel[m] = FALSE))
    m = m-1
if (m >= 0){
    i = NumberOfNodes[h] - 1; j = NumberOfNodes[m];
} else
    The node C_i at level h has no father, i.e. its the root node
NodeAtLevel[h] = FALSE; node->Attribute = attribute;
node->Status = Finalized; thisAttribute = attribute;
return (m)
```

Figure 2. The flooding function of Salembier's algorithm adopted for area openings. The parameters h and m are the current and child node gray levels while attribute is a pixel count at level h within the same connected component. The parameter thisAttribute is used to pass child areas to parent nodes.

form of a static array called *HQueue* segmented to a number of entries equal to the number of gray levels. Data are accessed and stored in a first in - first out approach by the main routine (Fig. 2) which re-assigns priority pixels to the Max-Tree structure and stores new pixels retrieved from the neighborhood of the one under study, to the appropriate entries. The Max-Tree structure consists of nodes corresponding to pixels of a given peak component P_h^k at level h. Each node is characterized by its level h and index k and contains information about its parent node id, the node status and the attribute value.

The two structures are managed with the aid of three arrays; the *STATUS[p]*, the *NumberOfNodes[h]* and the *NodeAtLevel[h]*. *STATUS* is an array of image size that keeps track of the pixel status. A pixel p can either be *NotAna-*

lyzed, InTheQueue or already assigned to node k at level h. In this case *STATUS[p]=k*. The *NumberOfNodes* is an array that stores the number of nodes created until that moment at level h. Last, *NodeAtLevel* is a boolean array that flags the presence of a node still being flooded at level h.

During initialization, the status of all image pixels is set to *NotAnalyzed*. Similarly the *NumberOfNodes* is set to zero while *NodeAtLevel* is set to *FALSE* for each gray level. After computing the image histogram, the *HQueue* and Max-Tree structures are allocated accordingly while the first pixel at level h_{min} is retrieved and placed in the appropriate queue. This pixel defines the root node and is passed on to the main routine (flood) as the initial parameter.

The flooding routine is a recursive function involved in the construction phase of the Max-Tree. It is initiated by accessing the first root pixel from the queue at level h_{min} and proceeds with flooding nodes along the different root paths that emerge during this process. The pseudo-code in Fig. 2 describes in detail the steps involved. Note that *ORI* is an image-size array that stores the pixel intensities. The construction phase terminates when all pixels have been assigned to their corresponding nodes and the Max-Tree structure is complete.

Constructing the Dual Input Max-Tree

Our implementation of the construction phase requires two input images. The first is the original image while the second is a copy modified by a clustering or partitioning operator. The idea can be summarized as follows; image data are loaded on the *HQueue* structure from the modified image to be mapped on the Max-Tree which is shaped by the histogram of the original image.

Upon finalizing the initialization process with both histograms computed, h_{min} is retrieved from the modified connectivity map $\psi(X)$ and placed in the corresponding queue while the three arrays are updated. The flooding function proceeds as described earlier by inspecting the neighbors of the starting pixel and distributes them to the appropriate queues. Within the *while* loop of Fig. 2 we add a test condition which checks for an intensity mismatch between the same pixel in the two images (see Fig. 3). Denoting with *P_ORI* the array storing the pixel intensity in the modified image if $ORI[p] < P\_ORI[p]$ where p is the pixel under study, the modified image is a result of an extensive operator ψ_c while if the opposite is true it is due to an anti-extensive operator ψ_p (see Fig. 3).

The first case involving clustering implies that p is a background pixel in the original image therefore it is regarded as connected to the current active node at level $ORI[p]$ through the connected component at level $P\_ORI[p]$; i.e. it defines a peak component at level $ORI[p]$ to which p in the modified image is connected. *NodeAtLevel[ORI[p]]* is set and the status of p is updated to the node id at level $ORI[p]$. Additionally the node area is increased by a unit.

```
/* flood(h, thisAttribute) : Flooding function at level h        */
attribute = thisAttribute + node->Attribute /* node->Attribute is */
 /* added to account for pixels found during other calls to flood */
while (not HQueue-empty(h))            /* First step: propagation   */
{ p = HQueue-first(h)                  /* retrieve priority pixel   */
  STATUS[p] = NumberOfNodes[h]         /* STATUS = the node index   */
  if(ORI[p]!=h){                       /* Detect intensity mismatch */
      NodeAtLevel[ORI[p]]=TRUE         /* Same for both cases       */
      node = Tree + NodeOffsetAtLevel[ORI[p]] + NumberOfNodes[ORI[p]]
      node->Attribute ++
      if(ORI[p]>h){                    /* Anti-extensive case       */
          node->Parent = NodeOffsetAtLevel[h] + NumberOfNodes[h]
          node->Status = Finalized; node->Level  = ORI[p]
          NumberOfNodes[ORI[p]] += 1; NodeAtLevel[ORI[p]] = FALSE
          attribute++ }          /* Finalizing the singleton node */
} else
      attribute++ /* If pixel intensity is the same in both images*/
/* The rest as in Figure 2 ...                                    */
return (m)
```

Figure 3.   The flooding function of the *Dual Input Max-Tree* algorithm.

In the case where partitioning is involved the detected mismatch between the same pixel in the two images is of the form $P\_ORI[p] < ORI[p]$. Pixel p is therefore part of a discarded component according to Definition 4, and consequently is treated as a singleton. Singletons define a node of unit area at level $ORI[p]$ hence upon detection the node must be finalized before retrieving the next priority pixel from the corresponding queue at level $P\_ORI[p]$. This involves setting the node status to the node index at level $ORI[p]$ and detecting the parent node id. The area is simply set to a unit and upon completion *NodeAtLevel*[$ORI[p]$] is set to *FALSE* indicating that this node is finalized.

The flooding function following this inspection proceeds with the neighboring pixels q updating the appropriate queues and setting the node flag for every pixel at the current level $P\_ORI[p]$. If a neighbor with a higher level $P\_ORI[q]$ is detected the process is halted at level $P\_ORI[p]$ and a recursive call to the flooding function initiates the same process at level $P\_ORI[q]$. This is repeated until reaching the regional maximum along the given root path. The Max-Tree structure is completed when all nodes are finalized.

Filtering

Once the Max-Tree structure is computed, filtering which forms a separate stage, is performed in a same way for both cases. Filtering the nodes based on the attribute value λ involves visiting all nodes of the tree once. If the node

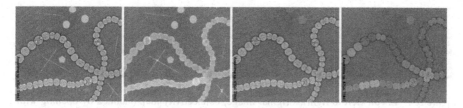

Figure 4. *Anabaena* colony (left to right): original image; connectivity map using closing by disc of radius 4; area opening ($\lambda = 900$) with Dual Input Max-Tree; area opening (same λ) with conventional Max-Tree. Image size 459×400 pixels.

attribute is less than λ the output gray level is set to that of the parent node and the comparison is repeated until the criterion is satisfied. The output image *Out* is generated by visiting all pixels p, retrieving their node ids from $ORI[p]$ and $STATUS[p]$ and assigning the output gray level of that node to $Out[p]$.

4. Discussion

The performance of the proposed algorithm was evaluated conducting a series of comparative experiments between the conventional and the dual input Max-Tree on sets of TEM images of bacteria. To demonstrate our results we chose two cases: one for clustering and one for partitioning. For a more extensive discussion of the utility of these filters the reader is refered to [1].

The first case involving clustering is demonstrated in Fig. 4 where artificial objects were added on an *Anabaena* colony to verify the filter's capability. Using a conventional Max-Tree representation, the attribute filter aiming at these additional objects, removes every connected component of area below the chosen criterion λ (set to 900) which includes parts of the colony too. In contrast to this, the same filter operated on the dual input Max-Tree considers the colony as a single object (as represented in the clustered connectivity map) and therefore removes only the unwanted objects of area less than λ. The second case considers partitioned objects and is demonstrated in Fig. 5. *Escherichia coli* cells in the original image are linked by filaments. Attribute filters on the conventional Max-Tree representation simply lower the intensity of these segments without eliminating the bridging effect. In the dual input Max-Tree however, the same object regions removed in the partitioned connectivity map (second from left), are converted to singletons in the original image hence any area filter with λ greater than the unit area discards them. This means that in the purely partitioning case, second-order connected attribute openings are equivalent to attribute openings of the connectivity map, as proven in [11].

The computational efficiency of our implementation has minimal difference from the conventional Max-Tree and this is due to loading two images and computing two histograms instead of one. The major computation takes place

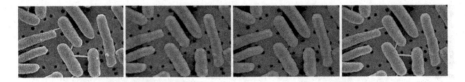

Figure 5.   E. coli (left to right): original image; connectivity map ontained by opening with disc of radius 2; area opening using the dual input Max-Tree ($\lambda = 100$); area opening using conventional Max-Tree (same λ). Image size 242×158 pixels.

within the flooding function which differs from the original implementation in the two test conditions that verify the type of generalization. In both cases the same number of input pixels have to mapped into the same size Max-Tree structure therefore if the same image is used twice our algorithm performs as a conventional Max-Tree. Our flooding function (Fig. 3) uses a single routine to handle both cases of generalization. This is primarily motivated by our current investigation on increasing operators that are neither extensive nor anti-extensive and the potential to manage image pairs in which the modified connectivity map comprises sets of both clustered and partitioned objects. We are studying the properties of such more general second-order connectivities.

References

[1] U. Braga-Neto and J. Goutsias. Connectivity on complete lattices: New results. *Comp. Vis. Image Understand.*, 85:22–53, 2002.

[2] U. Braga-Neto and J. Goutsias. A multiscale approach to connectivity. *Comp. Vis. Image Understand.*, 89:70–107, 2003.

[3] E. J. Breen and R. Jones. Attribute openings, thinnings and granulometries. *Comp. Vis. Image Understand.*, 64(3):377–389, 1996.

[4] P. Maragos and R. D. Ziff. Threshold superposition in morphological image analysis systems. *IEEE Trans. Pattern Anal. Mach. Intell.*, 12(5), 1990.

[5] A. Meijster and M. H. F. Wilkinson. A comparison of algorithms for connected set openings and closings. *IEEE Trans. Pattern Anal. Mach. Intell.*, 24(4):484–494, 2002.

[6] C. Ronse. Openings: Main properties, and how to construct them. Technical report, Université Louis Pasteur, Strasbourg, 1990.

[7] P. Salembier, A. Oliveras, and L. Garrido. Anti-extensive connected operators for image and sequence processing. *IEEE Trans. Image Proc.*, 7:555–570, 1998.

[8] J. Serra. Connectivity on complete lattices. *Mathematical Imaging and Vision*, 9:231–251, 1998.

[9] C. S. Tzafestas and P. Maragos. Shape connectivity: Multiscale analysis and application to generalized granulometries. *J. Math. Imag. Vis.*, 17:109–129, 2002.

[10] E. R. Urbach and M. H. F. Wilkinson. Shape-only granulometries and grey-scale shape filters. In *Proc. Int. Symp. Math. Morphology (ISMM) 2002*, pages 305–314, 2002.

[11] M. H. F. Wilkinson. Attribute-space connected filters. In *Proc. Int. Symp. Math. Morphology (ISMM) 2005*, 18-20 Apr 2005. These proceedings, pp. 85–94.

TRANSFORMATIONS WITH RECONSTRUCTION CRITERIA: IMAGE SEGMENTATION AND FILTERING

Iván R. Terol-Villalobos[1] and Jorge D. Mendiola-Santibañez[2]

[1]*CIDETEQ, S.C., Parque Tecnológico Querétaro S/N*
Sanfandila-Pedro Escobedo, C.P. 76700-APDO 064, Querétaro, México

[2]*Posgrado en Ingenieria, Universidad*
Autónoma de Queretaro, 76000, Queretaro, Mexico

Abstract   In this paper, a class of transformations with reconstruction criteria, derived from the reconstruction transformations, is investigated. The idea to build these transformations consists in stopping the reconstruction process according to a size criterion. This class of transformations was initially proposed for obtaining intermediate results between the morphological opening and the opening by reconstruction. Here, the transformations are presented in the general case, as in the reconstruction transformations case, by imposing some conditions on the marker. We show that the set of markers for the transformations with reconstruction criteria is given by the set of dilated images. The interest of these transformations in image segmentation is shown. Also the notion of granulometry and the alternating sequential filters are investigated.

Keywords: Opening and closing by reconstruction, opening and closing with reconstruction criteria, filtering, segmentation

1.   Introduction

In mathematical morphology (MM), the watershed-plus-markers approach is the traditional image segmentation method. This technique requires to correctly know the different morphological tools for extracting the markers. Among these tools, morphological filtering plays a fundamental role not only as a tool for symplifying the input image, but also for detecting markers. The basic morphological filters are the morphological opening and closing with a given structuring element. However, even if this type of filters permits the removal of undesirable regions, frequently, the remaining structures are modified. A way of attenuating this inconvinience is the use of the well-known filters by reconstruction. These filters that form a class of connected filters, process

75

C. Ronse et al. (eds.), Mathematical Morphology: 40 Years On, 75–84.
©2005 *Springer. Printed in the Netherlands.*

separately each connected component. Nevertheless, the main drawback of the filters by reconstruction is that they reconstruct *too much*, the so-called leakage problem, and sometimes it is not possible to extract the regions of interest. In other words, there is no way of controlling the reconstruction process. Several solutions have been proposed by Salembier and Oliveras [10], Tzafestas and Maragos [8], Serra [3], Terol-Villalobos and Vargas-Vázquez [4, 5], Vargas-Vázquez et *al.* [6] among others. In particular, Serra [3] characterizes the concept of viscous propagations by means of the notion of viscous lattices. In his work, Serra defines a connection on the viscous lattices which does not connect *too much* allowing to separate arc-wise connected components into a set of connected components in the viscous lattices sense. On the other hand, Terol-Villalobos and Vargas-Vázquez [4, 5] , introduce the notion of reconstruction criterion which allows the reconstruction to be stopped. In the present work, the reconstruction criterion will be used to introduce the transformations with reconstruction criteria. One shows that these transformations have a similar behavior than a class of transformations introduced by Serra in [3]. Also, the interest of building other transformations for segmenting and filtering images, is shown.

2.   Morphological Filtering

The basic morphological filters are the morphological opening $\gamma_{\mu B}$ and the morphological closing $\varphi_{\mu B}$ with a given structuring element. In this work, B is an elementary structuring element (3x3 pixels) containing its origin, $\check{B}$ is the transposed set ($\check{B} = \{-x : x \in B\}$) and μ is an homothetic parameter. The morphological opening and closing are given, respectively, by:

$$\gamma_{\mu B}(f)(x) = \delta_{\mu \check{B}}(\varepsilon_{\mu B}(f))(x) \quad \text{and} \quad \varphi_{\mu B}(f)(x) = \varepsilon_{\mu \check{B}}(\delta_{\mu B}(f))(x) \quad (1)$$

where $\varepsilon_{\mu B}(f)(x) = \wedge\{f(y) : y \in \mu\check{B}_x\}$ and $\delta_{\mu B}(f)(x) = \vee\{f(y) : y \in \mu\check{B}_x\}$ are the morphological erosion and dilation, respectively. $\wedge$ is the inf operator and $\vee$ is the sup operator. In the following, we will suppress the set B. The expressions $\gamma_\mu, \gamma_{\mu B}$ are equivalent (i.e. $\gamma_\mu = \gamma_{\mu B}$). When the parameter μ is equal to one, all parameters are suppressed (i.e. $\delta_B = \delta$).

Openings and Closings by Reconstruction

An interesting class of filters, called the filters by reconstruction, are built by means of the geodesic transformations [7]. In the binary case, the geodesic dilation (resp. erosion) of size 1 of a set Y (the marker) inside the set X is defined as $\delta_X^1(Y) = \delta(Y) \cap X$ (resp. $\varepsilon_X^1(Y) = \varepsilon(Y) \cup X$), while in the gray-level case is given by $\delta_f^1(g) = f \wedge \delta_B(g)$ (resp. $\varepsilon_f^1(g) = f \vee \varepsilon_B(g)$). When filters by reconstruction are built, the geodesic transformations are iterated until idempotence is reached. Consider two functions f and g, with $f \geq g$

$(f \leq g)$. Reconstruction transformations of the marker function g in f, using geodesic dilations and erosions, expressed by $R(f,g)$ and $R^*(f,g)$, respectively, are defined by:

$$R(f,g) = \lim_{n\to\infty} \delta_f^n(g) \qquad R^*(f,g) = \lim_{n\to\infty} \varepsilon_f^n(g)$$

When the marker function g is equal to the erosion or the dilation of the original function, the opening and the closing by reconstruction are obtained:

$$\tilde{\gamma}_\mu(f) = \lim_{n\to\infty} \delta_f^n(\varepsilon_\mu(f)) \quad \tilde{\varphi}_\mu(f) = \lim_{n\to\infty} \varepsilon_f^n(\delta_\mu(f)) \qquad (2)$$

Openings (Closings) with Reconstruction Criteria

It is well-known that the use of the opening by reconstruction does not enable the elimination of some structures of the image (this transformation reconstructs all connected regions during the reconstruction process). To attenuate this inconvenience, the openings and closings with reconstruction criteria were introduced in [4]. In [5], a modification in the criterion for building the transformations proposed in [4] not only permitted a better control of the reconstruction, but also generated connected transformations according to the notion of connectivity class [9]. This last class of openings and closings ([5]), are introduced in this section. Let us take the case of the opening with reconstruction criteria obtained by iterating the expression $\omega_{\lambda,f}^1(g) = f \wedge \delta\gamma_\lambda(g)$ using the marker image $g = \gamma_\mu(f)$. Observe that the only difference between this expression and the one for building the opening by reconstruction ($f \wedge \delta(g)$) is the opening γ_λ. The morphological opening γ_λ in the operator $\omega_{\lambda,f}^1$ plays the role of a reconstruction criterion by stopping the reconstruction of the regions where the criterion is not verified. Let γ_μ and φ_μ be the morphological opening and closing of size μ, respectively. The transformations given by:

$$\hat{\gamma}_{\lambda,\mu}(f) = \lim_{n\to\infty} \omega_{\lambda,f}^n(\gamma_\mu(f)) \qquad \hat{\varphi}_{\lambda,\mu}(f) = \lim_{n\to\infty} \alpha_{\lambda,f}^n(\varphi_\mu(f)) \qquad (3)$$

are an opening and a closing of size μ with $\lambda \leq \mu$, respectively, where $\omega_{\lambda,f}^1(g) = f \wedge \delta\gamma_\lambda(g)$ and $\alpha_{\lambda,f}^1(g) = f \vee \varepsilon\varphi_\lambda(g)$. The opening $\hat{\gamma}_{\lambda,\mu}$ (the closing $\hat{\varphi}_{\lambda,\mu}$) permits the obtention of intermediate results between the morphological opening (closing) and the opening (closing) by reconstruction. One has:

$$\gamma_\mu(f) \leq \hat{\gamma}_{\lambda,\mu}(f) \leq \tilde{\gamma}_\mu(f) \quad \varphi_\mu(f) \geq \hat{\varphi}_{\lambda,\mu}(f) \geq \tilde{\varphi}_\mu(f) \quad \forall \lambda \quad with \quad \lambda \leq \mu$$

However, the extreme values are not well-defined. One has for $\lambda = 0$ $\hat{\gamma}_{0,\mu}(f) = \tilde{\gamma}_\mu(f)$, but for $\lambda = \mu$, $\hat{\gamma}_{\mu,\mu}(f) = \delta_f^1\gamma_\mu(f)$.

3. Transformations with Reconstruction Criteria

One of the main problems of the transformations with reconstruction criteria above-described is that they are limited by the restriction of the use of only one type of marker function. In this section one studies the general condition that a marker must verify to be a marker. The following property expresses that the dilation permits the generation of invariants for the morphological opening [9].

PROPERTY 1 *For all pairs of parameters* λ_1, λ_2 *with* $\lambda_1 \leq \lambda_2$, $\delta_{\lambda_2}(g) = \gamma_{\lambda_1}(\delta_{\lambda_2}(g))$.

The dilation of a function g' ($g = \delta_{\lambda_2}(g')$), is an invariant of γ_{λ_1} ($\gamma_{\lambda_1}(g) = g$). This property is related to a new notion called viscous lattices, and recently proposed by Serra. In [3], Serra replaces the usual space $P(E)$ for the family $L(E) = \{\delta_\lambda(X), \quad \lambda > 0, \quad X \in P(E)\}$ that is both the image $P(E)$ under dilation δ_λ and under the opening $\delta_\lambda \varepsilon_\lambda$ and proposes the viscous lattice:

PROPOSITION 1 *The set $L(E)$ is a complete lattice regarding the inclusion ordering. In this lattice, the supremum coincides the set union, whereas the infimum $\wedge$ is the opening according to $\gamma_\lambda = \delta_\lambda \varepsilon_\lambda$ of the intersection.*

$$\wedge\{X_i, \quad i \in I\} = \gamma_\lambda(\cap\{X_i, \quad i \in I\}) \qquad \{X_i, \quad i \in I\} \in L(E)$$

The extreme elements of L are E and $\emptyset$. L is the viscous lattice of dilation δ_λ.

Using the elements of the viscous lattices as the markers ($g = \delta_{\lambda'}(g')$ with $\lambda' \geq \lambda$), the expression $\lim_{n \to \infty} \omega_{\lambda,f}^n(g)$ can be described in terms of geodesic dilations. Observe that at the first iteration of the operator $\omega_{\lambda,f}^1$, we have $\omega_{\lambda,f}^1(g) = f \wedge \delta\gamma_\lambda(g) = f \wedge \delta(g) = \delta_f^1(g)$ which is the geodesic dilation of size 1. At the second iteration, $\omega_{\lambda,f}^2(g) = f \wedge \delta\gamma_\lambda(\omega_{\lambda,f}^1(g)) = f \wedge \delta\gamma_\lambda(\delta_f^1(g)) = \delta_f^1\gamma_\lambda\delta_f^1(g)$. Thus, when stability is reached the transformation with reconstruction criteria can be established by:

$$R_{\lambda,f}(g) = \lim_{n \to \infty} \omega_{\lambda,f}^n(g) = \underbrace{\delta_f^1\gamma_\lambda\delta_f^1\gamma_\lambda \cdots \delta_f^1\gamma_\lambda\delta_f^1(g)}_{Until\ stability} \qquad (4)$$

and its dual transformation is given by: $R_{\lambda,f}^*(g) = \lim_{n \to \infty} \alpha_{\lambda,f}^n(g) = \underbrace{\epsilon_f^1\varphi_\lambda\epsilon_f^1\varphi_\lambda \cdots \epsilon_f^1\varphi_\lambda\epsilon_f^1(g)}_{Until\ stability}$. Where the marker g can be computed by means of the erosion; the erosion of a function g', $g = \epsilon_{\lambda_2}(g')$ with $\lambda_1 \leq \lambda_2$ is an invariant of the morphological closing size λ_1 ($\epsilon_{\lambda_2}(g') = \varphi_{\lambda_1}\epsilon_{\lambda_2}(g')$). Then, the working space of the marker images for the eqn. (4) is given by the viscous lattice $\{\delta_\lambda(g')\}$. However, the main problem of this transformation is that its output image does not belong to this set. Let us apply to this last equation an opening size λ,

$$R_{\lambda,f}'(g) = \gamma_\lambda \lim_{n \to \infty} \omega_{\lambda,f}^n(g) = \underbrace{\gamma_\lambda\delta_f^1\gamma_\lambda\delta_f^1 \cdots \gamma_\lambda\delta_f^1(g)}_{Until\ stability} \qquad (5)$$

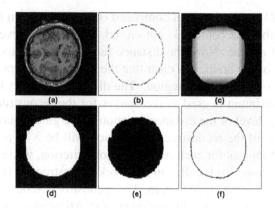

Figure 1. a) and b) Original image and its threshold, c) Reconstruction function, d) and e) Reconstructed binary image, f) Interpolation of images (d) and (e).

Now, the output image is an invariant for the morphological opening size λ and it is also an element of the viscous lattice. Furthermore, one observes that the reconstruction process is carried out inside the workspace of dilates images. In fact, observe that the reconstruction process is made by applying a sequence of a basic geodesic dilation followed by a morphological opening size λ. Thus, the output image at the *kth* iteration (before stability) of the sequence is an element of the viscous lattice. Now, this last reconstruction transformation and its dual transformation can be used to generate new openings and closings as defined by the update equations:

$$\gamma_{\lambda,\mu}(f) = R'_{\lambda,f}(\gamma_\mu(f)) \qquad \varphi_{\lambda,\mu}(f) = R'^*_{\lambda,f}(\varphi_\mu(f)) \qquad (6)$$

where R'^* is the dual transformation of R' computed by iterating until stability the sequence formed by a basic geodesic erosion followed by a morphological closing size λ. Observe that $\gamma_{\lambda,\mu}(f) \leq \hat{\gamma}_{\lambda,\mu}(f)$ and $\varphi_{\lambda,\mu}(f) \geq \hat{\varphi}_{\lambda,\mu}(f)$. Also, one can establish an order ($\gamma_\mu(f) \leq \gamma_{\lambda,\mu}(f) \leq \tilde{\gamma}_\mu(f)$ and $\varphi_\mu(f) \geq \varphi_{\lambda,\mu}(f) \geq \tilde{\varphi}_\mu(f)$, but in this case, the extreme values of the reconstruction criteria are strictly defined: for $\lambda = \mu$; $\gamma_{\mu,\mu}(f) = \gamma_\mu(f)$ and $\varphi_{\mu,\mu}(f) = \varphi_\mu(f)$ whereas for $\lambda = 0$; $\gamma_{0,\mu}(f) = \tilde{\gamma}_\mu(f)$ and $\varphi_{0,\mu}(f) = \tilde{\varphi}_\mu(f)$.

4. Image Segmentation

The idea for automatically selecting the parameter λ comes from the work of Serra [3]. Here, a procedure for selecting the parameter λ is proposed. This procedure consists in choosing the minimum value of λ such that it only permits the reconstruction of the region of interest. Consider the problem of detecting a contour on the skull of the image shown in Fig. 1(a). Since the highest gray-levels of the image are on the skull, thresholding the image between 221 and 255 will give a set composed only by points on the skull (Fig. 1(b)). Observe that the contour is not closed. Below gray-level 221 regions of the brain will appear. Now, the distance function is computed on this set. Due to the form of the brain, it is clear that the distance function will have the global maximum placed in this region. The dilation of the set formed by the points of this maximum is used as the marker for the reconstruction process. Let μ be the gray-level of the distance function in the global maximum. Then, the greatest value of the reconstruction criterion will be $\lambda = \mu - 1$. Now, in order to have all images for each reconstruction criterion, we use a gray level image Im. The image Im is called in this work reconstruction function image. Initially, this image is set equal to zero. The reconstruction begins with the parameter value $\lambda = \mu - 1$ on a binary image Is. All pixels x achieved by the reconstruction $(Is(x) = 1)$ are set at the gray-level value given by the λ parameter in the output image Im. When the reconstruction process of parameter λ stops, a second reconstruction process begins with parameter $\lambda - 1$. Each pixel x achieved by the second reconstruction process is set at gray-level $\lambda - 1$ in the output image Im. The procedure to build the reconstruction function image stops when we find a value λ', such that the reconstruction in Is touches the field borders. Figure 1(c) illustrates the reconstruction function. Then, the reconstruction criterion has the value $\lambda' + 1$ and its associated reconstruction image is computed by thresholding the gray-level image Im between $\lambda' + 1$ and 255. In Fig. 1(d) the binary image obtained with an automatically detected parameter is illustrated. Now, a second reconstruction is obtained from the image in Fig. 1(b), using the field borders as markers to reconstruct the image and the same value for the reconstruction criterion λ. An interpolation between both images (Figs. 1(d) and (e)) enables the computation of a better segmentation as illustrated in Fig. 1(f).

5.   Granulometry $\{\gamma_{\lambda,\mu}\}$

Let us study in this section the transformations with reconstruction criteria using the important concepts of granulometry (Serra [9]).

DEFINITION 1 *A family of openings $\{\gamma_{\mu_i}\}$ (or respectively of closings $\{\varphi_{\mu_i}\}$), where $\mu_i \in \{1, 2, \ldots n\}$, is a granulometry (respectively antigranulometry) if for all $\mu_i, \mu_j \in \{1, 2, \ldots n\}$ and for all function f, $\mu_i \leq \mu_j \quad \Rightarrow \quad \gamma_{\mu_i}(f) \geq \gamma_{\mu_j}(f)$ (resp. $\varphi_{\mu_i}(f) \leq \varphi_{\mu_j}(f)$)*

In practice, the granulometric curves can be computed from the granulometric residues between two different scales; $\gamma_{\mu_i}(f) - \gamma_{\mu_j}(f)$ with $\mu_i < \mu_j$. Then, one says that $\gamma_{\mu_i}(f) - \gamma_{\mu_j}(f)$ contains features of f that are larger than the scale μ_j, but smaller than scale μ_i. However, this is not strictly true. The image $\gamma_{\mu_i}(f) - \gamma_{\mu_j}(f)$ is not an invariant of the morphological opening γ_{μ_i}. On the contrary, the opening by reconstruction satisfies the following property.

PROPERTY 2 *Let $\widetilde{\gamma}_{\mu_i}$ and $\widetilde{\gamma}_{\mu_j}$ be two openings by reconstruction with $\mu_i < \mu_j$. Then, for all image f the difference $\widetilde{\gamma}_{\mu_i}(f) - \widetilde{\gamma}_{\mu_j}(f)$ is an invariant of $\widetilde{\gamma}_{\mu_i}$, i.e., $\widetilde{\gamma}_{\mu_i}[\widetilde{\gamma}_{\mu_i}(f) - \widetilde{\gamma}_{\mu_j}(f)] = \widetilde{\gamma}_{\mu_i}(f) - \widetilde{\gamma}_{\mu_j}(f)$.*

This condition ensures that no structure of size smaller than μ_i inside $\widetilde{\gamma}_{\mu_i}(f) - \widetilde{\gamma}_{\mu_j}(f)$ exists. Now, let us analyse the case of the opening with reconstruction criteria. By fixing the μ value one has that:

$$\gamma_{\lambda_i,\mu}[\gamma_{\lambda_i,\mu}(f) - \gamma_{\lambda_j,\mu}(f)] \neq \gamma_{\lambda_i,\mu}(f) - \gamma_{\lambda_j,\mu}(f) \qquad with \quad \lambda_i < \lambda_j \leq \mu$$

Observe that for this case (μ is fixed), the difference is carried out between the elements of the invariants set of γ_{λ_i} and those of γ_{λ_j}. The residues of the opening behave as in the morphological opening case. Now, take for λ a fixed value. It is clear that for $\lambda = 0$ property (2) is satisfied since $\gamma_{0,\mu_i} = \widetilde{\gamma}_{\mu_i}$

For other values of λ this equality is not verified. However, when λ is small the behavior of the residues practically verify this condition. The curves in Figs. 2(c) and (d) illustrate the granulometric density functions computed from the images in Figs. 2(a) and (b), using the residues of the openings $\widetilde{\gamma}_\mu$ and $\gamma_{\lambda,\mu}$, respectively, with $\lambda = 3$. Observe that the curves in Fig. 2(c) , correponding to $\widetilde{\gamma}_\mu$ (gray color) and $\gamma_{\lambda,\mu}$ (dark color) are very similar. This is not the case of granulometric functions computed from the image in Fig. 2(b). Let us illustrate under which conditions property (2) is satisfied by $\gamma_{\lambda,\mu}$. We know that the opening by reconstruction verifies the property; for all μ_1, μ_2 such that $\mu_1 < \mu_2$, and A a connected component; $\widetilde{\gamma}_{\mu_1}(A) \neq \emptyset$ and $\widetilde{\gamma}_{\mu_2}(A) \neq \emptyset$ implies that $\widetilde{\gamma}_{\mu_1}(A) = \widetilde{\gamma}_{\mu_2}(A) = A$. Then, for two given parameters $\mu_1 < \mu_2$, the difference $\widetilde{\gamma}_{\mu_1}(X) - \widetilde{\gamma}_{\mu_2}(X)$ is composed of connected components removed by $\widetilde{\gamma}_{\mu_2}$ from $\widetilde{\gamma}_{\mu_1}(X)$. In the binary case, a set X is arcwise connected if any pair of points x, y of X is linked by a path, entirely included in X. Where a path between two pixels of cardinal m is an m-tuple of pixels $x_0, x_1, ..., x_m$ such that $x = x_0$ and $x_m = y$ with x_k, x_{k+1} neighbors for all k. Similarly, a path to characterize some class of connected components A can be defined under some conditions. That is, a path between two pixels x, y with a given "thickness", formed by an m-tuple of pixels $x = x_0, x_1, ..., x_m = y$ with lambda $B_{x_k} \subset A$ and x_k, x_{k+1} neighbors for all k. Observe that the translates of λB, that hit the boundaries of the set $\omega^k_{\lambda,X}(\gamma_\mu(X))$ (inside this set) (see [5]):

$$\gamma_\lambda \omega^k_{\lambda,X}(\gamma_\mu(X)) = \underbrace{\gamma_\lambda \delta^1_f \gamma_\lambda \delta^1_f \cdots \gamma_\lambda \delta^1_f}_{k \ times}(\gamma_\mu(X)),$$

enable us to decide which points are added to form the reconstructed set. Then, for a class of connected components one has:

PROPERTY 3 *Let A be an arcwise connected component such that $\forall x, y \in A$ with $\lambda B_x \subset A$ and $\lambda B_y \subset A$, there exists a path $x = x_0, ..., x_m = y$, with*

x_k, x_{k+1} neighbors $\forall k$, and with $\lambda B_{x_k} \subset A$, $\forall k$. Then, $\forall \mu_1, \mu_2$ with $\lambda \leq \mu_1 \leq$ μ_2 such that $\gamma_{\lambda,\mu_1}(A) \neq \emptyset$ and $\gamma_{\lambda,\mu_2}(A) \neq \emptyset \Rightarrow \gamma_{\lambda,\mu_1}(A) = \gamma_{\lambda,\mu_2}(A) = A$

An arcwise connected component of the dilates set that verifies property (3) is removed by the opening $\gamma_{\lambda,\mu}$ if this component is completly removed by the morphological opening γ_μ, otherwise it remains intact.

6.   Alternating sequential filters

When the structures to be removed from the image have a wide range of scales, the use of a sequence of an opening (closing) followed by a closing (opening) does not lead to acceptable results. A solution to this problem is the use of the alternating sequential filters (ASF). Serra ([9]) defines and char-acterizes four operators $m_\mu(f) = \gamma_\mu \varphi_\mu(f)$, $n_\mu(f) = \varphi_\mu \gamma_\mu(f)$ $r_\mu(f) = \varphi_\mu \gamma_\mu \varphi_\mu(f)$, $s_\mu(f) = \gamma_\mu \varphi_\mu \gamma_\mu(f)$, where the size μ is indexed over a size distribution with $1 \leq \mu \leq \nu$. Let us take the operators defined by: $m_{\lambda,\mu}(f) = \gamma_{\lambda,\mu} \varphi_{\lambda,\mu}(f)$ and $n_{\lambda,\mu}(f) = \varphi_{\lambda,\mu} \gamma_{\lambda,\mu}(f)$. For the parameters λ, μ_1, μ_2 with $\lambda \leq \mu_1 \leq \mu_2$, these filters not only verify the following relationships:

$$m_{\lambda,\mu_2} m_{\lambda,\mu_1}(f) \leq m_{\lambda,\mu_2}(f) \leq m_{\lambda,\mu_1} m_{\lambda,\mu_2}(f)$$

$$n_{\lambda,\mu_1} n_{\lambda,\mu_2}(f) \leq n_{\lambda,\mu_2}(f) \leq n_{\lambda,\mu_2} n_{\lambda,\mu_1}(f)$$

But also, for the parameters μ, λ_1, λ_2 with $\lambda_1 \leq \lambda_2 \leq \mu$, one has that:

$$m_{\lambda_2,\mu} m_{\lambda_1,\mu}(f) \leq m_{\lambda_2,\mu}(f) \leq m_{\lambda_1,\mu} m_{\lambda_2,\mu}(f)$$

$$n_{\lambda_1,\mu} n_{\lambda_2,\mu}(f) \leq n_{\lambda_2,\mu}(f) \leq n_{\lambda_2,\mu} n_{\lambda_1,\mu}(f)$$

Let us use the μ parameter as the size of the structures of the image to be preserved and the λ parameter the size of those structures linked to the structures to be preserved. For a family $\{\lambda_i\}$ with $\lambda_j < \lambda_k$ if $j < k$ one has:

$$M_{\lambda_n,\mu}(f) = m_{\lambda_n,\mu} \ldots m_{\lambda_2,\mu} m_{\lambda_1,\mu}(f)$$

$$N_{\lambda_n,\mu}(f) = n_{\lambda_n,\mu} \ldots n_{\lambda_2,\mu} n_{\lambda_1,\mu}(f), \qquad (7)$$

with the condition $\lambda_n \leq \mu$. The image in Fig. 3(b) shows the output image computed from the image in Fig. 3(a) by the alternate filter using an open-ing and a closing by reconstruction $\tilde{\gamma}_\mu \tilde{\varphi}_\mu(f)$ with $\mu = 25$, while the image in Fig. 3(c) was computed by an ASF $\gamma_4 \varphi_4 \gamma_3 \varphi_3 \gamma_2 \varphi_2 \gamma_1 \varphi_1(f)$. Observe that the alternating filter, using reconstruction filters, enables us to extract the main structure of the image by removing all regions that are not connected to this structure, while the ASF, using morphological openings and closings, permits the removal of some structures linked to the main region (the cameraman). The idea of using openings and closings with reconstruction criteria consists in tak-ing into account both behaviors. Then, let us apply a sequence of openings and closings with reconstruction criteria. The images in Figs. 3(d) to (f) illustrate

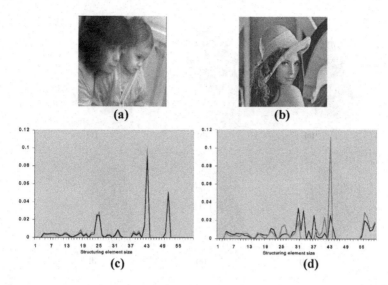

Figure 2. (a) and (b) Original images, (c) and (d) Granulometry curves computed from images 2(a) and 2(b), respectively, using the $\widetilde{\gamma}_\mu$ and $\gamma_{\lambda,\mu}$.

the use of ASF using openings and closings with reconstruction criteria. The image in Fig. 3(d) was computed by using the filter $\gamma_{\lambda,\mu}\varphi_{\lambda,\mu}(f)$ with $\mu = 25$ and $\lambda = 4$, while the image in Fig. 3(e) was computed by means of the ASF $\gamma_{4,\mu}\varphi_{4,\mu}\gamma_{3,\mu}\varphi_{3,\mu}\gamma_{2,\mu}\varphi_{2,\mu}\gamma_{1,\mu}\varphi_{1,\mu}(f)$ with $\mu = 25$. Finally, the image in Fig. 3(f) was computed with the ASF, $\gamma_{4,\mu}\varphi_{4,\mu}\gamma_{2,\mu}\varphi_{2,\mu}(f)$ with $\mu = 25$.

7. Conclusion

In this paper, a class of transformations with reconstruction criteria is investigated. In particular, the conditions required for selecting a marker for this type of transformations are studied. It is shown that a convenient framework for this type of transformations is the viscous lattice proposed by Serra [3]. The use of these transformations in the image segmentation problem is illustrated Also, the granulometry and the alternating sequential filters are analyzed.

Acknowledgments

The author I. Terol thanks Diego R. and Dario T.G. for their great encouragement. This work was funded by the government agency CONACyT (Mexico) under the grant 41170.

References

[1] Meyer F., and Beucher S.(1990) *Morphological segmentation.* J. Vis. Comm. Image Represent., **1**, 21-46.

Figure 3.   (a) Original image, (b) Alternating filter $\widetilde{\gamma}_\mu\widetilde{\varphi}_\mu(f)$ with $\mu = 25$, (c) ASF $\gamma_4\varphi_4\gamma_3\varphi_3\gamma_2\varphi_2\gamma_1\varphi_1(f)$, (d) Alternating filter $\gamma_{\lambda,\mu}\varphi_{\lambda,\mu}(f)$ with $\mu = 25$ and $\lambda = 4$, (e) ASF $\gamma_{4,\mu}\varphi_{4,\mu}\gamma_{3,\mu}\varphi_{3,\mu}\gamma_{2,\mu}\varphi_{2,\mu}\gamma_{1,\mu}\varphi_{1,\mu}(f)$ with $\mu = 25$, (f) ASF $\gamma_{4,\mu}\varphi_{4,\mu}\gamma_{2,\mu}\varphi_{2,\mu}(f)$.

[2] Crespo J., Serra J. and Schafer R.(1995) *Theoretical aspects of morphological filters by reconstruction.* Signal Process, **47**(2), 201–225.

[3] Serra J. (2002) *Viscous lattices.* In Mathematical Morphology, H. Talbot and R. Beare Eds., CSIRO (Australia), 79–89.

[4] Terol-Villalobos I. R., and Vargas-Vázquez D. (2001) *Openings and closings by reconstruction using propagation criteria.* Computer Analysis of Image and Patterns, W. Skarbek Ed., LNCS **2124**, Springer 502–509.

[5] Terol-Villalobos I. R., and Vargas-Vázquez D. (2002) *A study of openings and closing using reconstruction criteria.* In Mathematical Morphology, H. Talbot and R. Beare Eds., CSIRO (Australia), 413–423.

[6] Vargas-Vázquez, D., Crespo, J., Maojo, V., and Terol-Villalobos, I. R. (2003) *Medical image segmentation using openings and closings with reconstruction criteria.* International Conference on Image Processing, Vol. 1, 620–631.

[7] Vincent, L. (1993) *Morphological grayscale reconstruction: applications and efficient algorithms.*, IEEE Trans. on Image Processing, **2**(2), 176–201.

[8] Tzafestas, C.S., and Maragos, P. (2002) *Shape connectivity: multiscale analysis and application to generalized granulometries.* J. Mathematical Imaging and Vision, **17**, 109–129.

[9] Serra, J. (1998) *Image Analysis and Mathematical Morphology, Vol. II: Theoretical advances.* Academic Press.

[10] Salembier, Ph., and Oliveras A. (1996) *Practical extensions of connected operators.* In P. Maragos, R. W. Schafer and M.K. Butt, Eds., Mathematical Morphology and Its Aplications to Image and Signal Processing, pages 97–110. Kluwer.

ATTRIBUTE-SPACE CONNECTED FILTERS

Michael H.F. Wilkinson

Institute for Mathematics and Computing Science
University of Groningen, PO Box 800, 9700 AV Groningen, The Netherlands.
michael@cs.rug.nl

Abstract   In this paper connected operators from mathematical morphology are extended
to a wider class of operators, which are based on connectivities in higher dimension spaces, similar to scale spaces which will be called attribute spaces. Though
some properties of connected filters are lost, granulometries can be defined under certain conditions, and pattern spectra in most cases. The advantage of this
approach is that regions can be split into constituent parts before filtering more
naturally than by using partitioning connectivities.

Keywords: Mathematical morphology, multi-scale analysis, connected filters, perceptual
grouping.

1.   Introduction

Semantic analysis of images always involves grouping of pixels in some
way. The simplest form of grouping is modelled in digital image processing by
connectivity [4], which allows us to group pixels into connected components
or flat-zones in the grey-scale case. In mathematical morphology, *connected
operators* have been developed which perform filtering based on these kinds of
groupings [7][8][9]. However, the human observer may either interpret a single
connected component of a binary image as multiple visual entities, or group
multiple connected components into a single visual entity. These properties
have to some extent been encoded in second-order connectivities, which can
be either partitioning or clustering [1] [3][12].

In this paper I will demonstrate a problem with partitioning connectivities
when used for second-order connected attribute filters, due to the large numbers of singletons they produce in the image. This *over-segmentation* effect
is shown in Fig. 1. It will be shown that these attribute filters reduce to performing e.g. an opening with ball B followed by an application of the attribute
filter using the normal (4 or 8) connectivity. The approach presented here is
is different from second-order connectivities, in that it restates the connectiv-

85

C. Ronse et al. (eds.), Mathematical Morphology: 40 Years On, 85–94.
©2005 *Springer. Printed in the Netherlands.*

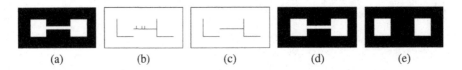

Figure 1.   Attribute-space compared to regular attribute filtering: (a) original image X; (b) the connected components of X according to $\mathcal{C}^{\psi}$, with ψ an opening by a 3×3 structuring element (see Section 3); (d) partitioning of X by attribute space method of Section 4; (e) regular attribute thinning $\Psi_{\psi}^{T}(X)$ with $T(C) = (I(C)/A^2(C) < 0.5)$; (f) attribute-space connected attribute thinning $\Psi_{A}^{T}(X)$ with the same T. T is designed to remove elongated structures. Note that only the attribute-space method removes the elongated bridge.

ity relationships in an image in terms of connectivity in higher-dimensional spaces, which I will call *attribute spaces* . As can be seen in Fig. 1, this leads to a more natural partitioning of the connected component into two squares and a single bridge. This effect is also shown in a practical application in Fig. 7.

This paper is organized as follows. First connected filters are described formally in Section 2, followed by second-order connectivities in Section 3. Problems with attribute filters using partitioning connectivities are dealt with in detail in this section. After this, attribute spaces are presented in section 4.

2.   Connectivity and Connected Filters

As is common in mathematical morphology binary images X are subsets of some universal set E (usually $E = \mathbb{Z}^n$). Let $\mathcal{P}(E)$ be the set of all subsets of E. Connectivity in E can be defined using *connectivity classes* [10].

DEFINITION 1 *A connectivity class* $\mathcal{C} \subseteq \mathcal{P}(E)$ *is a set of sets with the following three properties:*

1 $\emptyset \in \mathcal{C}$

2 $\{x\} \in \mathcal{C}$

3 for each family $\{C_i\} \subset \mathcal{C}$, $\cap C_i \neq \emptyset$ *implies* $\cup C_i \in \mathcal{C}$.

This means that both the empty set and singleton sets are connected, and any union of connected sets which have a nonempty intersection is connected.

Any image X is composed of a number of connected components or *grains* $C_i \in \mathcal{C}$, with i from some index set I. For each C_i there is no set $C \supset C_i$ such that $C \subseteq X$ and $C \in \mathcal{C}$. If a set C is a grain of X we denote this as $C \lessdot X$.

An alternative way to define connectivity is through *connected openings*, sometimes referred to as connectivity openings [1].

Figure 2. Binary attribute filters applied to an image of bacteria: (left) original; (middle) area opening using area threshold $\lambda = 150$; (right) elongation thinning using attribute $I/A^2 > 0.5$.

DEFINITION 2 *The binary connected opening* Γ_x *of* X *at point* $x \in \mathbf{M}$ *is given by*

$$\Gamma_x(X) = \begin{cases} C_i : x \in C_i \wedge C_i \lessdot X & \text{if } x \in X \\ \emptyset & \text{otherwise.} \end{cases} \tag{1}$$

Thus Γ_x extracts the grain C_i to which x belongs, discarding all others.

Attribute filters

Binary attribute openings are based on binary connected openings and *trivial openings*. A trivial opening Γ_T uses an increasing criterion T to accept or reject connected sets. A criterion T is increasing if the fact that C satisfies T implies that D satisfies T for all $D \supseteq C$. Usually T is of the form

$$T(C) = (Attr(C) \geq \lambda), \tag{2}$$

with $Attr(C)$ some real-valued attribute of C, and λ the attribute threshold. A trivial opening is defined as follows $\Gamma_T : \mathcal{C} \to \mathcal{C}$ operating on $C \in \mathcal{C}$ yields C if $T(C)$ is true, and $\emptyset$ otherwise. Note that $\Gamma_T(\emptyset) = \emptyset$. *Trivial thinnings* differ from trivial openings only in that the criterion T in non-increasing instead of increasing. An example is the scale-invariant elongation criterion of the form (2), in which $Attr(C) = I(C)/A^2(C)$, with $I(C)$ the moment of inertia of C and $A(C)$ the area [13]. The binary attribute opening is defined as follows.

DEFINITION 3 *The binary attribute opening* Γ^T *of set* X *with increasing criterion* T *is given by*

$$\Gamma^T(X) = \bigcup_{x \in X} \Gamma_T(\Gamma_x(X)) \tag{3}$$

The attribute opening is equivalent to performing a trivial opening on all grains in the image. Note that if the attribute T is non-increasing, we have an attribute thinning rather than an attribute opening [2, 8]. The grey-scale case can be derived through threshold decomposition [6]. An example in the binary case is shown in Figure 2.

3.   Second-Order Connectivities

Second-order connectivities are usually defined using an operator ψ which modifies X, and a base connectivity class $\mathcal{C}$ (4 or 8 connectivity)[1, 10]. The resulting connectivity class is referred to as $\mathcal{C}^{\psi}$. If ψ is extensive $\mathcal{C}^{\psi}$ is said to be *clustering*, if ψ is anti-extensive $\mathcal{C}^{\psi}$ is *partitioning* . In the general case, for any $x \in E$ three cases must be considered: (i) $x \in X \cap \psi(X)$, (ii) $x \in X \setminus \psi(X)$, and (iii) $x \notin X$. In the first case, the grain to which x belongs in $\psi(X)$ is computed according to $\mathcal{C}$, after which the intersection with X is taken to ensure that all grains $C_i \subseteq X$. In the second case, the x is considered to be a singleton grain. In the third case the connected opening returns $\emptyset$ as before.

DEFINITION 4 *The connected opening* Γ_x^{ψ} *for a second-order connectivity based on* ψ *of image* X *is*

$$\Gamma_x^{\psi}(X) = \begin{cases} \Gamma_x(\psi(X)) \cap X & \text{if } x \in X \cap \psi(X) \\ \{x\} & \text{if } x \in X \setminus \psi(X) \\ \emptyset & \text{otherwise,} \end{cases} \qquad (4)$$

in which Γ_x *is the connected opening based on* $\mathcal{C}$.

If $X \subset \psi(X)$ the second case of (4) never occurs. Conversely, if $\psi(X) \subset X$ we have $\psi(X) \cap X = \psi(X)$, simplifying the first condition in (4). An extensive discussion is given in [1, 10].

Attribute operators

Attribute operators can readily be defined for second-order connectivities by replacing the standard connected opening Γ_x by Γ_x^{ψ} in Definition 3.

DEFINITION 5 *The binary attribute opening* Γ_{ψ}^T *of set* X *with increasing criterion* T, *and connectivity class* $\mathcal{C}^{\psi}$ *is given by*

$$\Gamma_{\psi}^T(X) = \bigcup_{x \in X} \Gamma_T(\Gamma_x^{\psi}(X)) \qquad (5)$$

Though useful filters can be constructed in clustering case, and partition of grains in soil samples for computation of area pattern spectra has been used [12, 11], a problem emerges in the partitioning case.

PROPOSITION 1 *For partitioning connectivities based on* ψ *the attribute opening* Γ_{ψ}^T *with increasing, shift invariant criterion* T *is*

$$\Gamma_{\psi}^T(X) = \begin{cases} X & \text{if } T(\{x\}) \text{ is true} \\ \Gamma^T(\psi(X)) & \text{otherwise} \end{cases} \qquad (6)$$

with Γ^T *the underlying attribute opening from Definition 3.*

Proof If $T(\{x\})$ is true for any x, all $x \in X \setminus \psi(X)$ are preserved by Γ_ψ^T, because $\Gamma_x^\psi(X) = \{x\}$ for those pixels. Because T is increasing we have that $T(\{x\}) \Rightarrow T(C)$ for any $C \in \mathcal{C}$ with $C \neq \emptyset$. Thus, if $T(\{x\})$ is true for any x, all $x \in \psi(X)$ are also preserved, because $\Gamma_x(\psi(X)) \in \mathcal{C}$ and $\Gamma_x(\psi(X)) \neq \emptyset$ for those x. In other words if $T(\{x\})$ is true,

$$\Gamma_\psi^T(X) = X, \tag{7}$$

which proves (6) in the case that $T(\{x\})$ is true.

Conversely, if $T(\{x\})$ is false for any x, all $x \in X \setminus \psi(X)$ are rejected, i.e. $\Gamma_T(\{x\}) = \emptyset$. Therefore, if $T(\{x\})$ is false

$$\Gamma_\psi^T(X) = \bigcup_{x \in \psi(X)} \Gamma_T(\Gamma_x^\psi(X)). \tag{8}$$

Because all $x \in X \setminus \psi(X)$ are rejected, $\Gamma_x^\psi(X)$ can be rewritten as $\Gamma_x(\psi(X))$, and we have

$$\Gamma_\psi^T(X) = \bigcup_{x \in \psi(X)} \Gamma_T(\Gamma_x(\psi(X))) = \Gamma^T(\psi(X)). \tag{9}$$

The right-hand equality derives from Definition 3. □

Proposition 1 means that an attribute opening using a partitioning connectivity boils down to performing the standard attribute opening on $\psi(X)$, unless the criterion has been set such that Γ^T is the identity operator. The reason for this is the fact that the grains of $X \setminus \psi(X)$ according to the original connectivity are split up into singletons by Γ_x^ψ. Even if non-increasing criteria are used, singleton sets carry so little information that setting up meaningful filter criteria is not readily done. In Section 4 a comparison with the attribute-space alternative is given and illustrated in Figure 1.

4.   Attribute Spaces and Attribute-Space Filters

As was seen above, connectivities based on partitioning operators yield rather poor results in the attribute-filter case. To avoid this, I propose to transform the binary image image $X \subset E$ into a higher-dimensional *attribute-space* $E \times A$. Scale spaces are an examples of attribute spaces, but other attribute spaces will be explored here. Thus we can devise an operator $\Omega : \mathcal{P}(E) \rightarrow \mathcal{P}(E \times A)$. Thus $\Omega(X)$ is a binary image in $E \times A$. Typically $A \subseteq \mathbb{R}$ or $\mathbb{Z}$, although the theory presented here extends to cases such as $A \subseteq \mathbb{R}^n$. The inverse operator $\Omega^{-1} : \mathcal{P}(E \times A) \rightarrow \mathcal{P}(E)$, projects $\Omega(X)$ back onto X, i.e. $\Omega^{-1}(\Omega(X)) = X$ for all $X \in \mathcal{P}(E)$. Furthermore, Ω^{-1} must be increasing: $Y_1 \subseteq Y_2 \Rightarrow \Omega^{-1}(Y_1) \subseteq \Omega^{-1}(Y_2)$ for all $Y_1, Y_2 \in \mathcal{P}(E \times A)$. Attribute-space connected filters can now be defined as follows.

DEFINITION 6 *An attribute-space connected filter* $\Psi_A : \mathcal{P}(E) \to \mathcal{P}(E)$ *is defined as*

$$\Psi^A(X) = \Omega^{-1}(\Psi(\Omega(X))) \qquad (10)$$

with $X \in \mathcal{P}(E)$ *and* $\Psi : \mathcal{P}(E \times A) \to \mathcal{P}(E \times A)$ *a connected filter.*

Thus attribute-space connected filters work by first mapping the image to a higher dimensional space, applying a connected filter and projecting the result back. Note that the connected filter Ψ may use second-order connectivity rather the underlying connectivity in $E \times A$ (e.g. 26-connectivity in 3D). Note that if Ψ is anti-extensive (or extensive), so is Ψ^A due to the increasingness of Ω^{-1}. However, if Ψ is increasing, this property does not necessarily hold for Ψ^A, as will be shown on page 91 and following and Fig. 5. Similarly, idempotence of Ψ does not imply idempotence of Ψ^A. However, if

$$\Psi(\Omega(X)) = \Omega(\Psi^A(X)) = \Omega(\Omega^{-1}(\Psi(\Omega(X)))), \qquad (11)$$

for all $X \in \mathcal{P}(E)$, idempotence of Ψ does imply idempotence of Ψ^A, because Ω maps $\Psi^A(X)$ exactly back onto $\Psi(\Omega(X))$. Eqn. (11) obviously holds when $\Omega(\Omega^{-1}(Y)) = Y$ for all $Y \in \mathcal{P}(E \times A)$, but (11) is slightly more general.

We can also define attribute-space shape or size granulometries and spectra in analogy to connected shape or size granulometries [2, 13]. Let $\{\alpha_r\}$ be a granulometry, with each $\alpha_r : \mathcal{P}(E \times A) \to \mathcal{P}(E \times A)$ a connected filter, with r from some ordered set Λ. The set of attribute-space connected filters $\{\alpha_r^A\}$ defined as

$$\alpha_r^A = \Omega^{-1}(\alpha_r(\Omega(X))), \qquad (12)$$

has the following properties

$$\alpha_r^A(X) \subseteq X, \qquad (13)$$

$$s \leq r \Rightarrow \alpha_r^A(X) \subseteq \alpha_s^A(X) \qquad (14)$$

for all $X \subseteq E$. However, the stronger nesting property of granulometries, i.e.

$$\alpha_r^A(\alpha_s^A(X)) = \alpha_{\max(r,s)}^A(X) \qquad (15)$$

only holds if the condition on idempotence in (11) is true for all α_r in the granulometry. However, property (14) does lead to a nesting of the resulting images $\alpha_r^A(X)$ as a function of r, so a pattern spectra f_X^A based on these filters can be defined as

$$f_X^A(r) = \begin{cases} A(X \setminus \alpha_r^A(X)) & \text{if } r = 1 \\ A(\alpha_{r-1}(X) \setminus \alpha_r^A(X)) & \text{if } r > 1 \end{cases} \qquad (16)$$

with A the Lebesgue measure in E (area in 2-D), and $\Lambda = 1, 2, \ldots, N$, similar to [5]. Finally, note that connected filters form a special case of attribute-space connected filters, in which $\Omega = \Omega^{-1} = I$, with I the identity operator.

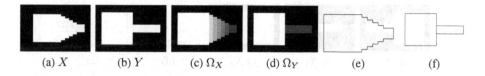

Figure 3. Attribute-space partitioning of two binary sets: (a) and (b) binary images X and Y each containing a single (classical) connected component (c) and (d) their respective opening transforms; (e) and (f) partitioning of X and Y using edge strength threshold $r = 1$. X is considered as one component due to the slow change in attribute value, whereas the abrupt change in width causes a split in Y.

Width-based attribute spaces

In the following $E = \mathbb{Z}^2$. As an example of mapping of a binary image $X \in \mathcal{P}(E)$ to binary image $Y \in \mathcal{P}(E \times A)$ we can use local width as an attribute to be assigned to each pixel $x \in X$, using an opening transform defined by granulometry $\{\beta_r\}$, in which each operator $\beta_r : E \rightarrow E$ is an opening with a structuring elements B_r. An opening transform is defined as

DEFINITION 7 *The opening transform Ω_X of a binary image X for a granulometry $\{\beta_r\}$ is* $\Omega_X(x) = \max\{r \in \Lambda | x \in \beta_r(X)\}$.   (17)

In the case that $\beta_r(X) = X \circ B_r$ with $\circ$ denoting structural openings and B_r ball-shaped structuring elements of radius r, an opening transform assigns the radius of the largest ball such that $x \in X \circ B_r$. An example is shown in Fig. 3. We can now devise a width-based attribute space by the mapping $\Omega_w : \mathcal{P}(E) \rightarrow \mathcal{P}(E \times \mathbb{Z})$ as

$$\Omega_w(X) = \{(x, \Omega_X(x)) | x \in X\} \quad (18)$$

The inverse is simply

$$\Omega_w^{-1}(Y) = \{x \in E | (x, y) \in Y\} \quad (19)$$

with $Y \in \mathcal{P}(E \times \mathbb{Z})$.

Let $C_i \subset E \times \mathbb{R}$ be the connected components of $\Omega_w(X)$ with i from some index set. Because a single attribute value is assigned to each pixel by Ω_X, it is obvious that the projections onto E of these sets $C_i^w = \Omega_w^{-1}(C_i)$ are disjoint as well. Thus they form a partition of the image plane in much the same way as classical connected components would do, as can be seen in Fig. 3. In this example we can work in a 2-D grey-scale image, rather than a 3-D binary image, for convenience. Connectivity in the attribute space is now partly encoded in the grey-level differences of adjacent flat zones in these images. In the simplest case, corresponding to 26-connectivity in the 3-D binary image, a grey-level difference of 1 means adjacent flat-zones are

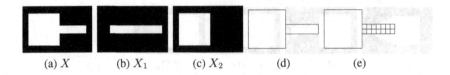

   (a) X (b) X_1 (c) X_2 (d)   (e)

Figure 4.   Attribute-space connectivity is not connectivity: (a) binary image X c is the union of two overlapping sets X_1 (b) and X_2 (c) each of which are considered connected in attribute space; however, X is partitioned into two sets (d) by the same attribute-space connectivity; (e) any partitioning connectivity which separates the square from the elongated part of X splits the elongated part into 14 singletons.

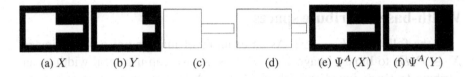

   (a) X (b) Y (c)   (d)   (e) $\Psi^A(X)$ (f) $\Psi^A(Y)$

Figure 5.   Non-increasingness of Ψ^A for increasing Ψ: (a) and (b) binary images X and Y, with $X \subseteq Y$; (c) and (d) partitions of X and Y in attribute space projection of Ω_w; (e) and (f) $Psi^A(X)$ and $Psi^A(Y)$, using for Psi an area opening with area threshold 10. Clearly $Psi^A(X) \not\subseteq Psi^A(Y)$, even though Ψ is increasing

connected in attribute space. More generally, we can use some threshold r on the grey level difference between adjacent flat zones. This corresponds to a second-order connectivity $\mathcal{C}^{\psi_r}$ with ψ_r a dilation in $\mathbb{Z}^3$, with structuring element $\{(0,0,-r),(0,0,-r+1),\ldots,(0,0,r)\}$. The effect of this can be seen in Fig. 3(f), in which abrupt changes in width lead to splitting of a connected component into two parts. Fig. 4 demonstrates that this splitting is different from caused by a partitioning connectivity . Fig. 5 shows the non-increasingness of an attribute-space area operator Ψ^A based on an area opening Ψ in $E \times A$. This effect occurs due to the fact that overlap of X_1 and X_2 in E does not imply overlap of $\Omega_w(X_1)$ and $\Omega_w(X_2)$ in $E \times A$.

A slightly different partitioning is obtained if we change (18)

$$\Omega_{\log w}(X) = \{(x, 1 + \log(\Omega_X(x)))| x \in X\} \tag{20}$$

with $\Omega_{\log w}^{-1} = \Omega_w^{-1}$. Note that one is added to the logarithm of the width to separate bridges of unity width from the background. Though very similar in behaviour to the attribute-space connectivity using Ω_w, attribute-space connectivity based on $\mathcal{C}^{\psi_r}$ is now scale-invariant, as is shown in Fig. 6. No second-order connectivity in E can achieve this, because they are all based on increasing operators [1][10], and scale-invariance and increasingness are incompatible [13].

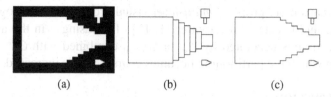

Figure 6. Scale invariant partitioning using 26-connectivity in 3-D: (a) Binary image in which the large and the bottom small connected component have identical shapes; (b) partitioning using Ω_w; (c) scale-invariant partitioning using $\Omega_{\log w}$, which splits the top small connected component, but regards the other two as single entities.

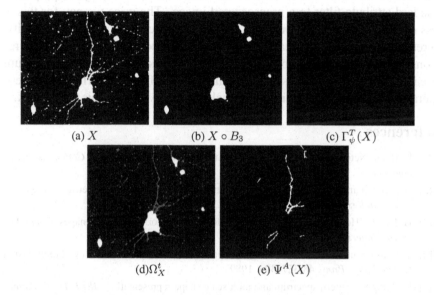

Figure 7. Elongation filtering of neurons: (a) binary image of neuron; (b) opening by B_3 to separate cell body from dendrites; (c) second-order connected attribute thinning *preserving* elongated features with $I(C)/A^2 > 0.5$; (e) Classification of pixels by thresholding Ω_X at the same value of $t = 3$; attribute-space connected filter result using same attribute as Γ_ψ^T.

Any nonlinear transformation on the attribute can be used to obtain different results, depending on the application. A simple method is to threshold the opening transform Ω_X assigning foreground pixels to different classes, denoted by Ω_X^t, allowing connectivity only within a class. A simple two-class classification is shown in Fig. 7, in which a second-order connected attribute filter is compared to the corresponding two-class pixel classification method. Only the attribute-space method recovers dendrites.

The first two attribute-space connectivities have a scale parameter, or rather a *scale-difference* or *scale-ratio* parameter. This means we can develop multiscale or perhaps more properly *multi-level* visual groupings in analogy to the well-defined multi-scale connectivities [1, 12]. Increasing r in the attribute-space connectivities generated by Ω_w or $\Omega_{\log w}$ combined with $\mathcal{C}^{\psi_r}$ yields a hierarchy, in which the partitioning becomes coarser as r is increased.

5.   Discussion

Attribute-space morphology solves the problems with attribute filters using partitioning connectivities as noted in Proposition 1. The fragmentation caused by splitting parts of connected components into singletons is absent. This means that attribute-space attribute filters are more than just applying a standard attribute filter to a preprocessed image. The price we pay for this is loss of the increasingness property, and increased computational complexity. In return we may achieve scale invariance, combined with a more intuitive response to, e.g., elongation-based attribute filters, as is seen in Fig. 1. Future research will focus on grey-scale generalizations, efficient algorithms for these operators, and on the possibilities of dealing with overlap in this framework.

References

[1] U. Braga-Neto and J. Goutsias. A multiscale approach to connectivity. *Comp. Vis. Image Understand.*, 89:70–107, 2003.

[2] E. J. Breen and R. Jones. Attribute openings, thinnings and granulometries. *Comp. Vis. Image Understand.*, 64(3):377–389, 1996.

[3] H. J. A. M. Heijmans. Connected morphological operators for binary images. *Comp. Vis. Image Understand.*, 73:99–120, 1999.

[4] T. Y. Kong and A. Rosenfeld. Digital topology: Introduction and survey. *Comp. Vision Graph. Image Proc.*, 48:357–393, 1989.

[5] P. Maragos. Pattern spectrum and multiscale shape representation. *IEEE Trans. Pattern Anal. Mach. Intell.*, 11:701–715, 1989.

[6] P. Maragos and R. D. Ziff. Threshold decomposition in morphological image analysis. *IEEE Trans. Pattern Anal. Mach. Intell.*, 12(5), 1990.

[7] P. Monasse and F. Guichard. Fast computation of a contrast invariant image representation. *IEEE Trans. Image Proc.*, 9:860–872, 2000.

[8] P. Salembier, A. Oliveras, and L. Garrido. Anti-extensive connected operators for image and sequence processing. *IEEE Trans. Image Proc.*, 7:555–570, 1998.

[9] P. Salembier and J. Serra. Flat zones filtering, connected operators, and filters by reconstruction. *IEEE Trans. Image Proc.*, 4:1153–1160, 1995.

[10] J. Serra. Connectivity on complete lattices. *J. Math. Imag. Vis.*, 9(3):231–251, 1998.

[11] A. Sofou, C. Tzafestas, and P. Maragos. Segmentation of soilsection images using connected operators. In *Int. Conf. Image Proc. 2001*, pages 1087–1090, 2001.

[12] C. S. Tzafestas and P. Maragos. Shape connectivity: Multiscale analysis and application to generalized granulometries. *J. Math. Imag. Vis.*, 17:109–129, 2002.

[13] E. R. Urbach and M. H. F. Wilkinson. Shape-only granulometries and grey-scale shape filters. In *Proc. Int. Symp. Math. Morphology (ISMM) 2002*, pages 305–314, 2002.

VECTOR-ATTRIBUTE FILTERS

Erik R. Urbach, Niek J. Boersma and Michael H.F. Wilkinson

Institute for Mathematics and Computing Science,
University of Groningen,
P.O. Box 800, 9700 AV Groningen, The Netherlands
{erik,michael}@cs.rug.nl

Abstract   A variant of morphological attribute filters is developed, in which the attribute on which filtering is based, is no longer a scalar, as is usual, but a vector. This leads to new granulometries and associated pattern spectra. When the vector-attribute used is a shape descriptor, the resulting granulometries filter an image based on a shape or shape family instead of one or more scalar values.

Keywords: Mathematical morphology, connected filters, multi-scale analysis, granulometries, pattern spectra, vector-attributes, shape filtering

Introduction

Attribute filters [2, 12], which preserve or remove components in an image based on the corresponding attribute value, are a comparatively new addition to the image processing toolbox of mathematical morphology. Besides binary and gray-scale 2-D images [2, 12], these filters have also been extended to handle vector images, like color images [5, 7] and tensor-valued data [3], and 3-D images. So far the attributes used in all of these cases have been scalars. Although the set of scalar attributes used in multi-variate filters and granulometries [14] can also be considered as a single vector-attribute, these multi-variate operators can always be written as a series of uni-variate scalar operators, which is not the case for vector-attribute filters.

In this paper vector-attribute filters and granulometries will be introduced, whose attributes consists of vectors instead of scalar values, followed by a discussion on their use as filters and in granulometries where the parameter is a single shape image or a family of shape images instead of a threshold value.

C. Ronse et al. (eds.), Mathematical Morphology: 40 Years On, 95–104.

1.   Theory

The theory of granulometries and attribute filters is presented only very briefly here. For more detail the reader is referred to [2, 9, 12, 16]. In the following discussion binary images X and Y are defined as subsets of the image domain $\mathbf{M} \subset \mathbb{R}^n$ (usually $n = 2$), and gray-scale images are mappings from $\mathbf{M}$ to $\mathbb{R}$.

Let us define a scaling X_λ of set X by a scalar factor $\lambda \in \mathbb{R}$ as

$$X_\lambda = \{x \in \mathbb{R}^n | \lambda^{-1}x \in X\}. \tag{1}$$

An operator ϕ is said to be *scale-invariant* if

$$\phi(X_\lambda) = (\phi(X))_\lambda \tag{2}$$

for all $\lambda > 0$. A scale-invariant operator is therefore sensitive to shape rather than to size. If an operator is scale, rotation and translation invariant, we call it a *shape operator*. A *shape filter* is simply an idempotent shape operator. In the digital case, pure scale invariance will be harder to achieve due to discretization artefacts, but a good approximation may be achieved.

Attribute openings and thinnings

Attribute filters, as introduced by Breen and Jones [2], use a criterion to remove or preserve connected components (or flat zones for the gray-scale case) based on their attributes. The concept of trivial thinnings Φ_T is used, which accepts or rejects connected sets based on a non-increasing criterion T. A criterion T is increasing if the fact that C satisfies T implies that D satisfies T for all $D \supset C$. The binary connected opening $\Gamma_x(X)$ of set X at point $x \in \mathbf{M}$ yields the connected component of X containing x if $x \in X$, and $\emptyset$ otherwise. Thus Γ_x extracts the connected component to which x belongs, discarding all others. The trivial thinning Φ_T of a connected set C with criterion T is just the set C *if C satisfies T*, and is empty otherwise. Furthermore, $\Phi_T(\emptyset) = \emptyset$.

DEFINITION 1 *The binary attribute thinning Φ^T of set X with criterion T is given by*

$$\Phi^T(X) = \bigcup_{x \in X} \Phi_T(\Gamma_x(X)) \tag{3}$$

It can be shown that this is a thinning because it is idempotent and anti-extensive [2]. The attribute thinning is equivalent to performing a trivial thinning on all connected components in the image, i.e., removing all connected components which do not meet the criterion. It is trivial to show that if criterion

T is scale-invariant:

$$T(C) = T(C_\lambda) \quad \forall \lambda > 0 \wedge C \subseteq \mathbf{M}, \tag{4}$$

so are Φ_T and Φ^T. Assume $T(C)$ can be written as $\tau(C) \geq r, r \in \Lambda$, with τ some scale-invariant attribute of the connected set C. Let the attribute thinnings formed by these T be denoted as Φ_r^τ. It can readily be shown that

$$\Phi_r^\tau(\Phi_s^\tau(X)) = \Phi_{\max(r,s)}^\tau(X). \tag{5}$$

Therefore, $\{\Phi_r^\tau\}$ is a shape granulometry, since attribute thinnings are anti-extensive, and scale invariance is provided by the scale invariance of $\tau(C)$. An attribute thinning with an increasing criterion is an attribute opening.

DEFINITION 2 *A binary shape granulometry is a set of operators $\{\beta_r\}$ with r from some totally ordered set Λ, with the following three properties*

$$\beta_r(X) \subset X \tag{6}$$
$$\beta_r(X_\lambda) = (\beta_r(X))_\lambda \tag{7}$$
$$\beta_r(\beta_s(X)) = \beta_{\max(r,s)}(X), \tag{8}$$

for all $r, s \in \Lambda$ and $\lambda > 0$.

Thus, a shape granulometry consists of operators which are anti-extensive, and idempotent, but not necessarily increasing. Therefore, the operators must be thinnings, rather than openings. To exclude any sensitivity to size, we add property (7), which is just scale invariance for all β_r.

Size and shape pattern spectra

Size pattern spectra were introduced by Maragos [8]. Essentially they are a histogram containing the number of pixels, or the amount of image detail over a range of size classes. If r is the scale parameter of a size granulometry, the size class of $x \in X$ is the smallest value of r for which $x \notin \alpha_r(X)$. Shape pattern spectra can be defined in a similar way [15]. The pattern spectra $s_\alpha(X)$ and $s_\beta(X)$ obtained by applying size and shape granulometries $\{\alpha_r\}$ and $\{\beta_r\}$ to a binary image X are defined as

$$(s_\alpha(X))(u) = -\left.\frac{\mathrm{d}A(\alpha_r(X))}{\mathrm{d}r}\right|_{r=u} \tag{9}$$

and

$$(s_\beta(X))(u) = -\left.\frac{\mathrm{d}A(\beta_r(X))}{\mathrm{d}r}\right|_{r=u} \tag{10}$$

in which $A(X)$ denotes the Lebesgue measure in $\mathbb{R}^n$, which is just the area if $n = 2$.

In the discrete case, a pattern spectrum can be computed by repeatedly filtering an image by each β_r, in ascending order of r. After each filter step, the sum of gray levels S_r of the resulting image $\beta_r(f)$ is computed. The pattern spectrum value at r is computed by subtracting S_r from S_{r-}, with r^- the scale immediately preceding r. In practice, faster methods for computing pattern spectra can be used [2, 10, 11]. These faster methods do not compute pattern spectra by filtering an image by each β_r. However, for methods using structuring elements this is usually unavoidable [1].

2.   Vector-attribute granulometries

Attribute filters as described by Breen and Jones [2] filter an image based on a criterion. Much work has been done since: uni- and multi-variate granulometries [1, 14] and their use on different types of images, such as binary, gray-scale, and vector images. Although the original definition of the attribute filters was not limited to scalar attributes, the attributes used so far have always been based on scalar values.

A multi-variate attribute thinning $\Phi^{\{T_i\}}(X)$ with scalar attributes $\{\tau_i\}$ and their corresponding criteria $\{T_i\}$, with $1 \leq i \leq N$, can be defined such that connected components are preserved if they satisfy at least one of the criteria $T_i = \tau_i(C) \geq r_i$ and are removed otherwise:

$$\Phi^{\{T_i\}}(X) = \bigcup_{i=1}^{N} \Phi^{T_i}(X). \tag{11}$$

The set of scalar attributes $\{\tau_i\}$ can also be considered as a single vector-attribute $\vec{\tau} = \{\tau_1, \tau_2, \ldots, \tau_N\}$, in which case a vector-attribute thinning is needed with a criterion:

$$T_{\vec{\tau}} = \exists i : \tau_i(C) \geq r_i \quad \text{for } 1 \leq i \leq N. \tag{12}$$

Although a thinning using this definition of $T_{\vec{\tau}}$ and $\vec{\tau}$ can be considered as a multi-variate thinning with scalar attributes, and thus be decomposed into a series of uni-variate thinnings (see definition 11), this is not the case with the vector-attributes and their corresponding filters for binary and gray-scale 2-D images that will be discussed below.

A binary vector-attribute thinning $\Phi_{\vec{r},\epsilon}^{\vec{\tau}}(X)$, with d-dimensional vectors from a space $\Upsilon \subseteq \mathbb{R}^d$, removes the connected components of a binary image X whose vector-attributes differ more than a given quantity from a reference vector $\vec{r} \in \Upsilon$. For this purpose we need to introduce some dissimilarity measure $d : \Upsilon \times \Upsilon \rightarrow \mathbb{R}$, which quantifies the difference between the attribute vector $\vec{\tau}(C)$ and $\vec{r}$. A connected component C is preserved if its vector-attribute

$\vec{\tau}(C) \in \Upsilon$ satisfies criterion $T^{\vec{\tau}}_{\vec{r},\epsilon}(C) = d(\vec{\tau}(C), \vec{r}) \geq \epsilon$ and is removed otherwise, with ϵ some threshold. Thus it satisfies $T^{\vec{\tau}}_{\vec{r},\epsilon}$ if the dissimilarity $d(\vec{\tau}(C), \vec{r})$ between vectors $\vec{\tau}(C)$ and $\vec{r}$ is at least ϵ. The simplest choice for d is the Euclidean distance: $d(\vec{u}, \vec{v}) = ||\vec{v} - \vec{u}||$, and any other distance measure (such as Mahalanobis) could be used. However, d need not be a distance, because the triangle inequality $d(a, c) \leq d(a, b) + d(b, c)$ is not required.

More formally, the vector-attribute thinning can be defined as:

DEFINITION 3 *The vector-attribute thinning $\Phi^{\vec{\tau}}_{\vec{r},\epsilon}$ of X with respect to a reference vector $\vec{r}$ and using vector-attribute $\vec{\tau}$ and scalar value ϵ is given by*

$$\Phi^{\vec{\tau}}_{\vec{r},\epsilon}(X) = \{x \in X \mid T^{\vec{\tau}}_{\vec{r},\epsilon}(\Gamma_x(X))\}. \tag{13}$$

This equation can be derived from definition 1 of the binary attribute thinning [2] by substituting T with $T^{\vec{\tau}}_{\vec{r},\epsilon}$ in the definition of the trivial thinning.

Although a multi-variate thinning $\Phi^{\{T_i\}}$ can be defined as a vector-attribute thinning $\Phi^{\vec{\tau}}_{\vec{r},\epsilon}$ with $T^{\vec{\tau}}_{\vec{r},\epsilon} = T^{\vec{\tau}}_{\vec{r}}$, equation 13 cannot be decomposed in a similar way, unless $d(\vec{\tau}(C), \vec{r})$ is the L_∞ norm.

It should be noted here that vector-attribute openings are vector-attribute thinnings with an increasing criterion $T^{\vec{\tau}}_{\vec{r},\epsilon}$. Although it is easy to define an increasing criterion based on scalar attributes, this is much harder for vector-attributes, i.e. a criterion using a vector-attribute consisting of only increasing scalar attributes is not necessarily increasing. Furthermore, since all of these scalar attributes are increasing, they will generally be strongly correlated. For this reason we restrict our attention to thinnings.

The reference vector $\vec{r}$ in the definition of vector-attribute thinnings can be computed using a given shape S: $\vec{r} = \vec{\tau}(S)$. This way a binary vector-attribute thinning with respect to a given shape S can be constructed:

DEFINITION 4 *The binary attribute thinning with respect to a shape $S \in C$ can be defined as:*

$$\Phi^{\vec{\tau}}_{S,\epsilon} = \Phi^{\vec{\tau}}_{\vec{\tau}(S),\epsilon} \tag{14}$$

In Fig. 1 the effect of ϵ in the criterion $T^{\vec{\tau}}_{\vec{r},\epsilon}(C) = d(\vec{\tau}(C), \vec{r}) \geq \epsilon$ is demonstrated, with $d(\vec{\tau}(C), \vec{r}) = ||\vec{r} - \vec{\tau}(C)||$. The reference vector $\vec{r}$ was computed from an image of the letter A. Three values were (manually) chosen for ϵ: the maximum (rounded) value that removes exactly one letter, one value that removes nearly all letters, and one value in between.

The extension of binary attribute filters and granulometries to gray-scale has been studied extensively [2, 10–12]. Extending our vector-attribute thinnings and granulometries can be done in a similar fashion. Gray-scale thinning with respect to a shape is demonstrated in Fig. 2.

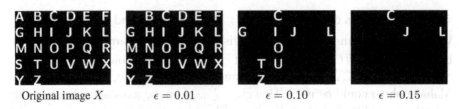

| Original image X | $\epsilon = 0.01$ | $\epsilon = 0.10$ | $\epsilon = 0.15$ |

Figure 1.   Filtering using a vector-attribute thinning $\Phi_{\vec{r},\epsilon}^{\vec{\tau}}(X)$ with increasing values of ϵ

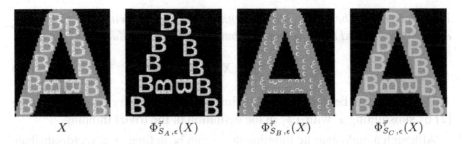

$$X \qquad \Phi_{S_A,\epsilon}^{\vec{\tau}}(X) \qquad \Phi_{S_B,\epsilon}^{\vec{\tau}}(X) \qquad \Phi_{S_C,\epsilon}^{\vec{\tau}}(X)$$

Figure 2.   Removal of letters using $\Phi_{S_i,\epsilon}^{\vec{\tau}}(X)$ in gray-scale image (left) of letters A, B, C with S_i being respectively the shapes S_A, S_B, and S_C

DEFINITION 5 *A granulometry with respect to reference vector $\vec{r} \in \Upsilon$, using scale, rotation and translation invariant vector-attribute $\vec{\tau} \in \Upsilon$ is given by the family of vector-attribute thinnings $\{\Phi_{\vec{r},\epsilon}^{\vec{\tau}}\}$ with ϵ from $\mathbb{R}$.*

It is obvious from (13) that $\Phi_{\vec{r},\epsilon}^{\vec{\tau}}$ is anti-extensive and idempotent, and more importantly that

$$\Phi_{\vec{r},\eta}^{\vec{\tau}}(\Phi_{\vec{r},\epsilon}^{\vec{\tau}}(X)) = \Phi_{\vec{r},\max(\epsilon,\eta)}^{\vec{\tau}}(X) \qquad \epsilon, \eta \in \mathbb{R} \qquad (15)$$

Furthermore, if $\vec{\tau}$ is scale, rotation, and translation invariant, $\Phi_{\vec{r},\epsilon}^{\vec{\tau}}$ is a shape filter and $\{\Phi_{\vec{r},\epsilon}^{\vec{\tau}}\}$ is a shape granulometry [15].

An example of a suitable vector-attribute for shape granulometries are moment invariants. Hu's moment invariants [6] are invariant to rotation, scaling and translation, and are therefore suitable as shape attribute. Recently, new sets of moment invariants have been presented, such as the Krawtchouk moment invariants [17], which form a set of discrete and orthogonal moment invariants, and a set of complete and independent moment invariants by Flusser and Suk [4]. A problem that occurs with Krawtchouk moment invariants when the reference shape is not rotationally symmetric, like most letters, is that the angle used in the definitions of these moment invariants is defined by the orientation instead of the direction of the shape, which means that a 180 degrees rotated version of a shape S will generate a different vector-attribute than S does. The sensitivity of the moment invariants of Hu and Krawtchouk to rotation and

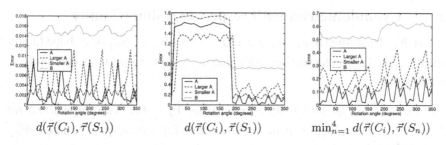

$$d(\vec{\tau}(C_i), \vec{\tau}(S_1)) \qquad d(\vec{\tau}(C_i), \vec{\tau}(S_1)) \qquad \min_{n=1}^{4} d(\vec{\tau}(C_i), \vec{\tau}(S_n))$$

Figure 3. Effect of orientation on the distance between the vector-attributes of a connected component C_i and a given reference image S_j for Hu (left) and Krawtchouk (middle and right) moment invariants, where C_i represents the letter A, double-sized A, half-sized A, and B for $i = 1, 2, 3, 4$ respectively; S_j represents the letter A for $j = 1, 2$ at 0 and 180 degrees rotation respectively

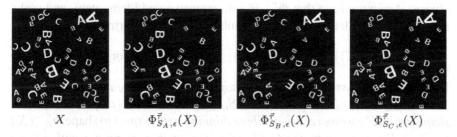

$$X \qquad \Phi_{S_A,\epsilon}^{\vec{\tau}}(X) \qquad \Phi_{S_B,\epsilon}^{\vec{\tau}}(X) \qquad \Phi_{S_C,\epsilon}^{\vec{\tau}}(X)$$

Figure 4. Left to right: original image and letters A, B, and C removed

scaling is demonstrated in Fig. 3, where one would expect the distance d between different orientations and sizes of the same letter A to be smaller than the distance between A and, according to the vector-attribute, the letter closest the A: the B. As can be seen, this is in both cases true for scaling, but it is clear that for Krawtchouk moment invariants rotation-invariance only holds for a certain range of orientations. This problem can be solved by using a filter that removes a connected component C if it matches any of the four orientations of a given shape S. This is demonstrated in Fig. 3(right). Furthermore, the Krawtchouk moments depend on the image size, which means that comparing two vectors requires that the same image size is used for the computation of both vectors and that some form of normalization is necessary. Considering these drawbacks of the Krawtchouk moment invariants we decided to use the well-known moment invariants of Hu for the other experiments described in this paper.

In Fig. 4 an image X consisting of the letters A, B, C, D, and E at different sizes and orientations is filtered with the goal of removing all instances of a certain letter in the image. As can be seen, especially the smallest letters in the image are not always removed when they should have been.

3.   Granulometries with respect to a shape family

Let $\Phi^{\vec{\tau}}_{S,\epsilon}$ be defined as above, and let $F = \{S_1, S_2, ..., S_n\}$ be a shape family with $F \subseteq C$. The vector-attribute thinning $\Phi^{\vec{\tau}}_{F,\epsilon}$ with respect to shape family F is defined as

DEFINITION 6 *The vector-attribute thinning $\Phi^{\vec{\tau}}_{F,\epsilon}$ of X with respect to a set F, with $F \subseteq C$ and using vector-attribute thinning with respect to shape $\Phi^{\vec{\tau}}_{S,\epsilon}$ is given by*

$$\Phi^{\vec{\tau}}_{F,\epsilon}(X) = \bigcap_{S \in F} \Phi^{\vec{\tau}}_{S,\epsilon} \qquad (16)$$

In other words, connected components are removed if they resemble any member of the shape family F closer than a given amount ϵ and are preserved otherwise. Again we have that $\Phi^{\vec{\tau}}_{F,\epsilon}$ is anti-extensive and idempotent, and scale, rotation, and translation invariance is inherited from $\vec{\tau}$. Furthermore,

$$\Phi^{\vec{\tau}}_{F,\epsilon}(\Phi^{\vec{\tau}}_{G,\epsilon}(X)) = \Phi^{\vec{\tau}}_{G,\epsilon}(\Phi^{\vec{\tau}}_{F,\epsilon}(X)) = \Phi^{\vec{\tau}}_{F,\epsilon}(X) \qquad \text{for } G \subseteq F. \qquad (17)$$

DEFINITION 7 *Assume we have N shapes $S_1, S_2, \ldots, S_N$ and let F_n be a set containing the $n \leq N$ shapes $S_1, \ldots, S_n$. A granulometry $\{\beta_n\}$ with respect to shape family F_N using vector-attribute thinning with respect to shape $\Phi^{\vec{\tau}}_{S_i,\epsilon}(X)$ for $S_i \in F_N$, is given by the family of vector-attribute thinnings with respect to shape family $\{\Phi^{\vec{\tau}}_{F_n,\epsilon}\}$ such that*

$$\beta_n = \Phi^{\vec{\tau}}_{F_n,\epsilon} \qquad (18)$$

It is easy to see that if all $\{\Phi^{\vec{\tau}}_{S_i,\epsilon}\}$ are a shape granulometry, then so is $\{\beta_n\}$.

The use of granulometries with respect to a shape family F for the computation of pattern spectra is demonstrated in Fig. 5, where a pattern spectrum of the input image in Fig. 4(left) is computed using a granulometry with respect to a family $F_n = \{S_1, \ldots, S_5\}$, with $S_1, \ldots, S_5$ representing the letters A till E respectively. As a comparison, a histogram was also computed representing the number of occurrences of each letter in the image.

4.   Conclusions

A new class of attribute filters was presented, whose attributes are vector instead of scalar values. These vector-attribute filtersvector-attribute filter are a subclass of the attribute filters defined by Breen and Jones. Using Hu's moment invariants, it was shown how thinnings and granulometries could be defined that filter images based on a given shape or a family of shapes.

For discrete images, the rotation- and scale-invariance of the moment invariant attributes is only by approximation. Furthermore, the rotation-invariance

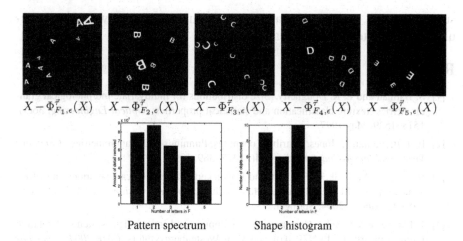

$$X - \Phi^{\vec{\tau}}_{F_1,\epsilon}(X) \quad X - \Phi^{\vec{\tau}}_{F_2,\epsilon}(X) \quad X - \Phi^{\vec{\tau}}_{F_3,\epsilon}(X) \quad X - \Phi^{\vec{\tau}}_{F_4,\epsilon}(X) \quad X - \Phi^{\vec{\tau}}_{F_5,\epsilon}(X)$$

Pattern spectrum   Shape histogram

Figure 5.   Pattern spectrum and shape histogram computed using $\Phi^{\vec{\tau}}_{F_n,\epsilon}(X)$ with $n = 1, 2, \ldots, 5$, resulting in filtering with family F_n, where F_n is the family of the first n letters in the alphabet. Each F_n includes one more shape to remove (top row)

of the Krawtchouk moment invariants does not hold for all orientations for shapes without rotational symmetry, due to the fact that the angle computed here refers to the orientation instead of the direction of the component. Although this problem can be solved by filtering using a few orientations of one shape, vector-attributes that do not have this problem, like Hu's moment invariants, are preferred. Future research will also investigate alternatives such as the complex moment invariants of Flusser and Suk [4]. More research is also needed to determine better ways for selecting the parameters like ϵ and the order and the choice of shape classes.

The dissimilarity measure d is also a critical choice. Other dissimilarity measures than the Euclidean distance should be investigated. If an adaptive system like a genetic algorithm would be used for d, an adaptive shape filter would be obtained. If multiple (reference) instances of the target class are available, the Mahalanobis distance is an option. This would lend more weight to directions in the attribute space Υ in which the class is compact, compared to directions in which the class is extended. Because we only use examples of the target class, the filtering problem resembles one-class classification [13]. This can be done with (kernel) density estimates to obtain a likelihood of class membership. The inverse of this probability would also yield a dissimilarity measure. Support-vector domain description could be used in a similar way [13].

An interesting approach would be the use of pattern spectra consisting of three dimensions: shape information from vector-attributes, size information such as the area, and the orientation of the components. This would be particularly useful in texture classification.

References

[1] S. Batman and E. R. Dougherty. Size distributions for multivariate morphological granulometries: texture classification and statistical properties. *Optical Engineering*, 36(5): 1518–1529, May 1997.

[2] E. J. Breen and R. Jones. Attribute openings, thinnings and granulometries. *Computer Vision and Image Understanding*, 64(3):377–389, 1996.

[3] B. Burgeth, M. Welk, C. Feddern, and J. Weickert. Morphological operations on matrix-valued images. In T. Pajdla and J. Matas, editors, *Computer Vision - ECCV 2004*, pages 155–167, Berlin, 2004. Springer.

[4] J. Flusser and T. Suk. Construction of complete and independent systems of rotation moment invariants. In N. Petkov and M.A. Westenberg, editors, *CAIP 2003: Computer Analysis of Images and Patterns*, pages 41–48, Berlin Heidelberg, 2003. Springer-Verlag.

[5] A. Hanbury and J. Serra. Mathematical morphology in the HLS colour space. In *12th British Mach. Vis. Conf.*, pages 451–460, Manchester, UK, September 2001.

[6] M. K. Hu. Visual pattern recognition by moment invariants. *IRE Transactions on Infomation Theory*, IT-8:179–187, 1962.

[7] R.A. Peters II. Mathematical morphology for angle-valued images. In *Nonlinear Image Processing VIII*, volume 3026 of *Proceedings of the SPIE*, pages 84–94, 1997.

[8] P. Maragos. Pattern spectrum and multiscale shape representation. *IEEE Trans. Patt. Anal. Mach. Intell.*, 11:701–715, 1989.

[9] G. Matheron. *Random Sets and Integral Geometry*. John Wiley, 1975.

[10] A. Meijster and M. H. F. Wilkinson. Fast computation of morphological area pattern spectra. In *Int. Conf. Image Proc. 2001*, pages 668–671, 2001.

[11] P. F. M. Nacken. Chamfer metrics, the medical axis and mathematical morphology. *Journal Mathematical Imaging and Vision*, 6:235–248, 1996.

[12] P. Salembier, A. Oliveras, and L. Garrido. Anti-extensive connected operators for image and sequence processing. *IEEE Transactions on Image Processing*, 7:555–570, 1998.

[13] D. M. J. Tax and R. P. W. Duin. Support vector domain description. *Pattern Recogn. Lett.*, 20:1191–1199, 1999.

[14] E. R. Urbach and J. B. T. M. Roerdink, and M. H. F. Wilkinson. Connected rotation-invariant size-shape granulometries. In *Proc. 17th Int. Conf. Pat. Rec.*, volume 1, pages 688–691, 2004.

[15] E. R. Urbach and M. H. F. Wilkinson. Shape-only granulometries and grey-scale shape filters. In *Proc. ISMM2002*, pages 305–314, 2002.

[16] L. Vincent. Granulometries and opening trees. *Fundamenta Informaticae*, 41:57–90, 2000.

[17] P. Yap, R. Paramesran, and S. Ong. Image analysis by Krawtchouk moments. *IEEE Trans. Image Proc.*, 12(11):1367–1377, November 2003.

RUMINATIONS ON TARJAN'S UNION-FIND ALGORITHM AND CONNECTED OPERATORS

Thierry Géraud

EPITA Research and Development Laboratory (LRDE)
14-16 rue Voltaire, F-94276 Le Kremlin-Bicêtre, France
Phone: +33 1 53 14 59 47, Fax: +33 1 53 14 59 22
thierry.geraud@lrde.epita.fr

Abstract   This papers presents a comprehensive and general form of the Tarjan's union-find algorithm dedicated to connected operators. An interesting feature of this form is to introduce the notion of separated domains. The properties of this form and its flexibility are discussed and highlighted with examples. In particular, we give clues to handle correctly the constraint of domain-disjointness preservation and, as a consequence, we show how we can rely on "union-find" to obtain algorithms for self-dual filters approaches and levelings with a marker function.

Keywords: Union-find algorithm, reconstructions, algebraic openings and closings, domain-disjointness preservation, self-dual filters, levelings.

Introduction

Connected operators have the important property of simplifying images while preserving contours. Several sub-classes of these operators have been formalized having stronger properties [8] and numerous applications have been derived from them, e.g., scale-space creation and feature analysis [17], video compression [14], or segmentation [10]. The behavior of connected operators is to merge most of the flat zones of an input image, thus delivering a partition which is much coarser than the input one. In that context, a relevant approach to implement such operators is to compute from an input image the resulting partition. The Tarjan's Union-Find Algorithm, *union-find* for short, computes a forest of disjoint sets while representing a set by a tree [16]. A connected component of points or a flat zone is thus encoded into a tree; a point becomes a node and a partition is a forest. union-find has been used to implement some connected operators; among them, connected component labeling [2], a watershed transform [6], algebraic closing and opening [19], and component tree

C. Ronse et al. (eds.), Mathematical Morphology: 40 Years On, 105–116.
©2005 *Springer. Printed in the Netherlands.*

computation [4, 11]. A tremendous advantage of union-find lies in its *simplicity*. However, the descriptions of morphological operators relying on this algorithm are usually spoiled by the presence of too many implementation details.

This paper intends to provide the image processing community with a simple and general form of union-find, which is highly adaptable to the large class of connected operators. We show that the description of a given operator with union-find is actually straightforward, comprehensive, and takes very few code. We also present how union-find can be used for the connected operators θ which verify a domain disjointness preservation property. Consequently we show that union-find is a simple way to get algorithms for folding induced self-dual filters [5], the inf-semilattice approach to self-dual morphology [3], and levelings defined on two functions [10].

In order to keep implementation details away from algorithmic considerations, we do not address any single optimization issue. Moreover, we do not enter into a comparison between union-find-based algorithms and other approaches; for those subjects, the reader can refer to [13, 7]. We claimed in [1] that our generic C++ image processing library, Olena [12], has been designed so that algorithms can easily be translated into programs while remaining very readable. To sustain this claim, programs given in this paper rely on our library and, thanks to it, they efficiently run on various image structures (signals, 2D and 3D images, graphs) whatever their data types (Boolean, different integer encodings, floating values, etc.)

In the present document we start from the simplest operator expressed with union-find, namely a connected component labeling, in order to bring to the fore the properties of union-find-based algorithms (Section 1). We stress on the notions of domains and of disjointness-preservation and we present a general formulation of union-find (Section 2). In the second part of this document we give a commented catalogue of connected operators with the help of that formulation (Section 3). Last we conclude (Section 4).

1.   Practicing on Connected Component Labeling

In union-find, a connected component is described as a tree. At any time of the algorithm computation, each existing component has a canonical element, a root point at the top of the tree. A link between a couple of nodes within a tree is expressed by a parent relationship. A convenient way to handle the notion of "parent" is to consider that parent is an image whose pixel values are points—given a point x, parent[x] is a point—and that a root point is its own parent. Finding the root point of a component recursively goes, starting from a point of this component, from parent to parent until reaching the root point.

Let us practice first on connected component labeling. The union-find algorithm is composed of an initialization followed by two passes. The first one aims at computing the tree forest and the second one aims at labeling.

```
void init() {
    is_processed = false; // that is, for all points
    cur_label = 0;
    O = false; // background is the default value
}

void first_pass() {
    bkd_scan p(D); // R_D is the bkd scan
    nbh_scan n(nbh);
    for_all (p)
        if ( is_in_I(p) ) { // "body 1.1"
            make_set(p); // so {p} is a set
            for_all_neighbors ( n, p )
                if ( D.holds(n) ) // n belongs to D
                    if ( is_processed[n] )
                        do_union(n, p);
            is_processed[p] = true;
        }
}

void second_pass() {
    fwd_scan p(I); // versus bkd in 1st pass
    for_all (p)
        if ( is_in_I(p) ) { // "body 2.1"
            if ( is_root(p) )
                set_O_value(p); // new label
            else
                O[p] = O[parent[p]]; // propagation
        }
}
```

In the first pass, points of the input set I are browsed in a given order. In the case of image processing, the domain D of the input image includes I. Practically, I is represented by its membership function defined over D and is encoded as a binary image. A convenient way to browse elements of I is thus performed by a classical forward or backward scan of the points p of D and testing if p is in I is required. Let us denote by $\mathcal{R}_D$ the ordering of points of D which corresponds to the way we browse D during the first pass. The current point p is first turned into a set: $\{p\}$. Neighbors of p are then inspected to eventually merge p with an existing tree. Let us denote by n a neighbor of p such as $n \in I$ (actually we have to ensure that n actually lies in the image domain D to prevent problems when p happens to belong to the internal boundary of D). If n has not been processed yet, it does not make sense to inspect n since it does not belong to a tree structure. In the other case, we proceed to the union of $\{p\}$ and of the set to which n belongs.

A key point of union-find lies in this task; performing the union of a couple of trees is a single simple operation: linking their respective root points, say r_1 and r_2 with $r_1 \mathcal{R}_D r_2$, so that the parent of r_1 is r_2. This rule ensures a very strong property: when processing p in the first loop, $\forall p'$ such as $p' \mathcal{R}_D p$, we have $p' \mathcal{R}_D parent(p')$ and $parent(p') \mathcal{R}_D p$. Adding the point p to the connected component that contains n is then as simple as setting to p the parent of r, where r is the root point of $\Gamma_I(n)$.

The second pass of union-find browses points of D in the reverse order, $\mathcal{R}_D^{-1}$, as compared to the first pass. Thanks to parenthood construction, the following two properties hold. 1) For any component (or flat zone) computed during the first pass, the first point of this component visited during the second pass is the root point. 2) In the second pass, a point is always visited after its parent. So, for each point $p \in I$ using $\mathcal{R}_D^{-1}$, we assign p a new label in the output image O if p is root, otherwise we assign p the same label as its parent.

```
bool is_in_I(point p) { return I[p] == true; }

void make_set(point p) { parent[p] = p; }

bool is_root(point p) { return parent[p] == p; }

point find_root(point x) {
  if ( is_root(x) ) return x;
  else return ( parent[x] = find_root(parent[x]) );
}

void set_parent(point r, point p) { parent[r] = p; }

void do_union(point n, point p) {
  point r = find_root(n);
  if ( r != p )
    set_parent(r, p);
}

L set_O_value(point p) { return ++cur_label; }
```

Both passes of connected component labeling using union-find are illustrated in Figure 1.

2.   Introducing Domains and Disjointness-Preservation

The main characteristic of union-find appears to be in its overall structure. First, let us take a partition of the image domain D into $m + 1$ disjoint sets defined by: $D = D' \cup D''$ with $D' = (\cup_{i=1}^{m} D_i)$. In this partition D'' is the set of points of D that are not subject to forest computation.

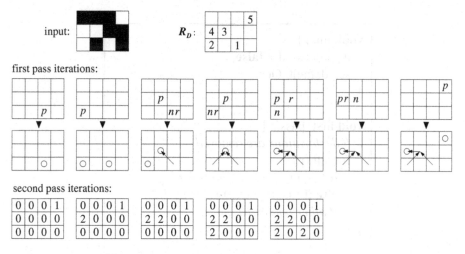

Figure 1. Connected component (8-connectivity) labeling using union-find: arrows represent parenthood and circles depict root points.

A requirement about D'' is therefore that, $\forall p \in D''$, the value $O[p]$ can be directly computed from the operator input. So we compute O over D'' as a whole in the initialization part of union-find (set_default_O routine). Also note that we will never have proper values for parenthood in D''.

A sub-domain D_i with $i = 1..m$ can have its own definition of forest computation, different from the ones of D_j with $j \neq i$. Consider for instance the simultaneous connected component labeling of both object and background in a binary 2D image; obviously we have $D'' = \emptyset$, $D_1 = I$, and $D_2 = I^c$. Processing two forests then allows us to rely on two distinct neighborhoods so that topological inconsistency is avoided. Keeping this idea in mind, the description of union-find mutes into a more general form depicted in Figure 2. Both first and second passes process in an independent way the domains D_i thanks to a test is_in_Di; that gives rise to the "*body 1.i*" and "*body 2.i*" sections. Furthermore, some variations between bodies in a same pass are now conceivable.

Preserving domain disjointness.   A strong assumption is implicitly managed in the writing of "*body 1.i*" due to the tests "is_in_Di(n)": we are preserving disjointness over domains, that is, each connected component (or flat zone) Γ created by such an algorithm cannot cross domains frontiers. We have:

$$\forall \Gamma \in O, \ \exists i \ \text{such as} \ \Gamma \subseteq D_i \ \text{and} \ \forall j \neq i, \ \Gamma \cap D_j = \emptyset.$$

Visiting domain boundaries.   During the first pass, while processing a neighbor n of p, if we do not enter component computation, that means that n does not belong to D_i. If $m \geq 2$, that also means that n can have already been

```
void init() {
  is_processed = false;
  set_default_O();
}

void first_pass() {
  // declarations are removed
  for_all (p)
    if ( is_in_D1(p) ) // body 1.1
    else if ( is_in_D2(p) ) // body 1.2
    // other bodies 1.i ...
}

void second_pass() {
  // declarations are removed
  for_all (p)
    if ( is_in_D1(p) ) // body 2.1
    else if ( is_in_D2(p) ) // body 2.2
    // other bodies 2.i ...
}

// with body 1.i being:
{
  make_set(p);
  for_all_neighbors(n, p)
    if ( D.holds(n) )
      if ( is_in_Di(n) and is_processed[n] )
        do_union(n, p);
      // optional:
      else visit_extB_of_Di(n, p);
  set_processed(p);
}

// with body 2.i being:
{
  if ( is_root(p) ) set_O_value_in_Di(p);
  else O[p] = O[parent[p]];
}
```

Figure 2.   General form of union-find with domains.

processed or not. Since $p \in D_i$, n belongs to the external boundary $\mathcal{B}^{ext}(D_i)$ of D_i and, when the first pass is completed, we have visited all the points of $\mathcal{B}^{ext}(D_i)$. This general version of union-find thus get us the ability of fetching information from $\mathcal{B}^{ext}(D_i)$ (through the visit_extB_of_Di routine).

Attaching auxiliary data to components.   A connected component encoded as a tree is represented by its root point. If one intends to implement with union-find a particular operator which requires information on components, some auxiliary data just have to be attached to every root points. Furthermore, we have the ability to attach to components a distinct type of data per domain. So we have to adapt the routines which deal with parenthood in the first pass:

```
void make_set_in_Di(point p) {
    parent[p] = p; // creation of {p}
    init_data_in_Di(p); // so p has data
}
void set_parent_in_Di(point r, point p) {
    parent[r] = p; // 2 components are now connected
    merge_data_in_Di(r, p); // so p carries data
}
```

Last, please remark that updating data is possible while visiting boundaries $\mathcal{B}^{ext}(D_i)$ (first pass) and that data are usually expected to influence the operator output (second pass, routine set_O_value_in_Di, called when p is root).

Extension to morphology on functions.   In the field of morphology on grey-level functions, given a function f, two trivial orderings between points of D can be derived from $\mathcal{R}_D$ and from the one of the complete lattice framework: $p\mathcal{R}^\uparrow p' \Leftrightarrow f(p) < f(p')$ or $(f(p) = f(p')$ and $p\mathcal{R}_D p')$ and its reverse ordering $\mathcal{R}^\downarrow$. If we choose $\mathcal{R}^\uparrow$ for the first pass of union-find, the evolution of connected components during this pass mimics a flooding of f and we get an extensive behavior of the connected operator. By duality, we obtain an anti-extensive behavior with $\mathcal{R}^\downarrow$. In the literature about implementing connected operators with union-find, namely algebraic openings and closings in [19, 7], the notion of domains is absent and the whole image domain is processed. We actually have $D = D'$ (so $D'' = \emptyset$) and at first glance that seems relevant. A novelty here appears during the first pass: connected components can grow even by merging flat zones of the input image. The expansion of a component is stopped when a given increasing criterion is no more satisfied; this component then turns "inactive". That leads us to:

```
void init_data_in_Di(point p) {
  is_active[p] = true; // and handle other data...
}
void do_union_in_Di(point n, point p) {
  point r = find_root(n);
  if ( r != p )
    if ( is_flat_in_Di(r, p) or equiv_in_Di(r, p) )
      set_parent_in_Di(r, p);
}
bool equiv_in_Di(point r, point p) {
  if ( not is_active[r] or not is_active[p] )
    return false;
  if ( not satisfies_criterion_in_Di(r, p) ) {
    is_active[p] = false;
    return false;
  }
  return true;
}
```

Connected operators relying on two functions.   So far, just notice that, changing the overall structure of union-find from the first way of browsing points of D to the second one (see below) does not affect the result of the algorithm thanks to the domain disjointness preservation property.

```
// first way of browsing points
for_all (p)
        if ( is_in_D1(p) ) // body *.1
    else if ( is_in_D2(p) ) // body *.2
    // ...

// second way of browsing points
    for_all (p_in_D1) // body *.1
    for_all (p_in_D2) // body *.2
    // ...
```

3.   A Catalogue of Union-Find-Based Connected Operators

In the previous section, we have presented a general form for union-find that relies on a separation between domains. We then just have to fill the holes of this form to get algorithms that implement some connected operators.

Morphology on Sets with Union-Find

Let us first take as an example the reconstruction by dilation operator, R^δ, which applies on a two sets, a marker F and a mask G such as $F \subseteq G$. In the following, given a set X and a point $x \in X$, we will denote by Γ_X a connected component of X. Denoting $O = R_G^\delta(F)$, we have the property $O \subset G$ and, since R^δ is a connected operator with respect to the mask, $\{\Gamma_O\} \subseteq \{\Gamma_G\}$. A first obvious idea for using union-find is then to compute the connected components of G and search for those that belong to O. So we have $D' = G$ and $\Gamma_{D'}$ are computed. However, we can consider the operator input as a four-valuated function $I_{F,G}$ defined on D, such as the value $I_{F,G}(p)$ indicates whether $p \in D$ is included in $F \cap G$, $F \cap G^c$, $F^c \cap G$, or $F^c \cap G^c$. That leads to different ways of using union-find to implement reconstructions depending on the choice of a partition of D into $D' \cup D''$.

DEFAULT	INIT(p)	BORDER(n, p)	MERGE(r, p)
R^δ obvious version: $D' = G \rightsquigarrow (\Gamma_{D'} \subseteq O \Leftrightarrow \exists p \in \Gamma_{D'}, p \in F)$			
O = false;	O[p] = F[p];		O[p] = O[r] **or** O[p];
R^δ alternative version: $D' = F^c \cap G \rightsquigarrow (\Gamma_{D'} \subseteq O \Leftrightarrow \exists n \in \mathcal{B}^{ext}(\Gamma_{D'}), n \in F)$			
O = F;	O[p] = **false**;	**if** (F[n]) O[p] = **true**;	O[p] = O[r] **or** O[p];
R^ϵ dual version of "R^δ with $D' = G$": $D' = G^c \rightsquigarrow (\Gamma_{D'} \subseteq O \Leftrightarrow \forall p \in \Gamma_{D'}, p \in F)$			
O = true;	O[p] = F[p];		O[p] = O[r] **and** O[p];
R^ϵ alternative version: $D' = F \cap G^c \rightsquigarrow (\Gamma_{D'} \subseteq O \Leftrightarrow \forall n \in \mathcal{B}^{ext}(\Gamma_{D'}), n \in F)$			
O = G;	O[p] = **true**;	**if** (**not** F[n]) O[p] = **false**;	O[p] = O[r] **and** O[p];

The table above presents for several operators the respective definitions of the following routines (from left to right): set_O_default_in_Di, init_data_in_Di, visit_extB_of_Di, and merge_data_in_Di. Last, when the only auxiliary data required to implement an operator represent the Boolean evaluation of $[\Gamma_{D'} \subseteq O]$, the output image can store these data as an attachment to root points and set_O_value has nothing left to perform. From our experiments, an appropriate choice for D'—depending on *a priori* knowledge about F and G—makes the union-find-based approach a serious competitor of the efficient hybrid algorithm proposed in [18]. Last, the case of regional extrema identification is summarized in the table below.

DEFAULT	INIT(p)	BORDER(n, p)	MERGE(r, p)
Regional minima identification of a function f with $D' = D$ (with either $\mathcal{R}^\downarrow$ or $\mathcal{R}^\uparrow$)			
	O[p] = **true**;	**if** (f[n] < f(p)) O[p] = **false**;	O[p] = O[r] **and** O[p];
Regional minima identification with $D' = D$ which relies on $\mathcal{R}^\downarrow$			
	O[p] = **true**;	**if** (is_processed[n] **and** f[n] != f(p)) O[p] = **false**;	O[p] = O[r] **and** O[p];

Extension to Functions

For some connected operators on functions that deliver functions, the table below recaps their corresponding definitions with union-find; the columns CRIT and VALUE respectively depict the result returned by satisfies_criterion and the body of set_O_value. For reconstructions, f and g being the input marker and mask functions, we compute flat zones from g, which is flattened by the operator, is_flat(r,p) returns g[r] == g[p].

INIT(p)	MERGE(r,p)	CRIT(r,p)	VALUE(p)
R^δ (f marker and g mask such as $f \leq g$); $\mathcal{R}^\downarrow$ is mandatory since g is lowered			
o[p] = f[p];	o[p] = max(o[r], o[p]);	g[p] >= o[r]	**if (not** is_active[p]) o[p] = g[p];
R^ϵ (f marker and g mask such as $f \geq g$); $\mathcal{R}^\uparrow$ is mandatory since g is "upper-ed"			
o[p] = f[p];	o[p] = min(o[r], o[p]);	g[p] <= o[r]	**if (not** is_active[p]) o[p] = g[p];
Area opening (resp. closing) of a function f; $\mathcal{R}^\downarrow$ (resp. $\mathcal{R}^\uparrow$) is mandatory			
area[p] = 1;	area[p] = area[r] + area[p];	area[r] < λ	o[p] = f[p];
Volume opening (resp. closing) of f; $\mathcal{R}^\downarrow$ (resp. $\mathcal{R}^\uparrow$) is mandatory			
area[p] = 1;	vol[p] = vol[r] + vol[p] + (area[p] $*$ abs(f[r]$-$f[p]))		
vol[p] = 1;	area[p] = area[r] + area[p];	vol[r] < λ	o[p] = f[p];

Operators Relying on Two Functions

Many morphological operators over two functions, f and g, have been defined from a couple of connected operators, φ extensive and ψ anti-extensive, following the general formulation [8, 5, 3]:

$$[\theta(f,g)](p) = \begin{cases} [\varphi(f,g)](p) & \text{if } p \in D^\uparrow(f,g) \\ [\psi(f,g)](p) & \text{if } p \in D^\downarrow(f,g) \\ f(p) & \text{otherwise.} \end{cases}$$

under the constraint of being disjointness-preservative regarding $D^\uparrow(f,g)$ and $D^\downarrow(f,g)$, the domains of D where θ is *expected* to be respectively extensive and anti-extensive. Let us denote by $D^\circ(f,g) = (D^\uparrow(f,g) \cup D^\downarrow(f,g))^c$ the domain where θ is expected to be constant. In the following we will elude in domain names the dependence upon f and g since that does not lead to any ambiguity. By extension, let us introduce $D^\Phi = D^\downarrow \cup D^\circ$ and $D^\Phi = D^\uparrow \cup D^\circ$.

Relying on union-find to get an algorithm starts with choosing domains such as $D = D' \cup D'' = (\cup_i D_i) \cup D''$ and $\forall j \neq i, D_j \cap D_i = \emptyset$. As a constraint we do not want to use the ability of union-find to visit domain boundaries. Since θ is *not* disjointness-preservative with respect to $D^\uparrow$ and D°, and to $D^\downarrow$ and D°, we cannot obtain a correct result with union-find when we consider setting the domains D_i with any combination of $D^\uparrow$, D°, and $D^\downarrow$. So far we are in a dead end.

Let us imagine that we relax the disjointness constraint of union-find (!) to form $D_1 = D^\Phi$ and $D_2 = D^\Phi$. The only weird aspect of this idea is that points of D° have to be processed twice during the first pass of union-find. For that, we just have to add a "refresh" step for the points of D° just after handling D^Φ during the first pass, so that D^Φ can be properly processed. This single modification is handled as follows:

```
for_all (p_in_D_upper_or_equal)
  // body 1.1
for_all (p_in_Do)
  is_processed[p] = false; // so p can be handled again
for_all (p_in_D_lower_or_equal)
  // body 1.2
```

and the second pass as described in Figure 2 remains unchanged. Although we have introduced a bond between domains (we have $D_1 \cap D_2 \neq \emptyset$), we do not have introduced any inconsistency in parenthood. Put differently, this modified version of union-find does not compute irrelevant components or flat zones.

We can now reuse the descriptions given previously of union-find-based operators to build levelings with markers as defined in [9], a domain-preserving self-dual reconstruction, and partial self-dual operators defined with the inf-semilattice approach in [3]. Results are summarized in the table below.

D^{Φ}			D^{Φ}		
$\mathcal{R}_D$	sub-domain	sub-operator	$\mathcal{R}_D$	sub-domain	sub-operator
some levelings g' of g given a function f such as $g' \in Inter(g, f)$ (see [10])					
$\mathcal{R}^{\downarrow}$ on g	$f(p) \leq g(p)$	γ, any lower leveling	$\mathcal{R}^{\top}$ on g	$f(p) \geq g(p)$	ϕ, any upper leveling
domain-preserving self-dual reconstruction with f marker and g mask					
$\mathcal{R}^{\downarrow}$ on g	$f(p) \leq g(p)$	$R_g^{\delta}(f)$	$\mathcal{R}^{\top}$ on g	$f(p) \geq g(p)$	$R_g^{\varepsilon}(f)$
inf-semilattice approach (input function is f) with any anti-extensive operator ψ					
$\mathcal{R}^{\downarrow}$ on $-f$	$f(p) \leq 0$	$-\psi(-f)$	$\mathcal{R}^{\downarrow}$ on f	$f(p) \geq 0$	$\psi(f)$

4.   Conclusion

We have presented a general formulation of Tarjan's union-find algorithm so that many connected operators can be straightforwardly mapped into algorithms; we definitely believe that this particular formulation can ease to express new segmentation methods using connected operators relying on union-find.

References

[1] J. Darbon, T. Géraud, and A. Duret-Lutz. Generic implementation of morphological image operators. In *Mathematical Morphology, Proc. of ISMM*, pages 175–184. Sciro, 2002.

[2] M. Dillencourt, H. Samet, and M. Tamminen. A general approach to connected-components labeling for arbitrary image representations. *Journal of the ACM*, 39(2):253–280, 1992.

[3] H. Heijmans and R. Keshet. Inf-semilattice approach to self-dual morphology. *Journal of Mathematical Imaging and Vision*, 17(1):55–80, 2002.

[4] W. H. Hesselink. Salembier's min-tree algorithm turned into breadth first search. *Information Processing Letters*, 88(1–2):225–229, 2003.

[5] A. Mehnert and P. Jackway. Folding induced self-dual filters. In *Mathematical Morphology and its Applications to Image and Signal Processing*, pages 99–108, 2000.

[6] A. Meijster and J. Roerdink. A disjoint set algorithm for the watershed transform. In *EUSIPCO IX European Signal Processing Conference*, pages 1665–1668, 1998.

[7] A. Meijster and M. Wilkinson. A comparison of algorithms for connected set openings and closings. *IEEE Trans. on PAMI*, 24(4):484–494, 2002.

[8] F. Meyer. From connected operators to levelings. In *Mathematical Morphology and its Applications to Image and Signal Processing*, pages 191–198. Kluwer, 1998.

[9] F. Meyer. The levelings. In *Mathematical Morphology and its Applications to Image and Signal Processing*, pages 199–206. Kluwer, 1998.

[10] F. Meyer. Levelings, image simplification filters for segmentation. *Journal of Mathematical Imaging and Vision*, 20(1–2):59–72, 2004.

[11] L. Najman and M. Couprie. Quasi-linear algorithm for the component tree. In *IS&T/SPIE Symposium on Electronic Imaging, In Vision Geometry XII*, pages 18–22, 2004.

[12] Olena. Generic C++ image processing library, `http://olena.lrde.epita.fr`, free software available under GNU Public Licence, EPITA Research and Development Laboratory, France, 2005.

[13] J. B. Roerdink and A. Meijster. The watershed transform: Definitions, algorithms and parallelization strategies. *Fundamenta Informaticae*, 41(1-2):187–228, 2000.

[14] P. Salembier and J. Ruiz. On filters by reconstruction for size and motion simplification. In *Mathematical Morphology, Proc. of ISMM*, pages 425–434. Sciro Publishing, 2002.

[15] P. Soille. *Morphological Image Analysis*. Springer-Verlag, 1999.

[16] R. E. Tarjan. Efficiency of a good but not linear set union algorithm. *Journal of the ACM*, 22(2):215–225, 1975.

[17] C. Vachier. Morphological Scale-Space Analysis and Feature Extraction. In *IEEE Intl. Conf. on Image Processing*, volume 3, pages 676–679, October 2001.

[18] L. Vincent. Morphological grayscale reconstruction in image analysis: Applications and efficient algorithms. *IEEE Trans. on Image Processing*, 2(2):176–201, 1993.

[19] M. Wilkinson and J. Roerdink. Fast morphological attribute operations using tarjan's union-find algorithm. In *Mathematical Morphology and its Applications to Image and Signal Processing, Proc. of ISMM*, pages 311–320, 2000.

LABELLED RECONSTRUCTION
OF BINARY OBJECTS:
A VECTOR PROPAGATION ALGORITHM

Michael Buckley[1] and Ryan Lagerstrom[1]

[1]*CSIRO Mathematical and Information Sciences*
Locked Bag 17, North Ryde, NSW 2113, Australia
Michael.Buckley,Ryan.Lagerstrom@csiro.au

Abstract   The *quench function* of a binary image is the distance transform of the image sampled on its skeleton. In principle the original image can be *reconstructed* from the quench function by drawing a disk at each point on the skeleton with radius given by the corresponding quench function value. This reconstruction process is of more than theoretical interest. One possible use is in coding of binary images, but our interest is in an applied image analysis context where the skeleton has been (1) *reduced* by, for example, deletion of barbs or other segments, and/or (2) *labelled* so that segments, or indeed individual pixels, have identifying labels. A useful reconstruction, or partial reconstruction, in such a case would be a labelled image, with labels propagated from the skeleton in some intuitive fashion, and the support of this labelled output would be the theoretical union of disks.

An algorithm which directly draws disks would, in many situations, be very inefficient. Moreover the label value for each pixel in the reconstruction is highly ambiguous in most cases where disks are highly overlapping. We propose a vector propagation algorithm based on Ragnelmalm's Euclidean distance transform algorithm which is both efficient and provides a natural label value for each pixel in the reconstruction. The algorithm is based on near-exact Euclidean distances in the sense that the reconstruction from a single-pixel skeleton is, to a very good approximation, a Euclidean digital disk. The method is illustrated using a biological example of neurite masks originating from images of neurons in culture.

Keywords: debarbing, Euclidean distance transform, object reconstruction, quench function, skeleton, vector propagation

C. Ronse et al. (eds.), Mathematical Morphology: 40 Years On, 117–128.

1.   Introduction

This paper deals with the process of *reconstruction* of a binary image using a sub-sample of the (Euclidean) *distance transform* of the image. Figure 1 shows part of a binary image representing neurites in an image of neurons in culture, while Figure 2 shows the Euclidean distance transform (EDT) of this image sampled on a *skeleton* of the image, with darker pixels corresponding to higher values of the distance transform. When, as here, a distance transform is sampled on a skeleton, the result is called a *quench function* [11, p. 159].

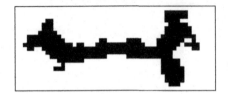

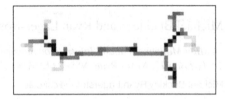

Figure 1.   Neurite Mask.   *Figure 2.*   Quench Function.

The original mask, Figure 1, can be *reconstructed* from the quench function provided the skeleton is, in some sense, sufficiently complete. This can, in principle, be achieved by drawing a binary disk at each point on the skeleton with radius given by the quench function value at that point. While this process may be useful for coding and decoding of binary image data, it is of theoretical interest only in a typical context of analysis of images where the original mask is already known anyway. A more interesting case is shown in Figures 3 and 4. The reduced quench function in Figure 3 has been obtained from the full quench function in Figure 2 by removing some of the smaller "barbs". The topology has slightly changed for complicated reasons in this particular example; the details are not relevant to this discussion.

The binary image in Figure 4 is a *partial reconstruction* of the original image produced by the new algorithm described in this paper using the reduced quench function. It may be seen that most but not all of the original image has been reconstructed. This is a first application of this kind of algorithm, namely filtering of binary masks of linear features such as neurite networks. Such networks can be complex and noisy. The process illustrated here cleans and simplifies a network mask essentially by (1) computing the quench function of the mask, (2) topologically simplifying (e.g. debarbing) the quench function, and finally (3) reconstructing a reduced mask from the reduced quench function.

Figures 5 and 6 illustrate another important aspect of the new algorithm, namely *label propagation*. Figure 5 shows a labelled skeleton corresponding to the (reduced) quench function in Figure 3. That is, Figures 3 and 5 are two positive valued functions defined on the same set of skeleton points, Figure 3

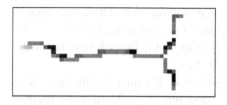

Figure 3. Reduced Quench Function.

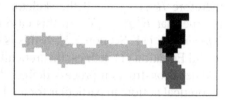

Figure 4. Partial Reconstruction.

giving a distance value for each pixel and Figure 5 giving a label value. In this case there are three label values, one for each of the three "branches" of the skeleton. The result of the labelled reconstruction is shown in Figure 6. Here each point in the reconstruction has been assigned one of the labels from the labelled skeleton. In this application a skeletonised neurite mask had been not

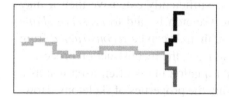

Figure 5. Labelled Skeleton.

Figure 6. Reconstruction with Labels.

only debarbed, but also each branch had been labelled as primary, secondary, tertiary etc. by another process. For each branching level of each tree, various measures such as width and brightness were needed, as well as masks for display purposes. Reduced and labelled masks such as that shown in Figure 6 fill these needs.

The remainder of this paper is organised as follows. Section 2 covers fundamental definitions of distance transforms, quench functions and reconstruction as a theoretical concept. Section 3 reviews the EDT algorithm of Ragnemalm [9] upon which the new algorithm is based. Finally Section 4 describes and discusses the new algorithm, and finally Section 5 shows some further examples.

2. Quench Functions and Reconstruction of Sets

The idea of a *quench function* is of both theoretical and practical interest in image analysis. For a set X, the *distance function*, D_X, is defined in terms of a metric d as

$$D_X(x) = \inf_{y \notin X} d(x, y);$$

see, for example, [11, Section 2.7]. The *skeleton*, S, of X is a subset of X which is defined in terms of the distance function. The quench function is defined simply as the restriction of the distance function to S. There are various possible definitions both of distance functions D_X and of skeletons $S \subseteq X$ (for example, see [5]), but given definitions of these the quench function may simply be defined as the restriction of D_X to S.

A closely related notion is that of *reconstruction*. If a quench function q has been defined from a distance function D_X and sets $S \subseteq X$, then the reconstruction $R(q)$ of q is

$$R(q) = \bigcup_{x \in S} \left(x + B_{q(x)} \right) \tag{1}$$

where B_r is the zero-centred ball of radius r with respect to metric d:

$$B_r = \{ y : d(y,0) < r \}.$$

It follows from the definitions that $x + B_{q(x)} \subseteq X$ for all $x \in X$, and therefore $R(q) \subseteq X$. If the skeleton S is sufficiently extensive then it may be true that $R(q) = X$. In this case the skeleton is said to *preserve shape information* [11, Section 5.5]. If this is true then X can be *reconstructed* from q, and Equation (1) defines, theoretically at least, the reconstruction process.

The reconstruction process defined by Equation (1) is often used just as a theoretical notion, in particular for analysing the properties of skeletons. However the actual computation of such a reconstruction is of practical interest in at least two contexts:

- coding of binary sets by their quench functions, and

- reconstruction of only a part of a set after deletion of part of its skeleton — for example, after pruning.

Here we are interested in the second of these two situations. If part of the skeleton has been deleted we would not in general expect the reconstruction via Equation (1) to produce all of X. In this paper we will in fact consider the computation of the reconstruction (1) given an arbitrary positive-valued function q defined on an arbitrary set S. That is, we will not assume that S is the skeleton of the set X, or that the values of $q(x)$ are actually sampled values of the distance function D_X for any set X.

In practice the quench function q will have the form of an image with non-negative values, and we will take $q(x) = 0$ to mean that x is outside the domain S of q, although strictly this is not necessary as $B_{q(x)} = \emptyset$ if $q(x) = 0$ and therefore such points contribute nothing to the reconstruction Equation (1).

As discussed in the Introduction we will also make use of an input *label* value for each point on the skeleton. This is an image of the same dimensions as q whose values are assumed to be non-zero wherever the corresponding value of q is non-zero. These label values on the skeleton are propagated in the reconstruction process to give a label value for each pixel in the reconstruction.

3. Euclidean Distance Transform (EDT) by Vector Propagation

In this section we describe a version of the algorithm proposed by Ingemar Ragnemalm in 1992 [9] for computation of a near-exact EDT via vector propagation with priority queues. The new algorithm described in this paper is based on this algorithm. In fact, the new algorithm is a kind of inverse of Ragnemalm's method which is a vector propagation method derived from the work of Danielsson [3].

Vector propagation methods for EDT's vector propagation are not exact, but when eight neighbours are considered at each propagation step, the great majority of distance values are exactly correct. All distance values up to 13 pixel units are guranteed to be exactly correct, and for those larger values which are not exactly correct, the maximum relative error is less than 0.3%; see [2, Section 2.2.2]. We therefore describe these methods, and the new method described in this paper, as "near-exact".

Simple versions of propagating reconstruction algorithms were given in [1] and [8]. These methods compute chamfer distances which, as is well-known, are relatively poor approximations to Euclidean distances, their contours typically being diamonds, octagons or other polygons rather than circles. Also the methods of [1] and [8] do not propagate labels.

Finally we note that several exact EDT methods have been developed, for example, [10, 4, 12, 7]. These exploit the separability of the squared Euclidean metric, but it does not seem easy to adapt these methods to reconstruction or label propagation. For surveys of EDT algorithms, see [2, Chapter 1] and [11, Section 2.7].

Our notation changes slightly here: whereas in the previous section X was a set, now X is a binary image with $X[j] = 1$ if and only if j is in the set X. We say j is a *foreground pixel* if $X[j] = 1$, and a *background pixel* if $X[j] = 0$. The value for each pixel i in the output image is the Euclidean distance to the nearest background pixel. This is the standard definition of the EDT.

As discussed in the Introduction above, the Ragnemalm method is not exact, but very close to exact. We note also that, as the squared EDT is always an integer for data on a rectangular grid, this form is sometimes more convenient. This algorithm can trivially be modified to produce a squared EDT.

The algorithms discussed in this paper make use of a *priority queue*. This is a data structure which allows an arbitrary number of elements to be added, in turn, each with a *priority*. Unless the queue is empty, the *first element* can be removed at any time, and this first element is guaranteed to have the smallest, or equal-smallest, priority value of all current elements on the queue. Priority queues usually support at least one identifier, or key, for each element. For these algorithms we require two such keys which we use to represent

- the index i of a pixel, and

- the index a of a *candidate ancestor* for pixel i – see below.

We use the word "index", but effectively i and a are two-dimensional co-ordinates of pixel locations.

The essential idea of the Ragnemalm EDT algorithm is to store index pairs, (pixel, ancestor) = (i, a), on a priority queue, ordered by $|i - a|$. Pixel i is always a foreground pixel and pixel a is a background pixel which is a *candidate ancestor* for i. The *true ancestor* of i is by definition the nearest background pixel to i in terms of Euclidean distance. Note that the ancestor of a foreground pixel i need not be unique, but the minimal distance $|i - a|$ is unique.

Initially, the output image output is set to zero values. Then all pairs (i, a) are placed on the queue where i and a are 8-neighbours, i is foreground and a is background. When the first (i_0, a_0) pair is removed from the queue, it is guaranteed that a_0 is the a true ancestor of i_0, and therefore we can assign output$[i_0] = |i_0 - a_0|$. This is because after the initialisation, while a foreground pixel i_0 may be, and usually is, enqueued with more than one neighbouring background pixel a,

- at least one of these is a true ancestor of i_0, and

- the nearest of these to i_0 is foremost in the queue as distance $|i - a|$ is used as the queue priority.

In general other pairs (i_0, a) will be removed from the queue later, but these cannot be closer pairs. We say that pixel i_0 has already been propagated, and this is indicated by the fact that output$[i_0] \neq 0$. Therefore any pair (i, a) coming from the queue with output$[i] \neq 0$ is ignored.

After such propagation of a pixel i_0, the neighbours j of i_0 are all processed. Neighbours j are ignored unless

- they are foreground pixels ($X[j] = 1$), and

- they have not been propagated already (output$[j] = 0$).

For any such neighbours j the pair (j, a) is placed on the queue. That is, the ancestor a of i_0 is used as a candidate ancestor for its neighbour j.

Then the process repeats: pairs (i, a) are repeatedly taken from the front of the queue and the distance values are propagated further and further from the background — i.e. from the outside in — and the process continues until the queue is empty. A nearly exact EDT is produced by this process because, except in very few cases, the true ancestor of each foreground pixel is also the true ancestor of one of its 8-neighbours with a smaller EDT value, and in the few cases when this is not true, the error in the distance value is very small. If

a 4-neighbourhood is used, a reasonable approximation only is obtained; see
[2].

Figure 7 shows pseudo-code for this algorithm. The add_to_queue(i, a)
procedure adds the pair of pixel locations (i, a) with priority $|i - a|$. In the
algorithm as shown here the queue is always longer than it needs to be, by a
factor of about three. This may be avoided by a process of assessment of each
(i, a) pair before queue insertion, but this is not shown here for the sake of
simplicity. The equivalent step in the reconstruction algorithm is described in
the next section, however.

```
1    output ≡ 0

2    For all neighbouring (foreground, background) pairs, (i, a)
3        add_to_queue(i, a)

4    while queue not empty
5        Remove first pair (i, a) from queue

6        if output[i] > 0   // Already propagated
7            continue   // i.e. Go to Line 4

8        output[i] = |i − a|   // Output the distance value for this pixel

9        for all neighbours j of i
10           if X[j] = 1 and output[j] = 0
11               add_to_queue(j, a)
```

Figure 7: Euclidean Distance Transform (EDT) Algorithm

4. Object Reconstruction Algorithm

Consider computing the EDT of the previous section for a binary image with
only one background pixel. The result is clearly a *conical* function. The cone
is downward-pointing, its point being at height zero at the single background
pixel. The slope of the cone is one. It has been observed that, more generally,
the EDT is the pointwise minimum of many such cones, one at each background point. In fact (see [10, 6]) the EDT is the morphological *erosion*, with
a conical structuring function, of an image which is $+\infty$ at foreground pixels
and zero at background pixels. We note that in similar fashion the squared
EDT may be computed as an erosion with a *paraboloid*. The separability of
a simple paraboloid is useful in this process, and this idea underlies several
algorithms for exact EDT calculation [7].

With this in mind, the Ragnemalm algorithm of the previous section can be
seen as a process of *competing cones*. Upward-pointing cones are "grown"

from the point upwards, beginning at each background point with at least one foreground neighbour. At each point the lowest of any competing cones dominates the others. It is the centre or source location a of the cone which dominates at point i which is the "ancestor" of i. Note that in principle a cone should be grown form all background points. However the cone at a background point with no foreground neighbours will always dominate at that location giving an EDT value of zero, and never dominate at any other location. Such cones therefore can be, and are, neglected in the interests of efficiency.

This "competing cones" process can also be seen as a *flooding* process, where the terrain, a minimum of cones, is computed as it is filled — from the bottom (zero-level) upwards. The catchment basins are characterised by the identity of the ancestor of each point.

The basic idea of the new reconstruction algorithm is the following. As for the Ragnemalm EDT we perform an erosion by a conical structuring function, but now with a different source function. The source function in the reconstruction algorithm is essentially *the negative of the quench function*. Specifically, at points outside the skeleton the source function is $+\infty$, and at points i on the skeleton its value is $-\text{EDT}[i]$. The result, f, of the cone-erosion of this source function has the property that $f[i] \leq 0$ if and only if, for some a on the skeleton, there exists a disk of radius $\text{EDT}[a]$ centred at a which includes i. In other words, the binary image $f \leq 0$ is exactly the reconstruction of the given skeleton with its corresponding EDT or quench function values.

Although there are advantages to using cones, and our algorithm uses cones, a similar statement is true for erosion using the squared EDT with a paraboloid structuring function. A reconstruction algorithm using this fact may be constructed using the separability of the paraboloid and the one-dimensional quadratic erosion algorithms of, for example, [12].

The new reconstruction algorithm begins by initiating a cone at each point on the skeleton. The source or ancestor point for each such point i is i itself, and the height of the function is $-\text{quench}[i]$. This initialisation amounts to placing pairs (i, i) on a priority queue with priorities $P = -\text{quench}[i]$.

Queue processing then proceeds as for the Ragnemalm EDT algorithm using, for pair (i, a), the priority function

$$P[i, a] = |i - a| - \text{quench}[i].$$

This is the height at i of a downward-pointing, unit-slope cone with point at $(a, -\text{quench}[a])$. The cone-erosion function $f[i]$ is

$$f[i] = \min_a P[i, a].$$

Pairs (i, a) with $P[i, a] > 0$ are not put on the queue. This effectively stops the flooding process at level zero – i.e. at $f[i] = 0$ – and ensures that the correct reconstruction is produced. An alternative would be to continue until the whole image is processed, writing the priority values to the output – i.e. $\text{output}[i] = P[i, a]$. This would produce the entire image f which could then be thresholded at zero to give the reconstruction. However the time required for this alternative procedure is proportional to the number of pixels in the image whereas our proposed algorithm has cost proportional to the number of pixels in the reconstructed set, and this set may be very much smaller than the whole image.

If the stopping condition $P[i, a] \leq 0$ is applied, then outputting the value 1 – i.e. $\text{output}[i] = 1$ – creates a binary mask of the reconstruction. We prefer a more general output, however: at each point i output the *label value* of the ancestor a of i – i.e. $\text{output}[i] = \text{label}[a]$. In this way each point in the reconstruction is labelled by a label value taken from the location of its ancestor. If the output is initially set to zero and all labels are strictly positive, then a threshold of this output image at zero gives the binary reconstruction.

Figure 8 shows the reconstruction algorithm in pseudo-code. Figure 9 shows the function add_to_queue which implements the priority function and the stopping rule.

```
1   output ≡ 0

2   for all skeleton pixels i — i.e. with quench[i] > 0
3       add_to_queue(i, i)

4   while queue not empty
5       Remove first pair (i, a) from queue

6       if output[i] > 0   // Already propagated
7           continue    // i.e. Go to Line 4

8       output[i] = labels[a]   // Output the label value for this pixel

9       for all neighbours j of i
10          if output[j] ≤ 0  // Not already propagated
11              add_to_queue(j, a)
```

Figure 8: Reconstruction Algorithm

Figure 10 shows an extended version of add_to_queue. This version avoids the unnecessary queue growth mentioned in Section 3 above. As each value of $\text{output}[i]$ is zero initially, the stopping condition, $P < 0$ is enforced at the first visit to each pixel i, and the current priority value, which is always negative, is written temporarily at output location i. Subsequently however, only pairs (i, a) with strictly lower priority can be added to the queue. This saves space and time, as any such pair would be ignored when dequeued.

```
1   add_to_queue(i, a)
2   {
3       P = |i - a| - quench[i]
4       if P < 0
5           Put pair (i, a) on queue with priority P
6   }
```

Figure 9: Function `add_to_queue`: Simple version

```
1   add_to_queue(i, a)
2   {
3       P = |i - a| - quench[i]
4       if P < output[i]
5           output[i] = P
6           Put pair (i, a) on queue with priority P
7   }
```

Figure 10: Function `add_to_queue`: Extended version

5.   Further Examples

Figures 11 and 12 are like Figures 5 and 6 but here skeleton points have much more varied labels. The quench function values leading to Figure 12 are the same as for Figure 6, namely those shown in Figure 3. This illustrates the fact that the algorithm will propagate an arbitrary collection of label values.

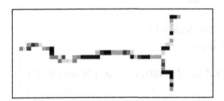

Figure 11.   Multi-Labelled Skeleton.   *Figure 12.*   Reconstruction with Labels.

Figures 13 and 14 illustrate the behavior of the reconstruction algorithm for simple artificial input. This input is indicated by Figure 13 which consists of four circles in different colours. This is purely illustrative of the data input to the reconstruction. This actual data consists of two images, both of which are zero-valued except at the four circle centres. The input "quench function" image contains the radii of the four circles at these points, while the input label image contains the corresponding four label values.

The result of the reconstruction with this input is shown in Figure 14. Firstly we can see that the reconstruction appears to be correct: the union of the la-belled regions in the reconstruction is equal to the union of the overlapping

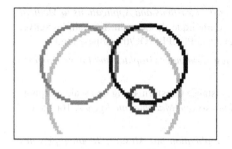

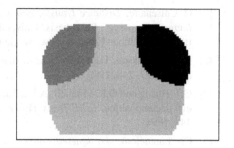

Figure 13. Artificial example.   *Figure 14.* Reconstruction.

input disks. Secondly we see the way in which labels are assigned to pixels which are in two or more of the input disks: region borders are at locations where the corresponding cones intersect.

Finally note how the smallest of the four input disks affects the reconstruction: it is not represented at all, not even its centre. This disk contains no unique points and is in this sense redundant. Moreover the quench function in this case is impossible or illegal: it cannot be a subset of the EDT of any set. This example illustrates that such anomalies are ignored by the reconstruction algorithm.

6.   Conclusion

We have described a new algorithm which produces a labelled reconstruction of a binary mask using sampled distance transform information and corresponding labels. The new algorithm inherits some of the key properties of the existing algorithm it is based on, namely Ragnemalm's vector propagating EDT algorithm. These properties include "near-exactness" and high speed. We have demonstrated useful behaviour of the algorithm for both artificial and real inputs.

Acknowledgments

The authors wish to thank Renovis, Inc., as well as Daniel Emerling, Nina Orike, and Krithika Ramamoorthy for permission to use masks derived from their neuron images. The original neuron images were acquired with an ImageXpress 5000A automated imaging system from Axon Instruments, Inc.

References

[1] P. Brigger, M. Kunt, and F. Meyer. The geodesic morphological skeleton and fast transformation algorithms. In J. Serra and P. Soille, editors, *Mathematical morphology and its applications to image processing*, pages 133–140. Kluwer Academic Publishers, 1994.

[2] O. Cuisenaire. *Distance Transformations: Fast Algorithms and Applications to Medical Image Processing.* PhD thesis, Université Catholique de Louvain, Louvain-la-Neuve, Belgium, October 1999. http://www.tele.ucl.ac.be/PEOPLE/OC/these/these.html.

[3] P.-E. Danielsson. Euclidean distance mapping. *Computer Graphics and Image Processing*, 14:227–248, 1980.

[4] C.T. Huang and O.R. Mitchell. A Euclidean distance transform using grayscale morphology decomposition. *IEEE Trans. Pattern Analysis and Machine Intelligence*, 16(4):443–448, 1994.

[5] Ch. Lantuéjoul. *La Squelettisation et son Application aux Mesures Topologiques des Mosaiques Polycrystalline.* PhD thesis, Ecole de Mines de Paris, 1978.

[6] Andrew J.H. Mehnert and Paul T. Jackway. On computing the exact Euclidean distance transform on rectangular and hexagonal grids. *J. Mathematical Imaging and Vision*, 11:223–230, 1999.

[7] A. Meijster, J.B.T.M. Roerdink, and W.H. Hesselink. A general algorithm for computing distance transforms in linear time. In J. Goutsias, L. Vincent, and D.S. Bloomberg, editors, *Mathematical Morphology and its Applications to Image and Signal Processing*, pages 331–340. Kluwer, 2000.

[8] Tun-Wen Pai and John H. L. Hansen. Boundary-constrained morphological skeleton minimization and skeleton reconstruction. *IEEE Trans. Pattern Analysis and Machine Intelligence*, pages 201–208, 1994.

[9] I. Ragnemalm. Fast erosion and dilation by contour processing and thresholding of distance maps. *Pattern Recognition Letters*, 13:161–166, 1992.

[10] F. Shih and O.R. Mitchell. A mathematical morphology approach to Euclidean distance transformation. *IEEE Trans. on Image Processing*, 2(1):197–204, 1992.

[11] P. Soille. *Morphological Image Analysis.* Springer-Verlag, 2003.

[12] R. van den Boomgaard, L. Dorst, S. Makram-Ebeid, and J. Schavemaker. Quadratic structuring functions in mathematical morphology. In P. Maragos, R. W. Schafer, and M. A. Butt, editors, *Mathematical Morphology and its Application to Image and Signal Processing*, pages 147–154. Kluwer Academic Publishers, Boston, 1996.

GRAYSCALE LEVEL MULTICONNECTIVITY

Ulisses Braga-Neto

Laboratório de Virologia e Terapia Experimental
Centro de Pesquisas Aggeu Magalhães - CPqAM/Fiocruz
Av. Moraes Rego s/n - Caixa Postal: 7472
Cidade Universitária, Recife, PE 50.670-420 BRAZIL
ulisses_braga@cpqam.fiocruz.br

Abstract   In [5], a novel concept of connectivity for grayscale images was introduced, which is called *grayscale level connectivity*. In that framework, a grayscale image is connected if all its threshold sets below a given level are connected. It was shown that grayscale level connectivity defines a *connection*, in the sense introduced by Jean Serra in [10]. In the present paper, we extend grayscale level connectivity to the case where different connectivities are used for different threshold sets, a concept we call *grayscale level multiconnectivity*. In particular, this leads to the definition of a new operator, called the multiconnected grayscale reconstruction operator. We show that grayscale level multiconnectivity defines a connection, provided that the connectivities used for the threshold sets obey a nesting condition. Multiconnected grayscale reconstruction is illustrated with an example of scale-space representation.

Keywords: Connectivity, Grayscale Images, Reconstruction, Mathematical Morphology Complete Lattice, Scale-Space.

Introduction

In [5], we introduced the notion of *grayscale level connectivity,* in which all threshold sets below a given level k are required to be connected according to a given binary connection. It was shown that a grayscale k-level connected image might have more than one regional maximum, but all regional maxima are at or above level k. It was also shown that grayscale level connectivity can be formulated as a *connection* [10] in an underlattice of the usual lattice of grayscale images. Grayscale level connectivities were shown to lead to effective tools for image segmentation, image filtering and multiscale image representation.

C. Ronse et al. (eds.), Mathematical Morphology: 40 Years On, 129–138.

However, it may be desirable to assign connectivities of varying strictness to different image levels. There may be details of interest at low levels of intensity that can be preserved if one uses a larger connection for lower levels, while at the same time employing a smaller connection at higher levels, in order to prevent relevant regional maxima belonging to different objects from fusing into a single grayscale connected component. In the present paper, we extend the concept of grayscale level connectivity to the case where different binary connections are assigned to different threshold sets of the image. In this case, each threshold set below a given level k is required to be connected according to the respective binary connection. The resulting connectivity is referred to as a *grayscale level multiconnectivity*. We will show that the crucial requirement to be made is that these binary connections be nested. We show that grayscale level multiconnectivity can be formulated as a connection. To compute grayscale level-k multiconnected components, we will employ a novel morphological operator, which we call *multiconnected grayscale reconstruction*; this is an extension of the usual grayscale reconstruction operator [11]. We illustrate the application of multiconnected grayscale reconstruction with an example of *skyline scale-space* [6].

1.   Review of Connectivity

We assume that the reader is familiar with basic notions of Lattice Theory and Mathematical Morphology [1, 7]. In this section, we review briefly the theory of connectivity on complete lattices; for a more detailed exposition, please see [10, 4, 2]. Consider a lattice $\mathcal{L}$, with a sup-generating family $\mathcal{S}$. A family $\mathcal{C} \subseteq \mathcal{L}$ is called a *connection* in $\mathcal{L}$ if (*i*) $O \in \mathcal{C}$, (*ii*) $\mathcal{S} \subseteq \mathcal{C}$, and (*iii*) if $\{C_\alpha\}$ in $\mathcal{C}$ with $\bigwedge C_\alpha \neq O$, then $\bigvee C_\alpha \in \mathcal{C}$. The family $\mathcal{C}$ generates a *connectivity* on $\mathcal{L}$, and the elements in $\mathcal{C}$ are said to be *connected*. We say that C is a *connected component* of $A \in \mathcal{L}$ if $C \in \mathcal{C}$, $C \leq A$ and there is no $C' \in \mathcal{C}$ different from C such that $C \leq C' \leq A$. In other words, a connected component of an object is a maximal connected part of the object. The set of connected components of A is denoted by $\mathcal{C}(A)$.

We can define an operator $\gamma_x(A)$ on $\mathcal{L}$ that extracts connected components from elements $A \in \mathcal{L}$, by

$$\gamma_x(A) = \bigvee\{C \in \mathcal{C} \mid x \leq C \leq A\}, \ x \in \mathcal{S}, \ A \in \mathcal{L}. \qquad (1)$$

It is easy to see that this operator is increasing, anti-extensive and idempotent, i.e., an opening [7]; it is called the *connectivity opening* associated with $\mathcal{C}$. It can be verified that $\gamma_x(A)$ is the connected component C of A *marked* by x (i.e., such that $x \leq C$).

For a marker $M \in \mathcal{L}$, the *reconstruction* $\rho(A \mid M)$ of a given $A \in \mathcal{L}$ from M is defined by

$$\rho(A \mid M) = \bigvee_{x \in \mathcal{S}(M)} \gamma_x(A) = \bigvee \{C \in \mathcal{C}(A) \mid C \wedge M \neq O\}. \qquad (2)$$

The second equality above can be easily verified [2]. Hence, the reconstruction operator $\rho(A \mid M)$ extracts the connected components of A that "intersect" marker M. It follows from the fact that the supremum of openings is an opening [7] that the operator $\rho(\cdot \mid M)$ is an opening on $\mathcal{L}$, for a fixed marker $M \in \mathcal{L}$ [7].

Given a connection in $\mathcal{P}(E)$, and the associated reconstruction operator $\rho: \mathcal{P}(E) \times \mathcal{P}(E) \rightarrow \mathcal{P}(E)$, we can define an operator $\tilde{\rho}: \operatorname{Fun}(E, \mathcal{T}) \times \operatorname{Fun}(E, \mathcal{T}) \rightarrow \operatorname{Fun}(E, \mathcal{T})$ by

$$\tilde{\rho}(f \mid g)(v) = \bigvee \{t \in \mathcal{T} \mid v \in \rho(X_t(f) \mid X_t(g))\}, \ v \in E, \qquad (3)$$

where $g \leq f$. It can be shown that $\tilde{\rho}(\cdot \mid g)$ is an opening on $\operatorname{Fun}(E, \mathcal{T})$, for a fixed marker $g \in \operatorname{Fun}(E, \mathcal{T})$ [7]. If we assume that $\mathcal{T}$ is a chain, then the operator $\tilde{\rho}(f \mid g)$ in (3) is known as the *grayscale reconstruction* of f from marker g, associated with the connection $\mathcal{C}$. Furthermore, if the chain $\mathcal{T}$ is finite, e.g., $\mathcal{T} = \mathcal{K} = \{0, 1, \dots, K\}$, then $X_k(\tilde{\rho}(f \mid g)) = \rho(X_k(f) \mid X_k(g))$ for $k \in \mathcal{K}$. The grayscale reconstruction is a well-known operator, frequently used in applications (e.g., see [11]).

2.   Grayscale Level Multiconnectivity

We consider discrete grayscale images modeled as elements of the lattice $\mathcal{L} = \operatorname{Fun}(E, \mathcal{K})$, consisting of all functions from an arbitrary domain of definition E into the discrete chain $\mathcal{K} = \{0, 1, \dots, K\}$ (the results presented here can be extended to the case of continuous-valued images, modeled as upper semi-continuous functions from a connected compact Hausdorff space E into $\overline{\mathbf{R}}$; e.g., see [6]).

Consider a family $\{\mathcal{C}_k \mid k \in \mathcal{K}\}$ of connections in $\mathcal{P}(E)$, such that $\mathcal{C}_k$ specifies the connectivity at level k. The only requirement made on the family $\{\mathcal{C}_k \mid k \in \mathcal{K}\}$ is that they satisfy the following *nesting condition*

$$\mathcal{C}_k \subseteq \mathcal{C}_l, \quad \text{for } k \geq l. \qquad (4)$$

The need to impose the nesting condition condition is made clear in the sequence. We remark that the family $\{\mathcal{C}_k \mid k \in \mathcal{K}\}$ with the nesting condition corresponds to a *connectivity pyramid*, in the terminology of [3].

For a particular example, let $E = \mathbf{Z}^d$ and take

$$\mathcal{C}_k = \{A \subseteq E \mid A \oplus B_k \in \mathcal{C}\}, \qquad (5)$$

where $\{B_k \mid k \in \mathcal{K}\}$ is a family of decreasing structuring elements, $A \oplus B_k$ denotes the dilation of A by B_k [7], and $\mathcal{C}$ is a connection in $\mathcal{P}(E)$. It can be shown that each $\mathcal{C}_k$ is a connection that contains $\mathcal{C}$, provided that each structuring element B_k contains the origin and is connected according to $\mathcal{C}$ [9, 2]. From this and the fact that $A \oplus B_k \oplus B_l = A \oplus B_{k+l}$ it follows easily that the nesting condition (4) is satisfied. For instance, B_k may be a digital disk of radius r_k centered at the origin, where $\{r_k\}$ is a decreasing sequence of integers. As the radius r_k varies, so does the number of connected sets in $\mathcal{C}_k$, which affords a varying degree of robustness against noise.

Given a function $f \in \mathrm{Fun}(E, \mathcal{K})$ and a value $k \in \mathcal{K}$, the *threshold set* of f at level k is the set $X_t(f) = \{v \in E \mid f(v) \geq t\}$. A set $R \subseteq E$ is a *regional maximum* of $f \in \mathrm{Fun}(E, \mathcal{K})$ at level $k \in \mathcal{K}$ if R is a connected component of $X_k(f)$, according to $\mathcal{C}_k$, and $R \cap X_l(f) = \emptyset$, for all $l \geq k + 1$. It is easy to see that a function $f \in \mathrm{Fun}(E, \mathcal{K})$ is constant over a regional maximum R; we denote this constant value by $f(R)$. In addition, we denote by $\mathcal{R}(f)$ the set of all regional maxima of a function f, and by $\mathcal{R}_k(f)$ the set of all regional maxima of f that are at or above level k.

A function $f \in \mathrm{Fun}(E, \mathcal{K})$ always has at least one regional maximum, which corresponds to the global maximum of f. We also have the following result (due to space constraints, proofs are ommitted).

Proposition 2.1 A function $f \in \mathrm{Fun}(E, \mathcal{K})$ has a single regional maximum if and only if $X_k(f) \in \mathcal{C}_k$, for all $k \in \mathcal{K}$.

The nesting condition (4) is essential in establishing this result. For an example, let

$$\mathcal{C}_k = \begin{cases} \mathcal{C}', & \text{for } k < k_0 \\ \mathcal{C}, & \text{for } k \geq k_0 \end{cases}, \tag{6}$$

where $k_0 \in \mathcal{K}$ and $\mathcal{C}, \mathcal{C}'$ are two connections in $\mathcal{P}(E)$ such that $\mathcal{C} \not\subseteq \mathcal{C}'$. See Fig. 1, where $\mathcal{C}$ and $\mathcal{C}'$ are chosen such that $X_{k_0}(f) = R \in \mathcal{C}$, but $X_k(f) \notin \mathcal{C}'$, for $k < k_0$. Here, f has a single regional maximum R at level k_0, even though this is the only connected image level. Anomalous situations like this are ruled out by the nesting condition.

We have the following definition.

Definition 2.1 For $k \in \mathcal{K}$, a function $f \in \mathrm{Fun}(E, \mathcal{K})$ is *level-k multiconnected* if $X_l(f) \in \mathcal{C}_l \setminus \{\emptyset\}$, for all $l \leq k$.

In other words, an image is level-k multiconnected if all its level sets at or below level k are non-empty and connected according to the respective connection. The parameter k reflects the "richness" of the connectivity. Smaller values of k allow more functions to be considered as connected. Clearly, if $k \geq l$, level-k multiconnectivity implies level-l multiconnectivity. See Fig. 2

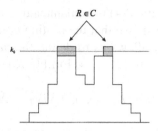

Figure 1. A function f with a single regional maximum R at level k_0, even though only the top image level is assumed to be connected. This anomalous situation is ruled out by the nesting condition.

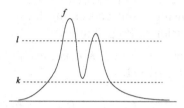

Figure 2. The function f is level-k connected but not level-l connected.

for an illustration in the case where all binary connections correspond to the ordinary connectivity on the real line.

A level-k multiconnected image is allowed to have more than one regional maximum, however, all of them must be above level k, as given by the following result.

Proposition 2.2 If $f \in \text{Fun}(E, \mathcal{K})$ is level-k multiconnected then $\mathcal{R}_k(f) = \mathcal{R}(f)$.

The converse to the previous result is not true (for a counter-example, see [6]).

For $k \in \mathcal{K}$, a function $g \in \text{Fun}(E, \mathcal{K})$ is a *level-k multiconnected component* of $f \in \text{Fun}(E, \mathcal{K})$ if g is level-k multiconnected, $g \leq f$, and there is no other level-k multiconnected function $g' \in \text{Fun}(E, \mathcal{K})$ different from g such that $g \leq g' \leq f$. The set of all level-k multiconnected components of $f \in \text{Fun}(E, \mathcal{K})$ is denoted by $\mathcal{N}_k(f)$.

Grayscale level-k connected components were computed in [6] using the standard grayscale reconstruction operator (3). In the multiconnected case, a new operator must be introduced to play the role of grayscale reconstruction. First we need the following result. Here, ρ_k denotes the reconstruction operator associated with $\mathcal{C}_k$, for $k \in \mathcal{K}$.

Proposition 2.3 The operators $\{\rho_k \mid k \in \mathcal{K}\}$ are ordered in a decreasing fashion: if $k \geq l$, then $\rho_k(A) \subseteq \rho_l(A)$, for all $A \subseteq E$.

Again, the nesting condition (4) is fundamental in establishing the property in Proposition 2 (it can be shown that the nesting condition is actually equivalent to it). This property allows us to define the *multiconnected grayscale reconstruction operator* $\widehat{\rho}$: $\mathrm{Fun}(E,\mathcal{K}) \times \mathrm{Fun}(E,\mathcal{K}) \to \mathrm{Fun}(E,\mathcal{K})$ by:

$$\widehat{\rho}(f \mid g)(v) = \bigvee \{k \in \mathcal{K} \mid v \in \rho_k(X_k(f) \mid X_k(g))\}, \ v \in E, \qquad (7)$$

As a supremum of openings, $\widehat{\rho}(\cdot \mid g)$ is an opening on $\mathrm{Fun}(E,\mathcal{K})$, for a fixed marker $g \in \mathrm{Fun}(E,\mathcal{K})$. The operator $\widehat{\rho}(\cdot \mid g)$ is the *semi-flat operator* generated by the family of operators $\{\rho_k(\cdot \mid X_k(g)) \mid k \in \mathcal{K}\}$, in the terminology of [7]. As a consequence of the nesting condition (4), it is possible to verify that $X_k(\widehat{\rho}(f \mid g)) = \rho_k(X_k(f) \mid X_k(g))$, for $k \in \mathcal{K}$.

We have the following result concerning level-k multiconnected components (this is the extension of Prop. 4 in [6]).

Proposition 2.4 Let $f \in \mathrm{Fun}(E,\mathcal{K})$. For each $C \in \mathcal{C}(X_k(f))$, there is an associated $g_C \in \mathcal{N}_k(f)$, given by

$$g_C = \bigvee \{\widehat{\rho}(f \mid h_{R,f(R)}) \mid R \in \mathcal{R}_k(f) \text{ s.t. } R \subseteq C\}, \qquad (8)$$

with $C = X_k(g_C)$, where $h_{A,r}$ is the cylinder of base $A \subseteq E$ and height $r \in \mathcal{K}$. Conversely, for each $g \in \mathcal{N}_k(f)$, there is an associated $C_g \in \mathcal{C}_k(X_k(f))$, given by $C_g = X_k(g)$, with $g = g_{C_g}$.

In other words, there is a bijection between $\mathcal{N}_k(f)$ and $\mathcal{C}_k(X_k(f))$; each level-k multiconnected component of f is associated with a connected component of $X_k(f)$ according to $\mathcal{C}_k$. Equation (8) provides a practical way to compute the level-k multiconnected components of an image, provided that one has available an implementation of the multiconnected grayscale reconstruction operator (more on this later).

Consider the subset of $\mathrm{Fun}(E,\mathcal{K})$ given by $\mathrm{Fun}_k(E,\mathcal{K}) = \{f \in \mathrm{Fun}(E,\mathcal{K}) \mid \mathcal{R}(f) = \mathcal{R}_k(f)\}$, for $k \in \mathcal{K}$. In other words, $\mathrm{Fun}_k(E,\mathcal{K})$ consists of the functions that have all regional maxima above level k.

Proposition 2.5 The set $\mathrm{Fun}_k(E,\mathcal{K})$ is a complete lattice under the pointwise partial order, with supremum $\bigvee^k$ and infimum $\bigwedge^k$, given by

$$\bigvee^k f_i = \bigvee f_i,$$
$$\bigwedge^k f_i = \widehat{\psi}_k \left(\bigwedge f_i \right)$$

where $\widehat{\psi}_k$ is an opening on $\mathrm{Fun}(E,\mathcal{K})$, given by

$$\widehat{\psi}_k(f) = \bigvee \{\widehat{\rho}(f \mid h_{R,f(R)}) \mid R \in \mathcal{R}_k(f)\}. \qquad (9)$$

Note that $\text{Fun}_k(E, \mathcal{K})$ is an *underlattice* of $\text{Fun}(E, \mathcal{K})$; i.e., it is a complete lattice under the partial of $\text{Fun}(E, \mathcal{K})$ [7]. In addition, $\text{Fun}_k(E, \mathcal{K})$ is an infinite $\vee$-distributive lattice. In this framework, $\bigwedge^k f_i = O$ if and only if $\bigwedge f_i$ has no regional maxima at level k or above, which happens if and only if $\bigcap_i X_k(f_i) = \emptyset$.

It is easy to verify that the family

$$\mathcal{S}_k = \{\delta_{v,l} \mid l \geq k\} \cup \{f \in \text{Fun}(E, \mathcal{K}) \mid \mathcal{R}(f) = \{R\}, \, f(R) = k\} \quad (10)$$

is sup-generating in $\text{Fun}_k(E, \mathcal{K})$. Note that $\mathcal{S}_k$ consists of all pulses of height at least k, along with the functions in $\text{Fun}_k(E, \mathcal{K})$ that have exactly one regional maximum at level k. This leads to the following fundamental result, which shows that grayscale level multiconnected images define a connection.

Proposition 2.6 For $k \in \mathcal{K}$, the family

$$\widehat{\mathcal{C}}_k = \{f \in \text{Fun}(E, \mathcal{K}) \mid f \text{ is level-}k \text{ multiconnected}\} \quad (11)$$

is a connection in lattice $\text{Fun}_k(E, \mathcal{K})$, with sup-generating family $\mathcal{S}_k$.

We now return to the issue of the implementation of the multiconnected grayscale reconstruction operator, which is required to compute grayscale level multiconnected components, by means of (8). Fast algorithms exist for the implementation of the standard grayscale reconstruction operator associated with usual adjacency connectivities on the digital plane [11]. If the underlying connectivity is instead given by the dilation-based connection $\mathcal{C}_k$ in (5), then we show in [2] that the grayscale reconstruction operator $\widetilde{\rho}_{B_k}$ associated with $\mathcal{C}_k$ is given by

$$\widetilde{\rho}_{B_k}(f \mid g) = f \wedge \widetilde{\rho}(f \oplus B_k \mid f \wedge g), \quad (12)$$

for $f, g \in \text{Fun}(E, \mathcal{K})$, where $\widetilde{\rho}$ is the grayscale reconstruction operator associated with the base connectivity $\mathcal{C}$ and $f \oplus B$ denotes the *flat grayscale dilation* of f by B [7]. Hence, the operators $\widetilde{\rho}_{B_k}$ can be easily implemented when $\mathcal{C}$ corresponds to a usual adjacency connectivity.

We will show next that $\widehat{\rho}$ can be written as a supremum of the standard component grayscale reconstruction operators. Given $k \in \mathcal{K}$, let w_k be the operator on $\text{Fun}(E, \mathcal{K})$ given by:

$$w_k(f)(v) = \begin{cases} f(v), & \text{if } f(v) \leq k \\ k, & \text{otherwise} \end{cases}, \quad v \in E, \quad (13)$$

The operator w_k "clamps" a function f at level k.

In the following, $\widetilde{\rho}_k$ denotes the grayscale reconstruction operator associated with $\mathcal{C}_k$, for $k \in \mathcal{K}$.

Proposition 2.7 The multiconnected grayscale reconstruction operator can be written as:

$$\widehat{\rho}(f \mid g) = \bigvee_{k \in \mathcal{K}} \widetilde{\rho}_k(w_k(f) \mid g), \tag{14}$$

for $f, g \in \mathrm{Fun}(E, \mathcal{K})$. Moreover, suppose that there are levels $\{k_i \mid i = 0, \ldots, L\}$, with $1 \leq L \leq K$ and $k_L = K$, such that

$$\mathcal{C}_k = \begin{cases} \mathcal{C}_0^*, & \text{for } 0 \leq k \leq k_0 \\ \mathcal{C}_i^*, & \text{for } k_{i-1} < k \leq k_i \end{cases}, \tag{15}$$

where $\{\mathcal{C}_i^* \mid i = 0, \ldots, L\}$ is a decreasing family of connections in $\mathcal{P}(E)$. Then (14) simplifies to

$$\widehat{\rho}(f \mid g) = \bigvee_{0 \leq i \leq L} \widetilde{\rho}_i^*(w_{k_i}(f) \mid g), \tag{16}$$

for $f, g \in \mathrm{Fun}(E, \mathcal{K})$, where $\widetilde{\rho}_i^* : \mathrm{Fun}(E, \mathcal{K}) \times \mathrm{Fun}(E, \mathcal{K}) \to \mathrm{Fun}(E, \mathcal{K})$ is the grayscale reconstruction operator associated with $\mathcal{C}_i^*$, for $1 \leq i \leq L$.

Hence, if one has available efficient implementations of the standard component grayscale reconstruction operators, such as in (12), then the multiconnected grayscale reconstruction operator can be efficiently implemented by means of (14) or (16).

Note that (16) realizes an economy with respect to (14), in the case where there is repetition among the connections $\{\mathcal{C}_k \mid k \in \mathcal{K}\}$. If $\mathcal{C}_k = \mathcal{C}$, for all $k \in \mathcal{K}$, then (14) and (16) reduce to $\widehat{\rho} = \widetilde{\rho}$; i.e., $\widehat{\rho}$ reduces to the usual grayscale reconstruction operator associated with $\mathcal{C}$.

As an example of application of Proposition 2.7, let $0 \leq k_0 < K$, let $B_0, B_1 \in \mathcal{P}(E)$ with $B_1 \subseteq B_0$, and consider two dilation-based connections $\mathcal{C}_0^* = \{A \subseteq E \mid A \oplus B_0 \in \mathcal{C}\}$ and $\mathcal{C}_1^* = \{A \subseteq E \mid A \oplus B_1 \in \mathcal{C}\}$, where $\mathcal{C}$ is a given connection in $\mathcal{P}(E)$, such that

$$\mathcal{C}_k = \begin{cases} \mathcal{C}_0^*, & \text{for } 0 \leq k \leq k_0 \\ \mathcal{C}_1^*, & \text{for } k_0 < k \leq K \end{cases}. \tag{17}$$

The *biconnected grayscale reconstruction operator* in this case is given simply by:

$$\begin{aligned} \widehat{\rho}(f \mid g) &= \widetilde{\rho}_{B_0}(w_{k_0}(f) \mid g) \vee \widetilde{\rho}_{B_1}(w_K(f) \mid g) \\ &= (f \wedge \widetilde{\rho}(w_{k_0}(f) \oplus B_0 \mid w_{k_0}(f) \wedge g) \vee (f \wedge \widetilde{\rho}(f \oplus B_1 \mid f \wedge g)) \\ &= f \wedge (\widetilde{\rho}(w_{k_0}(f) \oplus B_0 \mid w_{k_0}(f) \wedge g) \vee \widetilde{\rho}(f \oplus B_1 \mid f \wedge g)). \end{aligned} \tag{18}$$

One can easily implement (18), starting from an available efficient implementation of $\widetilde{\rho}$.

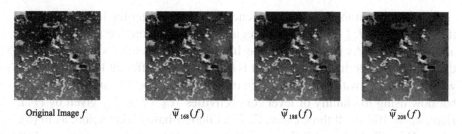

Original Image f $\tilde{\Psi}_{168}(f)$ $\tilde{\Psi}_{188}(f)$ $\tilde{\Psi}_{208}(f)$

Figure 3.   Supremal skyline scale-space. Here, K = 255, and the frames of the scale-space displayed correspond to k=168,188,208. The connectivity assumed at all levels is the usual 4-adjacency connectivity.

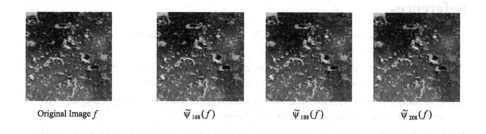

Original Image f $\tilde{\Psi}_{168}(f)$ $\tilde{\Psi}_{188}(f)$ $\tilde{\Psi}_{208}(f)$

Figure 4.   Supremal skyline scale-space in the multiconnected case, using the same image and values of k as in Fig. 3. Here we use a large dilation-based connectivity for levels below 160, and the usual 4-adjacency connectivity for levels above 160.

3.   Scale-Space Representation

In this section, we illustrate the application of multiconnected reconstruction in a skyline scale-space representation [6].

From (9), it is clear that $\widehat{\psi}_k \leq \widehat{\psi}_l$, for $k \geq l$. Since these operators are openings, it follows that $\{\widehat{\psi}_k \mid k \in \mathcal{K}\}$ is a *granulometry* [7]. Interpreting the value k as the *scale* of observation, the sequence $\{\widehat{\psi}_k(f) \mid k \in \mathcal{K}\}$ of approximations correspond to an "evolution" of f towards decreasing levels of "detail," in a *scale-space* representation [12, 8]. The evolution in this particular scale-space is akin to the view of a city skyline as one drives away from it; near the city, the shorter buildings can be seen, but far away, only the tallest buildings can be discerned. For this reason, this is called a *skyline scale-space* [6].

Fig. 3 illustrates the supremal skyline scale-space, using a image of blood sample containing filaria worms. The connectivity assumed at all levels is the usual 4-adjacency connectivity, so that the reconstruction operator used here is the standard one. Note that details are progressively removed as the value of k increases.

In the multiconnected case, as a general rule, the larger the level connectivities $\{C_k \mid k \in \mathcal{K}\}$ are, the fewer regional maxima there are to be flattened, and vice-versa. As mentioned in the Introduction, this allows one to control the amount of filtering of details that is introduced at different levels. In Fig. 4 we illustrate this with the same image used in Fig. 3, for the same values of k, but now using the family of level connectivities $\{C_k \mid k \in \mathcal{K}\}$ given by (17). Here, $k_0 = 160$ in all three cases, C_0^* is a dilation-based connection, as in (5), where B is a 7×7 box and C corresponds to 4-adjacency connectivity, and C_1^* corresponds to 4-adjacency connectivity. Note that more details at low levels of intensity are preserved than in the case of using 4-adjacency connectivity at all levels.

References

[1] G. Birkhoff. *Lattice Theory*, volume 25. American Mathematical Society, Providence, Rhode Island, 3rd edition, 1967.

[2] U.M. Braga-Neto and J. Goutsias. Connectivity on complete lattices: New results. *Computer Vision and Image Understanding*, 85(1):22–53, January 2002.

[3] U.M. Braga-Neto and J. Goutsias. A multiscale approach to connectivity. *Computer Vision and Image Understanding*, 89(1):70–107, January 2003.

[4] U.M. Braga-Neto and J. Goutsias. A theoretical tour of connectivity in image processing and analysis. *Journal of Mathematical Imaging and Vision*, 19(1):5–31, July 2003.

[5] U.M. Braga-Neto and J. Goutsias. Grayscale level connectivity. *IEEE Transactions on Image Processing*, 2004. In press.

[6] U.M. Braga-Neto and J. Goutsias. Supremal multiscale signal analysis. *SIAM Journal of Mathematical Analysis*, 36(1):94–120, 2004.

[7] H.J.A.M. Heijmans. *Morphological Image Operators*. Academic Press, Boston, Massachusetts, 1994.

[8] J.J. Koenderink. The structure of images. *Biological Cybernetics*, 50:363–370, 1984.

[9] J. Serra, editor. *Image Analysis and Mathematical Morphology. Volume 2: Theoretical Advances*. Academic Press, London, England, 1988.

[10] J. Serra. Connectivity on complete lattices. *Journal of Mathematical Imaging and Vision*, 9:231–251, 1998.

[11] L. Vincent. Morphological grayscale reconstruction in image analysis: Applications and efficient algorithms. *IEEE Transactions on Image Processing*, 2:176–201, 1993.

[12] A.P. Witkin. Scale-space filtering. In *Proceedings of 7th International Joint Conference on Artificial Intelligence*, pages 1019–1022. Karlsruhe, West Germany, 1983.

SHAPE-TREE SEMILATTICES

Variations and Implementation Schemes

Renato Keshet

Hewlett-Packard Labs—Israel
Technion City, Haifa 32000, Israel
renato.keshet@hp.com

Abstract   The shape-tree semilattice is a new framework for quasi-self-dual morphological processing, where eroded images have all shapes shrunk in a contrast-invariant way. This approach was recently introduced, and is further investigated here. Apart of reviewing their original definition, different algorithms for computing the shape-tree morphological operators are presented.

Keywords: Complete inf-semilattices, self-dual operators, tree of shapes, fillhole.

1.   Introduction

We have very recently introduced a new quasi-self-dual morphological approach for image processing [1]. The resulting flat morphological erosion, for instance, causes all shapes in an image to shrink, regardless their contrast (i.e., regardless to whether the are bright or dark). Motivation and connection to other works are described in [1].

For discrete binary images, the approach yields a set of morphological operators on the so-called *adjacency complete lattice*, which is defined by means of the adjacency tree representation (see Section 2 below). The scheme is generalized to discrete grayscale images by means of the *Tree of Shapes* (ToS), a recently-introduced grayscale image representation [7–10], which can be regarded as a grayscale generalization of the adjacency tree (see Section 3). In the grayscale case, however, the underlying space structure is not that of a complete lattice anymore, but of a complete inf-semilattice.

While the main motivation in the original article was on laying out the theoretical basis of the scheme, and investigating its mathematical soundness, the current article focuses on implementations. Different algorithms for implementing the desired basic operators are investigated; they slightly differ in general, but produce identical results for typical structuring elements.

C. Ronse et al. (eds.), Mathematical Morphology: 40 Years On, 139–148.

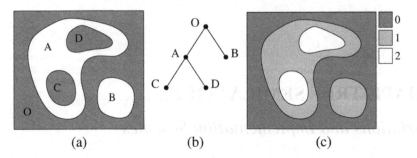

Figure 1.   (a) A binary image. Each letter corresponds to a connected component. O is the background component, A and B correspond to the two white connected components, and C and D correspond to the two dark connected components inside A. (b) The adjacency tree of (a). (c) Its unfolding transform.

2.   Binary Approach: Adjacency Lattice

The basic idea for our binary approach is to generate a quasi-self-dual complete lattice by using the data in the *adjacency tree*. In [2, page 89, Figure III.10], Serra describes the adjacency tree (which he called *homotopy tree*) as follows. If X is a bounded input binary image in an Euclidean space E, then the root of the adjacency tree is the infinite connected component of X^c. The first level nodes of the tree are those connected components of X that are adjacent to the root. The second level of nodes are the connected components of X^c that are adjacent to the first level of nodes, and so on. See an example in Fig. 1. The adjacency tree was thoroughly studied by Heijmans in [3].

Next, we review a characterization of the adjacency tree, which we presented in [1], and the subsequent derivation of the adjacency lattice. Let E be the 2D grid $\mathbb{Z}^2$, and R be a bounded region in E. Consider the set $\mathcal{P}(R)$ of all subsets of R (i.e., binary images within R). Define the *fillhole* $\phi(X)$ of a bounded binary image $X \in \mathcal{P}(R)$ as the complement of the morphological reconstruction (according to a given connectivity; e.g., either of the well-known 4- or 8-connectivities) of $E - X$ from the marker $E - R$. The fillhole operator is extended to discrete grayscale images by applying it to each level set:

$$\phi(f) = \theta^{-1}\{\phi[\theta_n(f)]\}, \tag{1}$$

where $\theta_n(f) \triangleq \{x \in E | f(x) \geq n\}$ is the level set of f of height n, and $\theta^{-1}\{T_n\} \triangleq \sup\{n \in \mathbb{N} | x \in T_n\}$ consolidates level sets back to a function. The grayscale fillhole operator ϕ is essentially the same as the FILL$(\cdot)$ operation, described in [4, page 208].

If a set or a function is invariant to the fillhole operator, then it is said to be *filled*.

The union $D_n(X)$ of all nodes of depth equal or larger than n on the adjacency tree of X can be calculated recursively according to

$$D_{n+1}(X) = \phi(D_n(X) \cap \mathsf{C}^n(X)), \quad n = 0, 1, 2, \ldots \quad (2)$$

where $D_0(X) \triangleq E$, and $\mathsf{C}(\cdot)$ is the *complement* operator, e.g., $\mathsf{C}(X) = X^c$ and $\mathsf{C}^2(X) = X$. The sequence of images $\{D_n(X)\}$ is decreasing w.r.t. the inclusion order, and $D_n(X)$ is filled, for all n.

Let $u_X(x)$ be the level of the connected component that a pixel $x \in E$ belongs to in the adjacency tree. We call the mapping $\mathcal{U} : X \to u_X$ the *unfolding transform*. E.g., see Fig. 1(c), for the unfolding transform of Fig. 1(a). The reason for this name is because $\mathcal{U}$ "unfolds" negative objects (those objects that are darker than its surroundings, i.e., belong to the background), providing a grayscale image where all objects are positive (brighter than the surroundings).

The unfolding transform can be calculated by:

$$u_X(x) \triangleq \theta^{-1}\{D_n(X)\}, \quad (3)$$

whereas the inverse mapping $\mathcal{U}^{-1} : u \to X$ is given by

$$X = \mathcal{U}^{-1}\{u\} = \{x \in E \mid u(x) \text{ is odd}\}. \quad (4)$$

We define the following partial ordering of binary images in $\mathcal{P}(R)$:

$$X \sqsubseteq Y \iff u_X(x) \le u_Y(x), \quad \forall x \in R. \quad (5)$$

In [1], we show that $(\mathcal{P}(R), \sqsubseteq)$ is a complete lattice, the infimum and supremum of which are given, respectively, by:

$$X \sqcap Y = \mathcal{U}^{-1}\{u_X \wedge u_Y\}, \quad \text{and} \quad X \sqcup Y = \mathcal{U}^{-1}\{\phi(u_X \vee u_Y)\}. \quad (6)$$

We call $(\mathcal{P}(R), \sqsubseteq)$ the *adjacency lattice*. The pair $(\varepsilon_B^A, \delta_B^A)$ of operators, defined by

$$\varepsilon_B^A \triangleq \mathcal{U}^{-1}\{u_X \ominus B\} \quad \text{and} \quad \delta_B^A \triangleq \mathcal{U}^{-1}\{\phi(u_X \oplus B)\}, \quad (7)$$

for some structuring element B, is an adjunction in the adjacency lattice. This erosion (resp. dilation) shrinks (resp. expands) all foreground *and* background "objects" in a binary image (except for the background component connected to the boundary). They are called *adjacency erosion and dilation*.

3. Shape-Tree Semilattice

The Tree of Shapes (ToS)

In order to generalize the above framework to grayscale images, we use the concept of image *shapes*. Monasse and Guichard define the *shapes* of an image

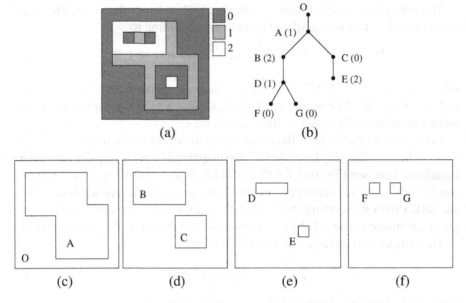

Figure 2.   Shape decomposition. (a) Original grayscale image f and (c)-(f) its shapes. (b) The tree of shapes associated to (a); the numbers in parenthesis are the shape graylevel.

f in [7, 8] as the collection $\mathcal{T}$ of sets given by:

$$\phi\left[\gamma_x\left(\theta_t\left(f\right)\right)\right] \text{ and } \phi\left[\gamma_x\left(\theta_t\left(f\right)^c\right)\right], \tag{8}$$

for all $x \in E$, and $t \in \mathbb{R}$, where $\gamma_x(X)$ extracts the connected component of X to which x belongs. Fig. 2 shows a simple grayscale image, and its associated shapes.

The above researchers show that every two shapes either contain one another or are disjoint. This provides $\mathcal{T}$ with a tree structure, where the parent of every shape $\tau \in \mathcal{T}$ is the smallest shape that contains τ. They call the resulting tree the *Tree of Shapes* of f (whose details were investigated by Ballester, Caselles, and Monasse in [10]), and they show that it is a contrast-invariant representation. Fig. 2(b) depicts the tree of shapes corresponding to the above example. In the sequel, we shall refer to the Tree of Shapes by the acronym ToS.

The ToS can be seen as a generalization of the adjacency tree. Indeed, if f is the indicator function of a given binary image X, then the ToS of f is identical to the adjacency tree of X.

Monasse and Guichard's work was developed for continuous 2D images ($E = \mathbb{R}^2$) and the usual topological connectivity, and extends naturally to the discrete case. However, they stress that, in the discrete case, one connectivity (8- or 4-connectivity) has to be used for the positive shapes, whereas the complementary connectivity (respectively, 4- or 8-connectivity) has to be used for

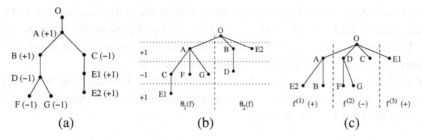

Figure 3. Alternative shape trees. (a) Replicated ToS (RToS), where the differences between the graylevels of each shape and its parent is either +1 or −1. Notice how shape E was replicated. The RToS is associated to Algorithm 1. (b) Threshold Adjacency Tree (TAT), associated with Algorithm 2. (c) Fillhole Tree, associated with Algorithm 3.

the negative shapes (see, e.g., [7]). Otherwise, the shape-inclusion property does not hold. These considerations are valid in our work as well.

The Semilattice

Now, we review the shape-tree semilattice, which we introduced in [1]. Let f be a function in $\text{Fun}(R, \mathbb{Z})$, the set of integer functions over R. Given the ToS of f, modify this tree by replicating each shape as many time as needed until the differences in graylevels between each shape and its parent is either +1 or −1. For example, if in the original ToS a shape τ has graylevel 7 and its parent 10, then τ is replicated twice with graylevels 8 and 9. Let $s(\tau)$ be the new difference in graylevel (either +1 or −1) between τ and its parent. Fig. 3(a) depicts the replicated ToS (which we call RToS) of the image in Fig. 2(a).

Now, for each pixel x, associate the binary sequence $s_f(x)$ of the differences $s(\tau_n(x))$ for all the shapes $\tau_n(x)$ that contain x, ordered from the largest to the smallest. E.g., regarding Fig. 2, for a pixel g in shape G, $s_f(g) = \{+1, +1, -1, -1\}$; a pixel d in shape D that is not in shapes F or G, $s_f(d) = \{+1, +1, -1\}$; and for e in E, $s_f(e) = \{+1, -1, +1, +1\}$. The mapping $f \mapsto s_f$ is called the grayscale unfolding transform, and denoted $\mathcal{U}$ as in the binary case. The inverse mapping is given by

$$f(x) = \mathcal{U}^{-1}\{s_f(x)\} = \sum_n [s_f(x)]_n, \qquad (9)$$

where $[s_f(x)]_n$ denotes the nth element of $s_f(x)$.

A partial ordering for grayscale images is defined as follows:

$$f \sqsubseteq g \iff s_f(x) \leq_4 s_g(x), \quad \forall x, \qquad (10)$$

where $\leq_4$ means here "is a prefix of" (e.g., $\{+1, +1\} \leq_4 \{+1, +1, -1, +1\}$). In fact, $\leq_4$ itself is a partial ordering. It provides the set of binary sequences with a complete inf-semilattice structure, where the null sequence $\{\}$ is the least element, and the infimum $s \bigtriangledown r$ is given by the greatest common prefix. The trivial supremum $\triangle$ is defined only when all operands are point-wise comparable w.r.t. $\leq_4$, in which case it returns the largest.

It is shown in [1] that $\sqsubseteq$ provides the set of discrete, bounded functions with a complete inf-semilattice structure. Flat erosion and dilation operators can be obtained on this inf-semilattice by:

$$\varepsilon_B^T(f) \;=\; \sqcap\{f_b \mid b \in B\} = \mathcal{U}^{-1}\left\{\bigtriangledown_{b \in B}(s_f)_b\right\}, \tag{11}$$

$$\delta_B^T(f) \;=\; \sqcup\{f_{-b} \mid b \in B\} = \mathcal{U}^{-1}\left\{\triangle_{b \in B}(s_f)_{-b}\right\}, \tag{12}$$

where B is a given structuring element, and f_b and $(s_f)_b$ denote the translation of f and s_f, respectively, by the vector b.

Because $(\mathrm{Fun}(R, \mathbb{Z}), \sqsubseteq)$ is a complete inf-semilattice, the erosion $\varepsilon_B^T(f)$ exists for all f in $\mathrm{Fun}(R, \mathbb{Z})$, but the dilation $\delta_B^T(f)$ does not. On the other hand, according to the semilattice theory presented in [5, 6], the dilation always exists when f is the erosion $\varepsilon_B^T(h)$ of some function $h \in \mathrm{Fun}(R, \mathbb{Z})$. Therefore, the operator $\gamma_B^T(\cdot) \stackrel{\triangle}{=} \delta_B^T \varepsilon_B^T(\cdot)$ is always well defined, and is indeed the opening operation associated to the above erosion.

Figure 4 illustrates the result of applying the erosion, its adjoint dilation (yielding opening), and the corresponding top-hat transform to a complex grayscale image.

4.   Erosion Algorithms

Consider the three algorithms presented in Fig. 5. Algorithm 1 is the straightforward implementation of the shape-tree semilattice erosion ε_B^T using (11), where $\varepsilon_B^S(s_f) \stackrel{\triangle}{=} \mathcal{U}^{-1}\left\{\bigtriangledown_{b \in B}(s_f)_b\right\}$. Algorithm 2 decomposes the input image into its threshold levels, erodes each threshold level with the (binary) adjacency erosion ε_B^A, and composes the result back into an image by summing up the (indicator χ of) the resulting threshold levels. An alternating-connectivity version of the (binary) unfolding transform should be used here, where the connectivity of ϕ is switched between 4 and 8 at each iteration of (2). One can show that the adjacency erosion is increasing, and therefore the eroded levels are still contained each in the one below. Algorithm 3 is slightly more complex. It decomposes the input image f into the series of grayscale images $\{f^{(n)}\}$, $n = 1, 2, \ldots$, obtained recursively by $f^{(n)} = \phi(z_n)$, where $z_n = f^{(n-1)} - z_{n-1}$ and $z_0 = f$. The connectivities of the fillhole operation should also alternate here. I.e., for each value of n, the connectivity is switched between 4 and 8. The image f is obtained from its fillhole tree by: $f(x) = \sum_n (-1)^{n-1} f^{(n)}(x)$.

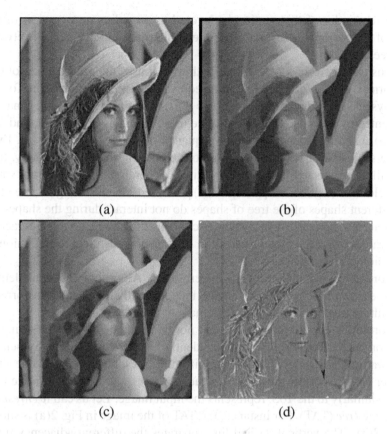

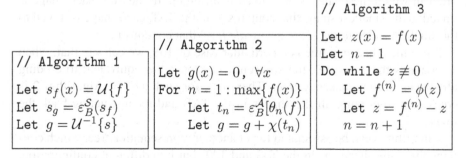

Figure 4. (a) Original grayscale image, (b) shape-tree erosion, (c) opening, and (d) top-hat. A cross structuring element of size 3 was used in all cases.

```
// Algorithm 1

Let  s_f(x) = U{f}
Let  s_g = ε_B^S(s_f)
Let  g = U^{-1}{s}
```

```
// Algorithm 2

Let  g(x) = 0,  ∀x
For  n = 1 : max{f(x)}
    Let  t_n = ε_B^A[θ_n(f)]
    Let  g = g + χ(t_n)
```

```
// Algorithm 3

Let  z(x) = f(x)
Let  n = 1
Do while  z ≢ 0
    Let  f^{(n)} = φ(z)
    Let  z = f^{(n)} − z
    n = n + 1
```

Figure 5. Erosion algorithms.

PROPOSITION 1 *If $B \supseteq \{(0,0),(1,0),(0,1),(-1,0),(0,-1)\}$, then all the three algorithms in Fig. 5 produce the same output g for a given grayscale image f.*

The remainder of this section is devoted to providing an overview of the proof of Proposition 1, by analyzing the similarities and differences between the three algorithms.

Let us take a closer look into Algorithm 1. It uses the grayscale unfolding transform, which, as noted before, is an equivalent representation of the ToS. In the grayscale-unfolding representation $s_f(x)$, two or more shapes of f may be represented by the same binary sequence (e.g., all pixels in shapes F, and G in Fig. 2(a) are both associated with the sequence $\{+1, +1, -1, -1\}$ in s_f). Even if these shapes are not connected one to the other in the 4-connectivity, they may still be connected in the 8-connectivity. Therefore, if an arbitrary structuring element (e.g., $B = \{(0,0), (1,1)\}$) is used, one cannot guarantee that two different shapes of the tree of shapes do not interact during the shape-tree semilattice erosion. However, when B contains the cross s.e., this interaction is neutralized, and the erosion of union of shapes is identical to the union of the erosion of these shapes.

In conclusion, when B contains the cross s.e., Algorithm 1 is equivalent to eroding each and every shape in the ToS independently by B, and transforming the resulting tree back to the image domain to obtain g.

Algorithm 2, on the other hand, can not be associated to the ToS, since it works on each threshold level independently. However, it can be associated to the various adjacency trees of the different threshold levels. If one links together all the roots of the various adjacency trees, one obtains a single tree, which, similarly to the ToS, represents the input image. Let us call it *Threshold Adjacency Tree* (TAT). For instance, the TAT of the image in Fig. 2(a) is shown in Fig. 3(b). The vertical dashed line separates the different adjacency trees, one for each threshold level.

Notice that the TAT contains the same shapes as in the ToS; only that the order in which these shapes are linked is different. In the ToS, each shape is linked to the smallest shape that contains it; in the TAT, each shape is linked to the smallest shape *in the same grayscale level* that contains it.

Since the adjacency tree is a particular case of ToS, we can say that, when B contains the cross s.e., the adjacency erosion ε_B^A is equivalent to eroding all shapes in the adjacency tree. Therefore, for such s.e. B, Algorithm 2 is equivalent to eroding all the shapes in the TAT, and reconstructing back the image.

Algorithm 3 can be associated to yet another representation tree, which contains the same shapes as in the ToS and TAT, but in a different configuration. We call this tree the *fillhole tree*, and it can be obtained by applying the threshold decomposition to each image $f^{(n)}$ of the fillhole decomposition. Each shape in a threshold level t of the image $f^{(n)}$ shall be linked to the smallest shape that contains it in the threshold level $t-1$ of $f^{(n)}$. The fillhole tree of Fig. 2(a) is shown in Fig. 3(c). One can verify that Algorithm 3 is equivalent to eroding each shape in the fillhole tree by B separately, and reconstructing the image back.

Since the shapes in all the three trees (ToS, TAT, and fillhole) are the same, and Algorithms 1 to 3 are all equivalent to eroding each of these shapes in these trees separately, then we conclude that the three Algorithms yield identical results (provided that B contains the cross s.e.).

5. Implementation Considerations

Even though Algorithm 2 provides insightful understanding of the geometrical relationship between the binary and grayscale operators, it is nevertheless significantly inefficient when the number of graylevels is high, which is the case regarding typical 8-bit images.

Algorithm 3 is the conceptually simplest and the easiest to implement, if one has a good implementation of the grayscale fillhole operator. For instance, it is a great option for Matlab simulations.

The most complex, but also the fastest, implementation of the shape-tree semilattice erosion can be obtained by Algorithm 1. In [7], a fast algorithm for calculating the ToS is presented. It is fairly simple to modify this algorithm to obtain the grayscale unfolding transform directly. One could consider a drawback of Algorithm 1 the size of the transformed data, which typically consist of a lengthy binary sequence for each pixel in the image. Here are a few ways to solve this problem. The first one is to consolidate the binary sequences $s_f(x)$ into alternating sequences of positive and negative numbers, representing the run-lengths of $+1$'s and -1's, respectively. E.g., $\{-1, -1, +1, +1, +1, +1, -1, -1, -1\}$ becomes $\{-2, 4, -3\}$. The infimum operation can be calculated directly on the alternating sequences by keeping their common prefix, followed by the weakest of the next elements, if they have the same sign. For instance, the infimum between $\{-2, 4, -3, 6\}$ and $\{-2, 4, -1, 5\}$ is $\{-2, 4, -1\}$, between $\{3, -6, 2, 7\}$ and $\{3, -6\}$ is the latter, and between $\{1, -2\}$ and $\{-3, 2, -9\}$ is $\{\}$ (empty sequence). A second approach to reduce data storage is to store a table that associates an index to each existing binary sequence in the image, and associate the corresponding index to each pixel. A third way is to associate to each pixel a pointer to one pixel in the parent shape. One has to take the care of pointing all pixels in a shape to the same parent pixel. Then, the infimum of two pixels is computed by following the pointers of each pixel recursively until both paths reach a common pixel.

6. Conclusion

We have reviewed the definition of adjacency lattice, which is the proposed framework for quasi-self-dual morphological processing of binary images. In this approach, each shape of the image is operated upon (erosion, dilation, etc.)

regardless to whether it belongs to the foreground or to the background. As a consequence, all shapes in the image suffer the same morphological modification in a contrast-invariant way.

The binary framework is generalized to grayscale images by means of the shape-tree semilattice. Three different algorithm for implementing the resulting morphological operators (erosion and adjoint dilation) are investigated.

References

[1] R. Keshet, "Shape-Tree Semilattice," *Journal of Mathematical Imaging and Vision, special issue of mathematical morphology after 40 years*, scheduled to March 2005.

[2] J. Serra, *Image Analysis and Mathematical Morphology, Vol. 1*, London: Academic Press, 1982.

[3] H.J.A.M. Heijmans, "Connected Morphological Operators for Binary Images," *Computer Vision and Image Understanding*, Vol. 73, No. 1, pp. 99-120, 1999.

[4] P. Soille, *Morphological Image Analysis: Principles and Applications*, NY: Springer, 2nd edition, 2003.

[5] R. Keshet (Kresch), "Mathematical Morphology on Complete Semilattices and its Applications to Image Processing," *Fundamenta Informaticae*, Vol. 41, Nos. 1-2, pp. 33-56, January 2000.

[6] H.J.A.M. Heijmans and R. Keshet, "Inf-semilattice approach to self-dual morphology," *Journal of Mathematical Imaging Vision*, Vol. 17, No. 1, pp. 55-80, July 2002.

[7] P. Monasse and F. Guichard, "Fast computation of a contrast-invariant image representation," *IEEE Trans. on Image Processing*, No. 9, Vol. 5, pp. 860-872, May 2000.

[8] P. Monasse and F. Guichard, "Scale-space from a level lines tree," *Journal of Visual Communication and Image Representation*, Vol. 11, pp. 224-236, 2000.

[9] V. Caselles and P. Monasse, "Grain filters," *Journal of Mathematical Imaging and Vision*, Vol. 17, No. 3, pp. 249-270, November 2002.

[10] C. Ballester, V. Caselles and P. Monasse, "The Tree of Shapes of an Image," *ESAIM: Control, Optimisation and Calculus of Variations*, Vol. 9, pp. 1-18, 2003.

III

SEGMENTATION

MORPHOLOGICAL SEGMENTATIONS OF COLOUR IMAGES

Jean Serra

Centre de Morphologie Mathématique - Ecole des Mines de Paris
35 rue Saint Honoré - 77300 Fontainebleau - France
Jean.Serra@cmm.ensmp.fr

Abstract   Colour images are multivariable functions, and for segmenting them one must go through a reducing step. It is classically obtained by calculating a gradient module, which is then segmented as a gray tone image. An alternative solution is proposed in the paper. It is based on separated segmentations, followed by a final merging into a unique partition. Three problems are treated this way. First, the search for alignments in the 2-D saturation/luminance histograms. It yields partial, but instructive results which suggest a model for the distribution of the light over the space. Second, the combination of luminance dominant and hue dominant regions in images. Third, the synthesis between colour and shape information in human bust tracking.

Keywords: colour, segmentation, saturation, norms, light propagation, connection, multivariate analysis

1.   Introduction

The present paper aims to analyse the way information is reduced when we go from multi-dimensional colour images to their segmentations, i.e. to final unique optimal partitions [15]. The problem is the following: sooner or later, the processing of such multi-dimensional data goes through a *scalar reduction*, which in turn yields the final partition. Usually, the scalar reduction arises rapidly, since in the most popular procedures it consists in replacing, from the beginning, the bunch of images by a sole gradient module on which the various minimizations hold (e.g. the watershed technique). When the scalar reduction occurs too soon, it risks to ignore specific features of each band, and to destroy them in the melting pot that generates the 1-D variable. The alternative approach we propose here works in the exactly opposite way. We will try and obtain first various intermediary partitions, and then make the final segmentation hold on them.

C. Ronse et al. (eds.), Mathematical Morphology: 40 Years On, 151–176.
©2005 *Springer. Printed in the Netherlands.*

This idea is developed below through three studies. The first one extends to colour images the simplest segmentation technique for numerical functions, which consists in histogram thresholding. How to extend it to the 2-D or 3-D histograms of the colour case? The question will be tackled in the framework of the *brightness-saturation-hue* representations. Such polar coordinates have to be defined in a suitable manner for image processing, as A. Hanbury and J. Serra did in [8]. We have also to check the pertinence of these new representations. J. Angulo [2] did it by analysing their 2-D histograms, which exhibit typical alignments. The physical interpretation of these structures leads to an original model for light reception, which is proposed below, in section 3.

The second study relies on the intuition that human vision exploits the hue for segmenting the highly saturated regions, and the luminance for the weakly saturated ones. This way of thinking already appears in literature with C.-H. Demarty and S. Beucher [6], and with P. Lambert and T. Carron [9]. But it is developed here differently, as we seek for an optimal partition by combining the three segmentations of the polar coordinates [1] (section 4).

The third variant enlarges the scope, and aims to synthesize segmentations according to both colour and shape. When we look at a bust, for example, the face of the person presents characteristic colours, whereas the shoulders are better described by their shape. How to mix together such heterogeneous sources? This sort of questions suggests a new model for multi-labelled connections that we will construct on the way (section 5).

The first two studies need a detour, as we have to justify the creation of new parameters (of saturation in particular). A brief remainder on the gamma correction is necessary. An excellent presentation of the theme may be found in Ch. Poynton's book [11], see also [18]. As for the notation, we follow Ch. Poynton, who differentiates by apostrophes the electronic colours (e.g. r') from the light intensities (e.g. r). Below, the rule is extended to the operations themselves; for example the arithmetic mean is written m for intensities and m' for video variables. Also, we adopt the convention of the CIE, which designates the absolute quantities by upper letters (e.g. X, Z) and the relative ones by lower case letters (e.g. x, z).

2.   The 3-D polar representations of the colour

Light intensities and gamma correction

Consider a television receiver. It uses three different colour representations. On the one side, the input Hertzian signal is coded as one grey image plus two other ones, associated to green-red and blue-yellow contrasts (i.e. one luminance and two chrominances). On the other side, the image on the monitor is obtained from three electrical signals, which excite three layers of green, red and blue photo-receivers. These two representations are quite different,

although technically sound for their respective purposes. However, the manu-
facturers take none of them for the user's interface, and prefer human adjust-
ments based on light (luminance), contrast (saturation), and, in case of an old
receiver, from hue. Hence, this last triplet turns out to be the simplest one for
human vision.

What are the relationships between these various representations? Do the
technological steps modify the initial light that enters a device? Colour image processing rests on a few basic operations (addition, comparison,...) and
properties (increasingness, distances...). Have these tools a physical meaning?
In colour imagery the basic notion is the spectral power distribution (SPD) of
the light radiating from or incident on a surface. This intensity has the dimension of an energy per unit area, such as watt per m^2. When the light arrives at a
photo-receiver, this sensor filters the intensities of each frequency by weighting
them according to fixed values. The sum of the resulting intensities generates
a signal that exhibits a certain "colour". The CIE (Commission Internationale
de l'Eclairage), in its *Rec 709*, has standardized the weights which yield the
triplet $R_{709}, G_{709}, B_{709}$ [4]. As energies, the intensities are additive, so that
all colours accessible from an RGB basis are obtain by sums of the primary
colours R, G, and B and by multiplications by non negative constants.

The exploration of the spectrum is lower bounded by $R = G = B = 0$
(zero energy) higher bounded by a maximum red R_0, green G_0 and blue B_0
that are given by the context (illumination, technological limits of the sensors,
or of the eye, etc.) in which we work. Generally, each technology fixes the
three bounds, which therefore define the reference white, and then introduces
the *reduced variables*

$$r = \frac{R}{R_0}, \qquad g = \frac{G}{G_0}, \qquad b = \frac{B}{B_0}. \tag{1}$$

The digital sensitive layers of cameras transform the light intensities into
voltages; conversely, the cathodic tubes (CRT) and the flat screens that display
images return photons from the electrical current. Now, their response is not
linear, but a power function of the input voltage whose exponent γ, (gamma),
varies around 2.5 according to the technologies. If we want the light intensities
of the CRT to be proportional to those of the scene itself, the gamma effect has
to be compensated. In video systems, this *gamma correction* is universally at
the camera. The *Rec. 709* of CIE proposes the following correction function

$$\begin{aligned} r' &= 4.5r & r \leq 0.018 \\ r' &= 1.099r^{0.45} - 0.099 & r > 0.018 \end{aligned} \tag{2}$$

that we write here for the reduced red intensity r, and where $1/\gamma = 0.45$. The
same transfer function is applied to both green and blue bands.

Fig. 1, drawn from [11] depicts the graph of Rel.(2). The variation domain
$[0, 1]$ is the same for the reduced intensities (r) as for the video colours (r'),

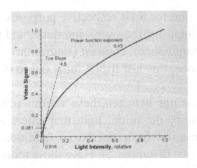

Figure 1. Gamma correction function.

which implies that the white point $R_0\ G_0\ B_0$ is left invariant. The linear beginning in Rel.(2) minimizes the effect of the sensor noise. An ideal monitor should invert the transform Rel.(2). Indeed, they generally have neither linear segment, nor gamma exponent equal to $1/0,45$ [11].

Fig. 1 shows that for r closed to1, the graph looks like a straight line. More precisely, the limited expansion

$$(1-u)^{1/\gamma} = 1 - \frac{u}{\gamma} + \epsilon(u) \qquad (3)$$

for small u, leads us to replace the second equation (2) by

$$r'^* = (0.55 + 0.45r)1.099 - 0.099 \qquad (4)$$

i.e., numerically

r	0.9	0.8	0.7	0.6	0.5
r'	0.949	0.895	0.837	0.774	0.705
r'^*	0.950	0.901	0.851	0.802	0.753
$\frac{r'-r'^*}{r'}$	0.1%	0.6%	1.4%	2.8%	4.8%

In comparison with the noise of the video systems, we can consider the approximation r'^* is perfect for $r \geq 0.8$ and excellent for $0.6 \leq r \leq 0.8$.

Colour Vector Spaces

Their linearity provide the intensities r, g, b with the structure of a 3 dimensions vector space, or rather of the part E which is limited to the unit cube $[0,1] \times [0,1] \times [0,1]$ of $\mathbb{R}^3$. For colour image processing purposes, it would be wise to go back from the video bands (r', g', b') to the reduced intensities (r, g, b) by the inverse transform of Rel.(2). When starting from the usual 3×8 bits (r', g', b') images, the best should probably be to code in 3×16 bits for

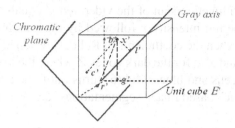

Figure 2. Chromatic plane and a-chromatic axis.

computation (or in floating variables). But as a matter of fact, people keeps the (r', g', b') video space, which is implicitly modelled as a part of a vector space, from which one builds arithmetic means, projections, histograms, Fourier transforms, etc... which often gives significant results.

What are the real consequences of the gamma correction Rel.(2) on the processing of colour data? Formally speaking, one can always consider the unit video cube (r', g', b') as a part, E' say, of a 3-dimensions vector space. This allows us to formulate operations, but their physical interpretations demand we come back to the intensities (r, g, b).

Fig. 2 depicts the unit cube E'. The vector $\vec{x}'$, of coordinates (r', g', b') can also be decomposed into two orthogonal vectors $\vec{c}$ and $\vec{l}'$ of the chromatic plane and the a-chromatic (or gray) axis respectively. The latter is the main diagonal of the cube going through the origin O and the chromatic plane is perpendicular to the gray axis in O. The two vectors $\vec{c}$ and $\vec{l}'$ have the following coordinates

$$3\vec{c} = (2r' - g' - b', 2g' - b' - r', 2b' - r' - g')$$
$$3\vec{l}' = (r' + g' + b', r' + g' + b', r' + g' + b') \tag{5}$$

Consider the red band $r'(z)$ over a zone Z in a colour image. What meaning can we give to the average red in Z? As we just saw, the only average that has a physical meaning is the quantity $\bar{r} = \frac{1}{Z} \int (r'(z))^\gamma \, dz$, which needs to be corrected into $\bar{r}^{1/\gamma}$ for display purposes (for the moment we neglect the constants 1,099 and 0,099 in Rel.(2)). On the other hand, the usual segmentations aim to split the space into regions Z where the colour is nearly constant. Then at each point $z \in Z$, we can approximate $r(z)$ by the limited expansion

$$r(z) = r'(z)^\gamma = \bar{r}'^\gamma \left[1 - \frac{\bar{r}' - r'(z)}{\bar{r}'}\right]^\gamma = \bar{r}'^\gamma \left[1 - \gamma \left(\frac{\bar{r}' - r'(z)}{\bar{r}'}\right) + \varepsilon(r')\right]$$

where $\bar{r}' = \frac{1}{Z} \int_Z r'(z) dz$. Under averaging in Z, the coefficient of the γ term in the right member becomes zero, so that

$$(\bar{r})^{1/\gamma} = \bar{r}' + \bar{\epsilon}(r') \tag{6}$$

Therefore, the arithmetic mean of the video red r' equals, at the second order, the mean of the red intensity r followed by the gamma correction. The result remains true when the coefficients of Rel.(2) are added, when the the average is weighted, and also for the dark zones Z where the first Rel.(2) applies. It extends to the greens and blues. Rel.(6) turns out to be a theoretical justification of the "mosaic" based image segmentations (e.g. waterfall algorithm).

Brightness

From the point of view of physics, brightness is nothing but the integral of the power spectrum, i.e., here, the sum of the three components r, g, and b, that stand for this spectrum. For colorimetric purposes, this sum has to be weighted relatively to the spectral sensitivity of the eye. The CIE *Rec. 709* defines a white point and three weighting functions of the spectrum which lead to the variables R_{709}, G_{709} and B_{709}, then to the *luminance*

$$Y_{709} = 0.212R_{709} + 0.715G_{709} + 0.072B_{709} \qquad (7)$$

and to the luminance Y_W of the associated white point. The three coefficients of Rel.(7) are related to the brightness sensitivity of the human vision and have been estimated by colorimetric measurements on a comprehensive population. The luminance Y_{709}, as a linear function of intensities, is an energy ($watts/m^2$).

Human vision responds to intensities in a logarithmic way, according to laws of the type $di/i = constant$. Just as we took into account the spectral sensitivity of the eye, we should not ignore its energetic sensitivity. Now, by an amazing coincidence vision response to intensity is closed to the gamma correction of Rel.(2): for example, when the luminance of a source is reduced to 20%, the eye perceives an intensity reduction of 50% only. Therefore, following many authors, we can consider the transforms

$$r' = r^{1/\gamma} \qquad g' = g^{1/\gamma} \qquad b' = b^{1/\gamma} \qquad (8)$$

for $\gamma \simeq 2.2$ as generating *perceptual intensities* . For example, the *Rec. BT 601-E* proposes the *luma* y'_{601} as a perceptual brightness measurement

$$y'_{601} = 0.299r' + 0.587g' + 0.144b'. \qquad (9)$$

However, this luma, as established from video values has not an energy dimension, and not any more the deriving additivity properties. The CIE follows the same direction, but defines the *lightness* l^* by taking a slightly different exponent

$$l^* = 116(\frac{Y_{709}}{Y_W})^{1/3} - 16 \qquad Y \geq 0.0089Y_W.$$

As regards the operations of segmentation in image processing, the situation is different. They do not hold on a *perceived* brightness, but on that of the *object under study*. In microscopy, the histological stainings usually range from blue to violet; the spectrum of a sunset, or that of a human face have nothing to do with the weights given to r, g, and b in Rel.(7) or (9). Thus in the absence of *a priori* information on the spectra of the objects under study, the purpose of segmentation leads us to take as brightness a *symmetrical function* of primary colours.

As regards the perceived energies now, consider, in the intensity space E, a vector x whose direction is given by $x_o = r_o, g_o, b_o$ but whose intensity varies, i.e.

$$x = (\lambda r_0, \lambda g_0, \lambda b_0) \qquad \lambda \in [0, \lambda_{\max}]$$

The point x describes the segment S which begins in O, goes through (r_o, g_o, b_o) and ends on the edge of cube E. In the video space E' there corresponds to x the point x' :

$$x' = \left((\lambda r_0)^{1/\gamma}, (\lambda g_0)^{1/\gamma}, (\lambda b_0)^{1/\gamma}\right) = \lambda^{1/\gamma} x_0' \qquad (10)$$

with $x_0' = r_0^{1/\gamma}, g_0^{1/\gamma}, b_0^{1/\gamma}$. Similarly, the point x' describes a segment S' in E'. When x varies, if we want its perceptual brightness to seem additive, then Rel.(10) implies that the corresponding brightness of x' is a linear function of the three primary components. Finally, since this "image processing brightness" has to vary from 0 to 1, as r and r' do, the only possibility is to take for it the arithmetic mean m' of the primary colours :

$$m' = \frac{1}{3}(r' + g' + b').\qquad (11)$$

Put $\lambda' = \lambda^{1/\gamma}$. The two expressions

$$|m(x_1) - m(x_2)| = |\lambda_1 - \lambda_2| \, m(x_0)$$
$$|m'(x_1') - m'(x_2')| = \left|\lambda_1^{1/\gamma} - \lambda_2^{1/\gamma}\right| m'(x_0')$$

turn out to be different distances in segments S and S' respectively. The exponent $1/\gamma$ provides the second one with a meaning of *perceptual homogeneity*. But image processing is more demanding, as we must be able to express that a colour point E' (or more generally a set of points) gets closer to another even when these two points are not aligned with the origin. Now, the mean (11) is nothing but the restriction to the cube E' of the L_1 norm, which is defined in the whole space $\mathbb{R}^3$ (i.e. for $r', g', h' \in [-\infty, +\infty]$) by taking $\alpha = 1$ in the relation

$$n(x') = \left(|r'(x)|^\alpha + |g'(x)|^\alpha + |b'(x)|^\alpha\right)^{1/\alpha} \qquad \alpha \geq 1 \qquad (12)$$

(This Rel.(12) introduces indeed a family of norms as soon as $\alpha \geq 1$. For $\alpha = 2$, we obtain the Euclidean norm L_2, and for $\alpha = \infty$, the "max" norm). In a vector space V, any norm n generates a distance d_n (see [5], section VII-1-4) by the relation

$$d_n\left(x'_1, x'_2\right) = n\left(x'_1 - x'_2\right) \qquad x'_1, x'_2 \in V \qquad (13)$$

Therefore L_1 is a distance, as well, of course, as its restriction to the unit cube E'.

For $\alpha = 1$, both brightness $m'(x')$ and distance $d\left(x'_1, x'_2\right) = m'\left(\left|x'_1 - x'_2\right|\right)$ in E' thus derive from a unique concept. This latter relation is important, as in segmentation a number of algorithms which were established for numerical functions extend to vector functions when a distance is provided (e.g. watershed).

Saturation

The CIE was more interested in various formulations of the brightness (luminance, lightness ...) than in saturation, that it defines as "*the colourfulness of an area judged in proportion to its brightness*". In other words, it is the concern of the part of uniform spectrum (i.e. of gray) in a colour spectrum, so that any maximal monochromatic colour has a unit saturation and so that any triplet $r = g = b$ has a zero saturation.

Intuitively, what the CIE means here is clear, but its definition of the saturation lends itself to various interpretations. From a given point $x \in E$, one can draw several paths along which the colour density varies in proportion to brightness. For example, in Fig. 2, supposed to represent cube E, we can take the perpendicular xc to the chromatic plane, or the perpendicular xl to the gray axis, or again the axis Ox, etc.. Which path to choose?

Indeed, these ambiguities vanish as soon as we set the context in the chromatic plane. The cube E is projected according to Hexagon H centered in O. Consider a point $x_o \in E$, of projection c_0 in H, and such that $c_0 \neq O$. Following the CIE, we define as a saturation any non negative function along the axis Oc_0 that *increases* from O; in O, it equals zero (pure gray) and has its maximum value when the edge of Hexagon H is reached, in $c_{\max}$ say (saturated colour). The hue remains constant along the segment $[0, c_{\max}]$, and the hue of the opposite segment $[0, \bar{c}_{\max}]$ is said to be *complementary* of that of segment $[0, c_{\max}]$. For a point $c \in [0, c_{\max}]$, we have $c = \lambda c_0, 0 \leq \lambda \leq 1$. Thus, given $c_0 \in H$, the saturation $s(c) = s(\lambda c_0)$ is a function of λ only, and this function is increasing.

We have to go back to the 3-D cube E, as point c_0, projection of x_0, is just an intermediary step (moreover $c_0 \notin E$). The saturation $s(x_0)$ of point $x_0 \in E$ is then defined by

$$s(x_0) = s(c_0)$$

Note that when a point $x \in E$ moves away from the chromatic plane along the perpendicular $c_0 x_0$ to this plane, its gray proportion increases, but its saturation $s(x)$ does not change: it is indeed a matter of *chromatism* and not of *energy* of the light intensity.

As point c describes the radius $[0, \bar{c}_{\max}]$ which is at the opposite of $[0, c_{\max}]$ in the chromatic plane, we have

$$c \in [0, \bar{c}_{\max}] \qquad \Longleftrightarrow \qquad c = \lambda c_0 \qquad \lambda(\bar{c}_{\max}) \leq \lambda \leq 0$$

where λ indicates the proportionality ratio, now negative, between c and $\bar{c}_0$. This purely vector equivalence admits a physical interpretation if we extend the definition of the saturation to all diameters $D(c_0) = [0, c_{\max}] \cup [0, \bar{c}_{\max}]$, $c_0 \in H$, of the hexagon H (saturation was previously introduced for radii only). This can be done by putting $c \in D(c_0)$, $s(c) = s(\lambda c_0) = s(|\lambda| c_0)$. Two opposite points have the same saturation, and more generally if $c_1 \in [0, c_{\max}]$ and $c_2 \in [0, \bar{c}_{\max}]$, then $c_1 + c_2 = (\lambda_1 + \lambda_2) c_0$, with $\lambda_1 \geq 0$ and $\lambda_2 \leq 0$. As s is increasing we have

$$c_1 \in [0, c_{\max}], \qquad c_2 \in [0, \bar{c}_{\max}] \qquad \Longrightarrow \qquad s(c_1 + c_2) \leq s(c_1) + s(c_2). \tag{14}$$

When $c_1 = c_{\max}$ and $c_2 = \bar{c}_{\max}$ we find in particular Newton's disc experiment, reduced to two complementary colours.

When considering the saturation in the video cube E', the conditions of increasingness of s' along the radii (now of H') and of its nullity on the gray axis are still valid. They must be completed by the two requirements of image processing, namely the symmetry w.r.t. r', g', b' and the fact that $s'(x_1' - x_2')$ must be a distance in E'.

We saw that the mean m', in Rel.(11), was the L_1 norm expressed in the unit cube E', and that $3m'(x')$ was both the norm of x' and of its projection l' on the gray axis, i.e.

$$L_1(x') = L_1(l') = 3m'(x')$$

It is tempting to keep the same norm for the hexagon H' of the chromatic plane. By using Rel.(5) we find

$$s'(x') = L_1(c') = \frac{1}{3} \left[|2r' - g' - b'| + |2g' - b' - r'| + |2b' - r' - g'| \right]. \tag{15}$$

By symmetry, $s'(x')$ depends on the three functions $max' = max(r', g', b,')$, $min' = min(r', g', b,')$, and $med' = mediane(r', g', b,')$ only, which gives

$$s' = \begin{cases} \frac{3}{2}(max' - m') & \text{if } m' \geq med' \\ \\ \frac{3}{2}(m' - min') & \text{if } m' \leq med' \end{cases} \tag{16}$$

On can find in [14] the derivation yielding s', and that of the following expression h' of the hue (which avoids to bring trigonometric terms into play),

$$h' = \frac{\pi}{3}\left[\lambda + \frac{1}{2} - (-1)^\lambda \frac{max' + min' - 2med'}{2s'}\right] \qquad (17)$$

with λ equals

$$0 \text{ if } r > g \geq b, \quad 1 \text{ if } g \geq r > b, \quad 2 \text{ if } g > b \geq r,$$
$$3 \text{ if } b \geq g > r, \quad 4 \text{ if } b > r \geq g, \quad 5 \text{ if } r \geq b > g \qquad (18)$$

The hue h', as a coordinate on the unit circle, is defined modulo 2π. The value $h' = 0$ in Eq.(17) corresponds to the red. For $s' = 0$, colour point lies on the gray axis, so that its hue is meaningless. The inverse formulae are given in [8], and the detailed proofs may be found in [14].

The relations (15) and (13) entail that $s'(c_1' - c_2') = L_1(c_1' - c_2')$ is a distance in the chromatic plane, which therefore brings into play both saturation and hue. On the other hand, as L_1 is a norm, Rel.(14) becomes true for all triplets c_1', c_2' and $c_1' + c_2'$ that are on a same diameter of H'. Remark that here the L_1 norm is the concern of the *projections* c', the norm of the vectors x' themselves being their arithmetic mean. Finally, the above comments apply also to the Euclidean norm and to the max-min, which both induce distances in the chromatic hexagon H'.

When passing from the video variables to the intensities, a first result is obtained by observing that the averaging of the saturation s' follows the same law than that of the brightness m', namely Rel.(6), in the zones Z where the colour varies weakly. Moreover, the mapping $x_0' = (r_0', g_0', b_0') \to x_0 = (r_0'^\gamma, g_0'^\gamma, b_0'^\gamma)$ shows that $c' = \lambda c_0'$ becomes $c = \lambda^\gamma c_0$, hence

$$s'(x') = \lambda L_1(c_0') = \lambda s'(x_0') \Leftrightarrow s(x) = \lambda^\gamma L_1(c_0) = \lambda s(x_0).$$

In other words, the L_1 norm is increasing on the radii, and is zero for the grays, on both chromatic hexagons H of the intensities and H' of the video variables. Thus it represents a saturation *in both spaces*. It seems difficult to go further, as two points $x_0', x_1' \in E'$ whose projections c_0' and c_1' lie on a same radius of H' may have homolog points x_0 and $x_1 \in E$ whose projections are not always aligned with O.

Two other norms

How to build polar representations which be not contradictory with the previous requirements? Besides the L_1 norm, we can think of two ones. Firstly, the Euclidean norm L_2. In practical image processing, it turns to be less convenient than the L_1 norm, which suits particularly well to linear and morphological operations, and provides nice inverses. In addition, the associated 2-D histograms are rather unclear [2].

Another possibility is to correct the classical HLS system, by replacing its saturation by $max(r,g,b) - min(r,g,b)$. In the whole space, the quantity $max - min$ is a semi-norm only: two distinct vectors c and c', whose difference c - c' is a gray have the same $max - min$ [8]. However, in the chromatic plane, $max - min$ becomes a norm. It can be used for the saturation in parallel with m' for the brightness. This is what we will do below each time $max - min$ norm is introduced.

Finally, the norm and distance based approach presents the significant advantage that it *separates the variables* : two points x_1' and $x_2' \in E'$ which have the same projection on the chromatic plane (resp. on the gray axis) have the same saturation (resp. the same brightness). However, the last property, on brightness, vanishes when the three bands are given different weights in the means m or m'.

The classical polar representations

Even though the transformation from RGB to hue, saturation and brightness coordinates is simply a transformation from a rectangular colour coordinate system (RGB) to a three-dimensional polar (cylindrical) coordinate system, one is faced with a bewildering array of such transformations described in the literature (HSI, HSB, HSV, HLS, etc.). Most of them date from the end of the seventies [17], and were conceived neither for processing purposes, nor for the current computing facilities. This results in a confusing choice between models which essentially all offer the same representation. The most popular one is the HLS triplet of System (19), which appears in many software packages. The comments which follow hold on this particular model, but they apply to the other ones. The HLS triplet derives from RGB by the following system

$$
\begin{cases}
l'_{HLS} = \dfrac{max(r',g',b')+min(r',g',b')}{2} \\[2ex]
s'_{HLS} = \begin{cases}
\dfrac{max(r',g',b')-min(r',g',b')}{max(r',g',b')+min(r',g',b')} & \text{if } \; l'_{HLS} \leq 0.5 \\[2ex]
\dfrac{max(r',g',b')-min(r',g',b')}{2-max(r',g',b')-min(r',g',b')} & \text{if } \; l'_{HLS} \geq 0.5
\end{cases}
\end{cases}
\tag{19}
$$

One easily checks that the HLS expressions do not preserve the above requirements of linearity (for the brightness), of increasingness (for the saturation) and of variables separation. The HLS luminance both RGB triplets

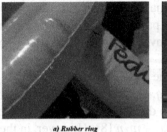 

a) Rubber ring   *b) Ana Blanco*

Figure 3.   Two test images.

$(1/2, 1/2, 0)$ and $(0, 1/2, 1/2)$ equals $1/4$, whereas that of their mean equals $3/8$, i.e. it is lighter than both terms of the mean. The HLS saturations of the RGB triplets $(4/6, 1/6, 1/6)$ and $(2/6, 3/6, 3/6)$ equals $3/5$ and $1/5$ respectively, whereas that of their sum is 1: it is just Newton's experiment denial! Finally the independence property is no more satisfied. Take the two RGB triplets $(1/2, 1/2, 0)$ and $(3/4, 3/4, 1/4)$. One passes from the first to the second by adding the gray $r' = g' = b' = 1/4$. Hence both triplets have the same projection on the chromatic plane. However, the HLS saturation of the first one equals 1 and that of the second $1/2$.

3.   2-D Histograms and linearly regionalized spectra

In practice, is it really worth deviating from beaten tracks, and lengthening the polar triplets list? What for? We may answer the question by comparing the *luminance/saturation* bi-dimensional histograms for HLS system and for various norms. J. Angulo and J. Serra did so on a dozen images [2] [3]. Two of them are depicted below, in Fig. 3.

Bi-dimensional histograms

In the first image, we observe strong reflections on the rubber ring, and various types of shadows. The corresponding L_1 and HLS histograms are reported in Fig. 4, with luminance on the x axis and saturation on y axis. No information can be drawn from HLS histogram, although alignments are visible on L_1 norm.

By coming back to the images, we can localize the pixels which give alignments, as depicted in Fig. 5. They correspond to three types of areas:

- shadows with steady hue,

- graduated shading on a plane,

- reflections with a partial saturation.

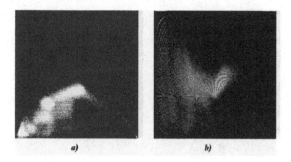

Figure 4.   Bi-dimensional histograms of the "rubber ring" image. The x-axis corresponds the luminance and the y-axis to the saturation. a) L_1 norm, b)HLS representation.

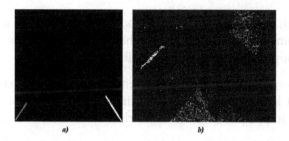

Figure 5.   Zones of "Rubber ring" associated with alignments. The left image a) show the supports of the alignments in Fig. 4 (in L_1 norm), and the right image indicate the locations of the aligned pixels in the space of the initial picture. The white (resp. gray) alignments of Fig. a) correspond to the white (resp. gray) pixels of Fig.b).

Consider now the more complex image of "Ana Blanco", in Fig.3b. It includes various sources light (television monitor, alpha-numerical incrustations...), and the light diffused by the background is piecewise uniform over the space. However, there are still alignments, which do not always go through points $(0,0)$, or $(1,0)$, and are sometimes parallel. In the lum/hue plane of the L_1 norm representation, several horizontal lines (constant hue) are located at different hue levels, and alternate with elongated clouds of points (Fig.6b).

All in all, we draw from the above histograms four main informations.

1 In the lum/sat histogram, there is no accumulation of pixels at point $(1,0)$. It means that the sensors we use are not physically saturated, which make realistic the proposed linear approach;

2 Still in the lum/sat histogram, some well drawn alignments can be extrapolated to point $(0,0)$ or point $(1,0)$. The others are parallels to the first ones;

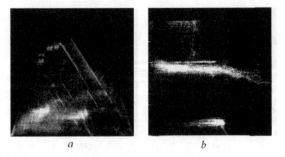

a　　　　　　　　　　　　　　　*b*

Figure 6.　　(a) and (b) the two histograms of "Ana Blanco", in the *luminance/saturation* and the *luminance/hue* plane respectively, both in L_1 norm.

3　However, most of the pixels form clouds in both lum/sat and lum/hue histograms are not aligned at all, whether the model does not apply, or the homogeneous zones are too small;

4　In the lum/hue histogram, most often the aligned pixels exhibit a (quasi) constant hue, i.e. draw horizontal lines. But sometimes, these "lines" turn out to be a narrow horizontal stripe.

Such characteristic structures, such distinct lines suggest we seek a physical explanation of the phenomenon. This is what we will do now. But besides any physical model, a first point is worth to be noticed: the only norm that enables us the extraction of reflection areas, of shadows and gradations is L_1. No other polar model results in such an achievement.

Linearly regionalized spectra (LR model)

If we assume that the alignments are a property of the spectrum, and not an artefact due to some particular representation, we have to express the spectrum in such a way that the sequence

$$(spectrum) \rightarrow (r'g'b') \rightarrow (m's'h') \rightarrow (m' = \alpha s' + \beta)$$

be true (in the alignments) whatever the weights generating r, g and b are, and also whatever the spectrum itself is. Consider a zone Z of the space whose all pixels yield an alignment in the L_1 histogram. Denote by $sp(\nu; z)$ the spectrum of the light intensity at point $z \in Z$. We will say that this spectrum is *linearly regionalized* in Z when for each point $z \in Z$ one can decompose $sp(\nu; z)$ into the sum of a first spectrum $sp_0(\nu)$, independent of point z, and of a second one, $\overline{\omega}(z)sp_1(\nu)$, which proportionally varies in Z from one point to another. For all $z \in Z$, we have

$$sp(\nu; z) = sp_0(\nu) + \overline{\omega}(z)sp_1(\nu) \tag{20}$$

where $\overline{\omega}(z)$ is a numerical function which depends on z only, and where sp_0 and sp_1 are two fixed spectra.

In the spectrum $sp(\nu; z)$, though sp_0 usually corresponds to diffuse light and sp_1 to specular one, we do not need to distinguish between the emitted and reflected components of the light. It can be the concern of the light transmitted through a net curtain, for example, or of that of a TV monitor; but it can also come from passive reflectance, such as those described by Shafer's dichromatic model [16], or by Obein et Al.'s model of glossiness [10]. But unlike these two models, the term $\overline{\omega}(z)sp_1$ may also represent an absorption, when it is negative. Similarly, we do not need to distinguish between diffuse and specular lights. The term sp_0 may describe a diffuse source over the zone Z, as well as a constant specular reflection stemming from the same zone. But above all, the emphasis is put here on the *space variation* of the spectrum. It is introduced by the weight $\overline{\omega}(z)$, that depends on point z, but not on spectrum sp_1. This weight may bring into play cosines, when the angle of the incident beam varies, or the normal to a glossy surface, etc...

The three spectra sp, sp_0 and sp_1 are known only through the weighting functions that generate a (R, G, B) triplet. We use here the notation (R, G, B) in a canonical manner, i.e. it may designate the (X, Y, Z) coordinates of the CIE, or the perceptual system (L, M, S) [18], as well as the (Y, U, V) and (Y, I, Q) TV standards. In all cases it is a matter of *scalar products* of the spectra by such or such frequency weighting. In particular, the white colour given by $r = g = b = 1$ can be obtained from a spectrum which is far from being uniform. We write

$$r(z) = \int \left[sp_0(\nu) + \overline{\omega}(z)sp_1(\nu)\right]\xi(\nu)\,d\nu = r_0 + r_1\overline{\omega}(z) \qquad (21)$$

$$g(z) = \int sp(\nu; z)\,\chi(\nu)\,d\nu = g_0 + g_1\overline{\omega}(z) \qquad (22)$$

and

$$b(z) = \int s(\nu; z)\,\psi(\nu)\,d\nu = b_0 + b_1\overline{\omega}(z) \qquad (23)$$

where ξ, χ and ψ are the three weighting functions that generate the primary colours r, g and b.

As sp_0 and sp_1 are power spectra, they induce *intensities* r, g, and b. Now, in the above histograms, the L_1 norm applies to the *video variables* $r' = r^{1/\gamma}$, $g' = g^{1/\gamma}$, and $b' = b^{1/\gamma}$ (if we neglect the behaviour near the origin). Then we draw from Rel.(21)

$$r'(z) = \left[r(z)\right]^{1/\gamma} = \left[r_0 + \overline{\omega}(z)\,r_1\right]^{1/\gamma}, \qquad (24)$$

with similar derivations for the video green and blue bands.

Is the linearly regionalized model able to explain the alignments in video histograms, despite the gamma correction? For the sake of simplicity, we will tackle this question by fixing the order of the video bands as $r' \geq g' \geq b'$, and $m' \geq g'$. Then we have

$$3m'(z) = r'(z) + g'(z) + b'(z)$$
$$2s'(z) = 2r'(z) - g'(z) - b'(z)$$

Alignments with the dark point In the *luminance/saturation* histograms in L_1 norm, several alignments are in the prolongation of the point $(0, 0)$, of zero luminance and saturation. The shadow regions of the "rubber ring" image illustrate this situation.

Suppose that, in the relation (20) which defines the LR spectrum, the term $sp_0(\nu; z)$ is identically zero. Then $r(z)$ reduces to $\overline{\omega}(z)r_1$, which gives

$$r'(z) = r^{1/\gamma} = \overline{\omega}^{1/\gamma} r_1^{1/\gamma} = \overline{\omega}^{1/\gamma}(z)r_1',$$

with similar derivations for two other bands. Therefore we have

$$3m'(z) = \overline{\omega}^{1/\gamma}(z)\left[r_1^{1/\gamma} + g_1^{1/\gamma} + b_1^{1/\gamma}\right] = 3\overline{\omega}^{1/\gamma}(z)m_1'$$

and

$$2s'(z) = 2r'(z) - g'(z) - b'(z) = \overline{\omega}^{1/\gamma}(z)\left[2r_1' - g_1' - b_1'\right]$$

hence $m'(z)s_1' = m_1's'(z)$. In the space E of the intensities, we find in the same way that $m(z)s_1 = m_1s(z)$. Therefore the nullity of the constant spectrum $sp_0(\nu)$ entails that both m' and s' on the one hand, and m and s on the other one, are proportional. Each *video* alignment indicates a zone where the *intensities* spectrum varies proportionally from one point to another.

Alignments with the white point The "rubber ring" image generates also an alignment along a line going through the point $(1, 0)$, i.e. the point with maximum luminance and zero saturation. That suggests to suppose the spectrum $sp_0(\nu; z)$ constant and equal to 1, and in addition that the three colors r_1, g_1, b_1 are not identical (if not, the saturation s' should be zero). We have

$$r(z) = 1 + \overline{\omega}(z)r_1 \tag{25}$$

and the two sister relations for $g(z)$ and $b(z)$. Under gamma correction, $r(z)$ becomes

$$r'(z) = (1 + \overline{\omega}(z)r_1)^{1/\gamma}.$$

Now, to say that the alignment is closed to a point of maximum luminance comes down to saying that $r_1, g_1,$ and b_1 are small with respect to 1, or again that

$$r'(z) = 1 + \frac{\overline{\omega}(z)}{\gamma} r_1 + \varepsilon(r_1), \tag{26}$$

hence $m'(z) = 1 + \frac{\overline{\omega}(z)}{\gamma} m_1$ and $s'(z) = \frac{\overline{\omega}(z)}{\gamma}[2r_1 - g_1 - b_1]$. We observe that the two conditions $r_1 \geq 0$ and $r'(z) \leq 1$, jointly with Rel.(26) imply that the coefficient $\overline{\omega}(z)$ is negative. Moreover, as the three colours r_1, g_1, b_1 are distinct, the condition $s'(z) \geq 0$ implies in turn that the quantity $2r_1 - g_1 - b_1$ is strictly negative. By putting $\sigma_1 = -(2r_1 - g_1 - b_1) > 0$ (σ_1 is not the saturation at point z_1), we obtain the following linear relation with positive coefficients

$$m'(z) = 1 - \frac{m_1}{\sigma_1} s'(z). \tag{27}$$

As in the previous case, but without approximations, the mean $m(z)$ and the saturation $s(z)$ of the intensities are linked by the same equation (27): it is a direct consequence of Eq.(25). Again, both video and intensity histograms carry the same information, and indicate the zones of almost white reflections.

Alignments with a gray point   There appears in some images, as "Ana Blanco", series of parallel alignments. Their supports go through points of (quasi) zero saturation but their luminance is strictly comprised between 0 and 1. The interpretation we just gave for the case of reflections extends to such a situation. It is still assumed that $r_0 = g_0 = b_0$, but with $0 < r_0 \leq 1$, and that the terms $\overline{\omega}(z)r_1, \overline{\omega}(z)g_1$, and $\overline{\omega}(z)b_1$ are small with respect to r_0. Then we have,

$$r'(z) = (r_0 + \overline{\omega}(z)r_1)^{1/\gamma} = r_0^{1/\gamma} + r_0^{1/\gamma - 1}\frac{\overline{\omega}(z)}{\gamma} r_1,$$

and the two sister relations for g' and b'. Hence

$$m'(z) = r_0^{1/\gamma} + r_0^{1/\gamma - 1}\frac{\overline{\omega}(z)}{\gamma} m_1,$$

$$s'(z) = -r_0^{1/\gamma - 1}\frac{\overline{\omega}(z)}{\gamma}\sigma_1,$$

so that, finally

$$m'(z) = r_0^{1/\gamma} - \frac{m_1}{\sigma_1} s'(z). \tag{28}$$

When the colour component (r_1, g_1, b_1) remains unchanged, but that the gray component (r_0, g_0, b_0) takes successively various values, then each of them induces an alignment of the same slope $\frac{m_1}{s_1}$. Rel.(28) extends, without approximation, to the histograms of the intensities themselves.

Finally, we derive from Eq.(17) that, in the three cases, the hue remains *constant* in each alignment zone.

Figure 7.   Initial image of parrots.

4.   Saturation weighted segmentations

The most radical change between the classical HLS system and those based on norms holds on the saturation equation. In system (19), when $\min(r, g, b) = 0$, (with $l \leq 0.5$), or when $\max(r, g, b) = 1$, (with $l \geq 0.5$), then the saturation equals 1. Now for human vision, the most significant parameter is the hue in high saturated areas, and it turns to luminance when saturation decreases. Any person whose reaction to colours is normal can easily check it. In the darkness, or, at the opposite, in white scenes (e.g. a landscape of snowy mountains), the eye grasps the contours by scrutinizing all small grey variations, whereas when the scene juxtaposes spots of saturated colours, then the eye localizes the frontiers at the changes of the hue. But how to transcribe quantitatively such a remark by a saturation function that takes its maxima precisely when the colours loose their saturation, as the classical HLS system does?

The norms based representations correct this drawback, so that their saturations may serve to split the space into hue-dominant versus grey-dominant regions. This very convenient key to entering the segmentation of colour images was initially proposed by C.Demarty and S.Beucher [6]. They introduce the function $max - min$ on the image under study, and threshold it at a level s_0 that depends on the context. Then they adopt the HSV representation, but they replace its saturation by 1 in the regions above s_0 and by 0 elsewhere. Their downstream segmentations become easier and more robust.

However, they did not take the plunge of a new representation, and they worked at the pixel level, which is not the most informative. In order to go further in the same way of thinking, J. Angulo and J. Serra propose, in [1], the following two steps segmentation procedure:

1 to *separately* segment the luminance, the saturation and the hue in a correct Newtonian representation;

2 to combine the obtained partitions of the luminance and of the hue by means of that of the saturation: the later is taken as a *criterion* for choosing at each place either the luminance class, or the hue one.

Figure 8. Representation of the "Parrots" image 7 in L_1 norm: a) luminance, b) saturation, c) hue.

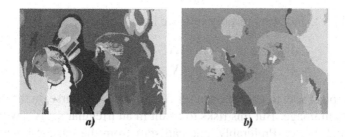

Figure 9. Grey segmentations of the luminance (a) and the hue (b). Both are depicted in false colour.

The three bands of the "parrots" image of Fig.7, in L_1 representation, are depicted in Fig.8(a-c). Each band is segmented by the *jump connection* algorithm [15] (one groups in same classes all points x where $f(x)$ differs by less than k of an extremum in the same connected component, these classes are then withdrawn from the image, and one iterates). The method depends only on the jump positive value k.

As the parameter k increases, the over-segmentations reduce, but in compensation heterogeneous regions appear. A satisfactory balance seems to be reached for $k = 20$ (for 8-bits images), up to the filtering of a few very small regions. We obtain the two segmentations depicted in Fig.9.

Synthetic partition

How to combine the two partitions of images 9a and 9b? The idea consists in splitting the saturation image into two sets X_s and X_s^c of high and low saturations respectively, and in assigning the hue partition to the first set, and the luminance one to the second. A class of the synthetic partition is either the intersection of a luminance class with the low saturation zone X_s^c, or the intersection of a hue class with the high saturation zone X_s. If the classes of the luminance, the hue, and the synthetic partition at point x are denoted by

Figure 10.   a) Segmentation of the saturation (presented in grey tones); b)optimal threshold
of a); c) final synthetic partition, superimposed to the initial image.

$A_m(x)$, $A_h(x)$, and $A(x)$ respectively, we have

$$A(x) = A_m(x) \cap X_s^c \quad \text{when} \quad x \in X_s^c$$
$$A(x) = A_h(x) \cap X_s \quad \text{when} \quad x \in X_s.$$

The simplest way to generate the set X_s consists, of course, in thresholding
the saturation image. But this risks to result in an irregular set X_s, with holes,
small particles, etc. Preferably, one can start from the mosaic image of the
saturation provided by the same segmentation algorithm as for the the hue and
the luminance (Fig.10a). An optimal threshold on the saturation histogram
determines the value for the a-chromatic/chromatic separation (Fig.10b). We
finally obtain the composite partition depicted in Fig.10c, which is excellent.

5.   Colour and mixed segmentations

Colour and shape

The discrimination we have just made between the zones of the space where
the hue is more significant than the luminance, and their complements, is a first
step towards the more precise discrimination between colour and shape that we
propose now. The video-image depicted in Fig. 11 illustrates the purpose. It
has been extracted from a test sequence for the image compression algorithm
proposed by C. Gomila in her Ph.D. Thesis [7] and described by steps a to f
below. The goal here consists in contouring the individual in the foreground,
in order to code him more finely than the background. This outline groups the
face and the bust. We observe that the former is better spotted by the colour of
the skin, which ranges in a specific domain, and the latter by the shape of the
shoulders, and by their location at the down part of the frame.

Beyond this example, the problem of combining two modes of description
into a merged segmentation is set. One may view to segment twice the image,
according to criteria associated with each mode separately, but then how to
manage the merging, and its overlapping? There is no referee, here, to play the
role given to the saturation in the "parrot" case. On the other hand, we cannot

a) *b)*

Figure 11. a) Initial image; b) final contouring.

afford just to take the supremum of the two partitions, as it is well known that in the classes of this supremum, both criteria may be satisfied : they do not generate an exclusive "or" (see for example [15], section 2). Nevertheless, if we provide a priority between the two modes of description, then an optimal bipartition of the space can be obtained, as shown below.

Quasi-connection

DEFINITION 1 *Let E be an arbitrary space. Any class $\mathcal{C}_1 \subseteq \mathcal{P}(E)$ such that*
(i) $\emptyset \in \mathcal{C}_1$
(ii) for each family $\{C_i \ i \in I\}$ in $\mathcal{C}_1$, $\cap C_i \neq \emptyset$ implies $\cup C_i \in \mathcal{C}_1$,
defines a quasi-connection *on E .*

Unlike a connection, a quasi-connection does not necessarily contain the set $\mathcal{S}$ of all singletons of $\mathcal{P}(E)$. What are the consequences of this missing axiom ? We still can associate a (quasi) connected opening $\gamma_{1,x}$ with each point $x \in E$ by putting for any $A \subseteq E$

$$\begin{aligned} &\gamma_{1,x}(A) = \emptyset \text{ when the family } \{C_1 \in \mathcal{C}_1, x \in C_1 \subseteq A\} \text{ is empty} \\ &\gamma_{1,x}(A) = \cup \{C_1 \in \mathcal{C}_1, x \in C_1 \subseteq A\} \text{ when not} \end{aligned} \qquad (29)$$

Operator $\gamma_{1,x}$ is obviously an opening since it coincides with that of the connection $\mathcal{C}_1 \cup \mathcal{S}$ when $\gamma_{1,x}(A) \neq \emptyset$. Moreover, for all $A \subseteq E$ and all $x, y \in E$, $\gamma_{1,x}(A)$ and $\gamma_{1,y}(A)$ are still equal or disjoint, i.e.

$$\gamma_{1,x}(A) \cap \gamma_{1,y}(A) \neq \emptyset \Rightarrow \gamma_{1,x}(A) = \gamma_{1,y}(A) \qquad (30)$$

and for all $A \subseteq E$ and all $x \in E$, we have $x \notin A \Rightarrow \gamma_{1,x}(A) = \emptyset$. The only change with the connection case is that now $\gamma_{1,x}(A)$ may equal $\emptyset$, even when $x \in A$. As a consequence, the supremum $\gamma_1 = \vee \{\gamma_{1,x}, x \in E\}$ generates an opening on $\mathcal{P}(E)$ whose residual $\rho_1(A) = A \setminus \gamma_1(A)$, for a set $A \subseteq E$ is not necessarily empty. In other words, to say that $x \in \rho_1(A)$ is equivalent to saying that the family $\{C_1 \in \mathcal{C}_1, x \in C_1 \subseteq A\}$ is empty. Remark that when $\mathcal{C}_1$ is a connection, then γ_1 turns out to be the identity operator.

Two levels mixed segmentations

Let $\mathcal{C}_1$ be a quasi-connection on $\mathcal{P}(E)$ of point openings $\{\gamma_{1,x}, x \in E\}$, and $\mathcal{C}_2$ be a connection on $\mathcal{P}(E)$ of point connected openings $\{\gamma_{2,x}, x \in E\}$. We introduce a hierarchy between them by restricting the classes according to $\mathcal{C}_2$ to the zones that are not reached by $\mathcal{C}_1$. This can be done via the operator

$$
\begin{aligned}
\chi_{2,x}(A) &= \gamma_{2,x}[\rho_1(A)] && \text{when } x \in \rho_1(A) && (31) \\
\chi_{2,x}(A) &= \emptyset && \text{when not}
\end{aligned}
$$

This operator $\chi_{2,x}$ is not increasing, as it acts on set $A \subseteq E$ via the residual of $\gamma_1(A)$. Nevertheless, it satisfies a few nice other properties.

PROPOSITION 2 *The operator $\chi_{2,x}$ defined by system (31) is anti-extensive, idempotent and disjunctive, i.e.*

$$
\chi_{2,x}(A) \cap \chi_{2,y}(A) \neq \emptyset \Rightarrow \chi_{2,x}(A) = \chi_{2,y}(A) \qquad \forall A \subseteq E \, ; \forall x, y \in E
$$

$$(32)$$

Proof. The anti-extensivity of $\chi_{2,x}$ is obvious. To prove its idempotence, suppose first that $\chi_{2,x}(A) \neq \emptyset$. Then, for all $z \in \chi_{2,x}(A)$, we have $\gamma_{1,z}[\chi_{2,x}(A)] \subseteq \gamma_{1,z}[\rho_1(A)] = \emptyset$, hence $\rho_1[\chi_{2,x}(A)] = \chi_{2,x}(A) \ni x$. As set $\chi_{2,x}(A)$ is an invariant of opening $\gamma_{2,x}$, we have

$$
\chi_{2,x}[\chi_{2,x}(A)] = \gamma_{2,x}[\rho_1(\chi_{2,x}(A))] = \gamma_{2,x}[\chi_{2,x}(A)] = \chi_{2,x}(A).
$$

Suppose now that $\chi_{2,x}(A) = \emptyset$. As $\gamma_{1,x}$ is an opening, we have $\gamma_{1,x}[\chi_{2,x}(A)] = \emptyset$, so $\rho_1[\chi_{2,x}(A)] = \emptyset$ hence $\chi_{2,x}[\chi_{2,x}(A)] = \emptyset$.

The disjunction implication remains to be proved (30). If the intersection $\chi_{2,x}(A) \cap \chi_{2,y}(A)$ is not empty, then it is equal to $\gamma_{2,x}[\rho_1(A)] \cap \gamma_{2,y}[\rho_1(A)]$. As $\gamma_{2,x}$ and $\gamma_{2,y}$ are two point openings of connection $\mathcal{C}_2$, we have $\chi_{2,x}(A) = \chi_{2,y}(A)$, which achieves the proof. ∎

Consider now the supremum $\chi_x = \gamma_{1,x} \vee \chi_{2,x}$ of the two operators $\gamma_{1,x}$ and $\chi_{2,x}$. We will prove that as x ranges over E, the family $\{\chi_x\}$ partitions all $A \subseteq E$ in an optimal way. More precisely, we can state

PROPOSITION 3 *Let $\mathcal{C}_1$ be a quasi-connection and $\mathcal{C}_2$ be a connection, both on $\mathcal{P}(E)$, where set E is arbitrary. Then,*

1 the union of the two families $\{\gamma_{1,x}\}$ and $\{\chi_{2,x}\}$, $x \in E$, of operators partition every set $A \subseteq E$ into two families of classes $\{A_{1,i}\}$ and $\{A_{2,j}\}$;

2 this partition is the greatest one with classes of $\mathcal{C}_1$ on $\gamma_1(A)$ and classes of $\mathcal{C}_2$ on $A \backslash \gamma_1(A)$.

Proof. 1/Let $A \subseteq E$. If $\gamma_1(A) \neq \emptyset$, then each point $x \in \gamma_1(A)$ belongs to the non empty A_1 class $\gamma_{1,x}(A)$ and if $A \neq \gamma_1(A)$, each point $y \in A \backslash \gamma_1(A)$

belongs to the non empty A_2 class $\gamma_{2,y} [A \backslash \gamma_1 (A)]$. If $\gamma_1 (A) = \emptyset$, then each point $y \in A$ belongs to class $\gamma_{2,y} (A)$. Therefore, the various classes cover set A. Moreover, Rel.(30) shows that the $\{A_{1,i}\}$ classes are disjunctive in the set $\gamma_1 (A)$ and Prop.2 that the $\{A_{2,j}\}$ classes are disjunctive in $A \backslash \gamma_1 (A)$.

2/ Let $\{A'_{1,r}\}$ and $\{A'_{2,s}\}$ be another partition of A into C_1 and C_2 components. Each point $x \in A$ belongs to one class of each partition. Suppose first that both classes are of type n°1, i.e. $x \in A_{1,i} \cap A'_{1,s}$, with $A_{1,i}, A'_{1,s} \in C_1$, for some i and some s. Therefore $A_{1,i} \cup A'_{1,s} \in C_1$ and $x \in A_{1,i} \cup A'_{1,s} \subseteq \gamma_{1,x} (A) = A_{1,i}$ hence $A'_{1,s} \subseteq A_{1,i}$. If we suppose now that both classes $A_{2,j}$ and $A'_{2,s}$ going through x are of type n°2, then the same proof, but for C_2, shows that $A'_{2,s} \subseteq A_{2,j}$. Finally, the combination at point x of a C_1-class of the first partition with a C_2-class of the second one is impossible. Indeed, we draw from the previous paragraph of this proof that $\cup \{A'_{1,r}\} \subseteq \cup \{A_{1,i}\} = \gamma_1 (A)$ and that $\cup \{A'_{2,s}\} \subseteq \cup \{A_{2,j}\} = A \backslash \gamma_1 (A)$; as $\cup \{A'_{1,r}\} \cup \{A'_{2,s}\} = A$ we have $\gamma_1 (A) = \cup \{A'_{1,r}\}$ and $A \backslash \gamma_1 (A) = \cup \{A'_{2,s}\}$, which achieves the proof. ∎

Proposition 3 allows us to partition A into a hierarchy of successive mixed segmentations. Clearly, it extends to three phases by replacing C_2 by a pseudo-connection and by adding a third connection C_3, ... and so on for n phases. The lack of increasingness entails that if $A \subseteq B$, then $\gamma_{1,x} (A) \subseteq \gamma_{1,x} (B)$ for all $x \in E$, but not $\chi_{2,x} (A) \subseteq \chi_{2,x} (B)$. Remark that when we look for segmenting the whole space E, the possible comparison of E with another set B becomes useless. An example of such an iteration based hierarchy is given by the jump connection [13] [15]. Given a continuous bounded function f the quasi one-jump criterion σ is defined by the following requirement: $\sigma [f, A] = 1$ iff for any point $x \in A$ there exists a minimum m of f in A such that $0 \leq f(x) - m < k$. The class C_1 of the A such that $\sigma [f, A] = 1$ generates a quasi-connection. Take $C_2 = C_1$ and iterate the process. The first step extracts the connected components of the space in which the function is less than k above the minima; the second step does the same on the function reduced to the residual space, and so on, which results in the so called "*jump connection from minima*".

Before illustrating Prop. 3 by an example, we would like to comment on the case when $C_2 = S \cup \emptyset$, i.e. when the second connection is the family of the singletons plus the empty set. Then, by taking $C = C_1 \cup C_2 = C_1 \cup S$, we obtain the smallest connection that contains C_1. However, the two phased approach (C_1, C_2) is more informative than the only use of C. For example, given the numerical function $f : \mathbb{R}^1 \to \mathbb{R}^1$, the segment along which $f (x) \geq 1$ forms a connection C. In particular, if we take

$$f = |x| \quad \text{for } x \neq 0 \quad \text{and} \quad f = 2 \quad \text{for } x = 0$$

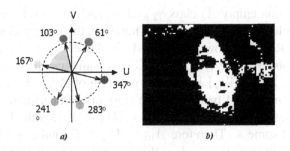

Figure 12.   a) Sector of the human skins in the (U, V) plane; b) Threshold of Fig. 11a according to the skins sector.

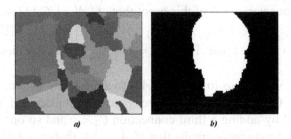

Figure 13.   a) Previous segmentation of Fig. 11a; b) Head reconstruction (after symmetry).

then all points of $]-1, +1[$ induce singleton classes, and there is no mean in connection for $\mathcal{C}$ distinguishing between $x = 0$, where f is ≥ 1, and the other points of $]-1, +1[$, where f is < 1. On the contrary, if we adopt the above twofold approach, then $\{0\} \in \mathcal{C}_1$, and the other singletons belong to $\mathcal{C}_2$.

Return to the colour and shape example

The colour/shape segmentation algorithm proposed by Ch. Gomila illustrates very well our twofold approach [7]. It proceeds as follows :

a/ the image under study (Fig. 11a) is given in the standard colour video representation YUV

$$y = 0.299r + 0.587g + 0.114b$$
$$u = 0.492(b - y) \tag{33}$$
$$v = 0.877(r - y)$$

b/ a previous segmentation resulted in the tesselation depicted in false colour in Fig. 13a or Fig. 14a. For the further steps, this mosaic becomes the working space E, whose "points" are the polygons of the mosaic;

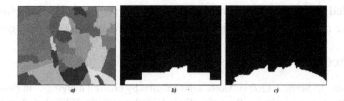

Figure 14.   a) Previous segmentation of Fig.11a; b) "shoulder/head" marker; c) bust reconstruction.

c/ classical studies have demonstrated that, for all types of human skins, two chrominances U and V practically lie in the sector region depicted in Fig. 12a. By thresholding the initial image Fig. 11a by this sector, we obtain the set Fig 12b, whose a small filtering by size suppresses the small regions, yielding a marker set;

d/ all "points" of E that contain at least a pixel of the marker set, or of its symmetrical w.r.t. a vertical axis, are kept, and the others are removed : this produces the opening $\gamma_1 (E)$, depicted in Fig. 13b;

e/ for the bust, an outside shape marker made of three superimposed rectangles is introduced. All their pixels that belong to a "point" of $\gamma_1 (E)$ are removed from the bust marker, since this second marker must hold on $E \backslash \gamma_1 (E)$ only. That is depicted in Fig. 14b, where one can notice how much the upper rectangle has been reduced; the associated opening $\gamma_2 [E \backslash \gamma_1 (E)]$ is depicted in Fig. 14c;

f/ the union $\gamma_1 (E) \cup \gamma_2 [E \backslash \gamma_1 (E)]$ defines the zone inside which the initial image Fig.11a is kept, as depicted in Fig.11b.

The example may seem simple; it holds on a rather poor discrete space and acts via two elementary quasi-connections and connections. However, it proved to be robust and well adapted to its goal, and its robustness is a direct consequence of the optimality of the involved segmentations.

6.   Conclusion

The three studies which compose this paper follow a certain order. The first two ones require imperatively a correct quantitative definition of the polar representation and of the saturation. When colour and shape are treated jointly, the physical meaning of the colour descriptors becomes less crucial, since the latter match with shape parameters whose physical meaning is totally different.

The above studies illustrate also a certain approach to segmentation. We have attempted to show that maximum partitions can be "added", conditioned by one another, can form hierarchies, etc... in order to express segmentations in the sense of [15], whose the main theorems underlay all the above examples.

Acknowledgements *The author gratefully thanks Prof Ch. Ronse and Dr J. Angulo for their valuable comments and the improvements they suggested.*

References

[1] Angulo, J., Serra, J. (2003) "Color segmentation by ordered mergings," in *Proc. of IEEE Int. Conference on Image Proc. (ICIP'03)*, Barcelona, Spain, IEEE, Vol. 2, p. 125–128.

[2] Angulo, J. (2003) *Morphologie mathématique et indexation d'images couleur. Application à la microscopie en biomédecine*, Thèse de Doctorat en Morphologie Mathématique, Ecole des Mines de Paris, Dec. 2003, 341 p.

[3] Angulo, J., Serra, J. (2004) "Traitements des images de couleur en représentation luminance/saturation/teinte par norme L_1," *Technical Report N-/04/MM*, Ecole des Mines de Paris, April 2004, 26 p.

[4] Commission Internationale de l'Eclairage (CIE) (1986) *Colorimetry*, Second Edition. *CIE Publication No. 15.2*, Vienna.

[5] Choquet, G. (1966) *Topologie*, Academic Press, New York.

[6] Demarty, C.-H., Beucher, S. (1998) "Color segmentation algorithm using an HLS transformation," in *Proc. of ISMM '98)*, Kluwer, p. 231–238.

[7] Gomila, C. (2001) *Mise en correspondance de partitions en vue du suivi d'objets*, Thèse de Doctorat en Morphologie Mathématique, Ecole des Mines de Paris, Sept. 2001, 242 p.

[8] Hanbury, A., Serra, J. (2003) "Colour Image Analysis in 3D-polar coordinates," in *Proc. of DAGM symposium*, Vienna, April 2003.

[9] Lambert, P., Carron, T. (1999) "Symbolic fusion of luminance-hue-chroma features for region segmentation," *Pattern Recognition*, Vol. 32, p. 1857–1872.

[10] Obein, G., Knoblauch, K., Viénot, F. (2002) "Perceptual scaling of the gloss of a one-dimensional series of painted black samples," *Perception*, Vol. 31, Suppl., p. 65.

[11] Poynton, Ch. (1996) *A technical Introduction to Digital Video*, Wiley, New York. Chapter 6, "Gamma" is available online at *http://www.poyton.com/PDFs/TIDV/Gamma.pdf*

[12] Salembier, P., Serra, J. (1995) "Flat Zones Filtering, Connected Operators, and Filters by Reconstruction," *IEEE Transactions on Image Processing*, Vol. 4, No. 8, p. 1153–1160.

[13] Serra, J. (2000) "Connectivity for sets and functions," *Fund. Inform.*, Vol. 41, 147–186.

[14] Serra, J. (2002) "Espaces couleur et traitement d'images," *Technical Report N-34/02/MM*, Ecole des Mines de Paris, Oct. 2002, 13 p.

[15] Serra, J. (2004) "A lattice approach to Image segmentation," To be published by *Journal of Mathematical Imaging and Vision*, 75 p.

[16] Shafer, S.A. (1985) "Using color to separate reflection components from a color image," *Color Research and Applications*, Vol. 10, No. 4, p. 210–218.

[17] Smith, A.R. (1978) "Color gammut transform pairs," *Computer Graphics*, Vol. 12, No. 3, p. 12–19.

[18] Tremeau, A., Fernandez-Maloigne, Ch., Bonton, P. (2004) *Image numérique couleur*, Ed. Dunod, Paris.

FAST IMPLEMENTATION OF WATERFALL BASED ON GRAPHS

Beatriz Marcotegui and Serge Beucher

Centre de Morphologie Mathématique.
Ecole des Mines de Paris
{marcotegui,beucher}@cmm.ensmp.fr

Abstract   The waterfall algorithm is a contrast-based hierarchical segmentation approach. In this paper we propose an efficient implementation based on the minimum spanning tree of the neighborhood graph. Furthermore, other hierarchies are proposed and compared to the original version of the algorithm.

Keywords: Hierarchical image segmentation, watershed, waterfall, minimum spanning tree, graphs.

Introduction

Segmentation, together with filtering is often the first step of image analysis or image interpretation. The success of the whole chain of treatment relies on the accuracy of the segmentation results. Important efforts have been devoted to segmentation during the last years and it still remains a key topic of research.

The watershed transformation [3, 1] is the paradigm of segmentation of Mathematical Morphology. It has proved to be a powerful tool used in the solution of multiple applications. Its main drawback is the over-segmentation produced. Two approaches are proposed in the literature to overcome this drawback:

- the selection of markers [9], which supposes that the characteristics of the interesting objects are known;

- hierarchical approaches, that are able to rank the importance of each region.

We focus on hierarchical approaches because of their ability to segment generic images.

177

C. Ronse et al. (eds.), Mathematical Morphology: 40 Years On, 177–186.

Several hierarchical approaches can be found in the literature. Grimaud [4] introduced the dynamics of minima that assign to each minimum a measure of its contrast. By thresholding this measure with increasing values, a hierarchy is obtained. Najman and Schmitt [10] showed that the same measure of dynamics may be assigned to a contour and introduced the geodesic saliency of watershed contours. Vachier and Meyer [11] generalized the concept of dynamics with the extinction values and proposed to assign to a minimum other measure than contrast such as area or volume. Volume extinction values result in a well adapted criterion for evaluating the visual significance of regions.

Meyer proposed a graph-based implementation of these hierarchies [7],[8]. Nodes correspond to the catchment basins of the topographic surface. If two catchment basins are neighbors, their corresponding nodes are linked by an edge. The valuation of this edge is the minimum pass point of the gradient along their common frontier. In the following we will refer to this graph as the neighborhood graph. Meyer found that all the information of a hierarchy may be stored in a very condensed structure: the minimum spanning tree (MST). This is due to the fact that the flooding always follows the path of minimum height, the same that chooses the MST of the neighborhood graph. This consideration leads to a very efficient algorithm of hierarchical segmentation [7] and has also been used for interactive segmentation [12].

In [1, 2] Beucher proposed a very interesting hierarchical segmentation approach: the waterfall. Starting from the watershed result, it consists in an iterative algorithm that at each step removes all the watershed contours completely surrounded by higher ones. Typically, less than 10 hierarchical levels are produced by iterating the waterfall algorithm. In [2] each step is implemented by a reconstruction process followed by a new flooding of the resulting image. Another implementation based on graphs is also proposed in [1, 2].

The hierarchies based on extinction values produce a different level for each merging of two regions. This is useful for interactive segmentation approaches because it offers flexibility. The waterfall generates several steps of the hierarchy with an autocalibrated number of regions. This autocalibration may be interesting for segmenting generic images without imposing a given number of regions, which can be a tricky parameter to fix.

In [5] an automatic track detection application is developed using the waterfall algorithm in the initialization step. Several waterfall iterations may be necessary until a region compatible with the track geometry is found. The existing implementation of the waterfall does not allow this application in real time.

In this paper we propose an implementation of the waterfall algorithm based on the MST. It allows to access to different levels of the hierarchy in a very efficient way. Furthermore it allows the possibility to obtain different hierarchies based on other criteria than the frontier height, used in the original version of the waterfall algorithm.

Section 1 describes the waterfall algorithm introduced by Beucher. The proposed efficient implementation based on graphs is presented in section 2. Other hierarchies are easily introduced in the new framework, as shown in section 3. Finally section 4 concludes.

1. Waterfall

Let's consider a partition P. It can be the outcome of the watershed of the gradient image (as proposed in [2]) or any other partition. The frontiers are valuated with a distance between regions (e.g. the minimum pass point of the gradient along the frontier, see figure 1(a)).

The waterfall algorithm removes from a partition all the frontiers completely surrounded by higher frontiers (see figure 1(b)). Thus, the importance of a frontier is measured with respect to its neighborhood. This process can be seen as a watershed applied not to the pixels of an image but to the frontiers of a partition. The iteration of the waterfall algorithm finishes with a partition of only one region.

Figure 2 illustrates the result of the waterfall process applied to a real image. Figure (a) shows the original image, (b) its gradient and (c) the watershed of the gradient. Figure (d), (e) and (f) are the different hierarchy levels produced by the iteration of the waterfall algorithm.

A first implementation of this algorithm based on graphs is proposed in [2]. The proposed graph contains a node for each arc frontier of the input partition and an edge between every pair of arcs (frontiers) delimiting the same catchment basin. (Note that this graph is different from the one we propose in the next section). The algorithm is considered as complex by the author and an image-based algorithm is proposed instead. The image-based algorithm consists in a reconstruction of the image from the watershed lines followed by the watershed transformation. Thus, a reconstruction and a new watershed computation are required to obtain a new level of the hierarchy.

In this paper, we propose an implementation of the waterfall algorithm that only requires a flooding of the image to obtain all levels of the hierarchy. The rest of the process is performed on the MST, that is much more efficient.

2. Waterfall based on the Minimum Spanning Tree

Given that the flooding always follows the path of minimum height, the MST obtained from the neighborhood graph, contains all the information required for the flooding process [6]. The MST is a very condensed way to store the information. Thus, it leads to very efficient implementation of hierarchical segmentation approaches.

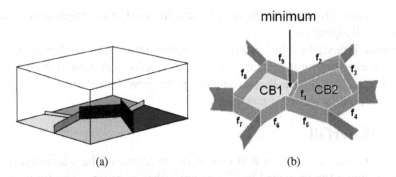

Figure 1. (a) Partition with valuated frontiers and (b) example of frontier : as the value of frontier f! is smaller than the values of its neighboring frontiers (f2 to f9) it will be removed by the waterfall algorithm.

We propose an implementation of the waterfall algorithm based on the MST. The algorithm is performed in two steps:

- Minimum spanning tree (MST) generation

- Waterfall from the MST

MST generation

We consider as input partition of the waterfall algorithm the result of the watershed. Thus, the gradient image is flooded to obtain the initial partition. The MST is obtained simultaneously to the flooding process [7, 12]. The graph is initialized with a node corresponding to each minimum of the image and without any edge. A lake is associated to each minimum. During the flooding each time that two regions of different lakes meet, an edge is added to the graph, linking both regions and the corresponding lakes are merged. Its valuation is the height of water when regions meet. At the end of the flooding process the graph has become the MST because:

- an edge is added only if regions that meet belong to different lakes (so the graph does not contain cycles),

- at the end of the flooding all the image belong to the same lake (it is an spanning tree)

- the flooding follows the path of minimum height (it is a minimum tree).

Note that the edges of the MST are valuated (not the nodes). We can define a regional minimum of the MST as a connected component of the graph, such as all the values of its edges are equal and it is surrounded by edges with strictly higher values. This definition will be used in the following subsection.

Figure 2. Waterfall iteration; First row: (a) original image and its (b) gradient. Second row: (c) Watershed segmentation and (d) first waterfall result Third row: (e),(f) Two more iterations of the waterfall algorithm

Figure 3 shows an example of partition (a) and its corresponding Minimum Spanning Tree (b).

In the following step we address the waterfall algorithm based on the MST.

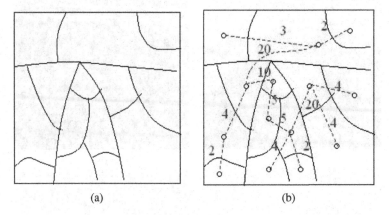

(a)   (b)

Figure 3.   Example of partition (a) and its associated Minimum Spanning Tree (b).

Waterfall from the MST

The waterfall algorithm removes from a partition the frontiers that are completely surrounded by higher frontiers, leading to a coarser partition. The waterfall can be implemented on the MST. The first step consists in identifying regional minima frontiers.

Let's consider the partition of figure 3(a) and its corresponding MST 3(b). If we take for example edge E linking regions V_1 and V_2 of figure 4(a) we should compare its valuation with the valuation of frontiers surrounding catchment basins $V_1 \bigcup V_2$. These frontiers are drawn in bold line in figure 4(b) (E1, E2, E3, F1, F2, F3 and F4). Edges named with an F do not belong to the MST, so by construction their valuations are higher than valuations of at least one edge named with Es. Therefore, in order to know if E is a regional minimum, it is enough to compare it with E1, E2 and E3. Thus, the MST has all the information required to identify regional minimum edges. In practice, we will compare the valuation of an edge E between V_1 and V_2 with edges of the MST having as one extremity V_1 or V_2. More generally, we are looking for all edges that belong to a regional minimum of the MST (defined in the previous subsection).

If E is a regional minimum edge, it corresponds to a frontier that should be removed by the waterfall. This is implemented by assigning the same label to both extremities (nodes) of the minimum edge. These labels identify regions in the output partition. Thus, a different label is assigned to each minimum edge. See figure 4(d). Regions in white (those that are not neighbors of a minimum edge) are not granted any label at this stage. Regions with a label are considered as markers. In order to obtain the final partition, markers are propagated following edges of the MST in increasing order of their valuation (fig 5). This process is a segmentation from markers on the MST [6].

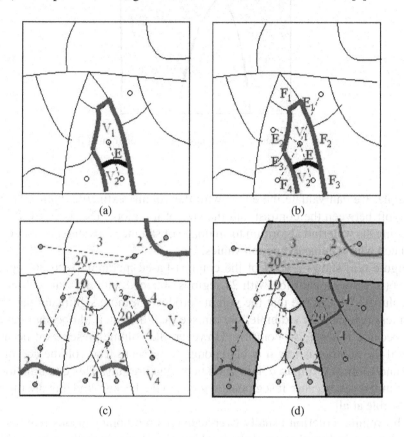

(a)   (b)

(c)   (d)

Figure 4.   Waterfall on the MST. First row : (a) Edge $E(V_1, V_2)$ and (b) comparison of E with its neighboring edges. Second row: (c) regional minimum edges. (d) Vertex labelled from minimal edges

If edges of the MST are valuated with the lowest pass point along frontiers, this algorithm is equivalent to the algorithm presented in [2].

3. Hierarchies with other criteria

The original waterfall algorithm removes edges according to their height. Thus it produces contrast-based hierarchies. The graph implementation of the algorithm, presented in the previous section, allows to easily produce other hierarchies, changing the edge valuation of the Minimum Spanning Tree. For

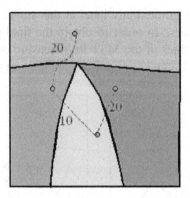

Figure 5.   Waterfall result (label propagation from 4(d)).

example, we can valuate the edges with the volume extinction values [11], a trade-off between the contrast and the size of a region. This is equivalent to applying the waterfall algorithm to an image of saliency of watershed contours, valuated with volume extinction values.

Figure 6(a) shows a level of the contrast-based hierarchy with 36 regions and 6(b) shows a partition with 24 regions obtained with the waterfall based on volume. We can see that the volume produces a more significant partition with less regions. For example, we can see that the waterfall does not get the hat, because it has a low contrast. However, the volume preserves it because even if the contrast is low it is big enough to be seen. The volume combines size and contrast trying to obtain good perceptual results. This combination is not optimal yet, because it segments regions in the background that are big but not visible at all.

The volume criterion usually over-segments big homogeneous regions. A last step may reduce this problem, just removing frontiers with contrast under a given (and small) threshold. Doing that at the end of the process is much more reliable because frontiers are longer and their contrast may be better estimated, reducing the effect of noise in a small frontier. Figure shows 7 the result of removing frontiers with contrast under 5 gray levels. The partition preserves the important regions (18) removing the over-segmentation of the background.

4.   Conclusion

In this paper we propose an efficient implementation of the waterfall algorithm. It consists in obtaining a Minimum Spanning Tree simultaneously to the flooding of the image. Then, all the process is performed on the graph. All the information required to manipulate the hierarchy is stored in a very condensed structure, leading to very efficient algorithms. Real-time applications such as

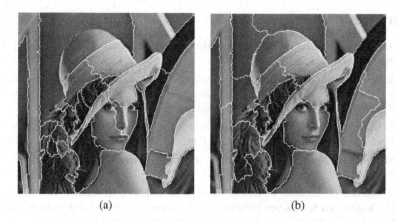

(a) (b)

Figure 6. Comparison between different hierarchies. (a) Contrast-based Waterfall, 36 regions. (b) Volume based Waterfall, 24 regions.

(a) (b)

Figure 7. Elimination of low contrasted frontiers. (a) volume based waterfall (24 regions). (b) From (a) low contrasted frontiers (contrast lower than 5) are removed (18 regions remain)

the one described in [5] may be addressed thanks to this new implementation. This implementation is based on the same data structure as the algorithms of volume extinction values, the MST. Thus, it opens the door to a combination of both approaches. For example, we have presented a waterfall based on volume extinction values and compared the results with the original version of the algorithm.

References

[1] S. Beucher. *Segmentation d'Images et Morphologie Mathématique*. PhD thesis, E.N.S. des Mines de Paris, 1990.

[2] S. Beucher. Watershed, hierarchical segmentation and waterfall algorithm. In *Mathematical Morphology and its Applications to Image Processing, Proc. ISMM'94*, pages 69–76, Fontainebleau, France, 1994. Kluwer Ac. Publ.

[3] S. Beucher and C. Lantuéjoul. Use of watersheds in contour detection. In *Proc. Int. Workshop Image Processing, Real-Time Edge and Motion Detection/Estimation*, 1979.

[4] M. Grimaud. New measure of contrast : dynamics. *Image Algebra and Morphological Processing III, San Diego CA, Proc. SPIE*, 1992.

[5] V. Marion, O. Lecoine, C. Lewandowski, J.G. Morillon, R. Aufrère, B. Marcotegui, R. Chapuis, and S. Beucher. Robust perception algorithm for road and track autonomous following. In *SPIE Defense & Security*, Orlando, USA, April 2004.

[6] F. Meyer. Minimum spanning forests for morphological segmentation. In *Mathematical Morphology and its Applications to Image Processing, Proc. ISMM'94*, pages 77–84, Fontainebleau, France, 1994. Kluwer Ac. Publ.

[7] F. Meyer. Graph based morphological segmentation. In *IAPR-TC-15 Workhop on Graph-Based Representation*, pages 51–61, Vienna, Austria, May 1999.

[8] F. Meyer. An overview of morphological segmentation. *International Journal of Pattern Recognition and Artificial Intelligence*, 15(7):1089–1118, 2001.

[9] F. Meyer and S. Beucher. Morphological segmentation. *Journal of Visual Communication and Image Representation*, 1(1):21–46, September 1990.

[10] L. Najman and M. Schmitt. Geodesic saliency of watershed contours and hierarchical segmentation. *IEEE Transactions on Pattern Analysis and Machine Intelligence*, 18(12), 1996.

[11] C. Vachier and F. Meyer. Extinction values: A new measurement of persistence. *IEEE Workshop on Non Linear Signal/Image Processing*, pages 254–257, June 1995.

[12] F. Zanoguera, B. Marcotegui, and F. Meyer. A tool-box for interactive image segmentation based on nested partions. In *IEEE International Conference on Image Processing*, Kobe, Japan, October 1999.

MOSAICS AND WATERSHEDS

Laurent Najman[1,2], Michel Couprie[1,2] and Gilles Bertrand[1,2]

[1] *Laboratoire A2SI, Groupe ESIEE*
Cité Descartes, BP99
93162 Noisy-le-Grand Cedex France

[2]*IGM, Unité Mixte de Recherche CNRS-UMLV-ESIEE UMR 8049*
{najmanl,coupriem,bertrang}@esiee.fr

Abstract   We investigate the effectiveness of the divide set produced by watershed algorithms. We introduce the mosaic to retrieve the altitude of points along the divide set. A desirable property is that, when two minima are separated by a crest in the original image, they are still separated by a crest of the same altitude in the mosaic. Our main result states that this is the case *if and only if* the mosaic is obtained through a topological thinning.

Keywords: segmentation, graph, mosaic, (topological) watershed, separation

Introduction

The watershed transform, introduced by S. Beucher and C. Lantuéjoul [4] for image segmentation, is now used as a fundamental step in many powerful segmentation procedures [8]. Watershed algorithms build a partition of the space by associating an influence zone to each minimum of the image, and by producing (in their "dividing" variant) a divide set which separates those influence zones; that is to say, they "extend" the minima.

In order to evaluate the effectiveness of the separation, we have to consider the altitude of points along the divide set. We call the greyscale image thus obtained a mosaic. The goal of this paper is to examine some properties of mosaics related to image segmentation. We say informally that a watershed algorithm produces a "separation" if the minima of the mosaic are of the same altitude as the ones of the original image and if, when two minima are separated by a crest in the original image, they are still separated by a crest of the same altitude in the mosaic. The formal definition relies on the altitude of the lowest pass which separates two minima, named pass value (see also [1, 9, 10]). Our main result states that a mosaic is a separation *if and only if* it is obtained through a topological thinning [5].

C. Ronse et al. (eds.), Mathematical Morphology: 40 Years On, 187–196.
©2005 *Springer. Printed in the Netherlands.*

1.   Basic notions and notations

Many fundamental notions related to watersheds in discrete spaces can be expressed in the framework of graphs. Let E be a finite set of vertices (or points), and let $\mathcal{P}(E)$ denote the set of all subsets of E. Throughout this paper, Γ denotes a binary relation on E, which is reflexive $((x, x) \in \Gamma)$ and symmetric $((x, y) \in \Gamma \Leftrightarrow (y, x) \in \Gamma)$. We say that the pair (E, Γ) is a *graph*. We also denote by Γ the map from E to $\mathcal{P}(E)$ such that, for all $x \in E$, $\Gamma(x) = \{y \in E | (x, y) \in \Gamma\}$. For any point x, the set $\Gamma(x)$ is called the *neighborhood of x*. If $y \in \Gamma(x)$ then we say that x and y are *adjacent*.

Let $X \subseteq E$. We denote by $\overline{X}$ the complement of X in E. Let $x_0, x_n \in X$. A *path from x_0 to x_n in X* is a sequence $\pi = (x_0, x_1, \ldots, x_n)$ of points of X such that $x_{i+1} \in \Gamma(x_i)$, with $i = 0 \ldots n - 1$. Let $x, y \in X$, we say that x and y are *linked for X* if there exists a path from x to y in X. We say that X is *connected* if any x and y in X are linked for X. We say that $Y \subseteq E$ is a *connected component of X* if $Y \subseteq X$, Y is connected, and Y is maximal for these two properties (*i.e.*, $Y = Z$ whenever $Y \subseteq Z \subseteq X$ and Z is connected). In the following, we assume that the graph (E, Γ) is connected, that is, E is made of exactly one connected component.

We denote by $\mathcal{F}(E)$ the set composed of all maps from E to $\mathbb{Z}$. A map $F \in \mathcal{F}(E)$ is also called an *image*, and if $x \in E$, $F(x)$ is called the *altitude of x (for F)*. Let $F \in \mathcal{F}(E)$. We write $F_k = \{x \in E | F(x) \geq k\}$ with $k \in \mathbb{Z}$, F_k is called an *upper section of F*, and $\overline{F_k}$ is called a *lower section of F*. A non-empty connected component of a lower section $\overline{F_k}$ is called a *(level k) lower-component of F*. A level k lower-component of F that does not contain a level $(k - 1)$ lower-component of F is called a *(regional) minimum of F*. We denote by $\mathcal{M}(F)$ the set of minima of F.

A subset X of E is *flat for F* if any two points x, y of X are such that $F(x) = F(y)$. If X is flat for F, we denote by $F(X)$ the altitude of any point of X for F.

2.   Minima extensions and mosaics

The result of most of the watershed algorithms is to associate an influence zone to each minimum of the image. We formalize this through the definition of a minima extension (see figure 1).

DEFINITION 1 *Let X be a subset of E, and let $F \in \mathcal{F}(E)$. We say that X is a minima extension of F if:*
- each connected component of X contains one and only one minimum of F, and
- each regional minimum of F is included in a connected component of X.
The complement of a minima extension of F in E is called a divide set of F.

Intuitively, for application to image analysis, the divide set represents the location of points which best separate the dark objects (regional minima), in terms of grey level difference (contrast). In order to evaluate the effectiveness of this separation, we have to consider the values of points along the divide set. This motivates the following definition.

DEFINITION 2 *Let $F \in \mathcal{F}(E)$ and let X be a minima extension of F. The mosaic of F associated with X is the map $F_X \in \mathcal{F}(E)$ such that*
- for any $x \notin X$, $F_X(x) = F(x)$; and
- for any $x \in X$, $F_X(x) = \min\{F(y)|y \in C_x\}$, where C_x denotes the connected component of X that contains x.

The term 'mosaic' for this kind of construction, was coined by S. Beucher [2].

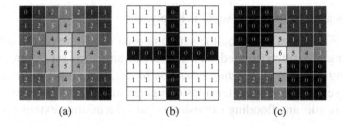

(a)   (b)   (c)

Figure 1.   (a) An image, (b) a minima extension of (a), and (c) the associated mosaic

Figure 1 shows a simple example of a minima extension and its associated mosaic. In all the examples of the paper, the graph (E, Γ) corresponds to the 4-adjacency relation on a subset $E \subset \mathbb{Z}^2$, *i.e.*, for all $x = (x_1, x_2) \in E$, $\Gamma(x) = \{(x_1, x_2), (x_1 + 1, x_2), (x_1 - 1, x_2), (x_1, x_2 + 1), (x_1, x_2 - 1)\} \cap E$.

3.   Mosaics and flooding extensions

A popular presentation of the watershed in the morphological community is based on a flooding paradigm. Let us consider the greyscale image as a topographical relief: the grey level of a pixel becomes the elevation of a point, the basins and valleys of the relief correspond to the dark areas, whereas the mountains and crest lines correspond to the light areas. Let us imagine the surface being immersed in a lake, with holes pierced in local minima. Water fills up basins starting at these local minima, and, at points where waters coming from different basins would meet, dams are built. As a result, the surface is partitioned into regions or basins separated by dams, called watershed divides.

Among the numerous algorithms [14, 12] that were developed following this idea, F. Meyer's algorithm [7] (called *flooding algorithm* in the sequel) is probably the simplest to describe and understand. Starting from an image $F \in \mathcal{F}(E)$ and the set M composed of all points belonging to the regional minima of F, the flooding algorithm expands as much as possible the set M, while preserving the connected components of M. It can be described as follows:

1. Attribute to each minimum a label, two distinct minima having distinct labels; mark each point belonging to a minimum with the label of the corresponding minimum. Initialize two sets Q and V to the empty set.

2. Insert every non-marked neighbor of every marked point in the set Q;

3. Extract from the set Q a point x which has the minimal altitude, that is, a point x such that $F(x) = \min\{F(y)|y \in Q\}$. Insert x in V. If all marked points in $\Gamma(x)$ have the same label, then

   - Mark x with this label; and
   - Insert in Q every $y \in \Gamma(x)$ such that $y \notin Q \cup V$;

4. Repeat step 3 until the set Q is empty.

Let $F \in \mathcal{F}(E)$, and let X be the set composed of all the points labeled by the flooding algorithm applied on F. We call any such set X produced by the flooding algorithm a *flooding extension (of F)*. Note that, in general, there may exist several flooding extensions of a given map F. It is easy to prove the following result: any flooding extension is indeed a minima extension of F.

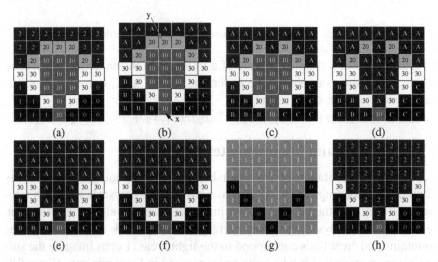

Figure 2. (a): Original image. (b-f): several steps of the extension algorithm. (g) The flooding extension of (a), and (h) the associated mosaic. One can note that the contour at altitude 20 in the original image (a) is not present in the mosaic (h).

The flooding algorithm, applied on figure 1.a, produces the minima extension 1.b, and the associated mosaic is the figure 1.c. Let us illustrate the behaviour of the algorithm on another example, the figure 2.a which presents an image with three minima at altitudes 0, 1 and 2.

- The minima at altitudes 2, 1, 0 are marked with the labels A, B, C respectively (figure 2.b). All the non-marked neighbors of the marked points are put into the set Q.
- The first point which is extracted from the set Q is the point x at altitude 10, which has points marked B and C among its neighbors (figure 2.b). This point cannot be marked.
- The next point to process is one of the points at altitude 20, for instance y (figure 2.b). The only marked points in the neighborhood of such a point are marked with the label A, and thus y is marked with the label A (figure 2.c), and the point at altitude 10 which is neighbor of y is put into the set Q.
- The next points to process are points at altitude 10. A few steps later, all points at altitude 10 but x are processed, and marked with the label A (figure 2.d).
- Then the other points at altitude 20 are processed. They are marked with the label A (figure 2.e). The next points to process are those at altitude 30, and we finally obtain the set of labeled points shown in figure 2.f.

Figure 2.g shows the flooding extension of figure 2.a, and figure 2.h is the associated mosaic.

Remark 1: we observe that (informally speaking) the algorithm does not preserve the "contrast" of the original image. In the original image, to go from, *e.g.*, the minimum at altitude 0 to the minimum at altitude 2, one has to climb to at least an altitude of 20. We observe that such a "contour" is not present in the mosaic produced by the algorithm. Similar configurations can be found for other adjacency relations, and in particular for the 6- and the 8-adjacency relations. Let us emphasize that configurations similar to the examples presented in this paper are found in real-world images.

4. Minima extensions and separations

This section introduces a formal framework that leads to a better understanding of the previous observation. In particular, two notions are pivotal in the sequel: greyscale minima extension, and separation.

Let F be a map and let F_X be the mosaic of F associated with a minima extension X of F. It is natural to try to associate any regional minimum of F_X to a connected component of X and conversely, and to compare the altitude of each minimum of F_X to the altitude of the corresponding minimum of F. We will see with forthcoming properties and examples, that both problems are in fact closely linked.

The following definition extends to greyscale maps the minima extension previously defined for sets.

DEFINITION 3 *Let F and G in $\mathcal{F}(E)$ such that $G \leq F$. We say that G is a (greyscale) minima extension (of F) if:*

i) the set composed by the union of all the minima of G is a minima extension of F.

ii) for any $X \in \mathcal{M}(F)$ and $Y \in \mathcal{M}(G)$ such that $X \subseteq Y$, we have $F(X) = G(Y)$.

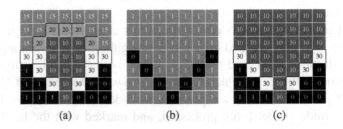

Figure 3.   (a) An image, (b) the flooding extension of (a), and (c) the mosaic of (a) associated to (b).

The image 2.h is an example of a mosaic that is a minima extension of the image 2.a. On the other hand, figure 3.a shows an image F and figure 3.c shows the mosaic F_X associated with the flooding extension X of F (figure 3.b). One can notice that the connected component of X which corresponds to the minimum of altitude 15 for F has an altitude of 10 for F_X, and is not a minimum of F_X. Thus, this mosaic F_X is not a minima extension of F.

We can now turn back to a more precise analysis of remark 1. To this aim, we present the pass value and the separation. Intuitively, the pass value between two points corresponds to the lowest altitude to which one has to climb to go from one of these points to the other one.

DEFINITION 4 *Let $F \in \mathcal{F}(E)$. Let $\pi = (x_0, \ldots, x_n)$ be a path in the graph (E, Γ), we set $F(\pi) = \max\{F(x_i)|i = 0, \ldots, n\}$.*
Let x, y be two points of E, the pass value for F between x and y is $F(x, y) = \min\{F(\pi)|\pi \in \Pi(x, y)\}$, where $\Pi(x, y)$ is the set of all paths from x to y.
Let X, Y be two subsets of E, the pass value for F between X and Y is defined by $F(X, Y) = \min\{F(x, y), \text{ for any } x \in X \text{ and any } y \in Y\}$.

A notion equivalent to the pass value up to an inversion of F (that is, replacing F by $-F$), has been introduced by A. Rosenfeld [13] under the name of *degree of connectivity* for studying connectivity in the framework of fuzzy sets. Figure 4 illustrates the pass value on the image F of figure 2.a.

Informally, a transformation "preserves the separation" if, when two points are separated by a crest in the original map, they are still separated by a crest of the same "height" in the transform.

DEFINITION 5 ([1]) *Let $F \in \mathcal{F}(E)$, let $x, y \in E$. We say that x and y are separated (for F) if $F(x, y) > \max\{F(x), F(y)\}$.*

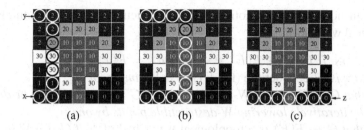

Figure 4. Illustration of paths and pass values on the image F of figure 2.a. (a) A path π_1 from the point x to the point y such that $F(\pi_1) = 30$. (b) A path π_2 from the point x to the point y such that $F(\pi_2) = 20$. It is not possible to find a path from x to y with a lower maximal altitude, hence $F(x,y) = 20$. (c) A path π_3 from the point x to the point z such that $F(\pi_3) = 10$, and we can easily check that $F(x,z) = 10$.

We say that x and y are $k-$separated (for F) if they are separated for F and if $k = F(x,y)$.
Let $G \in \mathcal{F}(E)$, with $G \leq F$. We say that G is a separation of F if, for all x and y in E, whenever x and y are k-separated for F, x and y are k-separated for G.
We say that G is a strong separation of F is G is both a separation of F and a minima extension of F.

Remark 2: we can now restate the remark 1 using the notions we have introduced in this section. Figure 3 shows that a mosaic produced by the flooding algorithm is not always a minima extension of the original map. Figure 2 shows that a mosaic produced by the flooding algorithm, even in the case where it is a minima extension, is not necessarily a separation of the original map.

5. Mosaics and topological watersheds

A different approach to the watershed was presented by M. Couprie and G. Bertrand [5]. The idea is to transform the image F into an image G while preserving some topological properties of F, namely the number of connected components of the lower cross-sections of F. A minima extension of F can then be obtained easily from G, by extracting the regional minima of G.

We begin by defining a "simple" point (in a graph), in a sense which is adapted to the watershed, then we extend this notion to weighted graphs through the use of lower sections [5].

DEFINITION 6 *Let $X \subseteq E$. The point $x \in X$ is W-simple (for X) if x is adjacent to one and only one connected component of $\overline{X}$.*

In other words, x is W-simple (for X) if the number of connected components of $\overline{X} \cup \{x\}$ equals the number of connected components of $\overline{X}$.

We can now define the notions of W-destructible point, W-thinning, and topological watershed.

DEFINITION 7 *Let $F \in \mathcal{F}(E)$, $x \in E$, and $k = F(x)$.*
The point x is W-destructible *(for F) if x is W-simple for F_k.*
We say that $G \in \mathcal{F}(E)$ is a W-thinning *of F if $G = F$ or if G may be derived from F by iteratively lowering W-destructible points by one.*
We say that $G \in \mathcal{F}(E)$ is a topological watershed *of F if G is a W-thinning of F and if there is no W-destructible point for G.*

As a consequence of the definition, a topological watershed G of a map F is a map which has the same number of regional minima as F. Furthermore, the number of connected components of any lower cross-section is preserved during this transformation. Quasi-linear algorithms for computing the topological watershed transform can be found in [6].

By the very definition of a W-destructible point, it may easily be proved that, if G is a W-thinning of F, then the union of all minima of G is a minima extension of F. This motivates the following definition.

DEFINITION 8 *Let $F \in \mathcal{F}(E)$ and let G be a W-thinning of F. The* mosaic *of F associated with G is the mosaic of F associated with the union of all minima of G.*

We have the following property.

PROPERTY 9 *Let $F \in \mathcal{F}(E)$, let G be a W-thinning of F, and let H be the mosaic of F associated with G. Then H is a minima extension of F.*

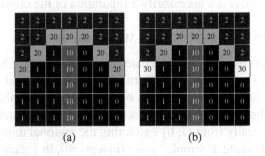

(a)   (b)

Figure 5.   Example of topological watershed. (a) A topological watershed of figure 2.a. (b) The associated mosaic.

Notice that in general, there exist different topological watersheds for a given map F. Figure 5.a presents one of the possible topological watersheds of figure 2.a, and figure 5.b shows the associated mosaic. One can note that both figure 5.a and figure 5.b are separations of figure 2.a.

6.   Mosaics and separations

Recently, G. Bertrand [1] showed that a mathematical key underlying the topological watershed is the *separation*. The following theorem states the equivalence between the notions of W-thinning and strong separation. The "if" part implies in particular that a topological watershed of an image F preserves the pass values between the minima of F. Furthermore, the "only if" part of the theorem mainly states that if one needs a transformation which is guaranteed to preserve the pass values between the minima of the original map, then this transformation is necessarily a W-thinning.

THEOREM 10 ([1]) *Let F and G be two elements of $\mathcal{F}(E)$. The map G is a W-thinning of F if and only if G is a strong separation of F.*

We can prove that the mosaic associated with any W-thinning of a map F is also a W-thinning of F (and thus, it is a separation of F).

PROPERTY 11 *Let $F \in \mathcal{F}(E)$, let G be a W-thinning of F, and H be the mosaic of F associated with G. Then H is necessarily a W-thinning of F.*

Furthermore, we prove that an arbitrary mosaic F_X of a map F is a separation of F *if and only if* F_X is a W-thinning of F. These strong results can be obtained thanks to the following property (the proofs can be found in [10]).

PROPERTY 12 *Let $F \in \mathcal{F}(E)$, let $X \subseteq E$ be a minima extension of F, and let F_X be the mosaic of F associated with X. Then, any regional minimum M of F_X is a connected component of X; furthermore $F_X(M) = F(m)$ where m denotes the unique regional minimum of F such that $m \subseteq M$.*

PROPERTY 13 *Let $F \in \mathcal{F}(E)$, let $X \subseteq E$ be a minima extension of F, and let F_X be the mosaic of F associated with X. If any connected component of X is a minimum for F_X, then F_X is a minima extension of F.*

PROPERTY 14 *Let $F \in \mathcal{F}(E)$, let $X \subseteq E$ be a minima extension of F, and let F_X be the mosaic of F associated with X. If F_X is a separation of F, then F_X is a minima extension of F.*

The following theorem is a straightforward consequence of Th. 10 and Prop. 14.

THEOREM 15 *Let $F \in \mathcal{F}(E)$, let $X \subseteq E$ be a minima extension of F, and let F_X be the mosaic of F associated with X. Then F_X is a separation of F if and only if F_X is a W-thinning of F.*

7.   Conclusion

The watershed transform is more and more used as a low-level operator in complex segmentation chains. Among those segmentation procedures, we can cite *hierarchical segmentation* [3] (waterfall) and *geodesic saliency of watershed contours* [11]. Such approaches need to compare several divides, or are based on neighborhood relationship between extended minima. It is thus important to be able to characterize some properties of the divides produced by watershed algorithms. This paper is a step in this direction. We introduced several notions that helped us to understand the watershed: minima extension, mosaic, and we also consider the pass values and separation. We show in particular that a mosaic is a separation if and only if it is a topological thinning.

Future work will build up on those results to revisit the contours saliency. We also aim at exploring criteria to choose among the various possible topological watersheds of a given image.

References

[1] G. Bertrand. On topological watersheds. *Journal of Mathematical Imaging and Vision*, 22, pp. 217-230, 2005. Special issue on Mathematical Morphology.

[2] S. Beucher. *Segmentation d'images et morphologie mathématique*. PhD thesis, École des Mines de Paris, France, 1990.

[3] S. Beucher. Watershed, hierarchical segmentation and waterfall algorithm. In J. Serra and P. Soille, editors, *Proc. Mathematical Morphology and its Applications to Image Processing*, pages 69–76, Fontainebleau, France, 1994. Kluwer.

[4] S. Beucher and C. Lantuéjoul. Use of watersheds in contour detection. In *Proc. Int. Workshop on Image Processing, Real-Time Edge and Motion Detection/Estimation*, Rennes, France, 1979.

[5] M. Couprie and G. Bertrand. Topological grayscale watershed transform. In *SPIE Vision Geometry V Proceedings*, volume 3168, pages 136–146, 1997.

[6] M. Couprie, L. Najman, and G. Bertrand. Quasi-linear algorithms for topological watershed. *Journal of Mathematical Imaging and Vision*, 22, pp. 231-249, 2005. Special issue on Mathematical Morphology.

[7] F. Meyer. Un algorithme optimal de ligne de partage des eaux. In *Actes du 8ème Congrès AFCET*, pages 847–859, Lyon-Villeurbanne, France, 1991.

[8] F. Meyer and S. Beucher. Morphological segmentation. *Journal of Visual Communication and Image Representation*, 1(1):21–46, 1990.

[9] L. Najman and M. Couprie. *DGCI'03*, volume 2886 of *LNCS*, chapter Watershed algorithms and contrast preservation, pages 62–71. Springer Verlag, 2003.

[10] L. Najman, M. Couprie, and G. Bertrand. Watersheds, mosaics, and the emergence paradigm. *Discrete Applied Mathematics*, 2005. In press.

[11] L. Najman and M. Schmitt. Geodesic saliency of watershed contours and hierarchical segmentation. *IEEE Trans. on PAMI*, 18(12):1163–1173, December 1996.

[12] J.B.T.M. Roerdink and A. Meijster. The watershed transform: Definitions, algorithms and parallelization strategies. *Fundamenta Informaticae*, 41:187–228, 2000.

[13] A. Rosenfeld. On connectivity properties of grayscale pictures. *Pattern Recognition*, 16:47–50, 1983.

[14] L. Vincent and P. Soille. Watersheds in digital spaces: An efficient algorithm based on immersion simulations. *IEEE Trans. on PAMI*, 13(6):583–598, June 1991.

A NEW DEFINITION FOR THE DYNAMICS

Gilles Bertrand [1,2]

[1]*Laboratoire A2SI, ESIEE, B.P. 99, 93162 Noisy-Le-Grand Cedex France*

[2]*Institut Gaspard Monge, Unité Mixte de Recherche CNRS-UMLV-ESIEE UMR 8049*
g.bertrand@esiee.fr

Abstract   We investigate the new definition of the ordered dynamics proposed in [4]. We
show that this definition leads to several properties. In particular we give necessary and sufficient conditions which indicate when a transformation preserves
the dynamics of the regional maxima. We also establish a link between the dynamics and minimum spanning trees.

Keywords: mathematical morphology, dynamics, graph, watershed, minimum spanning tree

Introduction

The dynamics, introduced by M. Grimaud [1, 2], allows to extract a measure
of a regional maximum (or a regional minimum) of a map. Such a measure may
be used for eliminating maxima which may be considered as "non significant".
In this paper we investigate a new definition of the dynamics. In particular we
establish some equivalence between a transformation which preserves the dynamics and a transformation which preserves some connection values (a kind
of measure of contrast) between pairs of points (Th. 11). Such an equivalence
is also given with topological watersheds (Prop. 19) and extensions (Prop. 16).
Furthermore we establish a link between the dynamics and minimum spanning
trees of a graph linking regional maxima, the cost of each arc being precisely
the connection value between the corresponding maxima (Th. 14).

1.   Basic definitions

Any function may be represented by its different threshold levels [5, 7, 8].
These levels constitute a "stack". In fact, the datum of a function is equivalent
to the datum of a stack. In this section, we introduce definitions for stacks and
related notions, this set of definitions allows to handle both the threshold levels
of a discrete function and the complements of these levels.

C. Ronse et al. (eds.), Mathematical Morphology: 40 Years On, 197–206.
©2005 *Springer. Printed in the Netherlands.*

Discrete maps and stacks

Here and subsequently E stands for a non-empty finite set and K stands for an element of $\mathbb{Z}$, with $K > 0$. If $X \subseteq E$, we write $\overline{X} = \{x \in E \mid x \notin X\}$. If k_1 and k_2 are elements of $\mathbb{Z}$, we define $[k_1, k_2] = \{k \in \mathbb{Z} \mid k_1 \leq k \leq k_2\}$. We set $\mathbb{K} = [-K, +K]$, and $\mathbb{K}° = [-K + 1, +K - 1]$.

Let $F = \{F[k] \subseteq E \mid k \in \mathbb{K}\}$ be a family of subsets of E with index set $\mathbb{K}$, such a family is said to be a $\mathbb{K}$-*family (on E)*. Any subset $F[k]$, $k \in \mathbb{K}$, is a *section of F (at level k)* or the k-*section of F*. We set:
$$\overline{F} = \{\overline{F[k]} \mid \overline{F[k]} = \overline{F[k]}, \ k \in \mathbb{K}\}, \text{ and}$$
$$F^{-1} = \{F^{-1}[k] \mid F^{-1}[k] = F[-k], \ k \in \mathbb{K}\},$$
which are, respectively, the *complement of F* and the *symmetric of F*.

We say that a $\mathbb{K}$-family F is an *upstack on E* if:
$$F[-K] = E, \ F[K] = \emptyset, \text{ and } F[j] \subseteq F[i] \text{ whenever } i < j.$$
We say that a $\mathbb{K}$-family F is a *downstack on E* if:
$$F[-K] = \emptyset, \ F[K] = E, \text{ and } F[i] \subseteq F[j] \text{ whenever } i < j.$$
A $\mathbb{K}$-family is a *stack* if it is either an upstack or a downstack.

We denote by $\mathcal{S}_E^+$ (resp. $\mathcal{S}_E^-$) the family composed of all upstacks on E (resp. downstacks on E). We also set $\mathcal{S}_E = \mathcal{S}_E^+ \cup \mathcal{S}_E^-$.

Let F, G be both in $\mathcal{S}_E^+$ or both in $\mathcal{S}_E^-$. We say that G is *under F*, written $G \subseteq F$ if, for all $k \in \mathbb{K}$, $G[k] \subseteq F[k]$.

Let $F \in \mathcal{S}_E^+$ and let $G \in \mathcal{S}_E^-$. We define two maps from E on $\mathbb{K}$, also denoted by F and G, such that, for any $x \in E$,
$$F(x) = \max\{k \in \mathbb{K} \mid x \in F[k]\} \text{ and } G(x) = \min\{k \in \mathbb{K} \mid x \in G[k]\},$$
which are, respectively, the *functions induced by the upstack F and the downstack G*, $F(x)$ and $G(x)$ are, respectively, the *altitudes of x for F and G*.

Let $F \in \mathcal{S}_E$ and let $x \in E$. We set $S(x, F) = F[k]$, with $k = F(x)$, $S(x, F)$ is the *section of x for F* (see illustration Fig. 1).

Graphs

Throughout this paper, Γ will denote a binary relation on E, which is reflexive and symmetric. We say that the pair (E, Γ) is a *graph*, each element of E is called a *vertex* or a *point*. We will also denote by Γ the map from E to 2^E, such that, for all $x \in E$, $\Gamma(x) = \{y \in E \mid (x, y) \in \Gamma\}$. If $y \in \Gamma(x)$, we say that y is *adjacent to x*. If $X \subseteq E$ and $y \in \Gamma(x)$ for some $x \in X$, we say that y *is adjacent to X*.

Let $X \subseteq E$, a *path in X* is a sequence $\pi = \langle x_0, ..., x_k \rangle$ such that $x_i \in X$, $i \in [0, k]$, and $x_i \in \Gamma(x_{i-1})$, $i \in [1, k]$. We also say that π is a *path from x_0 to x_k in X*. Let $x, y \in X$. We say that x and y are *linked for X* if there exists a path from x to y in X. We say that X is *connected* if any x and y in X are

F	a	b	c	d	e	f	g	h	i	F^{-1}
3	0	0	0	0	0	0	0	0	0	-3
2	0	0	0	1	1	0	0	0	0	-2
1	0	1	0	1	1	0	0	0	0	-1
0	0	1	1	1	1	0	0	1	0	0
-1	0	1	1	1	1	0	0	1	1	1
-2	1	1	1	1	1	0	0	1	1	2
-3	1	1	1	1	1	1	1	1	1	3

Figure 1. representation of an upstack F (the 1's, k on the left), the downstack F^{-1} (the 1's, k on the right), the downstack $\overline{F}$ (the 0's, k on the left), and the upstack $[\overline{F}]^{-1} = \overline{[F^{-1}]}$ (the 0's, k on the right). For example, we have $F[1] = F^{-1}[-1] = \{b,d,e\}$, $\overline{F}[1] = \{a,c,f,g,h,i\}$, $F(c) = 0$, $\overline{F}(c) = 1$, $S(c,F) = \{b,c,d,e,h\}$, $S(c,\overline{F}) = \{a,c,f,g,h,i\}$.

linked for X. We say that $Y \subseteq E$ is a *connected component of* $X \subseteq E$, if $Y \subseteq X$, Y is connected, and Y is maximal for these two properties.
Note: In the sequel of this paper, we assume that E is connected. All notions and properties may be easily extended for non-connected graphs.

Stacks and graphs

Let $F \in \mathcal{S}_E$ and let $k \in \mathbb{K}$. A connected component of a non-empty k-section of F is a *component of F (at level k)* or a *k-component of F*.
Let $x \in E$ and let $S(x,F)$ be the section of x for F. We denote by $C(x,F)$ the connected component of $S(x,F)$ which contains x, $C(x,F)$ is the *component of x for F*.
We say that $x \in E$ and $y \in E$ are *k-linked for F* if x and y are linked for $F[k]$, *i.e.*, if x and y belong to the same connected component of $F[k]$.

Let $F \in \mathcal{S}_E$. A subset $X \subseteq E$ is an *extremum of F* if X is a component of F and if X is minimal for this property (*i.e.*, no proper subset of X is a component of F). We denote by $\mathcal{E}_F$ the family composed of all extrema of F. If $F \in \mathcal{S}_E^+$, $G \in \mathcal{S}_E^-$, we also say that an extremum of F is a *maximum of F* and that an extremum of G is a *minimum of G*.
A subset $X \subseteq E$ is *flat for F* if $F(x) = F(y)$ for all x, y in X. If X is flat for F, the *altitude of X for F* is the value $F(X)$ such that $F(X) = F(x)$ for every $x \in X$.

Let $F \in \mathcal{S}_E^+$, $G \in \mathcal{S}_E^-$, and let x, y be two vertices in E. We define:
$$F(x,y) = \max\{k \mid x \text{ and } y \text{ are } k\text{-linked for } F\}, \text{ and}$$

$$G(x,y) = \min\{k \mid x \text{ and } y \text{ are } k\text{-linked for } G\},$$

$F(x,y)$ and $G(x,y)$ are the *connection values between x and y for F and G, respectively.*

If X and Y are two subsets of E, we set $F(X,Y) = \max\{F(x,y) \mid x \in X, y \in Y\}$, and $G(X,Y) = \min\{G(x,y) \mid x \in X, y \in Y\}$.

Let $F \in \mathcal{S}_E^+$, $G \in \mathcal{S}_E^-$, and let $\pi = \langle x_0, ..., x_k \rangle$ be a path in E, we set: $F(\pi) = \min\{F(x_i) \mid i \in [0,k]\}$ and $G(\pi) = \max\{G(x_i) \mid i \in [0,k]\}$. If $\Pi(x,y)$ is the set composed of all paths from x to y in E, we have: $F(x,y) = \max\{F(\pi) \mid \pi \in \Pi(x,y)\}$, $G(x,y) = \min\{G(\pi) \mid \pi \in \Pi(x,y)\}$.

2.   Separation and ordered extrema

We introduce the notion of separation (see also [4]) which plays a key role for the dynamics. Then, we show that, given an ordering on subsets of E, we may define connection values between these ordered subsets (Def. 4).

Definition 1.
Let $X \subseteq E$ and let x, y be in X. The points x and y are *separated for X* if x and y are not linked for X, i.e., if there is no path from x to y in X.
Let X, Y be subsets of E such that $X \subseteq Y$. We say that Y is a *separation of X* if any x and y in X which are separated for X, are separated for Y.
Let F and G be both in $\mathcal{S}_E^+$ (or both in $\mathcal{S}_E^-$) and such that $F \subseteq G$. We say that G *is a separation of F* if, for any $k \in \mathbb{K}$, $G[k]$ is a separation of $F[k]$.

Definition 2. Let $F \in \mathcal{S}_E^+$, $G \in \mathcal{S}_E^-$, and x, $y \in E$. The points x and y are *separated for F (resp. G)* if $F(x,y) < \min\{F(x), F(y)\}$ (resp. $G(x,y) > \max\{G(x), G(y)\}$). The points x and y are *k-separated for F (resp. G)* if they are separated and if $F(x,y) = k$ (resp. $G(x,y) = k$).

Proposition 3. *Let F, G be both in $\mathcal{S}_E^+$ (or both in $\mathcal{S}_E^-$) and such that $F \subseteq G$. The stack G is a separation of F if and only if any x and y in E which are k-separated for F, are k-separated for G.*

Definition 4. Let $\mathcal{O}$ be a family composed of non empty subsets of E and let $\prec$ be an *ordering on $\mathcal{O}$*, i.e., $\prec$ is a relation on $\mathcal{O}$ which is transitive and trichotomous (for any X, Y in $\mathcal{O}$, one and only one of $X \prec Y$, $Y \prec X$, $X = Y$ is true). We denote by $X_{max}^\prec$ the element of $\mathcal{O}$ such that, for all $Y \in \mathcal{O} \setminus \{X_{max}^\prec\}$, $Y \prec X_{max}^\prec$. Let $F \in \mathcal{S}_E^+$, $G \in \mathcal{S}_E^-$, and let $X \in \mathcal{O}$. The *connection value of X for $(F, \prec)$* is the number $F(X, \prec)$ such that:
- $F(X_{max}^\prec, \prec) = -\infty$; and
- $F(X, \prec) = \max\{F(X,Y) \mid Y \in \mathcal{O} \text{ and } X \prec Y\}$ if $X \neq X_{max}^\prec$.
The *connection value of X for $(G, \prec)$* is $G(X, \prec) = -G^{-1}(X, \prec)$.

Th. 5 shows that the connection values between ordered extrema are sufficient to decide whether or not a stack G is a separation of a stack F (see [4] for the proof).

Theorem 5 (ordered extrema). *Let F, G be both in $\mathcal{S}_E^+$ (or both in $\mathcal{S}_E^-$) such that $F \subseteq G$. Let $\prec$ be an ordering on $\mathcal{E}_F$. The stack G is a separation of F if and only if, for each X in $\mathcal{E}_F$, we have $F(X, \prec) = G(X, \prec)$.*

3. Dynamics

We will show now the connection between the previous notions and the notion of dynamics which was introduced as a measure of contrast of an extremum [1–3]. We first give a definition of the dynamics which may be proved to be equivalent to the original definition of Grimaud [1, 2].

Definition 6. Let $F \in \mathcal{S}_E^+$, and $X \in \mathcal{E}_F$. We define the number $\tilde{F}(X)$ by:
- $\tilde{F}(X) = -\infty$ if $F(X) = \max\{F(Y) \mid Y \in \mathcal{E}_F\}$;
- $\tilde{F}(X) = \max\{F(X, Y) \mid Y \in \mathcal{E}_F \text{ and } F(X) < F(Y)\}$ otherwise.
The *unordered dynamics of X for F* is the number $\Delta_F(X) = F(X) - \tilde{F}(X)$. If $G \in \mathcal{S}_E^-$ and $X \in \mathcal{E}_G$, we set $\tilde{G}(X) = -\tilde{G}^{-1}(X)$, the *unordered dynamics of X for G* is the number $\Delta_G(X) = \tilde{G}(X) - G(X)$.

If a stack G is a separation of a stack F, it may be seen that an extremum of G does not necessarily contain an extremum of F. In order to establish a link between the notion of separation and the dynamics, we introduce the following definition.

Definition 7. Let F, G be both in $\mathcal{S}_E^+$ (or both in $\mathcal{S}_E^-$) such that $F \subseteq G$. We say that G is an *extrema extension* or an *e-extension* of F if:
i) for each $X \in \mathcal{E}_F$, there exists $Y \in \mathcal{E}_G$, such that $X \subseteq Y$; and
ii) for each $Y \in \mathcal{E}_G$, there exists a unique $X \in \mathcal{E}_F$, such that $X \subseteq Y$; and
iii) for any $X \in \mathcal{E}_F$, $Y \in \mathcal{E}_G$, such that $X \subseteq Y$, we have $F(X) = G(Y)$.

It may be easily seen that two distinct extrema X, Y of a stack F are always separated for F (*i.e.*, any x in X and any y in Y are separated). In fact, since an e-extension "preserves the extrema", the following property is a direct consequence of Def. 1, 2, 6, 7 and Prop. 3.

Proposition 8. *Let G be an e-extension of F. If G is a separation of F, then for any $X \in \mathcal{E}_F$, $Y \in \mathcal{E}_G$, such that $X \subseteq Y$, we have $\Delta_F(X) = \Delta_G(Y)$.*

We observe that the converse of Prop. 8 is not true: Fig. 2 (a) shows a counter-example. The upstacks F and G have three maxima A, B, C. The upstack G is an e-extension of F but not a separation of F. Nevertheless $\Delta_F(A) = \Delta_G(A)$, $\Delta_F(B) = \Delta_G(B)$ and $\Delta_F(C) = \Delta_G(C) = \infty$.

Thus, separations (which are also e-extensions) "preserve the unordered dynamics of the extrema", but a transformation which preserves the unordered dynamics does not necessarily preserve k-separation between points and connection values between extrema. In this sense, we may say that the notion

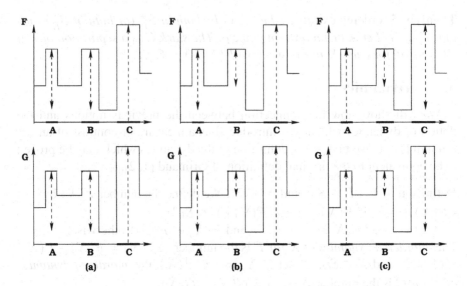

Figure 2. (a) the unordered dynamics of two upstacks F and G, (b) the ordered dynamics for the ordering $B \prec A \prec C$, (c) the ordered dynamics for the ordering $B \prec C \prec A$.

of separation conveys more information about the contrast of an image than the unordered dynamics. The following Th. 11 shows that it is possible to have a more powerful notion of dynamics if we introduce a definition based on extrema ordering.

Beforehand, it should be noted that it is not possible to obtain such a result by slightly changing the original definition of the dynamics. For example, if we replace the strict inequality appearing in the definition of the unordered dynamics by an inequality, some counter-examples for the converse of Prop. 8 may be easily found.

Definition 9. Let $F \in \mathcal{S}_E^+$ (resp. $F \in \mathcal{S}_E^-$) and let $\prec$ be an ordering on $\mathcal{E}_F$. Let X be an extremum for F. The *ordered dynamics of X for $(F, \prec)$* is the value $\Delta_F(X, \prec) = F(X) - F(X, \prec)$ (resp. $F(X, \prec) - F(X)$).

Remark 10. Let $F \in \mathcal{S}_E^+$ (resp. $F \in \mathcal{S}_E^-$) and let $\prec$ be an ordering on $\mathcal{E}_F$. We say that $\prec$ is an *altitude ordering of $\mathcal{E}_F$* if $X \prec Y$ (resp. $Y \prec X$) whenever $F(X) < F(Y)$. We observe that, if $\prec$ is an altitude ordering of $\mathcal{E}_F$ and if all extrema of F have distinct altitudes, then we have $\Delta_F(X, \prec) = \Delta_F(X)$.

In Fig. 2 (b) and (c) we can see that, with this new definition of the dynamics, the map G does not longer preserve the dynamics of the maxima of F. In fact, the following theorem, which is a direct consequence of Th. 5, shows that the ordered dynamics "encodes all information of a separation". The result holds whatever the choice of the ordering. Thus, an ordering which is not based on the altitudes of the extrema may be considered. For example, maxima may be ordered according to their areas.

Let $F \in \mathcal{S}_E$ and let $\prec$ be an ordering on $\mathcal{E}_F$. Let G be an e-extension of F. In the sequel of the paper, we will also denote by $\prec$ the ordering of $\mathcal{E}_G$ such that, for all X, Y in $\mathcal{E}_G$, $X \prec Y$ if and only if $X' \prec Y'$, X' and Y' being the extrema of F such that $X' \subseteq X$ and $Y' \subseteq Y$.

Theorem 11. *Let G be an e-extension of F and let $\prec$ be an ordering of $\mathcal{E}_F$. The stack G is a separation of F if and only if, for any $X \in \mathcal{E}_F$, $Y \in \mathcal{E}_G$, such that $X \subseteq Y$, we have $\Delta_F(X, \prec) = \Delta_G(Y, \prec)$.*

4.   Dynamics and maximum spanning tree

Theorem 11 invites us to try to recover, from the values $F(X, \prec)$, with $X \in \mathcal{E}_F$, the connection values between all extrema of a stack F. We will show, with Prop. 13, that this may be done provided we also know, for each extremum X of F (different from $X_{max}^{\prec}$), an extremum Y of F such that $X \prec Y$ and $F(X, Y) = F(X, \prec)$ (see Fig. 3 for illustrations of this section).

Definition 12. Let $F \in \mathcal{S}_E$ and let $\prec$ be an ordering on $\mathcal{E}_F$. Let Ψ be a map from $\mathcal{E}_F \setminus \{X_{max}^{\prec}\}$ on $\mathcal{E}_F$. We say that Ψ is a *connection map for* $\prec$ if, for each $X \in \mathcal{E}_F \setminus \{X_{max}^{\prec}\}$, we have $X \prec \Psi(X)$ and $F[X, \Psi(X)] = F(X, \prec)$.

Let S be a set, and let $T \subseteq S$. Let Ψ be a map from T on S, and let a, b, c be elements of S.
We say that b is *under* a, or that a *is over* b (for Ψ), if there is some $k \geq 0$ such that $a = \Psi^k(b)$. We say that c is a *common ancestor of a and b (for Ψ)* if a and b are under c. We say that c is the *least common ancestor of a and b (for Ψ)* if c is a common ancestor of a and b and no element under c distinct from c is a common ancestor of a and b.

Let $F \in \mathcal{S}_E$, let $\prec$ be an ordering on $\mathcal{E}(F)$, and let Ψ be a *connection map for* $\prec$. We observe that any extremum of F is under $X_{max}^{\prec}$. Furthermore any extrema X and Y of F admit a (unique) least common ancestor.

Proposition 13. *Let $F \in \mathcal{S}_E^+$ (resp. $G \in \mathcal{S}_E^-$), let $\prec$ be an ordering on $\mathcal{E}_F$ (resp. $\mathcal{E}_G$), and Ψ be a connection map for $\prec$. Let X, Y be distinct extrema of F (resp. G) and let Z be the least common ancestor of X and Y for Ψ. Then:*
- $F(X, Y) = \min\{F(S, \prec) \mid$ *for all S under Z and over X or Y, $S \neq Z\}$;*
- $G(X, Y) = \max\{G(S, \prec) \mid$ *for all S under Z and over X or Y, $S \neq Z\}$.*

Now, we will show the link between connection maps and maximum spanning trees (see [13] for maximum spanning trees).

Let $F \in \mathcal{S}_E$, let $\prec$ be an ordering on $\mathcal{E}_F$, and let Ψ be a connection map for $\prec$. We define the *graph induced by* Ψ as the graph $(\mathcal{E}_F, \tilde{\Psi})$ such that $\tilde{\Psi} = \{(X, \Psi(X)) \mid$ for all $X \in \mathcal{E}_F \setminus \{X_{max}^{\prec}\}\}$.

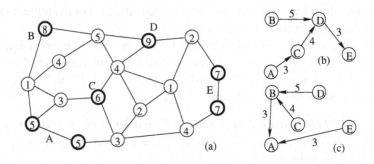

Figure 3.   (a) a graph (E, Γ) and an upstack F on E represented by its function, F has 5 maxima A, B, C, D, E, (b) a connection map for the ordering $A \prec B \prec C \prec D \prec E$, (c) a connection map for the ordering $E \prec D \prec C \prec B \prec A$.

Theorem 14. *Let $F \in \mathcal{S}_E^+$ (resp. $F \in \mathcal{S}_E^-$), let $\prec$ be an ordering on $\mathcal{E}_F$ and let Ψ be a connection map for $\prec$. The graph induced by Ψ is a maximum (resp. minimum) spanning tree of the graph $(\mathcal{E}_F, \mathcal{E}_F \times \mathcal{E}_F)$, the cost of an edge $(X, Y) \in \mathcal{E}_F \times \mathcal{E}_F$ being precisely the connection value between the extrema X and Y.*

5.   Dynamics, extension, and watersheds

We conclude this paper by outlining the deep link which exists between the ordered dynamics, the notion of topological watershed (see [9, 12, 4, 14]), and the notion of extension ([4]). The two following properties are direct consequences of two theorems presented in [4] (also in [10]).

Definition 15. *Let X, Y be non-empty subsets of E such that $X \subseteq Y$. We say that Y is an extension of X if each connected component of Y contains exactly one connected component of X. We also say that Y is an extension of X if X and Y are both 1empty.*
Let F, G be both in $\mathcal{S}_E^+$ (or both in $\mathcal{S}_E^-$). We say that G is an extension of F if, for any $k \in \mathbb{K}$, $G[k]$ is an extension of $F[k]$.

Proposition 16. *Let G be an e-extension of F and let $\prec$ be an ordering of $\mathcal{E}_F$. The stack G is an extension of F if and only if, for any $X \in \mathcal{E}_F$, $Y \in \mathcal{E}_G$, such that $X \subseteq Y$, we have $\Delta_F(X, \prec) = \Delta_G(Y, \prec)$.*

Definition 17. *Let $F \in \mathcal{S}_E$ and let $x \in E$ such that $F(x) \in \mathbb{K}^\circ$. We say that x is W-destructible for F if x is adjacent to exactly one connected component of $\overline{S(x, F)}$ (i.e., one connected component of $\overline{F[k]}$, with $k = F(x)$).*

For example, in Fig. 4 (a), the point x is W-destructible, but y is not W-destructible (y is adjacent to two connected components of $\overline{F[4]}$).

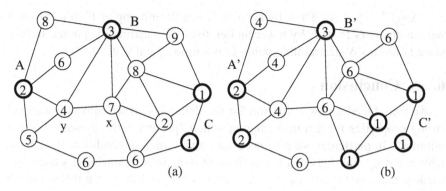

Figure 4. (a) An upstack F represented by its function, (b) an upstack $G \subseteq F$ which "preserves an ordered dynamics of the minima of F": by Prop. 8, the stack G is necessarily a W-thinning of F, in fact G is a topological watershed of F.

The following operation on stacks is the extension of the removal of a point from a set.

Let $F \in \mathcal{S}_E$ and let $x \in E$ such that $F(x) \in \mathbb{K}^\circ$. We denote by $F \setminus x$ the element of $\mathcal{S}_E$ such that $[F \setminus x][k] = F[k] \setminus \{x\}$ if $k = F(x)$, and $[F \setminus x][k] = F[k]$ otherwise.

Thus, if F is an upstack, then $[F \setminus x](x) = F(x) - 1$ and $[F \setminus x](y) = F(y)$ whenever $y \neq x$.

Definition 18. Let F, G be both in $\mathcal{S}_E^+$ (or both in $\mathcal{S}_E^-$). We say that G is a *W-thinning* of F, written $F \overset{W}{\searrow} G$, if:
i) $G = F$; or if
ii) there exists a W-thinning H of F and there exists a W-destructible point x for H, such that $G = H \setminus x$.
A W-thinning G of F is a *(topological) watershed* of F if $G \overset{W}{\searrow} H$ implies $H = G$.

Proposition 19. *Let $F, G \in \mathcal{S}_E$ such that $\overline{G}$ is an e-extension of $\overline{F}$ and let $\prec$ be an ordering of $\mathcal{E}_{\overline{F}}$. The stack G is a W-thinning of F if and only if, for any $X \in \mathcal{E}_{\overline{F}}$ and $Y \in \mathcal{E}_{\overline{G}}$ such that $X \subseteq Y$, we have $\Delta_{\overline{F}}(X, \prec) = \Delta_{\overline{G}}(Y, \prec)$.*

Fig. 4 provides an illustration of Prop. 19. Two upstacks F and G are represented. We have $G \subseteq F$, $\overline{F}$ and $\overline{G}$ have three minima A, B, C and A', B', C', respectively. We have $A \subseteq A'$, $B \subseteq B'$, $C \subseteq C'$. If F is an arbitrary upstack, we have $\overline{F}(x) = F(x) + 1$, for each $x \in E$ (see the subsection "Discrete maps and stacks"). Thus $\overline{F}(A) = \overline{G}(A') = 2+1$, $\overline{F}(B) = \overline{G}(B') = 3 + 1$, $\overline{F}(C) = \overline{G}(C') = 1 + 1$, it follows that $\overline{G}$ is an e-extension of $\overline{F}$. Let us consider the ordering $\prec$ such that $A \prec B \prec C$ and $A' \prec B' \prec C'$. We have $\Delta_{\overline{F}}(A, \prec) = \Delta_{\overline{G}}(A', \prec) = 4+1$, $\Delta_{\overline{F}}(B, \prec) = \Delta_{\overline{G}}(B', \prec) = 6+1$, $\Delta_{\overline{F}}(C, \prec$

$) = \Delta_{\overline{G}}(C', \prec) = \infty$. Thus, by Prop. 8, G is a W-thinning of F. It means that we can obtain G from F by lowering iteratively W-destructible points. In fact, since G has no W-destructible point, G is a topological watershed of F.

6.   Conclusion

We introduced a new definition for the dynamics which allows to recover strong properties for a transformation which "preserves the dynamics of the minima". In particular we give necessary and sufficient conditions which establish a deep link between the notions of dynamics, connection values, extension, and topological watersheds. Future work will show that this approach also provides a more powerful operator for filtering.

References

[1] M. Grimaud, *Géodésie numérique en morphologie mathématique. Application à la détection automatique de microcalcifications en mammographies numériques*, PhD Thesis, Ecole des Mines de Paris, December 1991.

[2] M. Grimaud, "A new measure of contrast: Dynamics", in *SPIE Vol. 1769, Image Algebra and Morphological Processing III*, pp. 292–305, 1992.

[3] F. Meyer, "The dynamics of minima and contours", in *ISMM 3rd*, R. S. P. Maragos and M. Butt, eds., *Computational Imaging and Vision*, pp. 329–336, Kluwer, 1996.

[4] G. Bertrand, "On topological watersheds", *Journal of Mathematical Imaging and Vision*, 22, pp. 217-230, 2005.

[5] P.D. Wendt, E.J. Coyle, and N.C. Gallagher, Jr., "Stack filters", *IEEE Trans. Acoust., Speech, Signal Processing*, Vol. ASSP-34, pp. 898-911, 1986.

[6] J. Serra, "Connections for sets and functions", *Fundamenta Informaticae*, 41, pp. 147–186, 2000.

[7] J. Serra, *Image analysis and mathematical morphology*, Vol. II, Th. Advances, Academic Press, 1988.

[8] H. Heijmans, "Theoretical aspects of gray-level morphology", *IEEE Trans. on Pattern Analysis and Machine Intelligence*, Vol. 13, No 6, pp. 568-592, 1991.

[9] M. Couprie and G. Bertrand, "Topological grayscale watershed transform," in *SPIE Vision Geometry V Proceedings*, **3168**, pp. 136–146, 1997.

[10] G. Bertrand, "Some properties of topological greyscale watersheds", *IS&T/SPIE Symposium on Electronic Imaging, Vision Geometry XII*, Vol. 5300, pp. 127–137, 2004.

[11] A. Rosenfeld, "Fuzzy digital topology", *Information and Control* **40**, pp. 76-87, 1979.

[12] L. Najman, M. Couprie, and G. Bertrand, "Watersheds, extension maps and the emergence paradigm", *Discrete Applied Mathematics*, to appear (Internal Report, IGM 2004-04, Institut Gaspard Monge, Université de Marne-La-Vallée, 2004).

[13] M. Gondran and M. Minoux, *Graphs and algorithms*, John Wiley & Sons, 1984.

[14] M. Couprie, L. Najman, and G. Bertrand, "Quasi-linear algorithms for the topological watershed", *Journal of Mathematical Imaging and Vision*, 22, pp. 231-249, 2005.

WATERSHED-DRIVEN REGION-BASED IMAGE RETRIEVAL

I. Pratikakis[1], I. Vanhamel[2], H. Sahli[2], B. Gatos[1] and S. Perantonis[1]

[1]*Institute of Informatics and Telecommunications*
National Centre for Scientific Research "Demokritos"
P.O. Box 60228, GR-15310 Athens, Greece
ipratika@iit.demokritos.gr, bgat@iit.demokritos.gr, sper@iit.demokritos.gr

[2]*ETRO - Vrije Universiteit Brussel*
Pleinlaan 2, B-1050 Brussels, Belgium
iuvanham@etro.vub.ac.be, hsahli@etro.vub.ac.be

Abstract   This paper presents a strategy for content-based image retrieval. It is based on a meaningful segmentation procedure that can provide proper distributions for matching via the Earth mover's distance as a similarity metric. The segmentation procedure is based on a hierarchical watershed-driven algorithm that extracts automatically meaningful regions. In this framework, the proposed robust feature extraction plays a major role along with a novel region weighting for enhancing feature discrimination. Experimental results demonstrate the performance of the proposed strategy.

Keywords: content-based image retrieval, image segmentation, Earth Mover's distance, Region weighting

Introduction

Increasing amounts of imagery due to advances in computer technologies and the advent of World Wide Web (WWW) have made apparent the need for effective and efficient imagery indexing and search of not only the metadata associated with it (eg. captions and annotations) but also retrieval directly on the visual content. During the evolution period of Content-Based Image Retrieval (CBIR) research the major bottleneck in any system is the gap between low level features and high level semantic concepts. Therefore, the obvious effort toward improving a CBIR system is to focus on methodologies that will enable a reduction or even, in the best case, bridging of the aforementioned gap. Im-

207

C. Ronse et al. (eds.), Mathematical Morphology: 40 Years On, 207–216.
©2005 *Springer. Printed in the Netherlands.*

age segmentation always plays a key role toward the semantic description of an image since it provides the delineation of the objects that are present in an image. Although, contemporary algorithms can not provide a perfect segmentation, some can produce a rich set of meaningful regions upon which robust discriminant regional features can be computed. In this paper, we present a strategy for content-based image retrieval. It is based on a meaningful segmentation procedure that can provide proper distributions for matching via the Earth mover's distance as a similarity metric. In the underlying framework, a major role plays the proposed robust feature extraction along with a novel region weighting for enhancing feature discrimination. The segmentation procedure is based on a hierarchical watershed-driven algorithm that extracts automatically meaningful regions. This paper is organized as follows: In Section 1, we provide the state-of-the-art in the region-based CBIR approaches. Section 2 describes the proposed segmentation scheme that guides the image representation along with the proposed feature set which is extracted out of each region. Section 3 is dedicated to the description of the selected similarity metric and a novel region weighting factor while in Section 4 experimental results demonstrate the performance of the proposed CBIR strategy.

1.   Related work

The fundamental aspects that characterize a region-based image retrieval system are the following : (i) the underlying segmentation scheme; (ii) the selected features for region representation; (iii) the region matching method and (iv) the user supervision.

In [8], the NeTra system is presented, where retrieval is based on segmented image regions. The segmentation scheme requires user supervision for parameter tuning and segmentation corrections. Furthermore, a one-to-one region matching is proposed after region selection by the user. In the same spirit, Blobworld system [1] is proposed, where a user is required to select important regions and features. As an extension to Blobworld, Greenspan *et al.* [4] compute blobs by using Gaussian mixture modeling and use Earth mover's distance (EMD) [12] to compute both the dissimilarity of the images and the flow-matrix of the blobs between the images. In [2], Fuh *et al.* use the idea of combining a color segmentation with relationship trees and a corresponding matching method. They use information concerning the hierarchical relationship of the regions along with the region features for a robust retrieval. In [16], an integrated matching algorithm is proposed that is based on region similarities with respect to a combination of color, shape and texture information. The proposed method enables one-to-many region matching. Hsieh and Grimson [5] propose a framework that supports a representation for a visual concept using regions of multiple images. They support one-to-many regions matching on two stages. First, a similarity comparison occurs followed by a region voting that leads to a final region matching. Finally, Jing *et al.* [6] propose an image retrieval framework that integrates efficient region-based representation and effective on-line learning capability. This approach is based on user's relevance feedback that makes user supervision an obligatory requirement.

In this paper, unlike the above approaches, we propose a strategy that does not require any supervision from the user rather than selecting an example image to be used as a query and permit a many-to-many region matching improving the robustness of the system. It is a region-based approach that takes advantage of the robustness of each subsequent module that consists of. More specifically, it is based on a watershed-driven hierarchical segmentation module which produces meaningful regions, and a feature extraction module that supports proper distributions for matching along with a robust similarity metric which is enhanced by a novel weighting factor.

2. Image representation

Automatic Multiscale Watershed Segmentation

The proposed multiscale hierarchical watershed-driven segmentation scheme for vector-valued images [15] [13], depicted in Fig.1, consists of three basic modules.

The first module (*Salient Measure Module*) is dedicated to a saliency measurement of the image partitions after a scale-space analysis which is supported by watershed segmentation and nonlinear diffusion filtering. The main goal of this module is to create a hierarchy among the gradient watersheds detected at the finest scale: the *localization scale*. To accomodate this hierarchy, we create a region adjacency graph (*RAG*), in which the nodes represent the detected gradient watersheds and the arcs represent the contours between two watershed segments, i.e. the adjacencies.

The entire process to retrieve the saliency measure for the gradient watersheds requires three basic steps: (i) *nonlinear scale-space filtering*; (ii) *Linking (deep image structure)*: At each scale the gradient magnitude of the transformed image is estimated. At the localization scale, the watershed transformation is performed to identify the position of all the contours in the image. At the higher scales, the duality between the regional minima of the gradient and the catchment basins of the watershed is exploited to make a robust region based parent-child linking scheme. The linking process is applied using the approach proposed in [11], where the linking of the minima in successive scales is applied by using the *proximity* criterion. The linking process produces a linkage list for all the regions detected at the localization scale. Inherently, the latter yields also a linkage list for each adjacency (contour) in the localization scale; (iii) *Contour valuation by downward projection*. To each contour, we attribute a saliency measure comprising the scale-space lifetime and the dynamics of contours in scale-space [11] [10]. The latter requires two types of information: (i) the dynamics of contours (DC) [9] at each scale, and (ii) the deep image structure, which relates the contours and regions detected at the different scales. The dynamics of contours in scale-space are used to valuate the contours detected at the localization scale. Let $\mathcal{L}(a_i) = \{a_i^{(t_0)}, a_i^{(t_1)}, \ldots, a_i^{(t_a)}\}$ be the linkage list for the contour a_i, where t_o is the localization scale, and the scale t_a is last scale in which the contour was detected (annihilation scale). Hence, the dynamics of contours in scale-space (DCS) are defined as:

$$\text{DCS}(a_i) = \sum_{b \in \mathcal{L}_{(a_i)}} \text{DC}(b) \tag{1}$$

The second module (*Hierarchical Level Retrieval Module*) identifies the different hierarchical levels through a hypothesis testing criterion. Starting from the watershed segmentation at the localization scale, a successive merging operation is performed until a stopping criterion which is based upon a color similarity measure is satisfied. The merging sequence is given by the multiscale saliency measure of the contours.

The last module (*Segmentation Evaluation Module*) concerns the extraction of the most suitable hierarchical level for further processing. For this purpose, we employ a criterion that is based on a measure that yields a global evaluation of the contrast between the segments and the segment uniformity: Contrast-Homogeneity criterion (*CH*) [14]. Additionally, we add a constraint on the amount of required segments. This allows us to exclude over-and under-segmented hierarchical levels.

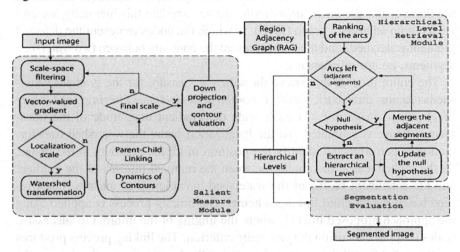

Figure 1. Schematic diagram for the automatic multi-scale segmentation scheme for vector-valued images.

Region features

After the image segmentation, a set of regions have been obtained. For each single segmented region we compute a set of features based mainly on color and spatial characteristics. Although, texture features are not explicitly part of the feature set, we have included texture information during region weighting, as it will be explained in Section 3. Finally, we have not used geometric properties since an image segmentation does not always provide a single region for each object in the image, and therefore, it is meaningless to compute representative shape features from such regions.

The color space that we use is the RGB color space. Although, it does not provide the color compaction of YCrCb and YIQ color space, neither the perceptual significance of Lab and YUV, our experimental results show a relative very good performance for retrieving. Other researchers in the area have confirmed our conclusions [5] [3].

Let R_i be a region in the segmented set $\{R\}$ with a set of adjacent regions $\{N(R_i)\}$. In our feature set, we do not only characterize each single region R_i but we also characterize its neighborhood by computing relational features. More specifically, the features we compute are the following :

• *mean Color component*

$$\mu C_k(R_i) = \frac{\sum_{j=1}^{A(R_i)} C_k(x_j, y_j)}{A(R_i)} \tag{2}$$

• *Area-weighted adjacent Color component*

$$\mu Adj C_k(R_i) = \frac{\sum_{j=1}^{Card(N(R_i))} A(R_j) * \mu C_k(R_j)}{\sum_{j=1}^{Card(N(R_i))} A(R_j)} \tag{3}$$

• *Area-weighted adjacent region contrast*

$$\mu Con(R_i) = \frac{\sum_{j=1}^{Card(N(R_i))} A(R_j) * (\| \mu C_k(R_i) - \mu C_k(R_j) \|)}{\sum_{j=1}^{Card(N(R_i))} A(R_j)} \tag{4}$$

• *Region geometric centroid*

$$G(R_i; \overline{x}, \overline{y}) = (\frac{\sum_{i=1}^{A(R_i)} x_i}{A(R_i)}, \frac{\sum_{i=1}^{A(R_i)} y_i}{A(R_i)}) \tag{5}$$

where C_k denotes the kth color component value with $k \in \{R, G, B\}$, $A(R_i)$ denotes the area of Region R_i, $Card(N(R_i))$ denotes cardinality of region's R_i neighborhood and (x_j, y_j) denotes the coordinates of a pixel that belongs to region R_j

3.   Image retrieval

Image Similarity Measure

The Earth Mover's Distance (EMD) [12] is originally introduced as a flexible similarity measure between multidimensional distributions. Intuitively, it measures the minimal cost that must be paid to transform one distribution into another. The EMD is based on the transportation problem and can be solved efficiently by linear optimization algorithms that take advantage of its special structure. Considering that EMD matches perceptually similarity well and can operate on variable-length representations of the distributions, it is suitable for region-based image similarity measure. Further motivation to use this distance as a similarity metric is based on the following properties of EMD : (i) It incorporates information from all segmented regions by allowing many-to-many relationship of the regions, thus information about an image is fully utilised; (ii) It is robust to inaccurate segmentation. For example, if a region is split into smaller ones, EMD will consider them as similar in the case that their distance is small and (iii) Concerning computational aspects, it is very efficient since the computational burden depends on the number of significant clusters rather than the dimension of the underlying feature space.

Formally, let $Q = \{(\mathbf{q}_1, w_{q_1}), (\mathbf{q}_2, w_{q_2}), \dots, (\mathbf{q}_m, w_{q_m})\}$ be the query image with m regions and $T = \{(\mathbf{t}_1, w_{t_1}), (\mathbf{t}_2, w_{t_2}), \dots, (\mathbf{t}_n, w_{t_n})\}$ be another image of the database with n regions, where $\mathbf{q}_i, \mathbf{t}_i$ denote the region feature set and w_{q_i}, w_{t_i} denote the corresponding weight of the region. Also, let $d(\mathbf{q}_i, \mathbf{t}_j)$ be the ground distance between $\mathbf{q}_i$ and $\mathbf{t}_j$. The EMD between Q and T is then:

$$EMD(Q,T) = \frac{\sum_{i=1}^m \sum_{j=1}^n f_{ij} d(\mathbf{q}_i, \mathbf{t}_j)}{\sum_{i=1}^m \sum_{j=1}^n f_{ij}} \tag{6}$$

where f_{ij} is the optimal admissible flow from $\mathbf{q}_i$ to $\mathbf{t}_j$ that minimizes the numerator of (6) subject to the following constraints:

$$\sum_{j=1}^n f_{ij} \le w_{q_i}, \sum_{i=1}^m f_{ij} \le w_{t_j}$$

$$\sum_{i=1}^m \sum_{j=1}^n f_{ij} = min(\sum_{i=1}^m w_{q_i}, \sum_{j=1}^n w_{t_j})$$

In the proposed approach, we define the ground distance as follows:

$$d(\mathbf{q}_i, \mathbf{t}_j) = (\sum_{k=1}^3 (\triangle \mu C_k)^2 + \sum_{k=1}^3 (\triangle \mu Adj C_k)^2 + (\triangle \mu Con)^2 + \beta(\triangle G(i; \overline{x}))^2 + \beta(\triangle G(i; \overline{y}))^2)^{\frac{1}{2}} \tag{7}$$

where β is a weighting parameter that enhances the importance of region's position relative to the remaining features.

Region weighting

We would like to identify and consequently, to attribute an importance in the regions produced by the selected segmentation scheme. Formally, we have to valuate the weighting factors w_{q_i} and w_{t_j} in Eq.(7). Most region-based approaches [16] [4] relate importance with the area size of a region. The larger the area is, the most important becomes the region. In our approach, we define an enhanced weighting factor which combines area with scale and global contrast in color/texture feature space, which can be all expressed by the valuation of dynamics of contours in scale-space (Eq. 1). More precisely, the weighting factor can be computed in the following way :

$$w_{q_i} = \frac{w_{DCS_i} * A(R_i)}{\sum_{j=1}^{Card(R)} w_{DCS_i} * A(R_i)} \tag{8}$$

$$w_{DCS_i} = \frac{\sum_{j=1}^{Card(N(R_i))} (\max DCS(\alpha_\kappa))}{Card(N(R_i))} \tag{9}$$

The term α_κ in Eq. 9 denotes the common border of two adjacent regions at the localization scale. One may observe that in Eq. 9, we compute the maximum value among the dynamics of contours in scale-space for each adjacency. This occurs because our final partitioning corresponds to a hierarchical segmentation level wherein a merging process has been applied. Due to merging, any common contour at the final partitioning may contain either a single or a set of contours which correspond to the localization scale.

4. Experimental results

The proposed strategy for content-based image retrieval has been evaluated with a general-purpose image database of 600 images from the Corel photo galleries that contain 6 categories (100 images per category). Evaluation is performed using precision versus recall (P/R) curves. Precision is the ratio of the number of relevant images to the number of retrieved images. Recall is the ratio of the number of relevant images to the total number of relevant images that exist in the database. They are defined as follows :

$$Precision(A) = \frac{R_a}{A}, Recall(A) = \frac{R_a}{S} \tag{10}$$

where A denotes the number of images shown to the user (the answer set), S denotes the number of images that belong to the class of the query and R_a denotes the number of relevant matches among A. To be objective, we have used

10 different queries for each category and we have averaged the precision/re-call values for each answer set. For comparison, we have tested our approach, denoted as "EMD hWSH", with two other approaches. All three approaches use as similarity metric the Earth Mover's Distance (EMD) which is adapted to the underlying feature set of each method. In Figure 2, we present representa-tive segmentation results of the corresponding partitioning methods which are discussed in our paper. The first approach is based on a k-means clustering [7] in the RGB color space which feeds the EMD with the produced distributions. In the presented (P/R) curves (Fig. 3), this approach appears as "EMD RGB". The second approach for comparison, that appears as "EMD JSEG", produces an image segmentation using the state-of-the-art JSEG algorithm [17]. The JSEG algorithm has been used before at the NeTRA CBIR system [8]. For each produced region we compute the feature set that is described in Section 1. We would like to notify that for "EMD JSEG", we compute region weights by taking into account the area of the region only. Furthermore, we have used optimal parameters for JSEG algorithm after a testing on a training set. In the produced P/R curves (Fig. 3), we can observe that both "EMD JSEG" and "EMD hWSH" (the proposed scheme) outperform the "EMD RGB". Each of methods, "EMD JSEG" and "EMD hWSH", has a very good performance after a severe testing of using 10 different queries. Examining the absolute values of the produced P/R curves for "EMD JSEG" and "EMD hWSH", we may draw the conclusion that the proposed strategy provides hiqh quality image retrievals. Furthermore, we may observe that although, JSEG provides worse quality segmentations than the proposed segmentation approach (see Figure 2), there is a sharing in marginal superiority of the one versus the other. At this point, we may note that although the final results do not provide an absolute superiority when using the proposed segmentation scheme, we have to seri-ously take into account that our proposed segmentation scheme is completely automatic while the JSEG algorithm is bound to parameter tuning for better performance. The general conclusion drawn upon this research is that the pro-posed CBIR strategy which does not require any user supervision can support image retrieval in a robust way. An enhancement of the feature set with ex-plicit texture information will further improve the current state of the proposed scheme.

(a)   (b)   (c)

Figure 2. Representative segmentation results using (a) k-means algorithm; (b) JSEG algorithm ; (c) proposed segmentation scheme

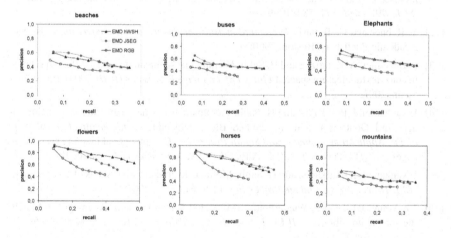

Figure 3. Precision / recall curves

References

[1] C. Carson, S. Belongie, H. Greenspan, and J. Malik. Blobworld: Image segmentation using E-M and its application to image querying. *IEEE Transactions on Pattern Analysis and Machine Intelligence*, 24:1026–1038, 2002.

[2] C-S Fuh, S-W Cho, and K. Essig. Hierarchical color image region segmentation for content-based image retrieval system. *IEEE Transactions on Image Processing*, 9(1):156–162, 2000.

[3] T. Gevers. Image segmentation and similarity of color-texture objects. *IEEE Transactions on Multimedia*, 4(4):509–516, 2002.

[4] H. Greenspan, G. Dvir, and Y. Rubner. Context-dependent segmentation and matching in image databases. *Computer Vision and Image Understanding*, 93:86–109, 2004.

[5] J-W. Hsieh and E. Grimson. Spatial template extraction for image retrieval by region matching. *IEEE Transactions on Image Processing*, 12(11):1404–1415, 2003.

[6] F. Jing, M. Li, H-J Zhang, and B. Zhang. An efficient and effective region-based image retrieval framework. *IEEE Transactions on Image Processing*, 13(5):699–709, 2004.

[7] T. Kanungo, D. Mount, C.D. Piatko N.S. Netanyahu, R. Silverman, and A.Y. Wu. An efficient k-means clustering algorithm: Analysis and implementation. *IEEE Transactions on Pattern Analysis and Machine Intelligence*, 24(7):881–892, 2002.

[8] W. Ma and B. Manjunath. NeTra: A toolbox for navigating large image databases. In *Proc. IEEE Int'l Conference Image Processing*, pages 568–571, 1997.

[9] L. Najman and M. Schmitt. Geodesic saliency of watershed contours and hierarchical segmentation. *IEEE Transactions on Pattern Analysis and Machine Intelligence*, 18(12):1163–1173, 1996.

[10] I. Pratikakis. *Watershed-driven image segmentation*. PhD thesis, Vrije Universiteit Brussel, 1998.

[11] I. Pratikakis, H. Sahli, and J. Cornelis. Hierarchical segmentaion using dynamics of multiscale gradient watersheds. In *11th Scandinavian Conference on Image Analysis (SCIA 99)*, pages 577–584, 1999.

[12] Y. Rubner and C. Tomasi. *Perceptual metrics for image database navigation*. Kluwer Academic Publishers, Boston, 2000.

[13] I. Vanhamel, A. Katartzis, and H. Sahli. Hierarchical segmentation via a diffusion scheme in color-texture feature space. In *Int. Conf. on Image Processing (ICIP-2003)*, Barcelona-Spain, 2003.

[14] I. Vanhamel, I. Pratikakis, and H. Sahli. Automatic watershed segmentation of color images. In J. Goutsias, L. Vincent, and D.S. Bloomberg, editors, *Mathematical Morphology and its Applications to Image and Signal Processing*, Computational imaging and vision, pages 207–214, Parc-Xerox, Palo Alto, CA-USA, 2000. Kluwer Academic Press.

[15] I. Vanhamel, I. Pratikakis, and H. Sahli. Multi-scale gradient watersheds of color images. *IEEE Transactions on Image Processing*, 12(6):617–626, 2003.

[16] J.Z. Wang, J. Li, and G. Wiederhold. SIMPLIcity: Semantics-Sensitive integrated Matching for picture libraries. *IEEE Transactions on Pattern Analysis and Machine Intelligence*, 23(9):947–963, 2001.

[17] Y.Deng and B.S.Manjunath. Unsupervised segmentation of color-texture regions in images and video. *IEEE Transactions on Pattern Analysis and Machine Intelligence*, 23(8):800–810, 2001.

EFFICIENT IMPLEMENTATION OF THE LOCALLY CONSTRAINED WATERSHED TRANSFORM AND SEEDED REGION GROWING

Richard Beare

CSIRO MIS
Locked Bag 17, North Ryde, Australia, 1670
Richard.Beare@csiro.au

Abstract   The watershed transform and seeded region growing are well known tools for image segmentation. They are members of a class of greedy region growing algorithms that are simple, fast and largely parameter free. The main control over these algorithms come from the selection of the marker image, which defines the number of regions and a starting position for each region.

Recently a number of alternative region segmentation approaches have been introduced that allow other types of constraints to be imposed on growing regions, such as limitations on border curvature. Examples of this type of algorithm include the geodesic active contour and classical PDEs.

This paper introduces an approach that allows similar sorts of border constraints to be applied to the watershed transform and seeded region growing. These constraints are imposed at all stages of the growing process and can therefore be used to restrict region leakage.

Keywords: watershed transform, region growing, constrained regions.

Introduction

Image segmentation aims to partition images into a number of disjoint regions according to some criterion, like color, edges or texture. The watershed transform [5] is a popular tool for performing region based segmentation of images. It is fast, flexible and parameter free. Seeded region growing [1] is a closely related approach that is usually applied to the raw image rather than the gradient image. Prior knowledge is usually provided to the watershed transform and seeded region growing algorithms by using a marker image [9] which defines the number of regions and the starting points for the growing process. However it is sometimes desirable to be able to impose additional constraints.

This paper will introduce a mechanism that can easily be included in the well known *hill-climbing* watershed transform implementation and the closely

C. Ronse et al. (eds.), Mathematical Morphology: 40 Years On, 217–226.
©2005 *Springer. Printed in the Netherlands.*

related seeded region growing implementation. The modification allows constraints to be applied to the curvature of region borders at all stages of the growing process. This makes some of the useful properties of other region segmentation approaches, like geodesic active contours and classical PDEs, available in more traditional region based segmentation frameworks. The modified algorithms are called *locally constrained* watershed transform and *locally constrained* seeded region growing.

Cost based frameworks for the modification have been developed elsewhere [4]. This paper will develop the modification from the point of view of a physical model and an efficient implementation of the algorithm.

The paper is structured as follows. Sections 1 and 2 introduce the watershed transform and previous work on constrained region growing. Sections 3 and 4 introduce the leakage problem that we are trying to correct and the physical model we are using to address it. Implementation, results and performance are discussed in Sections 5, 6 and 7.

1. Brief history of the watershed transform

A detailed description of the watershed transform's heritage is given in [12, 7]. Only a brief summary will be given here.

The watershed transform was first proposed as a segmentation method that modeled the progressive immersion of a topographical relief (an image) in a fluid [5]. Each regional minimum [1] in the surface corresponds to a different lake. Neighboring lakes meet at watershed lines as the level of flooding increases. Flooding continues until the entire relief is immersed.

An algorithmic definition of this model that allowed an efficient implementation employing priority queues was proposed by Vincent and Soille [14]. The algorithm defined a recursive relationship between gray levels of the image.

Meyer [8] defined the watershed in terms of a distance function called the topographical distance. This distance was defined in terms of the *lower slope* of *lower complete* images. images).

The catchment basin of a regional minimum is the set of pixels that are closer (in terms of the sum of value of regional minimum and the topographical distance) to that regional minimum than any other. The topographical distance watershed is the complement of the union of catchment basins.

These definitions produce a cost of zero in flat zones of an image, leading to a watershed that may be thicker than one pixel on a plateau. The usual solution to this problem is to transform images so that they are *lower complete*. This guarantees that there are no plateaus and that the watershed zones are thin.

More recent work by Nguyen, Worring and van den Boomgaard builds on the topographical distance framework and establishes the relationship between the watershed transform and energy-based minimization [11].

2. Previous work on constrained region based segmentation

A number of region segmentation techniques are able to include boundary constraints. In some cases the boundary constraints are essential for the sensible operation of the algorithms in real images.

Energy minimization based methods of region segmentation, such as those using classical PDEs [13], are able to constrain border curvature. This is done via a viscosity term in the energy function that modifies the rate of curve evolution. Careful selection of this term is often a critical factor in practical applications of these methods.

There has also been some recent work on other types of boundary constraints in traditional region growing contexts. One method modifies the image topology of the image using a viscous closing and then applies a traditional watershed transform [10]. The viscous closing is based on a geophysical model in which a fluid is subjected to a variety of pressures. Increasing the pressure decreases the viscosity of the fluid which is modeled by decreasing the size of the structuring element. A second method models the growing region as a polygon with the maximum edge length as the controlling parameter [3]. This algorithm is queue based, with polygon corners being placed on the queue.

A technique called *watersnakes* has also been described recently [11]. This work demonstrates the energy minimization nature of the conventional watershed transform and includes border related terms explicitly in the form of an approximation of local boundary length. This differs from the method described in this paper in which border constraints are imposed implicitly by the cost function.

3. Region leakage and segmentation stability

The aim of this work is to provide a mechanism for higher level knowledge to be included in a conventional region growing framework. The particular higher level knowledge we are interested in is the requirement that borders of regions should be, in some sense, smooth at all times during the growing process. This should make it possible to stabilize the growing process by preventing region leaks.

Let us consider how leaks may occur in the context of watershed transform from the point of view of the physical model from which it was derived. The physical model views an image as a terrain being progressively flooded by a fluid. Fluid enters through each marker and begins creating a lake. Adjacent lakes will eventually touch along ridges in the terrain, to form watershed lines. In the case of plateaus the lakes should meet halfway between the markers. Figure 1 illustrates a typical watershed scenario. In this example we are trying to segment a circular object (e.g. a cell). In the ideal case a pair of markers, one inside and one outside the object will result in the correct segmentation. In less ideal circumstances there may be breaks in the circle. In such cases the resulting segmentation will become strongly dependent on the placement of markers. In this example the external marker is not near the break and the watershed lines depart significantly from the circular contour.

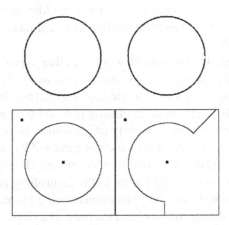

Figure 1.   Segmentation of a complete and incomplete circular object. The incomplete object has a small break on the right hand side. The lower images show markers and watershed lines.

It is important to note that the leak of the interior region through the gap in the circle is not "wrong" from the point of view of the definition of the watershed transform [2]. However it is not usually the result that would be desired. In this particular case we are actually aiming to create a closed object from one that isn't closed – i.e. part of the contour which we want the watershed to find is perceptual.

Obviously there are a number of ways of attacking this particular problem – we could use an approach designed to find closed contours [2] or could prefilter the image to close the contour. Geodesic active contours would also perform satisfactorily in this case.

The alternative we are examining in this paper is to impose a constraint on the boundary of the region at all stages of the growing process. In this example the nature of the boundary during the period in which it "escapes" from the object is obviously different to the result we finally expect – it will be much more highly curved at that time. Imposing a constraint that discourages high curvature should therefore help prevent this kind of problem.

4. Physical model

Classical PDEs include a viscosity term that imposes constraints that are very similar to the sort of thing we are after. Similar constraints have been introduced to the watershed transform [11], but these are not easily implemented using some of the standard approaches.

In this work our model is of flooding by a fluid represented by overlapping digital disks[3]. In the extreme case, where the disk diameter is a single pixel, the model is equivalent to the standard watershed implementation. If the diameter is greater than one the region will constrained to be a union of the selected structuring element or, in morphological terms, to be a set that is opened by the structuring element.

A region produced using this model will be a dilation of a sub region by the structuring element. This concept is used in the implementation.

This model is easily able to support the a different sized structuring element for each region. This can be useful in some circumstances.

We can think of solving the problem described in the previous section by selecting a fluid that cannot easily fit through the gap in the circle.

5. Implementation

The algorithm used here to implement the watershed transform and seeded region growing belongs to the *hill-climbing* class of algorithms. The implementation of both seeded region growing and watershed transform using the hill-climbing approach is essentially the same. Both exploit a priority queue and propagate labels. The main difference between them is the way the priorities are calculated. The inputs to both are a marker image and a control image. The priority queue contains pixel locations and label values. Pixels with the same priority are served in FIFO order. The steps in the procedure are:

1 **Initialize queue.**

2 **Exit** if queue is empty.

3 Take the pixel from the head of the queue.

4 Label that location if it isn't already labeled. Otherwise **Goto** step 2.

5 Compute the priority for each unlabeled neighbor and insert into the queue.

6 **Goto** step 2.

Queue initialization involves adding all pixels that are neighbors of markers in the marker image to the priority queue with the appropriate priority. A number of additions to the steps above can be made in the name of efficiency, but they are omitted for brevity. Marking watershed lines also requires some extra steps.

The way in which priority is calculated is the only difference between the hill-climbing implementation of watershed transform and seeded region growing. In the watershed transform the priority is simply the pixel gray level while in seeded region growing it is the difference between the color of the labeled region and the color of the candidate pixel [4]. The more complex priority function used by seeded region growing requires additional book keeping to implement it. This complicates the main loop. Details will not be discussed here.

The extension to locally constrained versions of the algorithms also requires modifications to the priority function. The priority function is now related to a region. Implementation of this priority function requires careful book keeping and is difficult to describe. The procedure will be illustrated using a walk through of the algorithm on some artificial data.

Figures 2(a) to 2(c) illustrate the concept. A 3×3 structuring element is being used in this example for clarity. In practice we tend to use a digital disk structuring element.

The basic idea behind the implementation is to maintain a pair of label images. The first is called the *center-map* and the second is called the *cover-map*. Label regions in each of these images grow concurrently. Regions in the cover-map are dilations of the corresponding region in the center-map.

Figure 2(a) shows the center map. The region is growing upward and is indicated by the cross patterned pixels. The black pixel has just been popped from the queue and is being processed. Figure 2(b) is the dilation of the center map by the 3×3 structuring element (including the black pixel - the top row of pixels in Figure 2(b) are labeled after the black pixel is popped from the queue). Pixels labeled A, B, C, D and E are neighbors of the black pixel and are unlabeled in the center map. Priority values need to be computed for each of these pixels so that they can be added to the priority queue. Figure 2(c) shows some pixels with a scale pattern that are used to compute the priority of pixel B. These pixels $\{P\}$ are defined mathematically as

$$\{P\} = \{B \oplus S\} \setminus L \qquad (1)$$

where S is the structuring element and L is the cover map region.

The set of pixels $\{P\}$ can be used to define the priority in a number of ways, which will be discussed in the next section. Once the priority has been computed pixel B is placed on the queue and the process is repeated for the other neighbors. The algorithm completes when the queue is empty.

This is the core of the algorithm. The cover map contains the final segmentation result. There are a number of complexities relating to efficiency and detection of borders between different regions in the cover map. Those details would severely complicate the description.

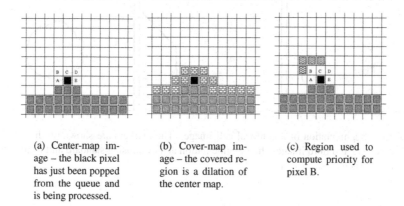

(a) Center-map image – the black pixel has just been popped from the queue and is being processed.

(b) Cover-map image – the covered region is a dilation of the center map.

(c) Region used to compute priority for pixel B.

Figure 2. Processing steps using center map and cover map (continued).

Details of priority function

The priority function for the locally constrained watershed transform should be an estimate of the gray level of the region defined by $\{P\}$. Obvious choices are the mean and median gray level, both of which may have advantages in different circumstances. Similar logic applies to the choice of priority function for seeded region growing.

The critical point to notice is that the priority is being computed using a set of pixels that belong to the border of the growing region, rather than a single pixel, and that the border pixels belong to a user defined structuring element.

This mechanism allows us to include border constraints, that are similar in character to those seen in geodesic active contour and classical PDE algorithms, in traditional region growing frameworks.

6. Real examples

Confocal cell image

Figure 3 shows a cell image taken with a confocal microscope. The image is a cross section through the cell and the wall of the cell is broken in many places. The standard watershed can easily leak through the broken cell wall, resulting in an undesirable segmentation. The locally constrained watershed, using a structuring element radius 5 for the cell, produces a much more useful segmentation that corresponds well to the cell outline.

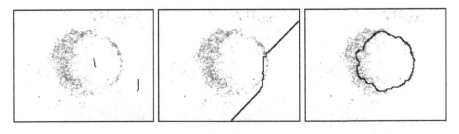

Figure 3.   Segmentation of a confocal cell image. The markers are shown on the first image. The second and third images show segmentation achieved using standard and locally constrained watersheds respectively.

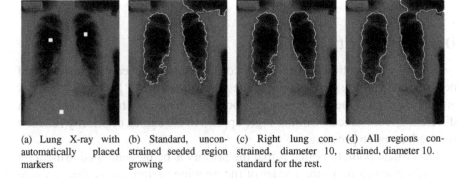

(a) Lung X-ray with automatically placed markers

(b) Standard, unconstrained seeded region growing

(c) Right lung constrained, diameter 10, standard for the rest.

(d) All regions constrained, diameter 10.

Figure 4.   Some segmentation results using seeded region growing.

Lung X-ray image

Figure 4 shows the result of a segmentation of a lung X-ray based on seeded region growing (SRG). SRG operates on the raw image rather than the gradient image and uses markers (shown in Figure 4(a)). These markers were found using the automated procedure based on the converging squares algorithm described in [1]. Figure 4(b) illustrates the results achieved by standard SRG, including a significant leakage through the faint region at the top of the right lung. Figure 4(c) shows the result that can be achieved when the lung is constrained and the body isn't. This is an example of the benefit of mixing constrained and unconstrained regions. If all regions are constrained (Figure 4(d)) the leak still occurs because the body region is unable to grow through the thin region near the top of the lung.

7. Performance

Execution times (using a mean based priority function) are shown below in Table 1. The times for the constrained watershed are slower than the standard watershed by factors of between 5 and 10. The standard watershed is implemented using the same framework and same queue structure (a splay queue as described in [6]). The constrained watershed executed with a single pixel structuring element produces the same result as the standard watershed, but takes nearly twice as long to execute. This is because some of the overheads associated with the use of constraints have an impact on performance even if the constraints are not being applied. These overheads include the step of looking up the constraint associated with the region currently being processed and the cost of calling the priority evaluation function. Neither of these steps is required by the standard watershed, leading to a much faster inner most loop.

Algorithm	circle radius	watershed line	time(s)	time ratio
standard watershed	NA	yes	0.44	1.0
constrained watershed	0	yes	0.7	1.59
constrained watershed	5	no	2.3	5.22
constrained watershed	5	yes	2.6	5.9
constrained watershed	10	no	2.8	6.36
constrained watershed	10	yes	3.15	7.15
constrained watershed	15	no	3.5	7.9
constrained watershed	15	yes	3.85	8.75

Table 1. Running times when applied to a 768×939 lung X-ray on a 1.7GHz Pentium 4 under Redhat Linux 7.2. The "time ratio" provides a comparison to the execution time of the standard watershed.

8. Conclusions

The report has introduced an efficient method of applying locally boundary constraints in a well known region growing framework. This advance makes some of the interesting properties of classical PDEs available to some greedy region growing approaches. The approaches described here offer another tool to stabilize some tricky segmentation problems.

Notes

1. A regional minimum is a connected set of pixels of equal value surrounded by pixels of higher value.

2. The final shape of the region illustrates artifacts of the non isotropic propagation across flat regions. This is a result of the implementation that is especially obvious in synthetic images.

3. In theory any structuring element can be used.

4. The color may be a vector quantity, which means that difference can be defined in a number of ways.

References

[1] R. Adams and L. Bischof. Seeded region growing. *IEEE Transactions on Pattern Analysis and Machine Intelligence*, 16:641–647, 1994.

[2] B. Appleton and H. Talbot. Globally optimal geodesic active contours. submitted to JMIV September 2002, revised October 2003.

[3] R. J. Beare. Regularized seeded region growing. In H. Talbot and R. Beare, editors, *Mathematical Morphology, Proceedings of the VIth International Symposium – ISMM 2002*, pages 91–99, 2002.

[4] R. J. Beare. A locally constrained watershed transform. Technical Report 04/187, CSIRO Mathematical and Information Sciences, 2004.

[5] S. Beucher and C. Lantuéjoul. Use of watersheds in contour detection. In *Int. Workshop on Image Processing*, Rennes, France, Sept. 1979. CCETT/IRISA.

[6] E. Breen and D. Monro. An evaluation of priority queues for mathematical morphology. In J. Serra and P. Soille, editors, *Mathematical morphology and its applications to image processing*, pages 249–256. Kluwer, 1994.

[7] A. Meijster. *Efficient Sequential and Parallel Algorithms for Morphological Image Processing*. PhD thesis, Institute for Mathematics and Computing Science, University of Groningen, 2004.

[8] F. Meyer. Topographic distance and watershed lines. *Signal Processing*, 38:113–125, 1994.

[9] F. Meyer and S. Beucher. Morphological segmentation. *Journal of Visual Communication and Image Representation*, 1(1):21–46, September 1990.

[10] F. Meyer and C. Vachier. Image segmentation based on viscous flooding simulation. In H. Talbot and R. Beare, editors, *Mathematical Morphology, Proceedings of the VIth International Symposium – ISMM 2002*, pages 69–77, 2002.

[11] H. Nguyen, M. Worring, and R. van den Boomgaard. Watersnakes: energy-driven watershed segmentation. *PAMI*, 25(3):330–342, 2003.

[12] J. B. T. M. Roerdink and A. Meijster. The watershed transform: definitions, algorithms, and parallelization strategies. *Fundamenta Informaticae*, 41:187–228, March 2000. ISBN 90-367-1977-1.

[13] J.A. Sethian. *Level Set Methods and Fast Marching Methods*. Cambridge University Press, 1999.

[14] L. Vincent and P. Soille. Watersheds in digital spaces: an efficient algorithm based on immersion simulations. *IEEE Transactions on Pattern Analysis and Machine Intelligence*, 13(6):583–598, June 1991.

IV

GEOMETRY AND TOPOLOGY

OPTIMAL SHAPE AND INCLUSION

open problems

Jean-Marc Chassery[1] and David Coeurjolly[2]

[1]*Laboratoire LIS, CNRS UMR 5083*
961, rue de la Houille Blanche - BP46,
F-38402 St Martin d'Hères, France
Jean-marc.Chassery@lis.inpg.fr

[2]*Laboratoire LIRIS, CNRS FRE 2672*
Université Claude Bernard Lyon 1,
43, Bd du 11 novembre 1918,
F-69622 Villeurbanne, France
david.coeurjolly@liris.cnrs.fr

Abstract   Access to the shape by its exterior is solved using convex hull. Many algorithms have been proposed in that way. This contribution addresses the open problem of the access of the shape by its interior also called convex skull. More precisely, we present approaches in discrete case. Furthermore, a simple algorithm to approximate the maximum convex subset of star-shaped polygons is described.

Keywords: shape approximation, convex hull, convex skull, potato peeling.

Introduction

In digital image processing we are often concerned with developing specialized algorithms that are dealing with the manipulation of shapes. A very classical and widely studied approach is the computation of the convex hull of an object. However, most of these studies focus only on exterior approaches for the computation of convexity, i.e., they are looking for the smallest convex set of points including a given shape. This contribution addresses the problem of the access of the shape by its interior. The computation of the best shape according to a criterion included in a given one has been studied in many few occasions in the continuous case involving convex skull and potato peeling.

In this article, we present successive configurations of fast approximation of the maximal convex subset of a polygon. First a discrete approach will illustrate an iterative process based on shrinking and convex hull. Second a

C. Ronse et al. (eds.), Mathematical Morphology: 40 Years On, 229–248.

region based approach will be proposed in specific case where the criterion is maximal horizontal-vertical convexity. Third study will be focused to the family of star-shaped polygons P. A simple algorithm extracts the maximal convex subset in $O(k \cdot n)$ in the worst case if n is the size of P and k its number of reflex points.

In section 1, we list classical problems of shape approximation. In section 2, we present the discrete approach followed by h-v convexity in section 3. In section 4, we introduce the Chang and Yap's optimal solution definitions that will be used in the rest of the presentation. In section 5, we present the proposed algorithm based on classical and simple geometric tools for star-shaped polygons. Finally, experiments are given.

1. Shape approximations

In this paper shapes are delimited by polygonal boundary. We suppose that polygons are simple in the sense that they are self-avoiding. Polygon inclusion problems are defined as follows: given a non-convex polygon, how to extract the maximum area subset included in that polygon ? In [7], Goodman calls this problem the *potato-peeling problem*. More generally, Chang and Yap [3] define the polygon *inclusion* problem class $Inc(\mathcal{P}, \mathcal{Q}, \mu)$: given a general polygon $P \in \mathcal{P}$, find the μ-largest $Q \in \mathcal{Q}$ contained in P, where $\mathcal{P}$ is a family of polygons, $\mathcal{Q}$ the set of solutions and μ a real function on $\mathcal{Q}$ elements such that

$$\forall Q' \in \mathcal{Q}, \quad Q' \subseteq Q \Rightarrow \mu(Q') \leq \mu(Q). \tag{1}$$

The maximum area convex subset is an inclusion problem where $\mathcal{Q}$ is the family of convex sets and μ gives the area of a solution Q in $\mathcal{Q}$. The inclusion problem arises in many applications where a quick internal approximation of the shape is needed [2, 5].

In a dual concept, we can mention enclosure problem presented as follows: given a non-convex polygon, how to extract the minimum area subset including that polygon ? We can define the polygon *enclosure* problem class $Enc(\mathcal{P}, \mathcal{Q}, \mu)$: given a general polygon $P \in \mathcal{P}$, find the μ-smallest $Q \in \mathcal{Q}$ containing P, where $\mathcal{P}$ is a family of polygons, $\mathcal{Q}$ the set of solutions and μ a real function on $\mathcal{Q}$ elements such that

$$\forall Q' \in \mathcal{Q}, \quad Q' \supseteq Q \Rightarrow \mu(Q') \geq \mu(Q). \tag{2}$$

Examples

The following list corresponds to examples of various situations depending on specifications of the $\mathcal{P}$ family, the $\mathcal{Q}$ family and the μ measure.

Example 1. $\mathcal{P}$ is a family of simple polygons,

$\mathcal{Q}$ is the family of convex polygons,

μ is the area measure,

$Enc(\mathcal{P}, \mathcal{Q}, \mu)$ is the problem of convex hull

Example 2. $\mathcal{P}$ is a family of simple polygons,

$\mathcal{Q}$ is the family of convex polygons,

μ is the area measure,

$Inc(\mathcal{P}, \mathcal{Q}, \mu)$ is the problem of potatoe pealing [7, 3]

Example 3. $\mathcal{P}$ is a family of n-sided convex polygons,

$\mathcal{Q}$ is the family of triangles,

μ is the area measure,

$Inc(\mathcal{P}, \mathcal{Q}, \mu)$ has a solution with complexity in O(n)

Example 4. [9] $\mathcal{P}$ is a family of n-sided simple polygons,

$\mathcal{Q}$ is the family of triangles,

μ is the area measure,

$Enc(\mathcal{P}, \mathcal{Q}, \mu)$ has a solution with complexity in O($nlog^2$n) O($nlog^2$n)

Example 5. [8] $\mathcal{P}$ is a family of n-sided simple polygons,

$\mathcal{Q}$ is the family of rectangles,

μ is the area measure,

$Enc(\mathcal{P}, \mathcal{Q}, \mu)$ has a solution with complexity in O(n) in 2D, O(n^3) in 3D.

Example 6. $\mathcal{P}$ is a family of n-sided orthogonal polygons,

$\mathcal{Q}$ is the family of convex orthogonal polygons,

μ is the area measure,

$Inc(\mathcal{P}, \mathcal{Q}, \mu)$ has a solution with complexity in $O(n^2)$ [12].

In the following we will focus on the problem 2.

2. Discrete approach

Let us consider a discrete shape defined as a connected set of pixels. The objective is to find a maximal convex set included in that discrete shape [4] .

The proposed algorithm is iterative, starting from the initial shape. Each iteration is composed of a shrinking and computation of the convex hull of the shrinked shape. If the convex hull is not included in the initial shape, we reiterate the shrinking process. Else we detect the side of the convex hull which contains concavity vertex of the initial connected component. We cut the initial component along this side and restart the algorithm with this new shape for initialization. The algorithm stops when the modified initial shape is convex. Figure 1 illustrates the different steps of this algorithm.

Figure 2 illustrates the result of this algorithm on various examples.

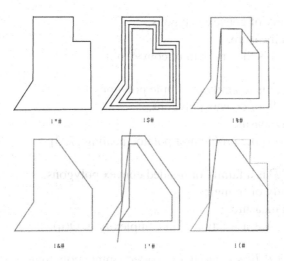

Figure 1. Iterative process by shrinking and convex hull computation: (a) initial shape, (b) successive shrinking, (c) the convex hull of the shrinked object is included in the initial shape, (d) cut of the initial shape for a second pass, (e) the convex hull of the shrinked object is included in the initial shape, (f) maximal solution.

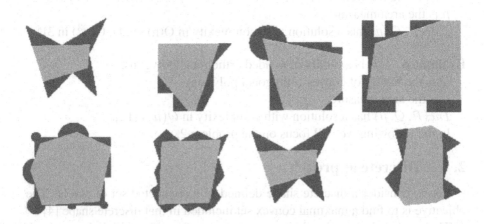

Figure 2. Illustration on different shapes: initial shape and result are superimposed.

In that case the μ measure is not the area but the Min-max distance between the boundaries of the initial shape and the solution. This corresponds to the Hausdorff distance between the two shapes P and Q.

$$\mu(P, Q) = -\max(\sup_{M \in P} \inf_{N \in Q} d(M, N), \sup_{M \in Q} \inf_{N \in P} d(M, N))$$

where d is the digital distance based on 4 or 8 connectivity.

3.   Limitation to h-v convexity

In that section we limit our study to the particular case of horizontal-vertical convexity noted by h-v convexity and illustrated on figure 3.

The search of the maximum ortho-convex polygon included into a simple orthogonal polygon has been solved in the continuous case by Wood and Yap with complexity in $O(n^2)$ [12].

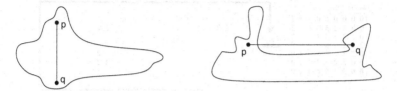

Figure 3.   h-v convexity: In left the shape is h-v convex, in right the shape is not h-v convex..

We will briefly present a method in the discrete case. This work is issued from an Erasmus project and a detailed version is in the students' report (see http://www.suerge.de/s/about/shape/shape.ps.gz).

Let us consider a given shape as a set of 4-connected component . The principle of this method is to assign a weight to all the pixels in the shape and then to partition the shape into h-v convex regions according to these weights.

Basically, the entire process consists of four steps which are: - *Horizontal/Vertical Sweeping for weight assignement* and - *Labeling of the shape into regions* and - *Construction of Adjacency and Incompatibility graphs between the labeled regions* and - *Finding the solution(s)*.

Illustration of these steps is given on the following example.

The weight being attached to each shape pixel indicates the number of sections (h-section and v-section) that include this pixel. The minimum weight of any shape pixel is equal to two. Figure 4 illustrates the weights assignation.

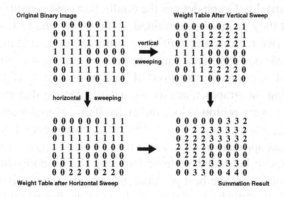

Figure 4.   Examples that simulate how the weights are being assigned.

A region is a maximal connected set of pixels having the same weight. The labeling process provides a different label to each region. It needs two steps:

1. Extract all the pixels with a same weight and generate a weighted image.
2. Perform labeling on this weighted image.

These two steps will be performed iteratively until all the different weights have been labeled (figure 5).

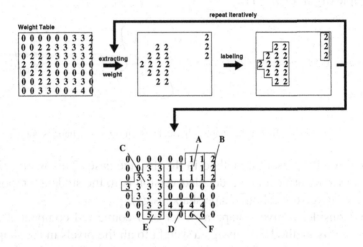

Figure 5.   An example that illustrates the labeling process.

Next step will illustrate relationships between regions in term of coexistence in a hv-convex set. Two graphs are introduce in which regions are vertices: Adjacency Graph and Incompatibility Graph.

The Adjacency Graph defines the connectivity between the different regions. It indicates that they can be merged to find the biggest hv-convex set. Two regions are adjacent to each other and their respective nodes will be linked if there is at least one pixel of each region neighbouring each other (figure 6).

The Incompatibility Graph defines the confliction between different regions. It indicates that they cannot be combined at the same time to find the biggest hv-convex set. Two regions are conflicting with each other and their respective nodes will be linked if there is at least one pixel of each region separated from each other by a h-section (or v-section) of background pixels (figure 6).

Before starting the graphs analysis we have to realise that every region is a solution, since every region is hv-convex itself and is included in the shape. However finding just a solution is not sufficient, since we are interested in the maximum or even optimal solution. A maximum solution is a solution in which no other region can be added without violating the hv-convexity constraint. The optimum solution is the best possible maximum solution. It does not have to be unique. To find the best solution we have to determine all the maximum solutions issued from the following steps.

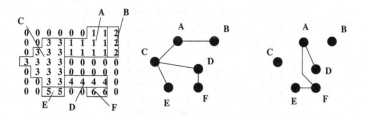

Figure 6. Adjacency (middle) and Incompatibility (right) graphs from the labeled shape (left).

1. Pick a region and put this in an initially empty set (let us call this set A).

2. Construct two lists, one called the adjacency list and the other one called the incompatibility list. These lists contain the labels of the regions which are respectively adjacent and conflicting with the regions in set A. If there is a region, which is in the adjacency list (and not in the incompatibility list), then add this to set A and update both lists. When this cannot be done anymore, we have found a maximum solution, since we do not have anymore candidates on the adjacency list.

3. Backtrack is needed to solve the problem of choice in the adjacency list. A region, which has not been picked before, should be placed into set A.

4. At the moment that all possibilities have been tried, we can change the initial region with which we started the search and go back to step 2 until every region has been the initial region of the set A.

If at every step the current solution is compared with the best solution so far, then at the end of the algorithm one of the optimal solutions is found.

The illustration in figure 7 shows the construction of maximal solutions by traversing the graphs of adjacency an incompatibility.

4. Preliminaries and Exact Polynomial Solution

For the rest of the presentation, we consider a polygon $P = (v_0, v_1, \ldots, v_{n-1})$ with n vertices. We denote by $R = (r_0, r_1, \ldots, r_{k-1})$ the k reflex vertices (or concave vertices) of P (maybe empty). The potato-peeling problem can be expressed as follows,

PROBLEM 1 *Find the maximum area convex subset (MACS for short) Q contained in P.*

In [7], Goodman proves that Q is a convex polygon.

He presents explicit solutions for $n \leq 5$ and leaves the problem unsolved in the general case.

In [3], Chang and Yap prove that the potato-peeling problem can be solved in polynomial time in the general case. More precisely, they detail an $O(n^7)$ time algorithm to extract Q from P. Since this algorithm uses complex geometric

Possible Solutions	Adjacency List	Incompatibility List
A	B C	D F
A B	C	D F
A B C	E	D F
A B C E ← Maximum Solution	D F
A C	B E	D F
A B C	E	D F
A B C E ← Maximum Solution	D F
A C E	B	D F
A B C E ← Maximum Solution	D F
. . . .	.	. .
C	A D E	
C D	E F	A
C D E ← Maximum Solution	A F
C D F ← Maximum Solution	A E
C A	B E	D F
C A B	E	D F
C A B E ← Maximum Solution	D F
. . . .	. .	. .

Figure 7. Table showing how to traverse the graphs of adjacency and incompatibility to get maximal solutions. ABCE is the optimal solution.

concepts and dynamic programming in several key steps, it is not tractable in practical applications.

Let us present some elements of the Chang and Yap's algorithm. First of all, we define a *chord* of P by a maximal segment fully contained in P. A chord is said to be *extremal* if it contains two or more vertices of P. In particular, an edge of P is always enclosed in an extremal chord. Let $C_1, C_2, \ldots, C_m$ be chords of P with $m \leq k$ such that each C_i passing through reflex vertices of P. We first consider the chords going through a unique reflex point. Let us associate to each such a chord C_i passing through u in R, the closed half-plane C_i^+ defined by C_i and such that at least one or two adjacent vertices to u does not belong to C_i^+. If C_i passes through more than one reflex vertex, the choice of the half-plane can be made in similar ways (see figure 8). They prove that the maximum area convex polygon Q is given by the intersection of P and a set of half-planes defined by a set of so-called optimal chords (see [3]). Hence, to solve the potato-peeling problem, we have to find the appropriate set of optimal chords associated to the reflex vertices.

If P has only one reflex vertex u, the optimal chord C_u that leads to the MACS Q can be easily found. First of all, Chang and Yap [3] define a *butterfly* as a sequence of points $[b', a, u, b, a']$ such that a, u and b are consecutive vertices in P with a and b adjacent vertices of u in P, and such that both $[a, u, a']$ and $[b, u, b']$ are extremal chords (see figure 9). Furthermore, the chord $[c, u, c']$ is said to be *balanced* if we have $|cu| = |uc'|$ (c and c' belonging to P). Based

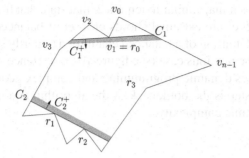

Figure 8. Notations and illustrations of chords and half-planes generated by these chords.

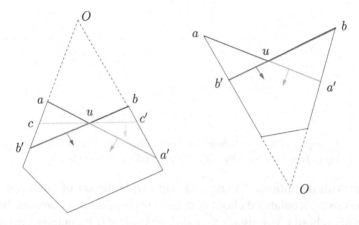

Figure 9. One reflex vertex case: an A-butterfly (left) and a V-butterfly (right). According to lemma 2, the optimal chord of each polygon is one of the gray segments.

on these definitions, two kinds of butterflies exist according to the position of the intersection O between the straight lines $(b'a)$ and (ba'), and the polygon. More precisely, a *A-butterfly* is such that the points b', a and O (or O, b and a') appear in this order (see figure 9-(left)) in the straight line (ab) (resp. $(a'b')$). Otherwise, it is called a *V-butterfly* (see figure 9-(right)). In this one reflex corner case, Chang and Yap prove the following lemma:

LEMMA 2 (BUTTERFLY LEMMA [3]) *Given a butterfly B and its optimal chord C_u, if B is a V-butterfly then C_u is an extremal chord. If B is a A-butterfly, C_u is either extremal or balanced.*

We notice that in case of V-butterfly we have two possible chords, in case of A-butterfly, three choices are possible. This lemma leads to a linear in time solution to the potato-peeling problem if $k = 1$.

In a general case, Chang and Yap [3] define other geometric objects such as series of A- and V-butterflies. Based on these definitions, they present an

A-lemma and a V-lemma, similar to lemma 2, that state that the optimal chords for a set of butterflies are extremal chords or a set of balanced chords. In the general case, the definition of a balanced chain for a butterfly sequence is more complex than in the previous case (see figure 10-(a)). Hence, the computation of such a chain uses dynamic programming and complex geometric concepts. Furthermore this step is the bottleneck of the algorithm and makes expensive the $O(n^7)$ global time complexity.

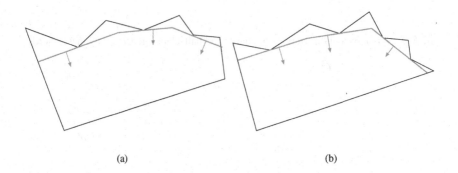

(a)   (b)

Figure 10. Examples of balanced chains of A-butterflies series: (a) a balanced chain given by single-pivot chords and (b) given by both single- and double-pivot chords.

To end with definitions, Chang and Yap divide the set of balanced chords into two classes: a balanced chord is called *single-pivot* if it contains only one reflex vertex (chord C_1 in figure 8) and *double-pivot* if it contains two distinct reflex points (chord C_2 in figure 8). Finally, the optimal set of chords that defines the MACS of P contains extremal, balanced single-pivot and balanced double-pivot chords. In figure 10-(a), the solution is composed by only single-pivot chords, in figure 10-(b) both single- and double-pivot chords. Note that for a reflex vertex r_i, from the set of extremal chords associated to r_i, we just have to consider the two extremal chords induced by the two adjacent edges to r_i. In the following, we restrict the definition of an extremal chord to a chord that contains an edge of P incident to a reflex vertex.

Our contribution starts here. In the next section, we present a fast algorithm to approximate the maximum area convex subset star-shaped polygons that uses Chang and Yap's analysis.

5.   Fast approximation algorithm

Kernel dilatation based heuristic

In this section, we assume that P is a star-shaped polygon. First of all, we remind basic definitions of such a polygon. P is a *star-shaped polygon* if there exist a point q in P such that $\overline{qv_i}$ lies inside P for all vertices v_i of P. The set of points q satisfying this property is called the *kernel* of P [11]. Using our definitions and [11], we have:

PROPOSITION 3 *The kernel of P is given by the intersection between P and the half-planes C_i^+ defined by all extremal chords C_i associated to all reflex vertices.*

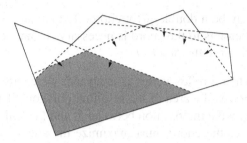

Figure 11.   Illustration of the kernel computation of the example in figure 10-(a).

Figure 11 is an illustration of such proposition. We have the theorem.

THEOREM 4 *Let P be a star-shaped polygon, then its kernel is a subset of the maximum area convex subset of P.*

PROOF. Let C_i be the optimal chord associated to a reflex point r_i of P. We consider the closed space K defined by the intersection between P and the two extremal chords of r_i. If C_i is an extremal chord, it is clear that $K \subseteq (C_i^+ \cap P)$. If C_i is a single or double-pivot balanced chord, the slope of C_i is strictly bounded by the slopes of the two extremal chords. Furthermore, since all the half-planes have the same orientation according to P and r_i, we also have $K \subseteq (C_i^+ \cap P)$ (see figure 9-(left) for example). Finally, since the maximum area convex polygon is the intersection between P and the set of half-planes defined by optimal chords, and since the two extremal chords always define a subset to the associated optimal chord, the intersection of all extremal chords is a subset of the MACS of P. With proposition 3, the kernel of P is a convex subset of the MACS. $\qquad\square$

In other words, there exists a continuous deformation that transforms the kernel to the MACS. In the following, the strategy we choose to approximate the MACS is to consider the deformation as an Euclidean dilatation of the kernel. Based on this heuristic, several observations can be made: the reflex vertices must be taken into account in the order in which they are reached by the dilatation wavefront. More formally, we consider the list $\mathcal{O}$ of reflex vertices such that the points are sorted according to their minimum distance to the kernel polygon. When a reflex vertex is analyzed, we fix the possible chords and introduce new definitions of chords as follows:

- the chord may be an extremal one as defined by Chang and Yap;

- the chord may be a single-pivot chord such that its slope is tangent to the wavefront (this point will be detailed in the next section);

- the chord may be a double-pivot chord. In that case, the second reflex vertex that belongs to the chord is necessary. It must correspond to the next reflex point in the order $\mathcal{O}$.

Furthermore, when a reflex vertex is analyzed, we choose the chord from this list that maximizes the area of the resulting polygon. If we denote by P' the polygon given by the intersection between P and the half-plane associated to the chosen chord, the chord must maximize the area of P'. In the algorithm, it is equivalent to minimize the area of the removed parts P/P'. Using these heuristics, the approximated MACS algorithm can be easily designed in a greedy process:

1:	Compute the kernel of P
2:	Compute the ordered list $\mathcal{O}$ of reflex vertices
3:	Extract the first point r_1 in $\mathcal{O}$
4:	**while** $\mathcal{O}$ is not empty **do**
5:	Extract the first point r_2 in $\mathcal{O}$
6:	Choose the best chord that maximizes the resulting polygon area with the chords (r_1, r_2)
7:	Modify the polygon P accordingly
8:	Update the list $\mathcal{O}$ removing reflex points excluded by the chord
9:	$r_1 \leftarrow r_2$
10:	**end while**

Algorithms and complexity analysis

In this section, we detail the algorithms and their computational costs. Keep in mind that n denotes the number of vertices of P and k the number of reflex vertices.

Distance-to-kernel computation.   First of all, the kernel of a polygon can be constructed in $O(n)$ time using the Preparata and Shamos's algorithm [11]. Note that the number of edges of the kernel is $O(k)$ (intersection of $2k$ half-planes). The problem is now to compute the distances between the reflex vertices and the kernel of P denoted $Kern(P)$. A first solution is given by the Edelsbrunner's algorithm that computes the extreme distances between two convex polygons [6].

However, we use another geometrical tool that will be reused in the next section. Let us consider the generalized Voronoi diagram of $Kern(P)$ [11, 1]. More precisely, we are interested in the exterior part to $Kern(P)$ of the diagram (see figure 12). As $Kern(P)$ is convex, such a diagram is defined by exterior angular bisectors $\{b_i\}_{i=1..5}$ of the kernel vertices. For example in figure 12, all exterior points to $Kern(P)$ located between the angular bisectors b_0 and b_1, are closer to the edge e (extremities are included) than to all other edges of $Kern(P)$. Hence, the minimum distance between such a point and $Kern(P)$ is equal to the distance between the point and the edge e.

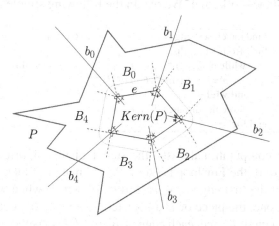

Figure 12. Euclidean growth of a convex polygon.

To efficiently compute the distance between all reflex vertices to the kernel, we first detail some notations. Let B_m be the open space defined by $Kern(P)$ and the exterior angular bisectors b_m and b_{m+1} (if $m+1$ is equal to $|Kern(P)|$ then b_0 is considered). Hence, the distance to kernel computation step can be viewed as a labelling of vertices of P according to the cell B_m they belong to. To solve this labelling problem, we have the following proposition.

PROPOSITION 5 *The vertices of P that belong to the cell B_m form only one connected piece of P.*

PROOF. Let us consider an half-line starting at a point of $Kern(P)$. By definition of $Kern(P)$, the intersection between such an half-line and P is either a point or an edge of P. The special case when the intersection is an edge of P only occurs when the starting point belongs to an edge of $Kern(P)$ and when the half-line is parallel to this edge (and thus parallel to an extremal chord of a reflex vertex of P). Hence, using the definition of the exterior angular bisectors and given a cell B_m, the half-line b_m (resp. b_{m+1}) crosses P at a point p_m (resp. p_{m+1}) of P. Furthermore, each vertex of P between p_m and p_{m+1} belongs to B_m, otherwise the number of intersections between P and b_m or b_{m+1} should be greater than one. Finally, only one connected piece of P belongs to B_m.   □

Then the proposition implies the corollary:

COROLLARY 6 *The order of the P vertex labels is given by the edges of $Kern(P)$.*

Hence, we can scan both the vertices of P and the edges of $Kern(P)$ to compute the distance-to-kernel. We obtain the following simple algorithm[1]:

1:	Find the closest edge e_j of $Kern(P)$ to the point $v_0 \in P$
2:	**for** i from 1 to n **do**
3:	**while** $d(v_i, e_j) > d(v_i, e_{j+1 \ (mod \ |Kern(P)|)})$ **do**
4:	$j := j + 1 \ (mod \ |Kern(P)|)$
5:	**end while**
6:	Store the distance $d(v_i, e_j)$ to v_i
7:	**end for**

To detail the computational costs, the step 1 of this algorithm is done in $O(k)$ and the cost of the **For** loop is $O(n)$. As the matter of fact, using Prop. 5 and excepted for the first edge $e_j \in Kern(P)$ of step 1, when we go from an edge $e_{j'}$ to next one, the piece of P associated to the cell B_m defined by $e_{j'}$ is completely computed. Hence, each edge of $Kern(P)$ is visited once during all the process. Note that the first edge e_j of step 1 may be used twice to complete the scan of P.

Finally, the minimum distances between the vertices of P and $Kern(P)$ are computed in $O(n)$. Note that since we are only interested in labelling reflex vertices, the above algorithm can transformed to have a $O(k)$ computational cost. However, a $O(n)$ scanning of P is still needed to identify reflex points. Furthermore, the sorted list $\mathcal{O}$ of reflex points according to such distance is computed in $O(k \cdot \log k)$.

Single-pivot chords computation.   Given a reflex point r_i of P, we have listed three possible classes of chord: extremal, single-pivot and double-pivot chords. The figure 13 reminds the possibles chords. The extremal and double-pivot chord computation is direct. However, we have to detail the single-pivot chord extraction. According to our heuristic, the single-pivot chord associated to r_i must be tangent to the wavefront propagation of the kernel dilatation.

Using the exterior angular bisector structure we have introduced above, we can efficiently compute the slopes of such chords. In figure 14-(a), let e_1 and e_2 be two adjacent edges of $Kern(P)$ (e_1 and e_2 are incident to the vertex v). Let p (resp. q) be a point in the plane that belongs to the cell generated by e_1 (resp. e_2). We can distinguish two cases: p is closer to e_1 than to one of

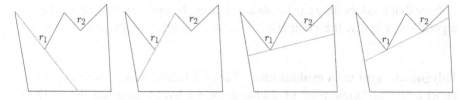

Figure 13. All possible chords that can be associated to the reflex point r_1 (the two extremal chords, a single-pivot balanced chord and the double-pivot chord).

its extremities and q is closer to v than to e_2 (without the extremities). Hence the straight line going through p and tangent to the wave-front propagation is parallel to e_1. In the second case, the tangent to wavefront straight line going through q is tangent to the circle of center v and radius $\|\vec{vq}\|$ (see figure 14). Moreover, two particular cases can be identified (see figure 14-(b)): the first one occurs when the distance between v and q is null, in that case the slope of the chord is the mean of the slopes of edges e_1 and e_2. The second case occurs when the single-pivot chord does not fulfill the maximal chord definition (see the right figure in 14-(b)). In that case, we do not consider this single-pivot chord in the optimal chord choice of line 6 in the main algorithm. Note that this last particular case can also occur with double-pivot chords. In such cases, the chord is not considered too.

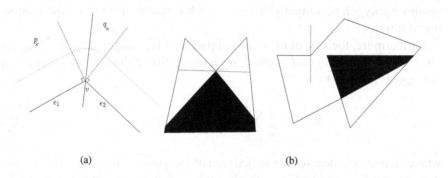

(a) (b)

Figure 14. Computing a chord parallel to the kernel dilatation wavefront: (a) slope computation in the general case, (b) illustration of the two particular cases: the distance between the reflex point and the kernel is null and the single-pivot chord is not a maximal chord.

Finally, if each reflex point r_i of P is labelled according to the closest edge e_i of $Kern(P)$ (extremities included), we can directly compute the single-pivot chord: if r_i is closer to e_j than one of its extremities, the chord is parallel to e_j, otherwise, the chord is tangent to a given circle.

Note that this labelling can be obtained using the previous distance-to-kernel algorithm. Finally, the computation of the k single-pivot chords is done in $O(k)$.

Polygon cut and area evaluation.   Given a reflex point r_i and a chord C_i either extremal, single-pivot of double-pivot, we have to compute the resulting polygon of the intersection between P and C_i^+. We suppose that the vertices of P are stored as a doubly-connected list.

The cutting algorithm is simple: starting from r_i, the vertices of P are scanned in the two directions until they belong to C_i^+ (see figure 15).

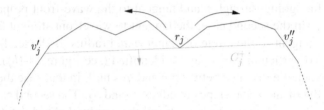

Figure 15.   Illustration of the vertex scan to compute the $(P \cap C_i^+)$ polygon.

During the process, let m be the number of removed vertices. Hence, the polygon cut according to C_i is done in $O(m)$. Furthermore, the resulting polygon has got $n - m$ vertices. Hence, given k reflex vertices and k chords, the resulting polygon is computed in $O(n)$: each vertex is visited a constant number of times.

Furthermore, the area of the removed parts can be computed without changing the computational cost. Indeed, let p be a point in the plane, the area of a polygon P is given by

$$\mathcal{A}(P) = \frac{1}{2} \sum_{i=0}^{n-1} A(p, v_i, v_{i+1 \ (mod \ n)}), \qquad (3)$$

where $A(p, v_i, v_j)$ denotes the signed area of the triangle (p, v_i, v_j) [10]. Hence, the area can be computed in a greedy process during the vertex removal process.

Overall computational analysis

Based on the previous analyses, we can detail the computational cost of the global algorithm presented in the section 5. First of all, step 1 (kernel determination) requires $O(n)$ computation using [11]. Then, we have presented a simple $O(n)$ algorithm to compute the distance-to-kernel of reflex vertices and thus the cost of step 2 (reflex vertices sorting) is $O(n + k \cdot \log k)$. The Step 3 and the step 5 in the **while** loop on reflex vertices requires $O(1)$ computations.

Given the two reflex points r_1 and r_2 of step 6, we have to decide which chord should be associated to r_1. We have two extremal chords, a single-pivot chord whose slope is computed in $O(1)$ and a double-pivot chord going through r_2. Using our heuristics, we choose the chord that minimizes the removed part of P. Hence, we compare the removed part area using the double chord and the 9 area measures of the 3×3 other possible choices for r_1 and r_2. Hence, at each step of the **while** loop, we compute 10 polygon cuts and we choose, for the r_1 chord, the chord that maximizes the resulting polygon area. Note that steps 7 and 8 are computed during the polygon cutting steps and do not change the computational cost.

In the worst case, each polygon cut step is done in $O(n)$. Hence, the overall complexity of the **while** loop is $O(k \cdot n)$. Finally, the cost of the approximated MACS extraction algorithm is $O(k \cdot n)$ in the worst case. Let N be the number of removed vertices while evaluating the optimal chord at r_1 in step 6. If we suppose that the reflex vertices of P are *uniformly distributed* in the sense that each other tested chord only visits the same amount $O(N)$ of vertices. Then, at each step, $O(N)$ vertices are visited and the modified polygon has got $O(n - N)$ vertices. Hence, the **while** loop complexity becomes $O(n)$. This leads to a global cost for the approximated MACS extraction in $O(n + k \cdot \log k)$. In practice, such *uniformity* has been observed in our examples.

To speed up the algorithm, several optimizations can be done without changing the worst case complexity. For example, when chords going through r_2 are tested, the obtained polygons are propagated to the next step of the while loop in order to reduce the number of polygon cut steps. Furthermore, since the area of the removed parts is computed during the vertex scan, the process can be stopped if the area is greater than the current minimum area already computed with other chords.

Experiments

In this section, we present some results of the proposed algorithm. First of all, the figure 16 compares the results between the optimal Chang and Yap's algorithm [3] and the approximated MACS extraction process. In practical experiments, the optimal $O(n^7)$ algorithm do not lead to a direct implementation. Indeed, many complex geometrical concepts are used and the overall algorithm is not really tractable. Hence, we use a doubly-exponential process to extract the optimal MACS. The main drawback of this implementation is that we cannot extract the optimal MACS if the number of the polygon points is important. In figure 16, the first column presents the polygon, its kernel and the distance labelling of all vertices, the second row contains the optimal MACS and the third one the approximated MACS. Note that the results of the last row are identical.

If we compute the area error between the optimal and the approximated MACS on these examples, the error is less than one percent.

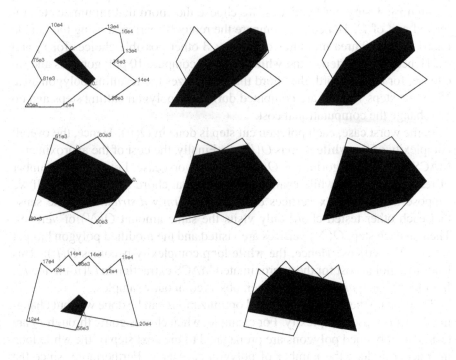

Figure 16. Comparisons between the optimal MACS and the proposed algorithm: the first column presents the input polygons, their kernels and the distance labelling, the second column shows the results of the Chang and Yap's algorithm. The last row present the result of the proposed algorithm.

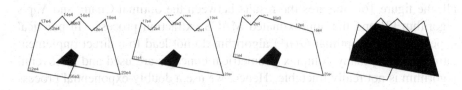

Figure 17. Some intermediate steps of the approximated MACS algorithm.

The figure 17 illustrates the intermediate steps of the approximated MACS algorithm and the figure 18 presents the result of the proposed algorithm on different shapes.

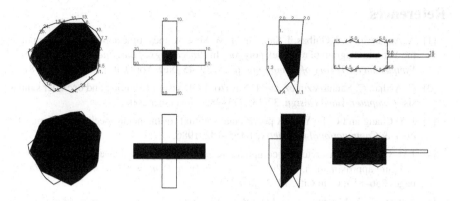

Figure 18. Results of the proposed approximated MACS algorithm on various shapes. The first row illustrates the shapes, their kernels and the distance labelling, the second row presents the obtained solutions.

6. Conclusion

In this article, we have proposed different approaches to extract a maximum convex subset of a polygon. In discrete representation space, we extract the maximum convex subset using the Hausdorff distance. A second illustration in discrete case was for finding the maximum h-v convex subset. In continous representation mode, for a star-shaped polygon, we propose a fast algorithm to extract an approximation of the maximum convex subset. The maximality has to be understand with the area measure. To do that, we have defined a kernel dilatation based heuristic that use classical tools in computational geometry. The computational worst case cost of the algorithm is $O(k \cdot n)$ where n is the number of points of the polygon and k is the number of reflex vertices. However, under some hypotheses on the reflex vertex distribution, the complexity can be bounded by $O(n + k \cdot \log k)$. In our experiments, the computational behavior of the algorithm is closer to the second complexity bound than to the first one.

In future works, a first task consists in a formal comparison between the proposed approximated solution and the optimal one, detailed by Chang and Yap [3]. However, heuristics choices make this comparison non-trivial. More generally, an optimization of the Chang and Yap's optimal algorithm is still an open problem. However, efforts should be made to extend this heuristic to general polygons.

Notes

1. $d(v_i, e_j)$ denotes the Euclidean distance between the point v_i and the segment e_j

References

[1] A. Aggarwal, L. J. Guibas, J. Saxe, and P. W. Shor. A linear time algorithm for computing the Voronoi diagram of a convex polygon. In *Proceedings of the Nineteenth Annual ACM Symposium on Theory of Computing*, pages 39–45, New York City, 25–27 May 1987.

[2] C. Andjar, C. Saona-Vzquez, and I. Navazo. LOD visibility culling and occluder synthesis. *Computer Aided Design*, 32(13):773–783, November 2000.

[3] J. S. Chang and C. K. Yap. A polynomial solution for the potato-peeling problem. *Discrete & Computational Geometry*, 1:155–182, 1986.

[4] J.M. Chassery. Discrete and computational geometry approaches applied to a problem of figure approximation. In *Proc. 6th Scandinzvian Conf on Image Analysis (SCIA 89)*, pages 856–859, Oulu City, 19–22 june 1989.

[5] K. Daniels, V. Milenkovic, and D. yRoth. Finding the largest area axis-parallel rectangle in a polygon. *Computational Geometry: Theory and Applications*, 7:125–148, 1997.

[6] H. Edelsbrunner. Computing the extreme distances between two convex polygons. *Journal of Algorithms*, 6:213–224, 1985.

[7] J. E. Goodman. On the largest convex polygon contained in a non-convex n-gon or how to peel a potato. *Geometricae Dedicata*, 11:99–106, 1981.

[8] D. Roth K. Daniels, V. Milekovic. Finding the largest area axis-parallel rectangle in a polygon. *Comput. Geom. Theory Appl*, 7:125–148, 1997.

[9] J. O'Rourke, A. Aggarwal, Sanjeev, R. Maddila, and M. Baldwin. An optimal algorithm for finding minimal enclosing triangles. *J. Algorithms*, 7(2):258–269, 1986.

[10] Joseph O'Rourke. *Computational Geometry in C*. Cambridge University Press, 1993. ISBN 0-521-44034-3.

[11] F. P. Preparata and M. I. Shamos. *Computational Geometry : An Introduction*. Springer-Verlag, 1985.

[12] D. Wood and C. K. Yap. The orthogonal convex skull problem. *Discrete and Computational Geometry*, 3:349–365, 1988.

REGULAR METRIC: DEFINITION AND CHARACTERIZATION IN THE DISCRETE PLANE

Gerald Jean Francis Banon[1]

[1]*National Institute for Space Research, INPE*
São José dos Campos, Brazil
banon@dpi.inpe.br

Abstract   We say that a metric space is regular if a straight-line (in the metric space sense) passing through the center of a sphere has at least two diametrically opposite points. The normed vector spaces have this property. Nevertheless, this property might not be satisfied in some metric spaces. In this work, we give a characterization of an integer-valued translation-invariant regular metric defined on the discrete plane, in terms of a symmetric subset B that induces through a recursive Minkowski sum, a chain of subsets that are morphologically closed with respect to B.

Keywords: Mathematical Morphology, symmetric subset, ball, lower regularity, upper regularity, regular metric space, integer-valued metric, translation-invariant metric, triangle inequality, recursive Minkowski sum, morphological closed subset, discrete plane, computational geometry, discrete geometry, digital geometry

Introduction

The continuous plane or, more precisely, the two-dimensional Euclidean vector space, has good geometrical properties. For example, in such space, a closed ball is included in another one only if the radius of the latter is greater than or equal to the sum of the distance between their centers and the radius of the former. Furthermore, in such space, two closed balls intersect each other if the sum of their radii is greater than or equal to the distance between their centers. Nevertheless, not all metric spaces have these properties.

In the first part of this work, we clarify the concept of regular metric space in which the above two geometrical properties are satisfied.

We say that a metric space is regular if its metric satisfies three regularity axioms or equivalently if a straight-line (in the metric space sense) passing through the center of a sphere has at least two diametrically opposite points. The Minkowski spaces (i.e., finite dimensional normed vector spaces) have this property.

C. Ronse et al. (eds.), Mathematical Morphology: 40 Years On, 249–258.

This regularity is generally lost when a metric on the continuous plane is restricted to the discrete plane, as it is the case of the Euclidean metric.

In the second part of this work, we study the characterization of the integer-valued translation-invariant regular metrics on the discrete plane in terms of some appropriate symmetric subsets.

We show that every such metric can be characterized in terms of a symmetric subset B that induces through a recursive Minkowski sum, a chain of subsets that are morphologically closed with respect to B.

Our characterization shows how to construct an integer-valued translation-invariant regular metric on the discrete plane.

This is an important issue in digital image analysis since the image domains are then discrete. In the sixties, Rosenfeld and Pfaltz [7] have already introduced a metric property and have used it to describe algorithms for computing some distance functions by performing repeated local operations. It appears that their property is precisely a necessary condition for a metric to be regular.

We came across the regularity property for a metric while we were trying to prove the one-pixel width of the skeleton of the "expanded" subsets proposed in [1]. Actually, what we needed at that time was just a "lower" regularity.

In one dimension, we observed that the (discrete) convexity is not a necessary condition to have the morphological closure property, so it was useless to solve our problem.

For the sake of simplicity of the presentation, in this work, we limit ourselves to the class of integer-valued metrics. More precisely, we consider the class of metrics that are mappings from the discrete plane *onto* the set of integers. This should not be a serious limitation because on the discrete plane the metrics assume only a countable number of values.

In Section 1, we give an axiomatic definition of regular metric spaces and we show that of the three axioms only two are sufficient to define the metric regularity. In order to get the regular metric characterization in the last section, we give in Section 2, independently of the metric definition, a definition of balls based on the notions of set translation and set transposition. In the same section, we recall the notions of recursive Minkowski sum, generated balls and border. In Section 3, we study the properties of the balls of a regular metric space. Conversely, in Section 4, we study the properties of the metric spaces constructed from the symmetric balls having a morphological closure property. Finally, in Section 5, we show the existence of a bijection between the set of integer-valued translation-invariant regular metrics defined on the discrete plane and the set of the symmetric balls satisfying the morphological closure property.

1. Regular Metric Space Definition

In a metric space we can define the concepts of straight-line and sphere. Let (E, d) be a metric space [4, Definition 7.10]. For any x and $y \in E$, and any $i \in d(\{x\} \times E)$ (i.e., the image of $\{x\} \times E$ through d), let us define the following subsets of E:

$$L_1(x, y) \overset{\triangle}{=} \{z \in E : d(x, z) = d(x, y) + d(y, z)\},$$
$$L_2(x, y) \overset{\triangle}{=} \{z \in E : d(x, y) = d(x, z) + d(z, y)\},$$
$$L_3(x, y) \overset{\triangle}{=} \{z \in E : d(z, y) = d(z, x) + d(x, y)\}, \text{ and}$$
$$S(x, i) \overset{\triangle}{=} \{z \in E : d(x, z) = i\}.$$

The subsets $L(x, y) = L_1(x, y) \cup L_2(x, y) \cup L_3(x, y)$ and $S(x, i)$ are called, respectively, the straight-line passing through the points x and y, and the sphere of center x and of radius i. Figure 1 shows the three parts of $L(x, y)$ in the Euclidean case.

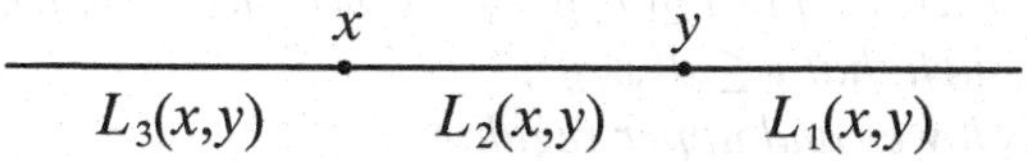

Figure 1. The three parts of the straight-line.

We must be aware that the above definition of straight-line is based on the concept of metric and the resulting object is generally different from the usual straight-line defined in the framework of linear vector space.

Because of the symmetry property of the distances, the straight-lines have some kind of symmetry as well (a detailed proof of all the propositions in this work can be found in [3]).

PROPOSITION 1 *(straight-line symmetry) - Let (E, d) be a metric space. For any x and $y \in E$,*
(i) $L_1(x, y) = L_3(y, x)$;
(ii) $L_2(x, y) = L_2(y, x)$;
(iii) $L(x, y) = L(y, x)$.

Based on the concepts of straight-line and sphere, we can define what we call regular metrics and regular metric spaces.

DEFINITION 1 *(regular metric spaces) - Let (E, d) be a metric space. The metric d on E is*
(i) lower regular of type 1 if $S(x, i) \cap L_1(x, y) \neq \emptyset$, for any x and $y \in E$, and any $i \in d(\{x\} \times E)$, such that $d(x, y) \leq i$;

(ii) lower regular of type 2 if $S(x,i) \cap L_2(x,y) \neq \emptyset$, for any x and $y \in E$, and any $i \in d(\{x\} \times E)$, such that $i \leq d(x,y)$;
(iii) upper regular if $S(x,i) \cap L_3(x,y) \neq \emptyset$, for any x and $y \in E$, and any $i \in d(\{x\} \times E)$;
(iv) regular *if it is lower regular (of type 1 and 2) and upper regular. A metric space (E,d) is* regular *if its metric is regular.*

Actually, because of Proposition 1, the three regularity axioms are not independent each other as we show in the next proposition.

PROPOSITION 2 *(axiom dependence) - Let (E,d) be a metric space. The metric d on E is lower regular of type 1 if and only if (iff) it is upper regular.*

The axiom dependence (Proposition 2) allows us to make an equivalent definition of regular metrics, but simpler with only two axioms, the lower regularity of type 2 being called simply lower regularity.

COROLLARY 2 *(first equivalent definition of regular metrics) - Let (E,d) be a metric space. The metric d on E is*
(i) lower regular *if $S(x,i) \cap L_2(x,y) \neq \emptyset$, for any x and $y \in E$, and any $i \in d(\{x\} \times E)$, such that $i \leq d(x,y)$;*
(ii) regular *iff it is lower and upper regular.*

In his lecture note [5], in order to compare balls with different centers, Kiselman has introduced for the translation-invariant metrics the properties of being upper and lower regular for the triangle inequality. Actually, for the translation-invariant metrics, his upper regular property is the same of our upper regularity. His lower regular property is our lower regularity of type 1 which is, by the above axiom dependence, equivalent to the upper regular property. Kiselman doesn't mention the independent axiom of lower regularity of type 2, which is our lower regularity.

The next proposition is another equivalent definition, which is more geometrical.

PROPOSITION 3 *(second equivalent definition of regular metrics) - A metric d on E is* regular *iff for any x and $y \in E$, and any $i \in d(\{x\} \times E)$, the intersection between the straight-line $L(x,y)$ and the sphere $S(x,i)$ have at least two diametrically opposite points in the sense that there exist u and $v \in S(x,i)$ such that $u \in (L_1(x,y) \cup L_2(x,y))$ and $v \in L_3(x,y)$.*

The next proposition shows that the lower (resp. upper) regularity property for a metric is a sufficient condition to have the usual ball intersection (resp. inclusion) property of the Euclidean vector spaces. We denote by $B_d(x,i) \stackrel{\triangle}{=} \{z \in E : d(x,z) \leq i\}$ the ball of center x and radius i in a metric space (E,d).

PROPOSITION 4 *(ball intersection and inclusion in a regular metric space) -*
Let (E, d) be a metric space then, for any x and $y \in E$, any $i \in d(\{x\} \times E)$
and any $j \in d(E \times \{y\})$,
(i) $B_d(x, i) \cap B_d(y, j) \neq \emptyset \Rightarrow d(x, y) \leq i + j$;
(ii) if (E, d) is lower regular, then $d(x, y) \leq i + j \Rightarrow B_d(x, i) \cap B_d(y, j) \neq \emptyset$;
(iii) $i + d(x, y) \leq j \Rightarrow B_d(x, i) \subset B_d(y, j)$;
(iv) if (E, d) is upper regular, then $B_d(x, i) \subset B_d(y, j) \Rightarrow i + d(x, y) \leq j$.

2. Balls, Recursice Minkowski sum, Generated Balls and Border

From the operations of translation and transposition, we can build the sub-collection of symmetric subsets and their translated versions, that we call here balls. The symmetry assumption is made in order to establish, in Sections 3 and 4, the relationship with the distances (which are symmetric mappings).

Because of our interest in digital image processing, the balls will be considered as subsets of the discrete plane $(\mathbf{Z}^2, +)$ (the Cartesian product of the set of integers by itself, provided with the usual integer pair addition).

We denote by X_u the translated version of a subset X of $\mathbf{Z}^2$ by a point u in $\mathbf{Z}^2$, that is, $X_u \overset{\triangle}{=} \{y \in \mathbf{Z}^2 : y - u \in X\}$.

We denote by o the unit element of the addition on $\mathbf{Z}^2$ and we call it *origin* of the discrete plane. A subset X of $\mathbf{Z}^2$ is *symmetric* (with respect to the origin o) if it is equal to its transpose, that is, $x \in X \Leftrightarrow -x \in X$. $\mathbf{Z}^2$ is an example of symmetric subset.

DEFINITION 3 *(balls) - A subset X of $\mathbf{Z}^2$ is a ball if $\exists u \in \mathbf{Z}^2$ such that X_u is a finite symmetric subset of $\mathbf{Z}^2$ or is $\mathbf{Z}^2$.*

Let $\mathbf{N}$ be the set of natural numbers and let $\mathbf{N}^+$ be the set of extended natural numbers (i.e., the natural numbers plus an element denoted ∞) with the usual addition extended in such a way that, for any $j \in \mathbf{N}^+$, $j + \infty = \infty + j = \infty$ and with the usual order extended in such a way that, for any $j \in \mathbf{N}^+$, $j \leq \infty$.

We denote by $X \oplus Y$ (resp. $X \ominus Y$) the Minkowski sum (resp. difference) of the subsets X and Y of $\mathbf{Z}^2$ [2][4][6]. From the Minkowski sum we can define a recursive Minkowski sum.

DEFINITION 4 *(recursive Minkowski sum) - Let $B \in \mathcal{P}(\mathbf{Z}^2)$ (all the parts of $\mathbf{Z}^2$) such that $B \neq \emptyset$, and let $j \in \mathbf{N}^+$, the recursive Minkowski sum of B times j is the subset jB of $\mathbf{Z}^2$ given by*

$$jB \overset{\triangle}{=} \begin{cases} \{o\} & \text{if } j = 0 \\ B & \text{if } j = 1 \\ ((j-1)B) \oplus B & \text{if } 1 < j < \infty \\ \mathbf{Z}^2 & \text{if } j = \infty \end{cases}$$

With a finite symmetric ball B we can associate, through the recursive Minkowski sum, a sub-collection of balls $\mathcal{B}_B \triangleq \{X \in \mathcal{P}(\mathbf{Z}^2) : \exists j \in \mathbf{N}^+$ and $\exists u \in \mathbf{Z}^2, X = (jB)_u\}$. We say that B induces or generates the sub-collection $\mathcal{B}_B$. We call the elements of $\mathcal{B}_B$ *generated balls*.

In order to be able to characterize the integer-valued translation-invariant regular metrics later on, we need the concept of sub-collection of *closed* generated balls. For convenience, we say that B *has the closure property* if every member of $\mathcal{B}_B$ is morphologically B-closed, that is, for every $X \in \mathcal{B}_B$, X satisfies the equation $X = (X \oplus B) \ominus B$.

Based on the mix distributivity of the recursive Minkowski sum: for any i and $j \in \mathbf{N}$, $(i+j)B = (iB) \oplus (jB)$, we can establish the next two propositions.

PROPOSITION 5 *(generated balls versus intersection) - Let B be a finite symmetric ball, then for any x and y in $\mathbf{Z}^2$ and any i and j in $\mathbf{N}$,*
$$y \in ((i + j)B)_x \Leftrightarrow (iB)_x \cap (jB)_y \neq \emptyset.$$

PROPOSITION 6 *(generated balls versus inclusion) - Let B be a finite symmetric ball, then for any x and y in $\mathbf{Z}^2$ and any i and j in $\mathbf{N}$,*
(i) $x \in (jB)_y \Rightarrow (iB)_x \subset ((i + j)B)_y$;
(ii) if B has the closure property, then $(iB)_x \subset ((i + j)B)_y \Rightarrow x \in (jB)_y$.

We now recall the definitions of an erosion by a structuring element and of a B-border. We will use the latter definition in the study of the ball border.

Let B be a subset of $\mathbf{Z}^2$, the *erosion by B* is the mappings from $\mathcal{P}(\mathbf{Z}^2)$ to $\mathcal{P}(\mathbf{Z}^2)$, $\varepsilon_B : X \mapsto X \ominus B$; the *$B$-border* of a finite subset X of $\mathbf{Z}^2$ is the subset $\partial_B(X) = X \setminus \varepsilon_B(X)$.

We observe that the B-border is an inner border: $\partial_B(X) \subset X$.

The following proposition about the erosion of a generated ball will be useful for the study of the generated ball border properties in Proposition 8.

PROPOSITION 7 *(erosion of a generated ball) - Let B be a subset of $\mathbf{Z}^2$. For any $x \in \mathbf{Z}^2$ and $j \in \mathbf{N} \setminus \{0\}$,*
(i) $\varepsilon_B((jB)_x) \supset ((j-1)B)_x$
(ii) if B has the closure property, then $\varepsilon_B((jB)_x) = ((j-1)B)_x$.

Proof - Property (i) follows from the closing extensivity. Property (ii) follows directly from the closure property of B.   □

PROPOSITION 8 *(generated balls versus B-border) - Let B be a finite symmetric ball having more than one element and having the closure property. Then for any x and y in $\mathbf{Z}^2$ and any i and j in $\mathbf{N}$,*
(i) $x \in \partial_B(((i + j)B)_y) \Rightarrow \partial_B((iB)_x) \cap \partial_B((jB)_y) \neq \emptyset$;
(ii) $x \in \partial_B((jB)_y) \Rightarrow \partial_B(((i + j)B)_y) \cap \partial_B((iB)_x) \neq \emptyset$.

Proof - Property (i) follows from Proposition 5 (generated balls versus intersection) and Proposition 7 (erosion of a generated ball). Property (ii) follows from Proposition 6 (generated balls versus inclusion) and again from Proposition 7. □

3.   From Metric to Symmetric Ball

With a translation-invariant (t.i.) metric [3][5], we can associate a ball of center at the origin.

Let $(\mathbf{Z}^2, d)$ be a t.i. metric space such that d is a mapping onto $\mathbf{N}$ or $\mathbf{N}^+$, that is, $d(\mathbf{Z}^2 \times \mathbf{Z}^2) = \mathbf{N}$ or, in the case of a generalized metric, $d(\mathbf{Z}^2 \times \mathbf{Z}^2) = \mathbf{N}^+$.

As usual the *unit ball of* $(\mathbf{Z}^2, d)$, denoted by B_d, is the set of all the points at a distance less than or equal to one from the origin o, that is,

$$B_d \triangleq \{u \in \mathbf{Z}^2 : d(u, o) \leq 1\}.$$

In the next proposition, we show a relationship between a t.i. lower regular metric and the recursive Minkowski sum of its unit ball.

PROPOSITION 9 *(property of the generated balls in a lower regular metric space) - Let* $(\mathbf{Z}^2, d)$ *be a t.i. metric space such that* $d(\mathbf{Z}^2 \times \mathbf{Z}^2) = \mathbf{N}$ *(resp.* $\mathbf{N}^+$*), for any* x *and* $y \in \mathbf{Z}^2$*, and any* $j \in \mathbf{N}$ *(resp.* $\mathbf{N}^+$*),*
(i) $x - y \in jB_d \Rightarrow d(x, y) \leq j$;
(ii) if d *is lower regular, then* $d(x, y) \leq j \Rightarrow x - y \in jB_d$.

Proof - We can prove recursively that Properties (i) and (ii) follow, respectively, from Properties (i) and (ii) of Proposition 4 (ball intersection and inclusion in a regular metric space) substituting i and j, by respectively, $j - 1$ and 1. □

The next proposition, which is a consequence of the previous one, will be used in Section 5 to characterize the regular metrics. It shows that in a regular metric space the recursive Minkowski sum of the unit ball B_d generates morphologically closed subsets with respect to B_d.

PROPOSITION 10 *(closure property of the unit ball of a regular metric space) - Let* $(\mathbf{Z}^2, d)$ *be a t.i. metric space such that* $d(\mathbf{Z}^2 \times \mathbf{Z}^2) = \mathbf{N}$ *(or* $\mathbf{N}^+$*). If* d *is regular, then* B_d *has the closure property.*

Proof - By using Property (iv) of Proposition 4 (ball intersection and inclusion in a regular metric space) and by applying Proposition 9 (property of the generated balls in a lower regular metric space), we can prove that, for any $j \in \mathbf{N}$, $(((jB_d) \oplus B_d) \ominus B_d) \subset jB_d$. Furthermore, by the closing extensivity, for any $j \in \mathbf{N}$, $jB_d \subset (((jB_d) \oplus B_d) \ominus B_d)$. That is, by the anti-reflexivity of the inclusion, for any $j \in \mathbf{N}$, jB_d is B_d-closed. □

4.   From Symmetric Ball to Metric

By using the recursive Minkowski sum, with a symmetric ball not reduced to but containing the origin, we can associate a t.i. metric.

Let B be a finite symmetric ball, such that $o \in B$ and $B \neq \{o\}$ and let $B_B(o) \stackrel{\triangle}{=} \{jB : j \in \mathbf{N}^+\}$, that is, $B_B(o)$ is the sub-collection of B_B consisting of all the generated balls of center at the origin.

Let $x \in \mathbf{Z}^2$ and let $M^x = \bigcap\{X \in B_B(o) : x \in X\}$. By a chain property of the generated balls with same center, M^x belongs to $B_B(o)$ and it is the smallest generated ball of center at the origin that contains the point x.

We denote by $\mathrm{radius}_B(X)$ the integer j such that for some $u \in \mathbf{Z}^2$, $X_u = jB$. Finally, we denote by f_B the mapping from $\mathbf{Z}^2$ to $\mathbf{N}^+$ given by, for any $x \in \mathbf{Z}^2$, $f_B(x) \stackrel{\triangle}{=} \mathrm{radius}_B(M^x)$, and by d_B the mapping from $\mathbf{Z}^2 \times \mathbf{Z}^2$ to $\mathbf{N}^+$ given by, for any x and $y \in \mathbf{Z}^2$,

$$d_B(x,y) \stackrel{\triangle}{=} f_B(x - y).$$

We observe that for any $x \in \mathbf{Z}^2$, $d_B(\{x\} \times \mathbf{Z}^2) = \mathbf{N}$ (resp. $\mathbf{N}^+$), that is d_B is onto $\mathbf{N}$ (resp. $\mathbf{N}^+$).

We now give the relationship between the border of a ball and its radius as follows.

PROPOSITION 11 *(ball border versus ball radius) - Let B be a finite symmetric ball, such that $o \in B$ and $B \neq \{o\}$. For any $x \in \mathbf{Z}^2$ and any finite $X \in B_B(o)$,*
(i) $x \in \partial_B(X) \Rightarrow f_B(x) = \mathrm{radius}_B(X)$;
(ii) if B has the closure property, then $f_B(x) = \mathrm{radius}_B(X) \Rightarrow x \in \partial_B(X)$.

Proof - Properties (i) and (ii) follow, respectively, from Properties (i) and (ii) of Proposition 7 (erosion of a generated ball).   □

The following proposition, which is a consequence of the ball border properties of Proposition 11 applied to the generated balls, will be used in the next section to characterize the regular metric.

PROPOSITION 12 *(regularity of a metric induced by a ball having the closure property) - Let B be a finite symmetric ball, such that $o \in B$ and $B \neq \{o\}$. If B has the closure property, then d_B is a regular metric.*

Proof - The lower regularity of d_B follows from Property (i) of Proposition 8 (generated balls versus B-border) and Proposition 11 (ball border versus ball radius). Its upper regularity follows from Property (ii) of Proposition 8 and Proposition 11.   □

5.   Regular Metrics Space Characterization

Let $\mathcal{B}_c$ be the sub-collection of finite symmetric balls having the closure property and such that $o \in B$ and $B \neq \{o\}$, and let $\mathcal{M}_r$ be the set of integer-valued t.i. regular metrics onto $\mathbf{N}$ (or $\mathbf{N}^+$) (i.e., $d(\mathbf{Z}^2 \times \mathbf{Z}^2) = \mathbf{N}$ (or $\mathbf{N}^+$).

THEOREM 5 *(characterization of integer-valued translation-invariant regular metrics) - The mapping $d \mapsto B_d$ from $\mathcal{M}_r$ to $\mathcal{B}_c$ is a bijection. Its inverse is the mapping $B \mapsto d_B$.*

Proof - Let us divide the proof in two parts.
(a) Let d be a regular metric such that $d(\mathbf{Z}^2 \times \mathbf{Z}^2) = \mathbf{N}$ (or $\mathbf{N}^+$), by Proposition 10 (closure property of the unit ball of a regular metric space), B has the closure property. Therefore, for any $d \in \mathcal{M}_r$, $B_d \in \mathcal{B}_c$. By using Proposition 9 (property of the generated balls in a lower regular metric space), we verify that the mapping $B \mapsto d_B$ is a left inverse for $d \mapsto B_d$.
(b) Let B be a symmetric ball satisfying the closure property, by Proposition 12 (regularity of a metric induced by a ball having the closure property) d_B is regular. Therefore, for any $B \in \mathcal{B}_c$, $d_B \in \mathcal{M}_r$. The sub-collection $\mathcal{B}_B(o)$ being a chain, we verify that $B \mapsto d_B$ is a right inverse for $d \mapsto B_d$. $\square$

Hence, from (a) and (b), the mapping $d \mapsto B_d$ from $\mathcal{M}_r$ to $\mathcal{B}_c$ is a bijection and its inverse is the mapping $B \mapsto d_B$.

The existence of a left inverse shows that every integer-valued t.i. regular metric (and lower regular metric as well) onto $\mathbf{N}$ (or $\mathbf{N}^+$), can be reconstructed from its unit ball by using the recursive Minkowski sum.

The city-block and chessboard distances on $\mathbf{Z}^2$ are examples of integer-valued t.i. regular metric onto $\mathbf{N}$. In [3] we give a detailed proof of the regularity of the chessboard distance.

Conclusion

In the first part of this work we have introduced a definition of regular metric spaces and commented its relation with the Kiselman's upper and lower regularity for the triangle inequality. We have pointed out, in particular, that the lower regularity of type 1 is a redundant axiom when the definition of regular metrics is based on the upper regularity axiom.

In the second part, we have established a one-to-one relationship between the set of integer-valued and translation-invariant regular metrics defined on the discrete plane, and the set of symmetric balls satisfying a special closure property.

From this result we now know how to construct a regular metric on the discrete plane. For this purpose, we choose in the discrete plane a symmetric ball B that has the closure property, i.e., that induces, through the recursive Minkowski sum, a chain of generated balls that are morphologically closed with respect to B. Then the distance of a point x to the origin is given by the radius (in the sense of the recursive Minkowski sum) of the smallest ball of the chain, that contains x.

This construction shows that to preserve in the discrete plane the regularity and isotropic properties of the Euclidean metric on the continuous plane, we have to reach a compromise between a good approximation of a continuous ball and thin contours. "Closer" B from an Euclidean continuous ball, bigger B and thicker the borders in the discrete plane.

Actually it will be interesting in a future work to give a proof that if B is the intersection of an Euclidean continuous ball with the discrete plane then the generated balls are morphologically closed with respect to B.

The proof should be based on a closure property of the convex subsets of the continuous plane ([4], Proposition 9.8).

Acknowledgments

We wish to thank Arley F. Souza for the compelling discussions about isotropic skeletons. We would also like to express our gratitude to the anonymous reviewers for their comments and suggestions. This work was partially supported by CNPq (Conselho Nacional de Desenvolvimento Científico e Tecnológico) under contract 300966/90-3.

References

[1] Banon, Gerald J. F. (2000). New insight on digital topology. In: International Symposium on Mathematical Morphology and its Applications to Image and Signal Processing, 5., 26-28 June 2000, Palo Alto, USA. *Proceedings...* p. 138-148. Published as: INPE-7884-PRE/3724.

[2] Banon, Gerald J. F. and Barrera, Junior (1998). *Bases da morfologia matemática para a análise de imagens binárias*. 2sd. ed. São José dos Campos: INPE, Posted at: <http://iris.sid.inpe.br:1912/rep/dpi.inpe.br/banon/1998/06.30.17.56>. Access in: 2003, Apr.17.

[3] Banon, Gerald J. F. (2004). *Characterization of the integer-valued translation-invariant regular metrics on the discrete plane*. São José dos Campos: INPE, Posted at: <http://iris.sid.inpe.br:1912/rep/dpi.inpe.br/banon/2004/03.08.12.12>. Access in: 2003, Apr.17.

[4] Heijmans, Henk J. A. M. (1994). *Morphological Image Operators*. Boston: Academic.

[5] Kiselman, Christer O. (2002). *Digital geometry and mathematical morphology*. S.n.t. Lecture Notes, Spring Semester. Posted at: <http://www.math.uu.se/kiselman/dgmm2002.pdf>. Access in 2002, Dec. 18.

[6] Serra, Jean (1982). *Image analysis and mathematical morphology*. London: Academic.

[7] Rosenfeld, Azriel and Pfaltz, John L. (1968). Digital functions on digital pictures. *Pattern Recognition*, v.1, n. 1, p. 33-61.

EUCLIDEAN SKELETONS OF 3D DATA SETS IN LINEAR TIME BY THE INTEGER MEDIAL AXIS TRANSFORM

Wim H. Hesselink, Menno Visser and Jos B.T.M. Roerdink[1]

[1]*Institute for Mathematics and Computing Science, University of Groningen, the Netherlands*
{wim,menno,roe}@cs.rug.nl

Abstract   A general algorithm for computing Euclidean skeletons of 3D data sets in linear time is presented. These skeletons are defined in terms of a new concept,
called the *integer medial axis* (*IMA*) transform. The algorithm is based upon the
computation of 3D feature transforms, using a modification of an algorithm for
Euclidean distance transforms. The skeletonization algorithm has a time complexity which is linear in the amount of voxels, and can be easily parallelized.
The relation of the *IMA* skeleton to the usual definition in terms of centers of
maximal disks is discussed.

Keywords: Feature transform, integer medial axis, 3-D Euclidean skeletonization.

1.   Introduction

In computer vision, skeleton generation is often one of the first steps in image description and analysis. Intuitively, a skeleton consists of the center lines
of an object, and therefore skeletons provide important structural information
about image objects by a relatively small number of pixels.

There are four main approaches to skeletonization: 1) thinning, i.e. iterative
removal of points from the boundary; 2) wave propagation from the boundary;
3) detection of crest points in the distance transformed image; 4) analytical
methods. A large number of skeletonization algorithms exist, see e.g. [15],
many of them based upon mathematical morphology [2, 10, 14, 17, 19, 20].
For a parallel 3D skeletonization algorithm based on thinning, see [9].

We note that in algorithms of type 3) one often restricts oneself to local
maxima of the distance transform [18], but the resulting skeleton is far from the
Euclidean one. The approach we present here is a variant of the third approach,
using a definition of skeletons based on Blum's medial axis transform [3].

Often, one is satisfied with approximations to the Euclidean metric (e.g.,
using chamfer metrics). In 1980, Danielsson [6] gave two good approximating

C. Ronse et al. (eds.), Mathematical Morphology: 40 Years On, 259–268.
©2005 *Springer. Printed in the Netherlands.*

Euclidean distance transform algorithms, and applied them to obtain the centers of maximal (integer) disks (*CMD*), see below. He notes (p. 243) that application of skeletons has been hampered by the lack of true Euclidean distance maps. Especially in the 3D case where data size can be very large, many existing algorithms for computing 3D Euclidean skeletons are computationally too expensive [4]. Ge and Fitzpatrick [7] clearly identified the problem to determine the *CMD*: "The problems with existing methods lie in the discrepancies between continuous and discrete image maps". The paper [7] also mentions the goal of linking the centers of maximal disks into *connected* skeletons.

The main contribution of the present work is that we present a simple and easily parallelizable linear time algorithm which computes a skeleton defined in terms of a new concept, called the *integer medial axis* (*IMA*) transform. The algorithm works in arbitrary dimensions, and is based upon the general linear time Euclidean distance transform (EDT) algorithm of Hirata [8], which has been rediscovered several times, i.e., by ourselves, see Meijster *et al.* [13], and later by Maurer *et al.* [11, 12]. The skeletonization algorithm has two phases. First, a feature transform is computed, which uses essentially the same algorithm as for the distance transform, the difference being that not only distances are computed, but also the boundary points which realize the closest distance. The actual skeletonization is performed in a second pass through the data, where the integer medial axis is computed by assigning points to the skeleton depending on their feature transform.

Our method does not aim at a minimal skeleton useful for image compression with exact reconstruction, but at a computation of connected skeletons directly from the Euclidean feature transform, thus avoiding the costly and complicated phase of removing centers of not-quite-maximal disks by the techniques of [16]. We establish a number of mathematical properties of the *IMA* and point out some relations to Blum's real medial axis (*RMA*) and to the *CMD* skeleton. More work is needed to establish the topological characteristics of the *IMA* skeleton.

Often, simplification or pruning of the skeleton is used as a postprocessing step to remove unwanted points, which arise especially in noisy data [1]. In our approach, skeleton pruning can be handled in the algorithm itself, by a single adjustable parameter through which one can prune the skeleton during the second pass of the algorithm.

In order to derive our algorithm, we first modify the *EDT* algorithm of Meijster *et al.* to calculate 3D feature transforms, from which the *IMA* skeletons are derived. For all program parts, explicit and compact pseudocode is given.

2.   Feature transform computation

We briefly describe extension of the Euclidean distance transform algorithm to the computation of feature transforms, closely adhering to the notation and approach given in [13]. The algorithm can deal with several types of distances (Manhattan, chessboard, or chamfer distances), but we will limit ourselves to the case of the Euclidean distance here, since we focus on Euclidean skeletons in this paper.

The length of a vector $\mathbf{r} \in \mathbb{R}^d$ is denoted by $||\mathbf{r}|| = \sqrt{\sum_i \mathbf{r}_i^2}$. We regard $\mathbb{Z}^d$ as a grid embeddded in $\mathbb{R}^d$. The elements of $\mathbb{Z}^d$ are called grid points.

Let B be the background, which is a given nonempty set of grid points. The Euclidean distance transform dt of B is the function that assigns to every grid point $\mathbf{r}$ the distance to the nearest background point, so $dt(\mathbf{r}, B) = \min\{||\mathbf{r} - \mathbf{y}|| \mid \mathbf{y} \in B\}$. The *feature transform FT* is defined as the set-valued function that assigns to $\mathbf{r}$ the set of closest boundary points. So we have $FT(\mathbf{r}, B) = \{\mathbf{y} \in B \mid ||\mathbf{r} - \mathbf{y}|| = dt(\mathbf{r}, B)\}$. The parameter B is omitted from dt and FT when it is clear from the context.

It is possible to compute FT, but it is computationally cheaper and sufficient for our purposes to compute, for every point $\mathbf{r}$, just a single feature transform point $ft(\mathbf{r})$. So, the function ft is incompletely specified by $ft(\mathbf{r}) \in FT(\mathbf{r})$. In fact, we compute $ft(\mathbf{r})$ as the first element of $FT(\mathbf{r})$ with respect to a lexical ordering.

The computation of ft proceeds in d phases. We specify the results of these phases as follows. For $0 < i \leq d$, let L_i be the i-dimensional subspace spanned by the first i standard basis vectors of $\mathbb{R}^d$. The i-th phase computes the i-dimensional feature transform ft_i which is characterized by $ft_i(\mathbf{r}) \in FT(\mathbf{r}, B \cap (\mathbf{r} + L_i))$. The result of the last phase is $ft = ft_d$. Since the components of $ft_i(\mathbf{r})$ orthogonal to L_i are always equal to the corresponding components of $\mathbf{r}$, we only compute and use the orthogonal projection of ft_i on L_i.

In Figures 1 and 2, we present the computation for the case $d = 3$ in a box of size (m, n, p). Since ft_i is a vector-valued function, the three components of $ft_i(\mathbf{r})$ are written $ft_i[\mathbf{r}].x, ft_i[\mathbf{r}].y$, and $ft_i[\mathbf{r}].z$.

The first phase is the computation of ft_1 given in Fig. 1. For every pair (y, z), it consists of two scans over the line $(0, y, z) + L_1$. The boundary B is represented here by a 3D boolean array b. In the first scan, $g[x]$ becomes the distance to the next boundary point along the line. The second scan collects ft_1.

The second and third phases are given in Fig. 2. In the body of the outer loop, the value of ft_i is computed from ft_{i-1} for a given scan line, again by two scans. The results of the forward scan are collected on stacks s and t, with common stack pointer q. The backward scan reaps ft_i as harvest. The auxiliary functions f and Sep are given by $f(i, u) = (i - u)^2 + g(u)$ and

$Sep(i, u) = (u^2 - i^2 + g(u) - g(i))$ **div** $(2(u - i))$, where the function g is the squared Euclidean distance transform of the previous phase. So, $g(i) = (x - ft_1[x, i, z].x)^2$ in phase 2, and $g(i) = (x - ft_2[x, y, i].x)^2 + (y - ft_2[x, y, i].y)^2$ in phase 3. Note that, in the body of the outer loop, we regard x and z as constants for phase 2, and x and y as constants for phase 3.

Since the algorithm is completely analogous to our algorithm for the Euclidean distance transform, we refer to paper [13] for further details.

```
forall y ∈ [0..n − 1], z ∈ [0..p − 1] do
  (∗ scan 1 ∗)
  if b[m − 1, y, z] then   g[m − 1] := 0
  else   g[m − 1] := ∞
  endif
  for x := m − 2 downto 0 do
    if b[x, y, z] then   g[x] := 0
    else   g[x] := 1 + g[x + 1]
    endif
  end for
  (∗ scan 2 ∗)
  ft₁[0, y, z].x := g[0]
  for x := 1 to m − 1 do
    if x − ft₁[x − 1, y, z].x ≤ g[x] then
      ft₁[x, y, z].x := ft₁[x − 1, y, z].x
    else
      ft₁[x, y, z].x := x + g[x]
    endif
  end forall
```

Figure 1.   Program fragment for the first phase - one dimensional feature transform in 3D.

3. Skeletonization

The feature transform of a data set can be used to compute its skeleton. We first examine the definition of the medial axis [3], see also [5–7, 16]. Actually, we present three possible formalizations: *CMD*, *RMA*, and *IMA*. Since *RMA* is not restricted to grid points, whereas *CMD* and *IMA* are, the latter two are the main contenders.

The real medial axis and *CMD* skeleton.   For the moment we assume that the boundary B is a closed subset of $\mathbb{R}^d$. For every point $x \in \mathbb{R}^d$, we can form the largest open disk $D(x, r) = \{y \in \mathbb{R}^d \mid \|x - y\| < r\}$ that is disjoint with B. This is called the *inscribed disk* of x. If an inscribed disk at point p is not contained in any other inscribed disk of B, we call it a *maximal disk* with center p. We define the *real medial axis RMA* to consist of the points $x \in \mathbb{R}^d \setminus B$ which are centers of maximal disks.

forall $x \in [0..m - 1], z \in [0..p - 1]$ **do**	**forall** $x \in [0..m - 1], y \in [0..n - 1]$ **do**
$\quad q := 0; s[0] := 0; t[0] := 0$	$\quad q := 0; s[0] := 0; t[0] := 0$
$\quad$ **for** $u := 1$ **to** $n - 1$ **do** ($*$ scan 1 $*$)	$\quad$ **for** $u := 1$ **to** $p - 1$ **do** ($*$ scan 1 $*$)
$\quad\quad$ **while** $q \geq 0 \wedge f(t[q], s[q]) > f(t[q], u)$ **do**	$\quad\quad$ **while** $q \geq 0 \wedge f(t[q], s[q]) > f(t[q], u)$ **do**
$\quad\quad\quad q := q - 1$	$\quad\quad\quad q := q - 1$
$\quad\quad$ **if** $q < 0$ **then**	$\quad\quad$ **if** $q < 0$ **then**
$\quad\quad\quad q := 0; s[0] := u$	$\quad\quad\quad q := 0; s[0] := u$
$\quad\quad$ **else**	$\quad\quad$ **else**
$\quad\quad\quad w := 1 + Sep(s[q], u)$	$\quad\quad\quad w := 1 + Sep(s[q], u)$
$\quad\quad\quad$ **if** $w < n$ **then**	$\quad\quad\quad$ **if** $w < p$ **then**
$\quad\quad\quad\quad q := q + 1; s[q] := u; t[q] := w$	$\quad\quad\quad\quad q := q + 1; s[q] := u; t[q] := w$
$\quad\quad\quad$ **endif**	$\quad\quad\quad$ **endif**
$\quad\quad$ **endif**	$\quad\quad$ **endif**
$\quad$ **end for**	$\quad$ **end for**
$\quad$ **for** $u := n - 1$ **downto** 0 **do** ($*$ scan 2 $*$)	$\quad$ **for** $u := p - 1$ **downto** 0 **do** ($*$ scan 2 $*$)
$\quad\quad ft_2[x, u, z].x := ft_1[x, s[q], z].x$	$\quad\quad ft_3[x, y, u].x := ft_2[x, y, s[q]].x$
$\quad\quad ft_2[x, u, z].y := s[q]$	$\quad\quad ft_3[x, y, u].y := ft_2[x, y, s[q]].y$
$\quad\quad$ **if** $u = t[q]$ **then** $q := q - 1$ **endif**	$\quad\quad ft_3[x, y, u].z := s[q]$
$\quad$ **end for**	$\quad\quad$ **if** $u = t[q]$ **then** $q := q - 1$ **endif**
end forall	$\quad$ **end for**
	end forall

(a) Second phase	(b) Third phase

Figure 2. Program fragments for the second and third phase.

For $x \in \mathbb{Z}^d$, the *inscribed integer disk* $M(x)$ is the intersection $D(x, r) \cap \mathbb{Z}^d$, where $D(x, r)$ is its inscribed disk. The set *CMD* (centers of maximal disks) consists of the points $x \in \mathbb{Z}^d$ for which $M(x)$ is not contained in any $M(y)$ with $y \neq x$, see also [7, 16]. As is presumably well known, it is not true that $CMD \subseteq RMA \cap \mathbb{Z}^d$.

EXAMPLE 1 *Let B consist of the four points $(0, 0)$, $(3, 0)$, $(0, 3)$, and $(3, 3)$. The intersection $RMA \cap \mathbb{Z}^d$ is empty, but CMD contains the points $(1, 1)$, $(1, 2)$, $(2, 1)$, and $(2, 2)$.*

Our aim is to define a skeleton that looks like the real medial axis of a smoothing of the boundary and tends to be connected when the complement of the boundary is connected, while still being computable in linear time.

Recall that $dt(x) = \min\{\|x - y\| \mid y \in B\}$ and $FT(x) = \{y \in B \mid \|x - y\| = dt(x)\}$. Clearly, $dt(x)$ is the radius of the inscribed disk of x (for $x \in B$, we regard the empty set as an open disk with radius 0). The function $ft : \mathbb{R}^d \to B$ is incompletely specified by $ft(x) \in FT(x)$.

The next lemma may not be surprising, but it seems to be new.

LEMMA 2 *Assume B is a discrete (i.e., locally finite) subset of $\mathbb{R}^d$. Let $x \in \mathbb{R}^d$. Then $x \in RMA$ if and only if $FT(x)$ has more than one element.*

This lemma is not true when B is not discrete. For example, in the case of an ellipse, the real medial axis is a segment of the long axis strictly inside of the ellipse; the two extremal points of the segment belong to *RMA* and yet have only one element in the feature transform set.

Henceforth, we assume that the boundary consists of grid points only, i.e. that $B \subseteq \mathbb{Z}^d$. It follows that B is discrete, so that Lemma 2 applies. The following result is almost trivial to verify, but it is quite useful.

LEMMA 3 *Let $x \in \mathbb{R}^d$ and let y, z be two different elements of $FT(x)$. Then $||y - z|| \geq 1$. If moreover $x \in \mathbb{Z}^d$, then $||y - z|| > 1$.*

The integer medial axis.   Since we assume the boundary now to consist of grid points only, *RMA* contains many points that would disappear when the boundary is smoothed to the curved (hyper)surface in $\mathbb{R}^d$ it is supposed to represent. For example, in the case of a boundary that consists of the grid points of a horizontal line in $\mathbb{R}^2$, the real medial axis consists of the vertical lines with odd-half-integer x coordinates. The following definition avoids these unwanted points.

DEFINITION 4 *Let $E = \{e \in \mathbb{Z}^d \mid ||e|| = 1\}$. The integer medial axis IMA consists of the points $p \in \mathbb{Z}^d$ such that for some $e \in E$ we have $||ft(p + e) - ft(p)|| > 1$ and $||m - ft(p + e)|| \leq ||m - ft(p)||$ where $m = p + \frac{1}{2}e$ is the midpoint of the line segment from p to $p + e$.*

The second condition on the pair $(p, p+e)$ in the definition of *IMA* is introduced to get one point, rather than two, and specifically the point that is closest to the perpendicular bisector of the line segment from $ft(p)$ to $ft(p+e)$. If p and $p+e$ have equal claims, both are included. The reason to use ft rather than FT is that ft is computationally cheaper, but also that the restriction of FT to $\mathbb{Z}^d$ may well be everywhere single-valued, so that consideration of neighbouring points is needed in any case.

We prefer *IMA* over *CMD* since it is easier to compute and seems to give more image information when the boundary is a discretization of a continuous boundary.

The following lemma is easy to prove.

LEMMA 5 *$IMA \cap B = \emptyset$.*

The definition of *IMA* is primarily motivated by the next result that shows that *IMA* has "enough" elements.

THEOREM 6 *Let p and q be points of the boundary B. Every Manhattan-shortest path from p to q that is not contained in B, contains a point of IMA.*

Proof: Let $r(i)$, $0 \leq i \leq k$ be a Manhattan-shortest path from p to q that is not contained in B. Since it is a Manhattan-shortest path from p to q, we have

$r(0) = p$, $r(k) = q$, and $||r(i+1) - r(i)|| = 1$ for all $0 \leq i < k$. Since the path is not contained in B, there is an index j with $0 < j < k$ and $r(j) \notin B$. Without loss of generality, we may assume $r(1) \notin B$.

Let $x(i) = ft(r(i))$ for all i. Then $x(0) = p$ and $x(k) = q$ and $x(1) \neq r(1)$. We have $||p - r(1)|| = 1$ and hence $dt(r(1)) = 1$. By Lemma 3, this implies that $x(1) = x(0)$ or $||x(1) - x(0)|| > 1$. It follows that function x represents a path from p to q in k steps that is not a Manhattan-shortest path. This implies that there is an index j with $0 \leq j < k$ and $||x(j+1) - x(j)|| > 1$. Put $m = \frac{1}{2}(r(j+1) + r(j))$. If $||m - x(j+1)|| \leq ||m - x(j)||$ then $r(j) \in IMA$. Otherwise $r(j+1) \in IMA$. In that case $j + 1 < k$ because of Lemma 5. $\square$

While the previous result can be interpreted as saying that *IMA* has enough elements, the next result shows that *IMA* has not too many elements, in the sense that every one of them is close to *RMA*.

THEOREM 7 *For every $p \in IMA$, there is $e \in E$ and $t \in \mathbb{R}$ with $0 \leq t \leq \frac{1}{2}$ and $p + te \in RMA$.*

Proof: Let $p \in IMA$. Then there is $e \in E$ with $||ft(p) - ft(p+e)|| > 1$ and $||m - ft(p)|| \geq ||m - ft(p+e)||$ where $m = p + \frac{1}{2}e$. First, assume that $ft(p) \in FT(m)$. Then $ft(p)$ is a closest point on B to m. So $||m - ft(p)|| \leq ||m - ft(p+e)||$. Since $||m - ft(p)|| \geq ||m - ft(p+e)||$, it follows that $||m - ft(p)|| = ||m - ft(p+e)||$ and that both $ft(p)$ and $ft(p+e)$ are elements of $FT(m)$. In view of lemma 2, this implies that $m \in RMA$ is a point as looked for. It remains to treat the case with $ft(p) \notin FT(m)$. Let point z be the last point of the line segment from p to m with $ft(p) \in FT(z)$. By continuity, this point exists. Since $ft(p) \notin FT(z')$ for points z' arbitrary close to z, the set $FT(z)$ consists of more than one element. So $z \in RMA$. $\square$

As illustrated by Theorem 6, *IMA* has some good connectivity properties. In that respect, it is better than *CMD*.

EXAMPLE 8 *Let B be the intersection of $\mathbb{Z}^2$ with the union of the x-axis and the y-axis. Then IMA consists of the points $(x, \pm x)$ for all $x \in \mathbb{Z} \setminus \{0\}$, and CMD is a subset of IMA that contains $(\pm 3, \pm 3)$ and $(\pm 4, \pm 4)$ but misses at least the points $(\pm 1, \pm 1)$, $(\pm 2, \pm 2)$, $(\pm 5, \pm 5)$.*

In general, it seems that, if the complement of B is bounded and connected, then *IMA* is connected (with respect to 8-connectivity in $\mathbb{Z}^2$, or more generally, L_∞-connectivity for $\mathbb{Z}^d$).

A disadvantage of *IMA* is that it can (weakly) depend on the choice of function ft within FT.

Implementation. The code for the skeletonization step is shown in Fig. 3. One may work with squared distances instead of distances, which avoids the computation of square roots and thus saves time.

When the medial axis is used for image analysis, it is often useful to prune it of disturbing details in some postprocessing phase. Our construction of the integer medial axis yields some information that is very useful for this purpose. The easiest pruning is to strengthen the condition $\|ft(p) - ft(p+e)\| > 1$ in the definition of *IMA* by replacing '> 1' by '$> \gamma$' for some pruning parameter γ. This removes some points of *IMA* that are due to irregularities of the boundary.

With the tunable parameter γ, skeletons may be computed according to a user's need. Unwanted skeleton points which still remain can be removed in a postprocessing step, if desired.

```
procedure IMA skeleton
for i := 0 to m − 1 do
  for j := 0 to n − 1 do
    for k := 0 to p − 1 do
      if i > 0 then compare(i,j,k,i-1,j,k) endif
      if j > 0 then compare(i,j,k,i,j-1,k) endif
      if k > 0 then compare(i,j,k,i,j,k-1) endif
    end for
  end for
end for
```

```
procedure compare(i,j,k,p,q,r)
x := [i, j, k]; y := [p, q, r]
x_f := ft3[x]; y_f := ft3[y]
if ||x_f − y_f|| > γ then
  crit := inprod(x_f − y_f, x_f + y_f − x − y)
  if crit ≥ 0 then  skel [x]:= 1
  endif
  if crit ≤ 0 then  skel [y]:= 1
  endif
endif
```

Figure 3. Program fragment for computing the IMA skeleton from the feature transform.

Table 1. Timing results (in seconds) for several data sets.

Data	Size	Feature transform	Skeleton	Total
angio	256x256x128	3	4	7
engine	256x256x128	4	4	8
tooth	256x256x256	7	7	14
vessels	256x256x256	10	6	16
head	256x256x256	9	7	16

4.   Results

We have run the skeletonization algorithm on several 3D data sets. Timing results are given for three 3D data sets, i.e. CT scans of a head, a tooth and a number of blood vessels. The size of these sets and their timing results are given in Table 1. These results were obtained on an 1.7 GHz Pentium M processor with 1024 Mb internal memory.

Since the 3D skeletons form surfaces, they are somewhat hard to visualize. Therefore, to get an idea of the quality of our skeletonization algorithm, we first give a number of examples of 2D skeletons, see Fig. 4. For the 3D case, some insight into the structure of the skeleton surfaces can be gained by using volume rendering techniques. An example for the tooth data set is given in Fig. 5. For a better impression a sequence of views from different viewpoints is desired, which can be played as a movie.

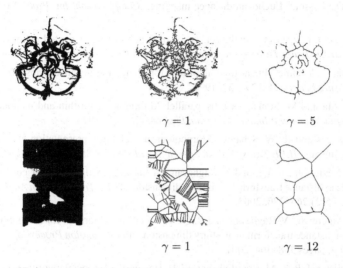

$\gamma = 1$ $\gamma = 5$

$\gamma = 1$ $\gamma = 12$

Figure 4.   2D images with their skeletons. Left: original images. Middle: IMA skeleton. Right: pruned IMA skeleton.

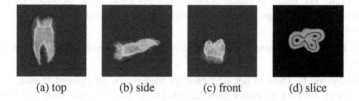

(a) top   (b) side   (c) front   (d) slice

Figure 5.   (a)-(c): Volume renderings of skeletons (white) inside the original data volumes. (d): Slice of the original tooth data combined with the skeleton.

References

[1] D. Attali and A. Montanvert. Computing and simplifying 2D and 3D continuous skeletons. *Computer Vision and Image Understanding*, 67(3):161–273, 1997.

[2] S. Beucher. Digital skeletons in Euclidean and geodesic spaces. *Signal Processing*, 38:127–141, 1994.

[3] H. Blum. A transformation for extracting new descriptors of shape. In W. Wathen-Dunn, editor, *Proc. Symposium Models for the perception of speech and visual form, Boston, November 1964*, pages 362–380. MIT Press, Cambridge, MA, 1967.

[4] G. Borgefors, I. Nystrom, and G. S. D. Baja. Computing skeletons in three dimensions. *Pattern Recognition*, 32(7):1225–1236, 1999.

[5] D. Coeurjolly. *d*-Dimensional reverse Euclidean distance transformation and Euclidean medial axis extraction in optimal time. In N. et al., editor, *DGCI 2003*, pages 327–337, New York, 2003. Springer. (LNCS 2886).

[6] P. E. Danielsson. Euclidean distance mapping. *Comp. Graph. Im. Proc.*, 14:227–248, 1980.

[7] Y. Ge and J. Fitzpatrick. On the generation of skeletons from discrete Euclidean distance maps. *IEEE Trans. Pattern Anal. Machine Intell.*, 18:1055–1066, 1996.

[8] T. Hirata. A unified linear-time algorithm for computing distance maps. *Information Processing Letters*, 58:129–133, 1996.

[9] C. M. Ma and M. Sonka. A fully parallel 3d thinning algorithm and its applications. *Computer Vision and Image Understanding*, 64:420–433, 1996.

[10] P. Maragos and R. W. Schafer. Morphological skeleton representation and coding of binary images. *IEEE Trans. Acoust. Speech Signal Proc.*, ASSP-34:1228–1244, 1986.

[11] C. R. Maurer Jr., R. Qi, and V. Raghavan. A linear time algorithm for computing the euclidean distance transform in arbitrary dimensions. *IEEE Trans. Pattern Anal. Machine Intell.*, 25(2):265–270, 2003.

[12] C. R. Maurer Jr., V. Raghavan, and R. Qi. A linear time algorithm for computing the euclidean distance transform in arbitrary dimensions. In *Information Processing in Medical Imaging*, pages 358–364, 2001.

[13] A. Meijster, J. B. T. M. Roerdink, and W. H. Hesselink. A general algorithm for computing distance transforms in linear time. In J. Goutsias, L. Vincent, and D. S. Bloomberg, editors, *Mathematical Morphology and its Applications to Image and Signal Processing*, pages 331–340. Kluwer Acad. Publ., Dordrecht, 2000.

[14] F. Meyer. The binary skeleton in three steps. In *Proc. IEEE Workshop on Computer Architecture and Image Database Management, IEEE Computer Society Press*, pages 477–483, 1985.

[15] J. R. Parker. *Algorithms for Image Processing and Computer Vision*. John Willey & Sons, 1996.

[16] E. Remy and E. Thiel. Look-up tables for medial axis on squared Euclidean distance transform. In N. et al., editor, *DGCI 2003*, pages 224–235, New York, 2003. Springer. (LNCS 2886).

[17] J. Serra. *Image Analysis and Mathematical Morphology*. Academic Press, New York, 1982.

[18] F. Y. Shih and C. C. Pu. A skeletonization algorithm by maxima tracking on Euclidean distance transform. *Pattern Recognition*, 28(3):331–341, 1995.

[19] H. Talbot and L. Vincent. Euclidean skeletons and conditional bisectors. In *Proc. SPIE Visual Communications and Image Processing'92, Boston (MA)*, volume 1818, pages 862–876, Nov. 1992.

[20] L. Vincent. Efficient computation of various types of skeletons. In *Proc. SPIE Symposium Medical Imaging V, San Jose, CA*, volume 1445, pages 297–311, Feb. 1991.

DIGITIZATION OF NON-REGULAR SHAPES

Peer Stelldinger

Cognitive Systems Group, University of Hamburg,
Vogt-Koelln-Str. 30, D-22527 Hamburg, Germany
stelldinger@informatik.uni-hamburg.de

Abstract   Only the very restricted class of r-regular shapes is proven not to change topology during digitization. Such shapes have a limited boundary curvature and cannot have corners. In this paper it is shown, how a much wider class of shapes, for which the morphological open-close and the close-open-operator with an r-disc lead to the same result, can be digitized correctly in a topological sense by using an additional repairing step. It is also shown that this class is very general and includes several commonly used shape descriptions. The repairing step is easy to compute and does not change as much pixels as a preprocessing regularization step. The results are applicable for arbitrary, even irregular, sampling grids.

Keywords: shape, digitization, repairing, topology, reconstruction, irregular grid

Introduction

The processing of images by a computer requires their prior digitization. But as Serra already stated in 1980 [5], "To digitize is not as easy as it looks." Shapes can be regarded as binary images and the simplest model for digitization is to take the image information only at some sampling points and to set the associated pixels to these values. Unfortunately even for this simple digitization model only a very restricted class of binary images is proven to preserve topological characteristics during digitization: Serra proved that the homotopy tree (i.e. the inclusion properties of foreground- and background components) of r-regular images (see Definition 1) does not change under digitization with a hexagonal grid of certain density [5]. Similarily Pavlidis showed that such images can be digitized with square grids of certain density without changing topology [4]. Latecki [3] also referred to this class of shapes. Recently the author proved together with Köthe that r-regular sets are not only sufficient but also necessary to be digitized topologically correctly with *any* sampling grid of a certain density [6]. But most shapes are not r-regular, e.g. have corners. To solve this problem Pavlidis said "Indeed suppose that we have a class of objects whose contours contain corners. We may choose a radius of curvature

269

C. Ronse et al. (eds.), Mathematical Morphology: 40 Years On, 269–278.
©2005 *Springer. Printed in the Netherlands.*

r and replace each corner by a circular arc with radius r" [4]. This approach to make shapes r-regular has two problems: (1) Pavlidis gives no algorithm how to do it exactly. He also does not say, for which shapes it is possible without changing the topology of the set. (2) It is a preprocessing step and thus cannot be computed by a computer, which only gets the digitized information. The aim of this paper is to solve both problems. After a short introduction in the definitions of r-regular images, sampling and reconstruction (section 1), the class of r-halfregular sets is defined in section 2, whose elements can be converted into r-regular sets by using a very simple morphological preprocessing step. In order to solve the second problem, it is shown how these shapes can be digitized topologically correctly by using a postprocessing algorithm instead of the preprocessing. These results are applicable for digitization with any type of sampling grid – only a certain density is needed. In section 3 it is shown that the concept of r-halfregular shapes includes several other shape descriptions. Finally in section 4 the postprocessing step is even more simplified in case of certain sampling grids. For square grids it simply means to delete all components and to fill all holes, which do not contain a 2x2 square of pixels. This is remarkably similar to the results of Giraldo et al. [2], who proved that finite polyhedra can be digitized with intersection digitization without changing their homotopy properties by filling all holes, which do not contain a 2x2 square. Unfortunately their approach was not applicable to other sets and was restricted to another digitization model.

1.   Regular Images, Sampling and Reconstruction

At first some basic notations are given: The Euclidean distance between two points x and y is noted as $d(x, y)$ and the Hausdorff distance between two sets is the maximal distance between one point of one set and the nearest point of the other. The Complement of a set A will be noted as A^c. The boundary ∂A is the set of all common accumulation points of A and A^c. A set A is open, if it does not intersect its boundary and it is closed if it contains the boundary, $A^0 := A \setminus \partial A$, $\overline{A} := A \cup \partial A$. $\mathcal{B}_r(c) := \{x \in \mathbb{R}^2 | d(x, c) \le r\}$ and $\mathcal{B}_r^0(c) := (\mathcal{B}_r(c))^0$ denote the closed and the open disc of radius r and center c. If $c = (0,0)$, write $\mathcal{B}_r$ and $\mathcal{B}_r^0$. The r-dilation $A \oplus \mathcal{B}_r^0$ of a set A is the union of all open r-discs with center in A and the r-erosion $A \ominus \mathcal{B}_r^0$ is the union of all center points of open r-discs lying inside of A. The morphological opening with an open r-disc is defined as $A \circ \mathcal{B}_r^0 := (A \ominus \mathcal{B}_r^0) \oplus \mathcal{B}_r^0$ and the respective closing as $A \bullet \mathcal{B}_r^0 := (A \oplus \mathcal{B}_r^0) \ominus \mathcal{B}_r^0$. The concept of r-regular images was introduced independently by Serra [5] and Pavlidis [4]. These sets are extremely well behaved – they are smooth, round and do not have any cusps (e.g. see Fig. 2B). Furthermore r-regular sets are invariant under morphological opening and closing, as already stated by Serra [5].

DEFINITION 1 *A set $A \subset \mathbb{R}^2$ is called r-regular if for each boundary point of A it is possible to find two osculating open discs of radius r, one lying entirely in A and the other lying entirely in A^c.*

Each outside or inside osculating disc at some boundary point x of a set A defines a tangent through x, which is unique if there exists both an outside and an inside osculating disc. The definitions of r-erosion and r-dilation imply that the boundary of a set does not change under opening or closing with an r-disc iff it is r-regular. In order to compare analog with digital images, a definition of the processes of sampling and reconstruction is needed. The most obvious approach for sampling is to restrict the domain of the image function to a set of sampling points, called sampling grid. In most approaches only special grids like square or hexagonal ones are taken into account [3][4, 5]. A more general approach only needs a grid to be a countable subset of $\mathbb{R}^2$, with the sampling points being not too sparse or too dense anywhere [6]. There the pixel shapes are introduced as Voronoi regions. Together with Köthe the author proved a sampling theorem, saying that a closed r-regular image is $\mathbb{R}^2$-homeomorphic to its reconstruction with an r'-grid if only $r' < r$ [6]. Two sets being $\mathbb{R}^2$-homeomorphic means that there exists a homeomorphism from $\mathbb{R}^2$ to $\mathbb{R}^2$, which maps the sets onto each other.

DEFINITION 2 *A countable set $S \subset \mathbb{R}^2$ of sampling points, where the Euclidean distance from each point $x \in \mathbb{R}^2$ to the nearest sampling point is at most $r \in \mathbb{R}$, is called an r-grid if $S \cap A$ is finite for any bounded set $A \in \mathbb{R}^2$. The pixel $\text{Pixel}_S(s)$ of a sampling point s is its Voronoi region, i.e. the set of all points lying at least as near to this point than to any other sampling point. The union of the pixels with sampling points lying in A is the reconstruction of A w.r.t. S: $\hat{A} := \bigcup_{s \in S \cap A} \text{Pixel}_S(s)$. Two pixels are adjacent if they share an edge. Two pixels of $\hat{A}$ are connected if there exists a chain of adjacent pixels in $\hat{A}$ between them. A component of $\hat{A}$ is a maximal set of connected pixels.*

2. Digitization of Halfregular Sets

Most shapes are not r-regular for any r. So if one wants to apply the above mentioned sampling theorem one at first has to construct an r-regular version of the shape before sampling, as suggested by Pavlidis. The question is how to define such a preprocessing step. Obviously a set, which is the result of an r-opening of another set, has an inside osculating open r-disc at any boundary point. Equivalently the r-closing of any set has an outside osculating open r-disc at any boundary point. So the straight forward idea is to combine these two operators. Unfortunately the r-closing of an r-open set does not need to be r-open anymore and as a consequence the open-close and the close-open operator with the same structuring element can have totally different results. The really interesting case is when this does not happen:

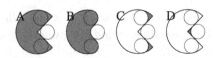

Figure 1.   The areas which change during regularization can be classified info three types: r-tips (left), r-waists (center) and r-spots (right).

Figure 2.   From left to right: An r-halfregular set A, its r-regularization B, and the changes of the fore- and background due to regularization C and D.

DEFINITION 3 *A set A is called r-regularizable if the open-close and the close-open-operator of radius r lead to the same (except of the boundary), $(A \circ \mathcal{B}_r^0) \bullet \mathcal{B}_r^0 = (A \bullet \mathcal{B}_r^0) \circ \mathcal{B}_r^0$. The r-regularization of A is $(A \circ \mathcal{B}_r^0) \bullet \mathcal{B}_r^0$.*

LEMMA 4 *The r-regularization of an r-regularizable set A is r-regular.*

PROOF Since opening and closing are idempotent, $(A \bullet \mathcal{B}_r^0) \circ \mathcal{B}_r^0$ is open and $(A \circ \mathcal{B}_r^0) \bullet \mathcal{B}_r^0)$ is closed w.r.t. $\mathcal{B}_r^0$. This implies r-regularity of A.   □

Note, that a shape and its regularization do not need to be $\mathbb{R}^2$-homeomorphic, since the topology can be totally changed during the regularization step. The changes can be classified into waists, tips and spots (see Fig. 1). The waists cause the biggest problems, because even big and thus important components can change their topology under regularization if they have waists. So if one wants to regularize a set it should have no waists.

DEFINITION 5 *For some set A let A' be a component of $(A \setminus (A \circ \mathcal{B}_r^0))^0$. Further let n be the number of open r-discs lying in A and touching A'. These discs are called* bounding discs *of A'. If n is zero, A' is called r-spot of A. If n is equal to 1, A' is called r-tip of A and if n is greater than 1, it is called r-waist of A (see Fig. 1). A set A is called r-halfregular if for each boundary point there exists an open inside or an open outside osculating disc of radius r, completely lying inside, respectively outside of A, and if neither A nor A^c has an s-waist for any $s \leq r$.*

Obviously an r-halfregular set is also s-halfregular for any $s < r$. For the rest of this section let A be an r-halfregular set with $r > 0$. Further let $B := (A \circ \mathcal{B}_r^0) \bullet \mathcal{B}_r^0$ be its r-regularization. $C := A \setminus (A \circ \mathcal{B}_r^0)$ shall be the difference between A and its opening with $\mathcal{B}_r^0$ and $D := (A \bullet \mathcal{B}_r^0) \setminus A$ the difference between A and its closing with $\mathcal{B}_r^0$ (see Fig. 2). The components of C and D are the r-spots, r-tips and r-waists, which change during the preprocessing regularization step.

LEMMA 6 *For each boundary point of C or of D there exists an outside osculating open r-disc and no component of $C \oplus \mathcal{B}_r^0$ or of $D \oplus \mathcal{B}_r^0$ contains an open disc of radius $2r$ as subset.*

PROOF C can have no open disc of radius r as subset, because due to the definition of C the center of such a disc is not in C. Now let x be a boundary point of C. Then x is either also boundary point of A or of $A \circ B_r^0$. In the first case there exists an osculating open r-disc lying completely outside of C since A is r-halfregular and C cannot include an inside osculating r-disc. In the second case there also exists an outside osculating disc for C, since $A \circ B_r^0$ is open w.r.t. B_r^0. Thus $\overline{C}$ is closed w.r.t. B_r^0 and $C \oplus B_r^0$ does not contain any open disc of radius $2r$ as subset. The proof for D is analog. □

LEMMA 7 *Let A be an r-halfregular set. Then every boundary point $y \in \partial A$ is also boundary point of $A \bullet B_r^0$ or $A \circ B_r^0$ and A is r-regularizable.*

PROOF Let y be some boundary point of A. If there exists an outside [inside] osculating r-disc, then y remains boundary point after r-closing [r-opening]. Now suppose A is not r-regularizable. Then $(A \circ B_r^0) \bullet B_r^0 \neq \overline{(A \bullet B_r^0) \circ B_r^0}$ and there either exists a point $x \in (A \circ B_r^0) \bullet B_r^0$, which is not in $\overline{(A \bullet B_r^0) \circ B_r^0}$ or there exists a point $x \in \overline{(A \bullet B_r^0) \circ B_r^0}$, which is not in $(A \circ B_r^0) \bullet B_r^0$. Such an x cannot lie inside or on the boundary of an r-disc being subset of A, because then x would be element of $\overline{A \circ B_r^0}$ and thus $x \in (A \circ B_r^0) \bullet B_r^0$, and – since closing is extensive and opening is increasing – $x \in \overline{(A \bullet B_r^0) \circ B_r^0}$ is true. If x lies inside or on the boundary of an r-disc in A^c, it cannot be in $(A \circ B_r^0) \bullet B_r^0$ or $\overline{(A \bullet B_r^0) \circ B_r^0}$ for analog reasons. Now suppose, x is in A, but not inside or on the boundary of some r-disc in A, thus $x \in C^0$. C has an outside osculating open r-disc at any boundary point y due to Lemma 6. Now let y be the boundary point of C being nearest to x. Then any y tangent has to be orthogonal to $\overline{xy}$. Thus there exists a unique tangent and also a unique outside osculating r-disc. Obviously the distance between x and y is smaller than r. Since y remains boundary point after r-closing of C, x cannot be in $\overline{(A \bullet B_r^0) \circ B_r^0}$. $x \oplus B_r^0$ can only intersect one outside osculating open r-disc of C lying inside $A \circ B_r^0$, because there exists at most one such disc for each component C' of C^0 due to the absense of r-waists. The center point of this disc is the only point of $A \ominus B_r^0$ having a distance of at most r to some point lying in C' or in $\partial C'$. Thus x cannot be element of $(A \circ B_r^0) \bullet B_r^0$. Analogously any x in D^0 is element of both $(A \circ B_r^0) \bullet B_r^0$ and $\overline{(A \bullet B_r^0) \circ B_r^0}$. Thus any $x \in \mathbb{R}^2$ is element of $(A \circ B_r^0) \bullet B_r^0$ iff it is element of $\overline{(A \bullet B_r^0) \circ B_r^0}$. □

As a consequence of Lemma 7 A can be constructed (except of its boundary) as $A^0 = (B \cup C)^0 \ominus D$. In the following the sampling theorem for halfregular sets is developed. Therefore one lemma needs to be proved before.

LEMMA 8 *No background [foreground] component in the reconstruction of $B \cup C$ [$B \setminus D$] w.r.t. an r'-grid, $r' < r$, is subset of $C \oplus B_r^0$ [$D \oplus B_r^0$].*

PROOF Let $c \in A^c$ be a background sampling point in $C \oplus \mathcal{B}_r^0$. Due to Lemma 6 there exists an open r-disc in C^c such that c lies in the disc. This disc can be chosen such that it lies either completely in B or completely outside of B. The center m of the disc is not in $C \oplus \mathcal{B}_r^0$. The halfline starting at c and going through m crosses $\partial \operatorname{Pixel}(c)$ at exactly one point c'. If $d(c, m) \leq d(c, c')$, m lies in $\operatorname{Pixel}(c)$ and thus the pixel is connected to the area outside of $C \oplus \mathcal{B}_r^0$, which implies that c cannot be part of a separate background component covered by $C \oplus \mathcal{B}_r^0$. If $d(c, m) > d(c, c')$, let g be the line defined by the edge of $\operatorname{Pixel}(c)$ going through c'. If there are two such lines (i.e. at a pixel corner), one is chosen arbitrarily.p he point c'' constructed by mirroring c on g is also a sampling point, and their pixels are adjacent. c'' lies on the circle of radius $d(d', c) = d(c', c'')$ with center c'. Among all points on this circle, c has the largest distance to m, and in particular $d(m, c'') < d(m, c)$. Thus, c'' lies outside of $C \oplus \mathcal{B}_r^0$, and is closer to m than c. By repeating this construction iteratively we obtain a chain of adjacent pixels whose sampling points successively get closer to m. Since $C \oplus \mathcal{B}_r^0$ contains only a finite number of sampling points, one such pixel will eventually not be covered by $C \oplus \mathcal{B}_r^0$. The constructed chain consists of pixels whose sampling points lie in a common r-disc outside of C. If this disc lies in B, they are not in the background, in contradiction to the supposition. Otherwise the pixels cannot be part of a separate background component in $C \oplus \mathcal{B}_r^0$. Since the chain is not infected by any sampling point lying in B they also cannot be part of a separate background component in the reconstruction of $B \cup C$ which is subset of $C \oplus \mathcal{B}_r^0$. Analogously there exists no foreground component in the reconstruction of $B \setminus D$, which is subset of $D \oplus \mathcal{B}_r^0$.   □

THEOREM 9 *Let A be a closed r-halfregular set with no $3r$-spot in A or in A^c, let $\hat{A}$ be the reconstruction of A with an r'-grid, $r' < r$, and let $\hat{A}'$ be the result of filling [deleting] all components of $\hat{A}^c$ [$\hat{A}$], which do not contain an open $2r$-disc. Then $\hat{A}'$ is $\mathbb{R}^2$-homeomorphic to A and the number of different pixels from $\hat{A}$ to $\hat{A}'$ is as most as high as from $\hat{A}$ to the reconstruction of the r-regularization B (see Fig. 3).*

PROOF $\hat{A}$ is equal to the union of the reconstructions of B and $C = A \setminus (A \circ \mathcal{B}_r^0)$ minus the reconstruction of $D = (A \bullet \mathcal{B}_r^0) \setminus A$. Due to Lemmas 4 and 7 is B r-regular and thus $\mathbb{R}^2$-homeomorphic to its reconstruction (see [6]). The components of $\hat{C}$ are either also separated components in $\hat{A}$ or connected with some component of $\hat{B}$. Lemma 6 states that no component of $C \oplus \mathcal{B}_r^0$ contains an open disc of radius $2r$. Due to $r' < r$, $\hat{C}$ is a subset of $C \oplus \mathcal{B}_r^0$. It follows that no component of $\hat{C}$ can contain an open disc of radius $2r$. Analogously no component of $\hat{D}$ can contain an open disc of radius $2r$. Due to Lemma 8 there cannot exist any background component in the reconstruction of $B \cup C$, which is subset of $C \oplus \mathcal{B}_r^0$, and there cannot exist any foreground component

Figure 3. The straightforward digitization may cause topological errors at r-tips (left) in contrast to the use of the regularization step (right) and the even better repairing step (center).

in the reconstruction of $B \setminus D$, which is subset of $D \oplus \mathcal{B}_r^0$. This implies that any separate component of $\hat{C}$ or of $\hat{D}$ is surrounded by pixels belonging to components which do not vanish under dilation or erosion with an open $3r$-disc. This also implies that the resulting image is independent of the order of the filling and deleting of components, which are subsets of $C \oplus \mathcal{B}_r^0$ or of $D \oplus \mathcal{B}_r^0$. So by filling all components of $\hat{A}^c$ and deleting all components of $\hat{A}$, which do not contain an open disc of radius $2r$, any component caused by C and D is affected, which is not part of bigger components in $\hat{A}$ and $\hat{A}^c$, respectively. Any component of $\hat{B}$ $[\hat{B}^c]$ is not deleted [filled], because it contains an open disc of radius $2r$ due to the fact that the corresponding component in B $[B^c]$ contains a disc of radius $3r$. It follows that the components of $\hat{A}'$ and $\hat{B}$ differ only in a way, which does not affect the topology or the neighborhood relations. Thus $\hat{A}'$ is $\mathbb{R}^2$-homeomorphic to B. Moreover they differ only in pixels lying in C or D. Since $\hat{A}$ and $\hat{A}'$ differ in all these pixels, the number of pixels, which changes due to the postprocessing repairing step is at most as big as the number of pixels, which changes due to the preprocessing regularization step. Since B can be constructed from A by removing r-tips from A and A^c, which is an $\mathbb{R}^2$-homeomorphic operation, $\hat{A}'$ is also $\mathbb{R}^2$-homeomorphic to A. $\square$

3. Examples for r-halfregular sets

In the last section it was shown, that digitization with repairing is sufficient to get a topologically correct digital version of some r-halfregular set, if each of the components of the set and its complement has a certain size (no $3r$-spots). Note that this is only a restriction to the sampling density and not a restriction to the class of correctly digitizable sets, since for each r-halfregular set there exists an $s \leq r$ such that the set has no $3s$-spots and is s-halfregular. Surprisingly the concept of r-halfregular sets is very general and several commonly used shape descriptions imply halfregularity. One example are polygonal shapes. The first part of Theorem 10 shows that any shape with polygonal boundary description is r-halfregular for some r. This is in particular interesting since any shape bounded by simple curves can be approximated by a polygonal shape, as shown by Bing[1]. Giraldo et al. already proved that such polygonal shapes can be reconstructed topologically correctly by using a simple repairing step, which is very similar to the one of this paper [2]. But their approach is restricted to polygonal shapes and square grids and uses intersection digitization. Intersection digitization can be simulated by the digitization model of this paper on r-dilated shapes. The second part of Theorem 10 states that any convex set is halfregular. Since convexity is stable under projection from 3D to 2D the shape of any image of a convex object, like a ball, a cylinder or a box, can be topologically correctly digitized by using the repairing step. This implies that also sets whose complement is convex, are halfregular. Since halfregularity bases only on local properties, sets are also halfregular, if they or their complements are convex in each local area of certain size, as can be seen in the third part of the theorem. These are only some examples for halfregular shapes. Another example are shapes which are bounded by spline curves, i.e. true type fonts. There it is not difficult to determine the minimal size of the spots and waists and the maximal curvature at non-corner points, which can be used to compute the minimal sampling density in order to digitize a text printout, such that topology is preserved (see Fig. 3).

THEOREM 10 *(a) Let A be a set, where the boundary components are polygonal Jordan curves. Further choose $r > 0$ such that the Hausdorff distance from each line segment to each non-adjacent corner point is at least $4r$ and the Hausdorff distance between each two non-adjacent line segments is at least $2r$. Then A is r-halfregular. (b) Each convex set is r-halfregular for any $r \in \mathbb{R}_+$. (c) Each set A, where for each boundary point $x \in \partial A$ either $A \cap \mathcal{B}_{2r}^0(x)$ or $A^c \cap \mathcal{B}_{2r}^0(x)$ is convex, is r-halfregular.*

PROOF (a) Since any waist has to be bounded by at least two non-adjacent line segments and since the distance between any two non-adjacent line segments is at least $2r$, there cannot exist any s-waist for $s \leq r$. Let a be an arbitrary boundary point of A and let $B = \mathcal{B}_{2r}^0(a)$ be the disc of radius $2r$ centered in a. a lies on some line segment L of the boundary, whose endpoints shall be called l_1 and l_2. If a is a corner point, choose either of the two adjacent line segments. Since the Hausdorff distance of non-adjacent line segments is at least $2r$, only line segments, which are adjacent to L can lie in B. W.l.o.g. let the distance from a to l_1 be at most as big as the distance from a to l_2. Then the line segment L' being adjacent to L by meeting in l_2 has a Hausdorff distance of at least $4r$ to l_1. If the angle between L and L' is at least $\pi/2$, L' cannot go through B. Otherwise suppose the foot of the perpendicular of L' going through a is in L'. Then due to the theorem on intersecting lines the Hausdorff distance from L' to a is $d(L', a) = d(a, l_2)/d(l_1, l_2) \cdot d(L', l_1) \geq \frac{1}{2} \cdot 4r$. If the foot of the perpendicular is not in L', L' is shorter than L. Then L' also cannot intersect

B, because otherwise L would intersect such a disc B' with radius $2r$ and center in L'. Thus the only line segments of the boundary, which can intersect B, are L and the adjacent line segment meeting L in l_1. The straight line which covers L cuts B into two open halfs such that L' intersects at most one of these halfs. The other half disc contains an open disc of radius r osculating a. This disc does not intersect the boundary, which implies r-halfregularity. □

(b) A convex set can be described as the intersection of halfplanes. Any outside osculating r-disc of some halfplane is completely outside of this intersection. Thus for any boundary point there exists an outside osculating r-disc. Obviously convex sets cannot have waists, which implies r-halfregularity. □

(c) First suppose there exists an s-waist A' of A with $s \leq r$. Now let x be a boundary point of A which is also boundary point of A'. If $A^c \cap \mathcal{B}_{2r}^0(x)$ is convex, then it is subset of a half of the $2r$-disc and the other half contains an r-disc lying in A. Thus $A \cap \mathcal{B}_{2r}^0(x)$ has to be convex. Since this has to be true for any such boundary point, A must be convex in the area $A' \oplus \mathcal{B}_{2r}^0$ of radius $2r$ around the waist. This implies that the union of the waist and its bounding discs also has to be convex. But this cannot be since the convex hull of this union is morphologically open w.r.t. $\mathcal{B}_s^0$ in contrast to the set itself. Thus there cannot exist any s-waist in A and analogously in A^c. For each boundary point x there exists an outside osculating r-disc if $A \cap \mathcal{B}_{2r}^0(x)$ is convex and an inside osculating disc if $A^c \cap \mathcal{B}_{2r}^0(x)$ is convex. Thus A is r-halfregular. □

4. Discrete Repairing

Although the repairing process is very simple – you only have to find components which do not contain a disc of a certain size – an implementation into a discrete algorithm is not straightforward since subpixel positions of such discs have to be considered. In this section it is shown that there are even better ways to find such components. The idea is that for any regular r'-grid there are only finitely many patterns which cover an r-disc ($r = r' + \varepsilon$) such that each pixel intersects the disc, and some patterns include others. So if one has a set of patterns such that any possible pattern is superset of an element of the set, one has only to look for components which do not include any of these elements. The following theorem shows this in detail for square and hexagonal grids.

THEOREM 11 *Let A be a closed r-halfregular set with no $3r$-spot in A or in A^c, let $\hat{A}$ be the reconstruction of A with a square grid [hexagonal grid] which is an r'-grid, $r' < r$, and let $\hat{A}'$ be the result of filling/deleting all components of $\hat{A}^c$ and $\hat{A}$, which do not contain the highlighted configuration shown in Fig. 4(a) [which do not contain one of the highlighted configurations in Fig. 1(b)]. Then $\hat{A}'$ is $\mathbb{R}^2$-homeomorphic to A and the number of different pixels between $\hat{A}$ and $\hat{A}'$ and $\hat{A}$ is as most as high as between $\hat{A}$ and the reconstruction of the r-regularization B.*

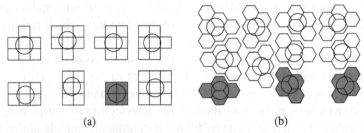

Figure 4.   There is only a finite number of pixel configurations which cover an r-disc when using square (a) or hexagonal (b) grids and a minimal set of subconfigurations (highlighted).

PROOF If a component of the reconstruction contains an r-disc, then it also contains one of the configurations of Fig. 4(a) [4(b)]. The highlighted one [At least one of the highlighted configurations in Fig. 1(b)] is subset of any of these configurations. Otherwise if a component of the reconstruction contains such a highlighted configuration, it also contains an $(r' + \varepsilon)$-disc for a sufficiently small ε. This is all to show since A is also $(r' + \varepsilon)$-halfregular.   □

5.   Conclusions

The new class of r-halfregular shapes was firstly introduced and it was shown that this class can be digitized topologically correctly by using a simple postprocessing step. The main result simply says that the digitization of an r-halfregular shape with an arbitrary sampling grid of sampling density $r' < r$ is topologically undistinguishable from the original shape after applying a post-processing step which simply removes all components, which do not exceed a certain size. This is much more general than the restriction to r-regular shapes, which was used in literature before. Is was also shown that the postprocessing step leads to better results than a morphological preprocessing step, which makes an r-halfregular shape r-regular. Further on it was proven that the class of r-halfregular shapes subsumes other shape classes like polygonal or convex shapes. Finally the postprocessing step was even more simplified in case of using regular sampling grids like square or hexagonal ones.

References

[1] Bing, R.H.: *Approx. surfaces with polyhedral ones.*. Ann. Math. **65**, pp. 456-483, 1957.

[2] Giraldo, A., Gross, A. Latecki, L.J.: *Dig. Preserving Shape.* PR **32**, pp. 365-376, 1999.

[3] Latecki, L.J.: *Discrete Rep. of Spatial Objects in Computer Vision.* Kluwer, 1998.

[4] Pavlidis, T.: *Alg. for Graphics and Image Processing.* Computer Science Press, 1982.

[5] Serra, J.: *Image Analysis and Math. Morphology.* Academic Press, 1982.

[6] Stelldinger, P., Köthe U.: *Shape Preservation During Digitization: Tight Bounds Based on the Morphing Distance.* Pattern Recognition, LNCS **2781**, pp. 108-115, Springer, 2003.

DOWNSAMPLING OF BINARY IMAGES USING ADAPTIVE CROSSING NUMBERS

Etienne Decencière[1] and Michel Bilodeau[1]

[1]*Centre de Morphologie Mathématique, Ecole des Mines de Paris*
35, rue Saint Honoré, 77300 Fontainebleau, France
Etienne.Decenciere@ensmp.fr

Abstract   A downsampling method for binary images is presented, which aims at preserving the topology of the image. It uses a general reference sampling structure. The reference image is computed through the analysis of the connected components of the neighborhood of each pixel. The resulting downsampling operator is auto-dual, which ensures that white and black structures are treated in the same way. Experiments show that the image topology is indeed preserved, when there is enough space, satisfactorily.

Keywords: Digital topology, binary downsampling, reference downsampling

1.   Introduction

In this era of expanding mobile multimedia devices, small screens will soon be in every pocket. Their relatively small resolutions (a PDA screen is typically 320 by 320 pixels) pose display problems, worsened by the fact that visual digital documents are often thought for high resolution displays. For example, how can a faxed document, or a tourist brochure, which has been scanned with a 200 dpi resolution, be conveniently displayed on a PDA screen?

As it can be seen, we are confronted with a severe downsampling problem. Moreover, these images often are binary or nearly so, like faxes, diagrams, maps, etc. In these particular cases, classical downsampling methods work very badly, because they aim at removing from the image those structures which cannot be represented at the lower resolution level. For example, depict a thin black line on a white background. If downsampled with a classical linear method (i.e. high frequencies are filtered out before downsampling), this line will be smoothed away. If we require that the resulting image is binary, thin structures might be simply erased. In many application domains this is a normal, and welcome, feature. However, when displaying graphical data on small displays, the opposite might be more interesting, that is, preserving small

279

C. Ronse et al. (eds.), Mathematical Morphology: 40 Years On, 279–288.
©2005 *Springer. Printed in the Netherlands.*

structures when there is enough place in the image. In other words, we want, if possible, to preserve the homotopy of the initial image.

In this paper, focus is on the problem of nice downsampling of binary images. After this introduction, we will define the problem, and review existing methods. Then, in section 3 we will introduce a general adaptive downsampling scheme which will be used as basis in the following section for a binary downsampling method which aims at preserving homotopy. In the next section results are presented and commented. Finally, conclusions are drawn.

Demonstrations, which are quite simple, are not given for lack of space. They are included in a technical report [4].

2.   Framework and objectives

Only binary images will be considered in this paper. They typically correspond to text, diagrams, graphics, or maps.

Thin and small structures in binary images are often semantically very important. Therefore, we want to preserve them through the downsampling procedure as long as it is possible. In other words, and borrowing vocabulary from the image compression world, we can say that we want to achieve *graceful degradation* of the information.

Of course, the detection of what is important is not trivial, nor is easy to know how long it is possible to preserve data which is considered meaningful. On the other hand, we are not subject to one important constraint that most downsampling methods try to satisfy : reversibility. Indeed, most downsampling methods propose an up-sampling operator such that the reconstructed image is as close as possible to the original one. The idea behind reversibility is to limit the loss of data. In our case, we do not aim at this characteristic because we pretend to preserve semantic information, and we suppose that the image topology is directly related to this information.

We do not know beforehand if the important structures of a binary image are black or white. Therefore, we will treat them in the same way. In other words, the downsampling method should be *auto-dual*.

3.   State of the art

The classical linear downsampling approach is based on the removal from the original image of those frequencies which are too high to be represented at the lower resolution level. This is clearly not adapted to our framework, where high frequencies convey often important semantic information. For example, a thin line would be blurred or erased by such methods.

Morphological downsampling methods are also based on the same idea [8, 9, 7] : first, they remove those structures which are considered too small to be represented at a lower resolution level, and then a point downsampling is applied.

In a series of articles ([1, 3, 2]), Borgefors et al. propose a multiscale representation of binary images. Their aim is to preserve the shape of the objects. Even if these methods tend to preserve the topology of the image, this is not their main objective. Futhermore, the proposed downsampling method is not auto-dual, an essential property in our framework.

Adaptive downsampling methods analyse the image contents before downsampling in order to preserve meaningful details when possible. A method based on the morphological tophat transformation was proposed for downsampling grey level and binary images [6, 5]. It takes into account the size of the structures, by comparison with a structuring element (i.e. a reference set), in order to favour those pixels which are considered more interesting. In this paper, we will improve these results in the case of binary images, thanks to the use of topology information.

4.   Reference downsampling

A general reference downsampling method has been introduced by Decencière et al. [6, 5]. We present below a version adapted to binary images.

We model a binary image I as binary function of $\mathbb{Z}^2$:

$$I : \quad \mathbb{Z}^2 \longrightarrow \quad \{0, 1\} \tag{1}$$
$$(x, y) \longmapsto \quad I(x, y) . \tag{2}$$

The set of binary images is denoted $\mathcal{I}$. We will often identify I to the set $\{P \mid I(P) = 1\}$. For instance, when we say that a point M of $\mathbb{Z}^2$ belongs to I, we mean : $M \in \{P \mid I(P) = 1\}$. The complement image of I will be denoted $\bar{I} : \bar{I} = 1 - I$.

A grey level image R is a function of $\mathbb{Z}^2$ into $\{0, \ldots, 255\}$:

$$R : \quad \mathbb{Z}^2 \longrightarrow \quad \{0, \ldots, 255\} \tag{3}$$
$$(x, y) \longmapsto \quad R(x, y). \tag{4}$$

The set of grey level images is denoted $\mathcal{R}$. We partition $\mathbb{Z}^2$ into 2×2 blocks :

$$B(x, y) = \{(2x, 2y), (2x + 1, 2y), (2x, 2y + 1), (2x + 1, 2y + 1)\}, \tag{5}$$

and we define $index\_max(R, B(x, y))$ as the element of $B(x, y)$ where R takes its maximal value. If there were two or more elements of $B(x, y)$ where R takes its maximal value, then the first in lexicographic order is taken.

The binary reference downsampling operator Ω is defined as :

$$\Omega : \quad \mathcal{I} \times \mathcal{R} \longrightarrow \mathcal{I} \tag{6}$$
$$(I, R) \longmapsto I_R = \Omega(I, R) , \tag{7}$$

with

$$I_R(x, y) = I(index\_max(R, B(x, y))). \tag{8}$$

I_R is called the refence downsampling of I with reference R.

THEOREM 1 Ω *is auto-dual with respect to its second parameter :*

$$\Omega(R, I) = \overline{\Omega(R, \bar{I})}. \tag{9}$$

The simplest binary downsampling method, called point sampling, which consists in taking the first pixel of each $B(x, y)$, is equivalent to applying a reference downsampling operator with a constant reference image. Needless to say, this method gives very poor results.

The choice of R is essential to build interesting sampling operators. The objective of this approach is to build R from I, in such a way that the value of $R(x, y)$ corresponds to the importance of pixel (x, y) in image I.

For example, we have presented in previous work a method to build the reference image using a tophat operator [6, 5]. If wth_B and bth_B are respectively the white tophat and black tophat operators with structuring element B, then the reference image, in the binary case, is built in the following way :

$$R_{th} = wth_B(I) \bigvee bth_B(I). \tag{10}$$

The reference downsampling operator $\Omega(I, R_{th})$ built from R_{th} will give the priority to points belonging to thin structures. This downsampling operator is auto-dual.

However, being thin is not a criterion discrimant enough. For example, figure 1(a) shows an example where three pixels a, b and c would be considered equally important using the thinness criterion described above. In this case, the value of pixel a would be kept instead of the value of pixel c, because it is first in the lexicographic order. It is necessary to give to pixel c more importance. Figure 1(b) shows the result obtained when doing a reference downsampling using the tophat. Notice that this result is considerably better than the result obtained with point sampling (which, in this case would produce a void image). The tophat sampling gives an interesting result, but has introduced a topological modification of the object. Such modifications are often annoying when dealing with binary data. For example, in this case image (a) would be interpreted as a letter "C", whereas image (b) would be misunderstood as a letter "O". We would like to compute a reference image that would give the third result shown by image (c) through reference downsampling.

5.   Topological downsampling

We recall the main digital topology notions that will be used in the following. For a complete introduction to digital topology, the reader may consult the article by Kong and Rosenfeld [11].

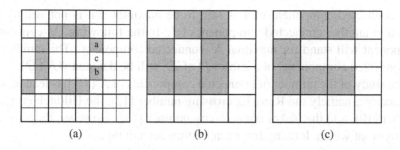

Figure 1. (a) Example of binary image containing three pixels (marked a, b, c) that would be considered equally important by a tophat-based reference image. (b) Tophat sampling ; (c) Aimed result.

Several approaches to characterize the importance of an image pixel from a topological point of view have been proposed.

Let $\mathcal{N}$ be a neighbor relation on $\mathbb{Z}^2$, i.e. a binary relation on $\mathbb{Z}^2$ which is symmetric. When points P and M of $\mathbb{Z}^2$ are in relation through $\mathcal{N}$, we say that they are neighbors and we write $P\mathcal{N}M$. Moreover, we adopt the following convention : we take $\mathcal{N}$ such that a point P is never in relation with itself through $\mathcal{N}$. We will denote $\mathcal{N}(P)$ the set of neighbors of P. As $P\mathcal{N}P$ is always false, P never belongs to $\mathcal{N}(P)$.

Typical neighbor relations used in image processing are the 4-, 6- and 8-neighborhoods, respectively denoted $\mathcal{N}_4$, $\mathcal{N}_8$ and $\mathcal{N}_6$. Among these, $\mathcal{N}_6$ has the best topological properties, as it is the only one that verifies the digital Jordan curve theorem. But, when the image has been digitized following a square grid, 6-connectivity causes some unwelcome phenomena. For example, some diagonal lines will not be connected.

In order to palliate the defects of 4- and 8- neighborhoods, neighborhood relations which depend on the image have been proposed, and widely used. For example, the (4,8)- neighborhood relation $\mathcal{N}^I_{(4,8)}$ is defined as :

$$P\mathcal{N}^I_{(4,8)}M \;=\; \begin{cases} P\mathcal{N}_4M \text{ if } I(P)=1 \text{ or } I(M)=1 \\ P\mathcal{N}_8M \text{ otherwise.} \end{cases} \qquad (11)$$

The (8,4)- neighborhood relation, $\mathcal{N}^I_{(8,4)}$, is defined analogously. These image-dependant neighborhoods verify the digital Jordan curve theorem (see [11] for references to the various demonstrations).

Once equipped with a neighborhood relation, the points of an image can be aggregated into larger structures. A subset D of I is a $\mathcal{N}$-connected component of I if for each pair (M,P) of D, there is a sequence of points of D $(Q_k)_{0 \le k \le k_{max}}$ such that $Q_0 = M$, $Q_{k_{max}} = P$ and :

$$\forall k, \quad 0 \le k < k_{max}, \quad Q_k\mathcal{N}Q_{k+1} \qquad (12)$$

A connected component of I is said to be maximal if it is not strictly included in another connected component of I. In the following, $\mathcal{N}$-connected component will stand for maximal $\mathcal{N}$-connected component. The number of $\mathcal{N}$-connected components of a subset D of $\mathbb{Z}^2$ will be denoted $\#\mathcal{N}(D)$.

The study of the number of connected components of $\mathcal{N}(P)$ has lead to several notions, namely the Rutovitz crossing number [12], the Hilditch crossing number [10], and the Yokoi connectivity number [13]. However, they are not exactly what we are looking for, because they are not dual.

In fact, connectivities such as $\mathcal{N}_{(4,8)}^I$ are practically never invariant with respect to image inversion : we may have, $P\mathcal{N}_{(4,8)}^I M$ but not $P\mathcal{N}_{(4,8)}^{\bar{I}} M$. Therefore, we propose a new neighborhood, which is invariant with respect to image inversion.

Consider a pixel P and a binary image I. In order to answer the question "does P belong to the object", we compute the number $n^I(P)$ of 8-neighbors of P where I takes the same value as P. This is given by :

$$n^I(P) = Card(\{M \in \mathcal{N}_8(P) \mid I(M) = I(P)\}) \tag{13}$$

whose values are included between 0 ($\mathcal{N}_8$-isolated point) and 8 ($\mathcal{N}_8$-interior point). If this value is equal or greater than 4, then we will consider P as a background pixel, otherwise, as an object point.

THEOREM 2 *The operator n^I is invariant with respect to image inversion :*

$$\forall I \in \mathcal{I}, \forall P \in \mathbb{Z}^2, n^I(P) = n^{\bar{I}}(P). \tag{14}$$

As we want a strong connectivity for object pixels, we choose an 8- neighborhood for them. For background pixels, we take 4- neighborhood. The resulting neighbor relation, $\mathcal{N}_n^I$ is therefore :

$$P\mathcal{N}_n^I M = \begin{cases} P\mathcal{N}_8 M & \text{if } n^I(P) < 4 \text{ or } n^I(M) < 4 \\ P\mathcal{N}_4 M & \text{otherwise.} \end{cases} \tag{15}$$

Inspired by the crossing and connectivity numbers, we now study the neighborhood $\mathcal{N}_n^I(P)$ of P.

Let us consider a pixel P. It is either an object pixel of I, or a background pixel of I.

If P is an object pixel of I (i.e. $n^I(P) < 4$), then it has 8 neighbours through $\mathcal{N}_n^I$ (see Figure 2 (a)). On some of these 8 neighbours, I takes the same value as on P. We call the number of $\mathcal{N}_4$-connected components of this subset the *adaptive crossing number* of an object pixel.

Similarly, if P is a background pixel of I (i.e. $n^I(P) \geq 4$), then it has 4 neighbours through $\mathcal{N}_n^I$ (see Figure 2 (b)). On some of these 4 neighbours, I takes the same value as on P. We call the number of $\mathcal{N}_8$-connected components of this subset the *adaptive crossing number* of a background pixel.

For example, in Figure 2 (a), the number of $\mathcal{N}_4$-connected components of the set $\{M \in \mathcal{N}_n^I(a) \mid I(M) = I(a)\}$ is 1, and in Figure 2 (b), the number of $\mathcal{N}_8$-connected components of the set $\{M \in \mathcal{N}_n^I(b) \mid I(M) = I(b)\}$ is 2. Formally, we define the adaptive crossing number $X^I(P)$ as :

$$X^I(P) = \begin{cases} \#\mathcal{N}_4(\{M \in \mathcal{N}_8(P) \mid I(M) = I(P)\} & \text{if } n^I(P) < 4 \\ \#\mathcal{N}_8(\{M \in \mathcal{N}_4(P) \mid I(M) = I(P)\} & \text{if } n^I(P) \geq 4 . \end{cases}$$
(16)

THEOREM 3 X^I *is invariant with respect to image inversion :*

$$\forall I \in \mathcal{I}, \forall P \in \mathbb{Z}^2, X^I(P) = X^{\bar{I}}(P). \tag{17}$$

As a consequence, the reference image R_n^I built from X^I is also invariant with respect to image inversion :

$$R_n^I(P) = \begin{cases} X^I(P) & \text{if } X^I(P) > 0 \\ 5 & \text{otherwise.} \end{cases} \tag{18}$$

The particular case for $X^I(P) = 0$, i.e. for isolated points, is necessary if we want these pixels to be preserved. The value 5 is arbitrary ; it has to be higher than the other values of $X^I(P)$.

Finally, we obtain the following downsampling operator

$$\Omega_n(I) = \Omega(R_n^I, I), \tag{19}$$

which has the property we were seeking for :

THEOREM 4 Ω_n *is is auto-dual :*

$$\forall I \in \mathcal{I}, \Omega_n(I) = \overline{\Omega_n(\bar{I})} \tag{20}$$

6. Results

First of all, in figure 2(c) we give the values of $X^I(P)$ for some pixels of the test image. Notice that the value associated to pixel c is now higher than the values of its neighbors a and b. Thanks to this, the resulting downsampled image with the reference image R_n we have just defined is the one given by figure 1(c).

Figure 3 shows a more complex test image, containing geometric structures and text. Its size is 512 x 512. Figure 4 shows some reference downsampling results. In positions (a) and (b) we respectively find the results of tophat and topological downsampling.

Notice that topological downsampling has done a better work in preserving some important structures. For example, in many cases topological downsampling has avoided the fusion between letters, whereas tophat sampling has not.

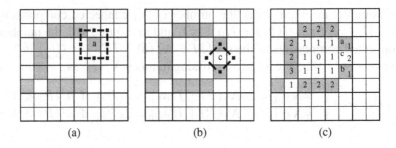

Figure 2.   (a) $\mathcal{N}_n^I$-neighbors of pixel a. Neighborhood relation on them, used to compute $X^I(a)$, indicated by segments. (b) $\mathcal{N}_n^I$-neighbors of pixel c marked with dots. Neighborhood relation on them, used to compute $X^I(c)$, indicated by segments. (c) Test image with some values of the adaptive crossing number. Notice that the value associated to pixel c is higher than those given to a and b.

Figure 3.   Test image (512 x 512).

We have iterated the procedure. In position (c) (resp. (d)) we have the result of downsampling image (a) (resp. (b)) again with tophat (resp. topological) sampling. Notice again that topological sampling has better preserved the image topology. However, tophat sampling has preserved some geometric details that are also important (look for instance at letters "t", "j" or "r"). In some other cases (see for example letter "V" in image (d)) the topology of some structures has not been preserved. The main reason for this behavior is the lack of space (pixels per letter) in the resulting image. In some other cases, the local analysis does not correctly evaluate the value of some pixels (letters "h" or "y" in image (b)).

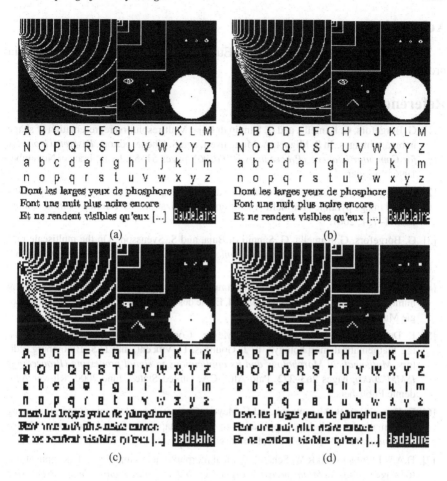

Figure 4. (a) Tophat sampling. (b) Topological sampling. (c) Tophat sampling, iterated. (d) Topological sampling, iterated.

7. Conclusion and future developments

A binary downsampling method which aims at preserving the image topology has been presented. It does a good job preserving structures from a topology point of view. However, it could be improved by incorporating into the reference image some geometric information, that could be computed by means of a tophat, for example.

It should be noted that the operations involved in the computation of the reference image are not computationally greedy. The implementation of this method on mobile processors should not be a problem.

The next step in this work will be to extend this sampling method to grey level images.

Acknowledgments

This work has been financed by the European MEDEA+ Pocket Multimedia project.

References

[1] G. Borgefors, G. Ramella, and G. Sanniti di Baja. Multiresolution representation of shape in binary images. In S. Miguet, A. Montanvert, and S. Ubéda, editors, *Discrete Geometry for Computer Imagery, (Proceedings DCGI'96)*, pages 51–58, Lyon, France, November 1996. Springer.

[2] G. Borgefors, G. Ramella, and G. Sanniti di Baja. Shape and topology preserving multi-valued image pyramids for multi-resolution skeletonization. *Pattern Recognition Letters*, 22:741–751, 2001.

[3] G. Borgefors, G. Ramella, G. Sanniti di Baja, and S. Svensson. On the multiscale representation of 2d and 3d shapes. *Graphical Models and Image Processing*, 61:44–62, 1999.

[4] E. Decencière and M. Bilodeau. Downsampling of binary imagesusing adaptive crossing numbers. Technical Report N 01/05/MM, Ecole des Mines de Paris, Centre de Morphologie Mathématique, January 2005.

[5] E. Decencière, B. Marcotegui, and F. Meyer. Content dependent image sampling using mathematical morphology. In J. Goutsias, L. Vincent, and D.S. Bloomberg, editors, *Mathematical Morphology and its applications to signal processing (Proceedings ISMM'2000)*, pages 263–272, Palo Alto, CA, United States, June 2000. Kluwer Academic Publishers.

[6] E. Decencière, B. Marcotegui, and F. Meyer. Content-dependent image sampling using mathematical morphology: application to texture mapping. *Signal Processing: Image Communication*, 16(6):567–584, February 2001.

[7] D.A.F. Florêncio and R.W. Schafer. Critical morphological sampling and its applications to image coding. In *Mathematical Morphology and its Applications to Image Processing (Proceedings ISMM'94)*, pages 109–116, September 1994.

[8] R.M. Haralick, X. Zhuang, C. Lin, and J.S.J. Lee. The digital morphological sampling theorem. *IEEE Transactions on Acoustics, Speech and Signal Processing*, 37:2067–2090, December 1989.

[9] H.J.A.M. Heijmans and A. Toet. Morphological sampling. *Computer Vision, Graphics, and Image Processing: Image Understanding*, 54:384–400, November 1991.

[10] C.J. Hilditch. Linear skeletons from square cupboards. In B. Meltzer and D. Michie, editors, *Machine Intelligence, Vol. 4*, pages 403–420. Edinburgh Univ. Press, 1969.

[11] T.Y. Kong and A. Rosenfeld. Digital topology: introduction and survey. *Computer Vision, Graphics, and Image Processing*, 48:357–393, 1989.

[12] D. Rutovitz. Pattern recognition. *J. Royal Statist. Soc.*, 129:504–530, 1966.

[13] S. Yokoi, J. Toriwaki, and T. Fukumura. Topological properties in digital binary pictures. *Systems Comput. Controls*, 4:32–40, 1973.

GREY-WEIGHTED, ULTRAMETRIC AND LEXICOGRAPHIC DISTANCES

Fernand Meyer

Centre de Morphologie Mathématique, École des Mines de Paris
35 rue Saint-Honoré, 77305 Fontainebleau, France
fernand.meyer@ensmp.fr

Abstract   Shortest distances, grey weighted distances and ultrametric distance are classically used in mathematical morphology. We introduce a lexicographic distance,
for which any segmentation with markers becomes a Voronoï tessellation.

Keywords: Grey-weighted distances, ultrametric distances, lexicographic distances, marker
segmentation, path algebra

1.   Introduction

Mathematical morphology makes a great use of distances. The classical distance $d(x, y)$ between two pixels x and y is defined as the length of the shortest
path linking these two pixels. If the family of admissible paths is constrained
to remain within a set Y or on the contrary to miss a set Z we obtain geodesic
distances. Grey weighted distances are obtained by assigning a weight to each
edge of a graph. The length of a path being the sum of the weights of all
its edges. Among them are the Chamfer distances [2], approximating the Euclidean distances on a grid and topographic distances for the construction of
the watershed line [10, 7].

Ultrametric distances govern morphological segmentation. This is due to
the fact that watershed segmentation is linked to the flooding of topographic
surfaces: the minimal level of flooding for which two pixels belong to a same
lake precisely is an ultrametric distance. The key contribution of this paper
is the introduction of a lexicographic distance, for which a segmentation with
markers becomes a Voronoï tessellation of SKIZ of the markers.

The paper is organized as follows. Restricting ourselves to distances defined on graphs, we present the various distances encountered in morphology.
We then analyze the segmentation with markers and show that the flooding
ultrametric distance is myopic: only a lexicographic distance has sufficient
discriminant power for correctly describing the segmentation with markers as

C. Ronse et al. (eds.), Mathematical Morphology: 40 Years On, 289–298.

a SKIZ of the markers. In the last part we introduce the "path algebra", which unifies all shortest distance algorithms whatever their type.

2. Graphs and distances

Graphs encountered in morphology

Graphs are the good framework for dealing with distances. A non oriented graph $G = [X, E]$ is a collection of a set X whose elements are called vertices or nodes and of a set E whose elements $e_{ij} = (i, j) \in E$ are pairs of vertices called edges. Two edges that share one or several nodes are said to be *adjacent*. A *path* of length n is a sequence of n edges $L = \{e_{12}, u_{23}, \ldots, e_{n-1n}\}$, such that successive edges are adjacent. Any partition $\mathcal{A}$ for which a dissimilarity between adjacent regions has been defined can be represented as a region adjacency graph $G = (X, E)$, where X is the set of nodes and E is the set of edges. The nodes represent regions of the partition. Adjacent regions i and j are linked by an edge $e_{ij} = (i, j)$ with a weight s_{ij} expressing the dissimilarity between them. The adjacency matrix $A = (a_{ij})$ of the graph is defined by $a_{ij} = s_{ij}$ if $(i, j) \in E$ and $a_{ij} = \infty$ if not. By convention $a_{ii} = 0$ as a_{ii} represents the dissimilarity between i and i itself.

In the simplest case, A represents the pixels of the image, the edges linking neighboring pixels. In case of a topographic surface, the watershed graph is obtained by taking the catchment basins as nodes, the dissimilarity between two adjacent basins being the altitude of the pass separating them.

Grey weighted distances

The "weighted length"of a path is defined as the sum of the weights of its edges. If the weight of an edge is equal to its length, we obtain the length of the path. The distance $d(x, y)$ between two nodes x and y is the minimal length of all paths between x and y. If there is no path between them, the distance is equal to ∞. In fig.1A the shortest path between x and y is a bold line and has a length of 4.

This classical distance is well known and allows to define the distance between a pixel and a set, the Hausdorff distance between sets, distance functions, crest and saddle points, ultimate erosions, skeletons, Voronoï tesselations, extremities of particles etc. A recent review on the use of distance functions in mathematical morphology may be found in [11]. Grey weighted distances [12] have been used for finding the shortest paths in an image, the weight of the edges being an increasing function of the grey tones of their extremities. This allows doing virtual endoscopy [3] or detecting fibers on a noisy background [13]. The watershed line itself is the skeleton of influence of the regional min-

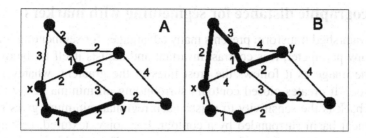

Figure 1. Different types of distance functions:
- A: shortest path : the length of a path is equal to the sum of the lengths of its edges
- B: flooding distance: each path is characterised by the highest weight assigned to one of its edges. The easiest path between two nodes is the path for which this highest weight is the lowest.

ima of the image for the topographic distance, where the weights equal the modulus of the gradient [10, 7].

Ultrametric distances

Consider the case where the graph G represents a region adjacency graph. For any $\lambda \geq 0$, one defines a derived graph $G_\lambda = [X, E_\lambda]$ with the same nodes but only a subset of edges : $E_\lambda = \{(i,j) \mid \alpha_{ij} \leq \lambda\}$. The connected components of this graph create a partition of the objects X into classes. Two nodes x and y belong to the same connected component of G_λ if and only if there exists a path $L = (i, i_2, ...i_p, j)$ linking these two nodes and verifying: $\max(\alpha_{ii_1}, \alpha_{i_1 i_2}, ...\alpha_{i_p j}) \leq \lambda$. Such a path exists among all paths linking x and y in G if and only if $\alpha_{xy}^* = \min_{L \in C_{xy}} (\max(\alpha_{xi_1}, \alpha_{i_1 i_2}, ...\alpha_{i_p y})) \leq \lambda$ where C_{xy} is the set of all paths between x and y. In case of a watershed graph, α_{xy}^* represents the lowest level of flooding for which x and y belong to a same lake. In fig.1B, the shortest path between x and y is bold and has a maximal altitude of 3. Obviously the level of the minimal lake containing x and y is below the level of the minimal lake containing not only x and y but also an additional pixel z ; the level of the lake containing all three pixels is $\max\left[\text{flood}(x, z), \text{flood}(z, y)\right]$, hence the ultrametric inequality : $\text{flood}(x, y) \leq \max\left[\text{flood}(x, z), \text{flood}(z, y)\right]$. The closed balls $\text{Ball}(i, \rho) = \left\{j \in X \mid \alpha_{ij}^* \leq \rho\right\}$ form the classes of the partition at level ρ. For increasing levels of ρ one obtains coarser classes. This is the simplest type of segmentation associated to hierarchies : select a level of the hierarchy as the result of the segmentation [8].

A lexicographic distance for segmenting with markers

The watershed transform presents many advantages for segmenting images: free of any parameter, it is contrast invariant and adjusts itself to the dynamics of the image as it follows the crest lines of the gradient, whatever their magnitude. It creates closed contours surrounding all minima of the image. Its drawback is the sensitivity to noise: each regional minimum gives rise to a catchment basin surrounded by a contour. For noisy, textured, or complex images, this leads to a dramatic oversegmentation. The solution consists in introducing markers in order to regularize the result [9, 1]. Using the markers as sources, it is possible to flood the gradient image, finding the strongest contours surrounding the markers and producing a closed contour for each marker.

Segmenting with markers is fast, as it requires only one flooding. However, in some situations of interactive segmentation one wishes to play with the markers, adding, suppressing or modifying markers. The same operation has to be repeated for each new set of markers, which is time consuming. In such a situation it is interesting to use the "watershed graph" introduced earlier. Some of the nodes will be markers. Segmenting with markers amounts to construct a minimum spanning forest, where each tree is rooted within one marker [6].

Consider now a flooding process organized on the watershed graph: a source is placed at each marker ; as the level of the flooding increases, edges are crossed and new nodes are flooded, if they were not already flooded earlier from another source. Each flooded node is assigned to the source where the flood comes from. The flooding distance seems to govern the flooding : if m_1 and m_2 are markers and their flooding distances to a given node x verify $\mathrm{flood}(x, m_1) < \mathrm{flood}(x, m_2)$, then the flood coming from m_1 will reach x before the flood coming from m_2. So it seems, at first sight, that the minimum spanning forest can also be interpreted as the skeleton by zone of influence of the markers for the flooding distance. But this is not so, as we will establish now.

The competition between floodings.   Consider the graph of fig.2, where the nodes a, b, c and d are markers. If a flood starts from the markers, which flood will attain the node x first ? In our case it is obviously the flood coming from the source b ; hence x will be assigned to the marker b. Is there a distance function for which the distance between x and b is shorter than the distance from x to any other marker ? The flooding distances from x to the markers are ranked as follows : $(\mathrm{flood}(x, d) = 7) > (\mathrm{flood}(x, a) = \mathrm{flood}(x, b) = \mathrm{flood}(x, c) = 6)$. From the first inequality we can discard the flood from d, as it has an edge of altitude 7 to pass, whereas the highest edge for the other floods is only 6. The ultrametric distance is myopic and is unable to discriminate further

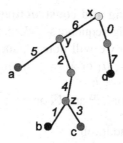

Figure 2. Following the path from x towards the sources, one sees that :
$(\text{ult}(x,d) = 7) > (\text{ult}(x,a) = \text{ult}(x,b) = \text{ult}(x,c) = 6)$
$(\text{ult}(y,a) = 5) > (\text{ult}(y,b) = \text{ult}(y,c) = 4)$
$(\text{ult}(z,c) = 3) > (\text{ult}(z,b) = 1)$

between the markers a, b and c. The floods from a, b and c have all to pass through the edge (y, x) of altitude 6 ; the first reaching the node y will be the first crossing the edge (y, x), hence the first reaching x. Applying the same reasoning to the node y, we can discard a as $\text{flood}(y, a) = 5 > \text{flood}(y, b) = \text{flood}(y, c) = 4$. Among the remaining candidates b and c, the flood coming from b reaches z and crosses the edge of altitude 4 before the flood coming from c, as $\text{flood}(z, b) = 1 < \text{flood}(z, c) = 3$. Finally the flood coming from b ultimately wins because it arrives first all along the unique path between b and x. The distance which is minimal along the winning path of flooding is a lexicographic distance. The lexicographic length $\Lambda(A)$ of a path A between a node x and a marker m_1 is defined as a sequence $(\lambda_1, \lambda_2, ..., \lambda_k)$ of decreasing weights $(\lambda_1 \geq \lambda_2 \geq .... \geq \lambda_k)$: the largest weight λ_1 on the path between x and m_1, then the largest weight λ_2 on the remaining part of the path and so on until one reaches the marker m_1. Let $\pi = \{e_{12}, u_{23}, ..., e_{n-1n}\}$ be a path. We define $\max(\pi) = \max(a_{ii+1} \mid 1 \leq i \leq n - 1)$. The lexicographic length of π is the sequence $\Lambda(\pi)$ such that $\Lambda(\pi) = \max(\pi) \triangleright \Lambda(e_{ii+1}, ...e_{n-1n})$, where $\triangleright$ is the concatenation and i is the smallest index such that $a_{i-1i} = \max(\pi)$. The lexicographic distance $\widehat{\Lambda}(x, y)$ between a node x and a node y is defined as the lexicographic length of the shortest path from x to y. By convention $\widehat{\Lambda}(x, x) = 0$. In our case we see that the lexicographic distance correctly ranks the floodings reaching x : $\widehat{\Lambda}(x, d) = 7 > \widehat{\Lambda}(x, a) = (6, 5) > \widehat{\Lambda}(x, c) = (6, 4, 3) > \widehat{\Lambda}(x, b) = (6, 4, 1)$.

A node will be flooded by a marker a if and only if its lexicographic distance to this marker is smaller than the lexicographic distances to all other markers. In [5] Ch. Lenart proposes a more complex distance, made of the set of all paths belonging to the minimum spanning trees of the graph. The minimum spanning forests also are skeletons of influence for Lenart's distance.

Two operators for comparing and constructing lexicographic distances.
Let S be the set of decreasing sequences. We define on S the usual lexico-
graphic order relation, which we will note $\prec$, such that: $(\lambda_1, \lambda_2, \ldots, \lambda_k) \prec$
$(\mu_1, \mu_2, \ldots, \mu_k)$ if either $\lambda_1 < \mu_1$ or $\lambda_i = \mu_i$ until rank s where $\lambda_{s+1} < \mu_{s+1}$.
We define $a \preceq b$ as $a \prec b$ or $a = b$.

We now define two operators, $\boxplus$ and $\boxtimes$ on S:

- an operator $\boxplus$ called "addition"

$$a \boxplus b = \begin{array}{l} a \text{ if } a \preceq b \\ b \text{ if } b \preceq a \end{array} \quad \forall a, b \in S, \text{ i.e. the shortest lexicographic length.}$$

  The $\boxplus$ operation is associative, commutative and has a neutral element
  ∞ called the zero element: $a \boxplus \infty = a$

- an operator $\boxtimes$ called "multiplication" permits to compute the lexicographic length $\Lambda(A)$ of a path $A = e_{12}e_{23}\ldots e_{n-1n}$ obtained by concatenating two adjacent paths $B = e_{12}e_{23}\ldots e_{k-1k}$ and $C = e_{kk+1}\ldots e_{n-1n}$.
  The operator $\boxtimes$ is designed such that $\Lambda(A) = \Lambda(B) \boxtimes \Lambda(C)$. Let
  $a = (\lambda_1, \lambda_2, \ldots, \lambda_k)$ and $b = (\mu_1, \mu_2, \ldots, \mu_k)$; we will define $a \boxtimes b$
  by:

  - if $\mu_1 > \lambda_1$ then $a \boxtimes b = b$
  - if $\lambda_k \geq \mu_1$ then $a \boxtimes b = a \triangleright b$.
  - else let j be the highest index verifying $\lambda_{j-1} \geq \mu_1 \geq \lambda_j$ then
    $a \boxtimes b = (\lambda_1, \lambda_2, \ldots, \lambda_{j-1}, \mu_1, \mu_2, \ldots, \mu_l)$

This algorithm guarantees that $a \boxtimes \infty = \infty$. We also define $\infty \boxtimes a = \infty$,
so that the zero element is an absorbing element for $\boxtimes$. The operator $\boxtimes$ is
associative, has a neutral element 0 called unit element: $a \boxtimes 0 = 0 \boxtimes a = a$. The
multiplication is distributive with respect to the addition both to the left and to
the right. The lexicographic length of the path $A = e_{12}e_{23}\ldots e_{n-1n}$ is equal to
the product by $\boxtimes$ of the weights of all its edges: $\Lambda(A) = a_{12} \boxtimes a_{23} \boxtimes \ldots \boxtimes a_{n-1n}$.
Remark that the lexicographic length of a path is generally not equal to the
length of this path in reverse order. Hence the lexicographic distance is not
symmetrical.

3. Paths algebra : unifying shortest distance algorithms

Three algebras for three distances

The (min, plus) algebra. Shortest-path algorithms on graphs can be expressed as linear algebra operations on dioids [4]. The three types of distances
presented above can be computed within the same formalism. In order to understand how, we illustrate the ground ideas on a simple example.

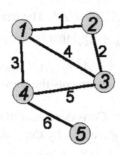

Figure 3. A weighted graph

Its incidence matrix is $A = [a_{ij}] = \begin{bmatrix} 0 & 1 & 4 & 3 & \infty \\ 1 & 0 & 2 & \infty & \infty \\ 4 & 2 & 0 & 5 & \infty \\ 3 & \infty & 5 & 0 & 6 \\ \infty & \infty & \infty & 6 & 0 \end{bmatrix}$. Consider

the ordinary matrix product $A^2 = \left[a_{ij}^2\right]$. If we replace in the formal expression of $a_{ij}^{(2)} = a_{i1} * a_{1j} + a_{i2} * a_{2j} + ... + a_{i5} * a_{5j}$ the operator $*$ by the operator $+$ and simultaneously the operator $+$ by the operator minimum $\wedge$, then we get $a_{ij}^{(2)} = (a_{i1} + a_{1j}) \wedge (a_{i2} + a_{2j}) \wedge ... \wedge (a_{i5} + a_{5j})$, the meaning of which is the following: $a_{ik} + a_{kj}$ is the length of the path $(e_{ik} e_{kj})$; this length is infinite if the weight of one of the edges e_{ik} or e_{kj} is infinite. On the other hand $a_{ii} + a_{ij} = 0 + a_{ij}$ and $a_{ij} + a_{jj} = a_{ij} + 0$ represent the path (i, j) restricted to one edge. Since k takes all values, $a_{ij}^{(2)}$ represents the shortest path comprising one or two edges between i and j.

Consider now $A^2 A^2 = \left[a_{ij}^{(4)}\right]$. The element $a_{ij}^{(4)} = \bigwedge_k \left(a_{ik}^{(2)} + a_{kj}^{(2)}\right)$ where $a_{ik}^{(2)}$ is the shortest path of length 1 or 2 edges between i and k and $a_{kj}^{(2)}$ is the shortest path of length 1 or 2 edges between k and j. Hence $a_{ik}^{(2)} + a_{kj}^{(2)}$ is the shortest path of length 2, 3 or 4 passing through k between i and j. On the other hand $a_{ii}^{(2)} = 0$ and $a_{ii}^{(2)} + a_{ij}^{(2)}$ is the shortest path of length 1 or 2 between i and j. Hence $A^2 A^2 = \left[a_{ij}^{(4)}\right]$ records the shortest paths of length 1,2,3 or 4 between any pair of nodes.

By successive multiplications, one finally gets an idempotent state where $A^n A^n = A^n$ represents the shortest path of any length between all pairs of nodes, that is the shortest distance between any pair of nodes. We call the matrix obtained at idempotence $A^* = A^n$. The triangular inequality is verified as $a_{ij}^* = \bigwedge_k \left(a_{ik}^* + a_{kj}^* \right) \leq a_{ik}^* + a_{kj}^*$, that is $d(i,j) \leq d(i,k) + d(k,j)$.

We also remark that $A^* = A \wedge A^2 \wedge \dots A^n$, since A^k represents the shortest paths of length up to k and A^n represents the shortest paths of any length.

The (min, max) algebra.   Consider again the same graph represented in fig.3 but we now replace the operator $*$ by the $\vee$ and the operator $+$ by $\wedge$. We get $a_{ij}^{(2)} = (a_{i1} \vee a_{1j}) \wedge (a_{i2} \vee a_{2j}) \wedge \dots \wedge (a_{i5} \vee a_{5j})$, which may be interpreted as follows: $a_{ik} \vee a_{kj}$ is the highest weight along the path $(e_{ik}e_{kj})$. Calling the highest weight of the edges in a path the altitude of this path, $a_{ij}^{(2)}$ can be interpreted as the largest weight of the path of lowest altitude and of length 1 or 2 between the nodes i and j.. As previously it is possible to show that there exists an m such that $A^m A^m = A^m = A^*$. a_{ij}^* can be interpreted as the largest weight of the path of lowest altitude between the nodes i and j ; it is identical with the flooding distance defined previously. The triangular inequality obtained in the algebra $(min, +)$ becomes the ultrametric inequality in this new algebra (min, max) : $a_{ij}^{(n)} = \bigwedge_k \left(a_{ik}^{(n)} \vee a_{kj}^{(n)} \right) \leq a_{ik}^{(n)} \vee a_{kj}^{(n)}$.

The $(\boxplus, \boxtimes)$ algebra.   Let S be the set of all lexicographic distances, that is decreasing n-tuples $(\lambda_1, \lambda_2, \dots, \lambda_k)$ defined earlier.

Consider again the same graph represented in fig.3 but with a last interpretation to $+$ and $*$ in the expression of $a_{ij}^{(2)} = a_{i1} * a_{1j} + a_{i2} * a_{2j} + \dots + a_{i5} * a_{5j}$. If we replace the operator $*$ by the $\boxtimes$ and the operator $+$ by $\boxplus$, then we get $a_{ij}^{(2)} = (a_{i1} \boxtimes a_{1j}) \boxplus (a_{i2} \boxtimes a_{2j}) \boxplus \dots \boxplus (a_{i5} \boxtimes a_{5j})$, which may be interpreted as follows: $a_{ik} \boxtimes a_{kj}$ is the lexicographic length of the path $(e_{ik}e_{kj})$. We get $A^2 =$

$$
\begin{bmatrix}
0 & 1 & 2 & 3 & 6 \\
1 & 0 & 2 & 3 & \infty \\
2,1 & 2 & 0 & 4,3 & 6 \\
3 & 3,1 & 4 & 0 & 6 \\
6,3 & \infty & 6,5 & 6 & 0
\end{bmatrix}
\text{ and } A^4 =
\begin{bmatrix}
0 & 1 & 2 & 3 & 6 \\
1 & 0 & 2 & 3 & 6 \\
2,1 & 2 & 0 & 3 & 6 \\
3 & 3,1 & 3,2 & 0 & 6 \\
6,3 & 6,3,1 & 6,3,2 & 6 & 0
\end{bmatrix}
$$

A_{ij}^k the length of the shortest path with at most $k+1$ nodes between i and j. There exists an index for which $A^n A^n = A^n = A^*$, which represents then the shortest lexicographic distance between all pairs of nodes. The triangular inequality is replaced by $a_{ij}^* = \boxplus \left(a_{ik}^* \boxtimes a_{kj}^* \right) \leq a_{ik}^* \boxtimes a_{kj}^*$.

Segmenting with markers: We have established earlier that a node will be flooded by a marker a if and only if its lexicographic distance to this marker is smaller than the lexicographic distances to all other markers. It is now immediate to construct a segmentation with markers, by comparing the columns of the markers. For instance if the nodes 1 and 2 are markers, one assigns to 1 all nodes for which the weight in column 1 is smaller than the corresponding weights in column 2. In our case, nodes 1,4 and 5 are assigned to the node 1 and node 2 and 3 assigned to node 2.

Algebraic solutions of a linear system

The algebraic treatment presented now applies to all three algebras met above: $(\wedge, +)$ for the ordinary distance, $(\wedge, \vee)$ for the flooding distance, $(\boxplus, \boxtimes)$ for the lexicographic distance. We will now introduce it for $(\boxplus, \boxtimes)$, knowing that it is also valid for $(\wedge, +)$ and $(\wedge, \vee)$. All elements of the identity matrix are equal to ∞ except the diagonal elements equal to 0. The matrix multiplication is denoted $A \boxtimes B$ or AB. As we have seen above, there exists an n such that A^n becomes idempotent and $A^n = A^*$ verifies $A^* = E \boxplus A^2 \boxplus A^3 \boxplus ... \boxplus A^n = E \boxplus AA^*$.

Multiplying by a matrix B yields $A^*B = B \boxplus AA^*B$. Defining $Y = A^*B$, one sees that A^*B is solution to the equation $Y = B \boxplus AY$. Moreover it is the smallest solution.

With the right choice for B, various problems may be solved:

- For finding A^* one may solve $Y = E \boxplus AY$ yielding $A^*E = A^*$ as solution

- with $B = \begin{bmatrix} \infty & ... & 0 & ... & \infty \end{bmatrix}^T$, one obtains A^*B, the i-th column of the matrix A^*, that is the distance of all nodes to node i.

For finding the solution to such linear systems, the classical linear algebra algorithms are still valid, such as the Jacobi, Gauss-Seidel or Jordan algorithm. We give here two others.

The step algorithm. The step algorithm for inverting a matrix is interesting in the case where A^* is already known for a graph G with N nodes, to which a new node has to be added.

Are estimated first the lengths of the paths between a node $i \leq N$ and the node $N + 1$:

$$a^*_{iN+1} = \sum_{j \in \Gamma^{-1}(N+1)}^{\boxplus} a^*_{ij} * a_{jN+1} \text{ and } a^*_{N+1,i} = \sum_{j \in \Gamma(N+1)}^{\boxplus} a^*_{N+1j} * a_{ji}$$

One then considers the possibility for a shortest path between i and i to pass through the node $N + 1$: $a^*_{ij} = a_{ij} \boxplus a^*_{iN+1} * a^*_{N+1,j}$.

The greedy algorithm (Dijkstra). The greedy algorithm [4] is the algebraic counterpart of the Dijkstra algorithm of shortest path. If $\overline{Y} = A^*B$ is the solution of $Y = AY \boxplus B$, for a column vector B then there exists and index i_0 such that $\overline{y_{i_0}} = \sum^{\boxplus} b_i$. Hence the smallest b is solution : $\overline{y_{i_0}} = b_{i_0}$. Each element of $Y = AY \boxplus B$ can then be written $y_k = \overset{\boxplus}{\underset{j \neq k}{\sum}} a_{kj} * y_j \boxplus b_k =$

$\overset{\boxplus}{\underset{j \neq k, i_0}{\sum}} a_{kj} * y_j \boxplus a_{ki_0} y_{i_0} \boxplus b_k$. Suppressing the line and the column of rank i_0

and taking for B the vector $b_k^{(1)} = a_{ki_0} y_{i_0} \boxplus b_k$, one obtains a new system of size $N - 1$ to solve.

4.   Conclusion

Segmenting with markers being equivalent to constructing the skeleton of influence of the marker for a particular lexicographic distance, segmenting becomes equivalent to solving a linear system. According the particular situation, one then choses the best adapted linear algebra algorithm.

References

[1] S. Beucher and F. Meyer. The morphological approach to segmentation: the watershed transformation. In E. Dougherty, editor, *Mathematical morphology in image processing*, chapter 12, pages 433–481. Marcel Dekker, 1993.

[2] G. Borgefors. Distance transforms in arbitrary dimension. *Comput. Vision, Graphics, Image Processing*, 27:321–345, 1984.

[3] T. Deschamps and L. Cohen. Fast extraction of minimal paths in 3D images and applications to virtual endoscopy. *Medical Image Analysis*, 5(4), Dec. 2001.

[4] M. Gondran and M. Minoux. *Graphes et Algorithmes*. Eyrolles, 1995.

[5] C. Lenart. A generalized distance in graphs. *Siam J. of discrete mathematics*, 11(2):293–304, 1998.

[6] F. Meyer. Minimal spanning forests for morphological segmentation. *ISMM94 : Mathematical Morphology and its applications to Signal Processing*, pages 77–84, 1994.

[7] F. Meyer. Topographic distance and watershed lines. *Signal Processing*, pages 113–125, 1994.

[8] F. Meyer. An overview of morphological segmentation. *International Journal of Pattern Recognition and Artificial Intelligence*, 17(7):1089–1118, 2001.

[9] F. Meyer and S. Beucher. Morphological segmentation. *jvcip*, 1(1):21–46, Sept. 1990.

[10] L. Najman and M. Schmitt. Watershed of a continuous function. *Signal Processing*, 38:99–112, July 1994.

[11] M. Schmitt. Ubiquity of the distance function in mathematical morphology. In F. Meyer and M. Schmitt, editors, *Geostatistics, Mathematical Morphology, Random Sets..., a Tribute to Georges Matheron*. Springer, 2005.

[12] P. Verbeek and B. Verwer. Shading from shape, the eikonal equation solved by grey-weighted distance transform. *Pattern Recogn. Lett.*, 11:618–690, 1990.

[13] L. Vincent. Minimal paths algorithms for the robust detection of linear features in images. In H. Heijmans and J.Roerdink, editors, *Mathematical Morphology and Its Applications to Image Processing*, pages 331–338. Kluwer, 1998.

MATHEMATICAL MODELING OF THE RELATIONSHIP "BETWEEN" BASED ON MORPHOLOGICAL OPERATORS

Isabelle Bloch[1], Olivier Colliot[2] and Roberto M. Cesar[3]

[1]*Ecole Nationale Supérieure des Télécommunications, Département TSI, CNRS UMR 5141, 46 rue Barrault, 75634 Paris Cedex 13, France*
Isabelle.Bloch@enst.fr

[2]*McConnell Brain Imaging Center, MNI, McGill University, 3801 University, Montréal, Québec, H3A2B4, Canada*
colliot@bic.mni.mcgill.ca

[3]*Department of Computer Science, Institute of Mathematics and Statistics - IME, University of São Paulo - USP Rua do Matão, 1010, São Paulo, SP CEP 05508-090 - Brazil*
cesar@ime.usp.br

Abstract   The spatial relationship "between" is a notion which is intrinsically both fuzzy and contextual, and depends in particular on the shape of the objects. The few existing definitions do not take into account these aspects. We propose here definitions which are based on morphological operators and a fuzzy notion of visibility in order to model the main intuitive acceptions of the relation. We distinguish between cases where objects have similar spatial extensions and cases where one object is much more extended than the other. These definitions are illustrated on real data from brain images.

Keywords: Relationship "between", spatial reasoning, fuzzy dilation, visibility.

Introduction

Spatial reasoning and structural object recognition in images rely on characteristics or features of objects, but also on spatial relationships between these objects, which are often more stable and less prone to variability. Several of these relationships (like set theoretical ones, adjacency, distances) are mathematically well defined. Other ones are intrinsically vague and imprecise. Their modeling in the framework of fuzzy sets proved to be well adapted. This is the case for instance for directional relative direction, which can be adequately defined using fuzzy directional dilations [2]. Interestingly enough, these basic

299

C. Ronse et al. (eds.), Mathematical Morphology: 40 Years On, 299–308.
©2005 *Springer. Printed in the Netherlands.*

relationships can be expressed in terms of mathematical morphology, which endows this framework with a unifying feature [3]. More complex relationships have received very little attention until now. In this paper, we deal with the "between" relation, and propose to model it based on simple morphological operators.

Definitions of "between" in dictionaries involve the notion of separation ("in the space that separates"). From a cognitive point of view, two factors appear to play a role: the convex hull of the union of both objects and the notion of visibility. Several difficulties arise when trying to model this relationship. First, it is intrinsically vague and imprecise, even if objects are precise. For instance, in Figure 1 (a), we would like to consider that B is not completely between A_1 and A_2 but that it is between them to some degree. Moreover, the relation has several meanings and may vary depending on shape. The definitions should therefore be contextual rather than absolute. For instance, the between relation cannot be defined in the same way whether the objects have similar spatial extensions or not (Figure 1 (b)), hence the necessary dependence on the context of the definitions. The semantics of "between" change depending on whether we consider a person between two buildings, a fountain between a house and a road, or a road passing between two houses. These differences have been exhibited in cognitive and linguistic studies [13].

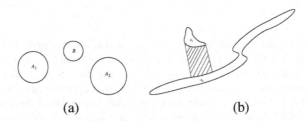

(a)   (b)

Figure 1.   (a) Is the object B between A_1 and A_2 and to which degree? (b) An example of objects with different spatial extension where a contextual definition is appropriate [13].

The primary aim of this paper is to propose some definitions of the relationship "between", modeling mathematically these intuitive ideas. More precisely, we try to answer the following question: *Which is the region of space, denoted by $\beta(A_1, A_2)$, located between two objects A_1 and A_2?* From the answer, we can then assess the degree to which an object B is between A_1 and A_2 by defining an appropriate measure of comparison between $\beta(A_1, A_2)$ and B [4].

Although this problem received very little attention, it was addressed by different communities. Approaches found in the domain of spatial logics and qualitative spatial reasoning rely on colinearity between points [1] or between centers of spheres [11]. They do not take into account the shape of objects nor the fuzziness of the "between" relationship. To our knowledge, only two approaches take the fuzziness into account. One applies on one-dimensional fuzzy sets [6] and relies on the definition of fuzzy ordering which makes its extension to higher dimensions an uneasy task. The only one which is close to our aim is the definition of [9]. The degree to which an object B is between two objects A_1 and A_2 in the 2D space is defined as a function of the average angle between segments linking points of B to points of A_1 and A_2. This approach has a major drawback, even in quite simple situations, due to the averaging and may lead to counter-intuitive results. Some linguistic and cognitive researches link the "between" relation to the mathematical definition of convex hull. In [10], the area between two objects is cognitively understood as "the minimal space bounded by the pair of reference objects". In [13], the region between A_1 and A_2 is defined as the strict interior of the convex hull $CH(A_1 \cup A_2)$ of $A_1 \cup A_2$ from which A_1 and A_2 are suppressed. Obviously this applies only in simple situations. For more complex objects, the connected components of $CH(A_1 \cup A_2) \setminus (A_1 \cup A_2)$ which are not adjacent to both A_1 and A_2 should be suppressed [4], but this does not solve all problems. In [13] an idea is also briefly mentioned, to deal with objects having different spatial extensions. In that case, the definition using the convex hull is not meaningful and more contextual definitions can be proposed, such as the area issued from the projection of the small object on the large one (Figure 1 (b)). It should be noted that, to our knowledge, no definition deals appropriately with such cases.

In this paper, we assume that, in continuous space, the considered objects are compact sets (enabling an easy link with the digital case), and that they have only one connected component. Extensions to objects having several connected components will be proposed based on a distributivity property. We also assume that objects A_1 and A_2 are not connected to each other.

The paper is organized as follows. Definitions based on morphological operators, in particular dilations and separation tools, are presented in Section 1; definitions based on the notion of visibility and fuzzy dilation are then proposed, in Section 2. In order to illustrate our work, in Section 3, we apply some of the proposed definitions on real objects, namely brain structures, and show that the results correspond to what is intuitively expected.

1.   Morphological dilations and directional dilations

We now try to implement the notion of separation that is found in standard dictionary definitions. Morphological dilation provides a good basis to this aim. If both objects are dilated until they meet, the ultimate intersection can be considered as being between both objects, and therefore constitutes a "seed" for constructing $\beta(A_1, A_2)$. Formally, we define:

$$\beta(A_1, A_2) = D^n[D^n(A_1) \cap D^n(A_2)] \cap A_1^C \cap A_2^C \qquad (1)$$

where D^n denotes the dilation by a disk of radius n, and where n is defined as: $n = \min\{k/D^k(A_1) \cap D^k(A_2) \neq \emptyset\}$ i.e. n is the half of the minimum distance between both sets. This definition applies for convex sets, but is clearly not adapted to non convex sets. Even for convex sets, this definition can be considered as too restricted, since it may exclude some parts of the convex hull of the union of both sets, as illustrated in Figure 2 (a). This effect can be even stronger in case of non convex sets, as shown in Figure 2 (b). But the result can also be too extended, typically if the distance between objects is much larger than their size.

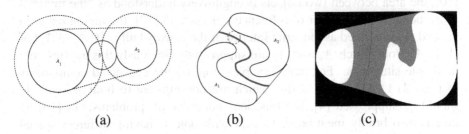

(a)   (b)   (c)

Figure 2.   (a) Dilation of the intersection of the dilations of A_1 and A_2 by a size equal to their half minimum distance. (b) Definition based on watershed or SKIZ (thick line), in a case where the definition based on simple dilation leads only to the disk limited by the dashed line. (c) Definition based on watershed, applied on the white objects and providing the grey area.

A line that "best" separates the two sets can also be considered as a seed of $\beta(A_1, A_2)$. This line can be implemented in mathematical morphology as the watersheds of the distance function to $A_1 \cup A_2$ (see Figure 2), or equivalently (with the assumptions made on the objects) the SKIZ (skeleton by influence zones). From this seed, the space that separates both sets can be defined as the geodesic dilation until convergence (reconstruction) of the watershed lines (or the SKIZ) in $CH'(A_1 \cup A_2)$, where $CH'(A_1 \cup A_2)$ denotes $CH(A_1 \cup A_2) \setminus (A_1 \cup A_2)$ from which the connected components not adjacent to both sets are suppressed. Actually, in the cases we consider (A_1 and A_2 compact sets, not connected to each other, and each having only one connected component), the reconstruction provides $CH'(A_1 \cup A_2)$. Another way to implement the notion of separation is to use the definition of [12], where a compact set is said to separate two compact sets A_1 and A_2 if any segment with extremities in A_1 and A_2 respectively hits this set. Unfortunately this definition does not solve the problem of non visible concavities and does not prevent the separating set to intersect A_1 or A_2.

Note that cases like in Figure 2 (c) could be handled in a satisfactory way by considering $CH(A_1 \cup A_2) \setminus (CH(A_1) \cup CH(A_2))$, i.e. by working on the convex hull of both objects, but this approach is not general enough and cannot deal with imbricated objects for instance.

In order to improve the dilation based approach, we develop an idea similar to the one proposed in 1D in [6], by replacing the ordering by the directional relative position of both objects. Directional dilation is then performed, using directional fuzzy structuring elements, with a similar approach as in [2]. The main direction between two objects can be determined from the angle histogram [14]. Given an axis of reference, say the x axis, denoted $\vec{u_x}$, the histogram of angles $h_{(A_1,A_2)}$ is defined as: $h_{(A_1,A_2)}(\theta) = |\{(a_1, a_2), a_1 \in A_1, a_2 \in A_2, \angle(\vec{a_1 a_2}, \vec{u_x}) = \theta\}|$. The maximum or the average value α of this histogram can be chosen as the main direction between A_1 and A_2. Let D_α denote the dilation in direction α. The structuring element can be either a crisp segment in the direction α, or a fuzzy structuring element where the membership function at a point (r, θ) (in polar coordinates) is a decreasing function of $|\theta - \alpha|$ [2]. From this dilation, we define $\beta_\alpha = D_\alpha(A_1) \cap D_{\pi+\alpha}(A_2) \cap A_1^C \cap A_2^C$. Since it can be difficult to find only one main direction (from histogram of angles for instance) we can use several values for α and define β as: $\beta = \cup_\alpha \beta_\alpha$ or $\beta = \cup_\alpha(\beta_\alpha \cup \beta_{\alpha+\pi})$. We can also use the histogram of angles directly as a fuzzy structuring element. For instance, let us define two fuzzy structuring elements ν_1 and ν_2 from the angle histogram $h_{(A_1,A_2)}(\theta)$ (Figure 5 (a)) as: $\nu_1(r, \theta) = h_{(A_1,A_2)}(\theta)$ (see Figure 5 (b)), and $\nu_2(r, \theta) = h_{(A_1,A_2)}(\theta + \pi) = \nu_1(r, \theta + \pi)$. Several definitions of the between region can be envisaged, the simplest being:

$$\beta(A_1, A_2) = D_{\nu_2}(A_1) \cap D_{\nu_1}(A_2) \cap A_1^C \cap A_2^C, \tag{2}$$

which is illustrated in Figure 5 (c). Another definition, inspired by [6], allows to remove the concavities which are not "facing each other":

$$\beta(A_1, A_2) = D_{\nu_2}(A_1) \cap D_{\nu_1}(A_2) \cap A_1^C \cap A_2^C \cap$$
$$[D_{\nu_1}(A_1) \cap D_{\nu_1}(A_2)]^C \cap [D_{\nu_2}(A_1) \cap D_{\nu_2}(A_2)]^C, \tag{3}$$

which is illustrated in Figure 5 (d). For instance, if A_2 is approximately to the right of A_1, this definition suppresses the concavities which are to the left (respectively right) of both objects.

2.   Visibility, fuzzy visibility and myopic vision

Let us consider again the situation in Figure 2 (c). If we assume that the two objects are buildings and that someone is supposed to meet another person between these buildings, then he would probably expect the person to wait in an area where he can surely be seen, not in the concavity. Another interpretation would be that B hides to A_1 a part of A_2. The notion of visibility has to play an important role: although $CH(A_1 \cup A_2) \setminus (A_1 \cup A_2)$ has only one connected component, which is adjacent to both A_1 and A_2, object A_2 has a concavity which is not visible from A_1 and should probably not be included in the between area (Figure 2 (c)). To take such situations into account, we propose to base the notion of visibility on admissible segments as introduced in [15]. A segment $]x_1, x_2[$, with x_1 in A_1 and x_2 in A_2 (A_1 and A_2 are still supposed to be compact sets), is said admissible if it is included in $A_1^C \cap A_2^C$. Note that x_1 and x_2 then necessarily belong to the boundary of A_1 and A_2, respectively (A_1 and A_2 are compact). This has interesting consequences from an algorithmic point of view, since it considerably reduces the size of the set of points to be explored. The visible points are those which belong to admissible segments. The region between A_1 and A_2 can then be defined as the union of admissible segments. It corresponds to the set $CH(A_1 \cup A_2) \setminus (A_1 \cup A_2)$ from which all points not belonging to admissible segments are suppressed.

However, the definition of admissible segments can be too strict in some cases. In order to get more flexibility, we introduce the notion of approximate (or fuzzy) visibility. It extends both the crisp definition of visibility and the definition proposed in [9] in the sense that the information is not reduced to an average angle. This is achieved by relaxing the admissibility to semi-admissibility by introducing an intermediary point P on the segments. A segment $]a_1, P]$ with $a_1 \in A_1$ (respectively $[P, a_2[$ with $a_2 \in A_2$) is said semi-admissible if it is included in $A_1^C \cap A_2^C$. At each point P of space, we compute the angle the closest to π between two semi-admissible segments from P to A_1 and A_2 respectively. This is formally defined as:

$$\theta_{min}(P) = \min\{|\pi - \theta|, \theta = \angle([a_1, P], [P, a_2]),$$
$$a_1 \in A_1, a_2 \in A_2,]a_1, P] \text{ and } [P, a_2[\text{ semi-admissible}\}. \quad (4)$$

The region between A_1 and A_2 is then defined as the fuzzy region of space with membership function $\beta(P) = f(\theta_{min}(P))$, where f is a function from $[0, \pi]$ to $[0, 1]$ such that $f(0) = 1$, f is decreasing, and becomes 0 at the largest acceptable distance to π (this value can be tuned according to the context). This idea is illustrated in Figure 3.

Now, we assume that one of the objects, say A_2, can be considered to have infinite size with respect to the other (we assume this to be known in advance). This is the case for instance when one says that a fountain is between the house and the road, or that the sport area is between the city hall and the beach. None of the previous definitions applies in such cases, since they consider objects globally. Intuitively, the between area should be considered between A_1 and the only part of A_2 which is the closest to A_1, instead of considering A_2 globally. Hence the idea of projecting A_1 onto A_2 in some sense, and to consider the "umbra" of A_1. Here we make an additional assumption, largely verified in most situations, by approximating by a segment the part the closest to A_1 (if this appear to be too restrictive, the part the closest to A_1 can be approximated by several segments, of different directions, and the orthogonal direction is then locally defined). Let us denote the segment direction by $\vec{u}$. The between

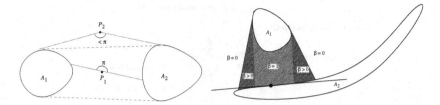

Figure 3.   Left: Illustration of the fuzzy visibility concept. For point P_1, we have $\theta_{min}(P_1) = 0$ and therefore $\beta(P_1) = 1$, while for point P_2, it is not possible to find two colinear semi-admissible segments from A_1 (respectively A_2) to P_2, thus $\theta_{min}(P_2) > 0$ and $\beta(P_2) < 1$, expressing that P_2 is not completely between A_1 and A_2. Right: Illustration of the definition of region β in case on an extended object (myopic vision). In the areas indicated by $\beta > 0$, the relation is satisfied to some degree between 0 and 1. They can be more or less spread depending on the structuring element, i.e. on the semantics of the relation.

region can then be defined by dilating A_1 by a structuring element defined as a segment orthogonal to $\vec{u}$ and limiting this dilation to the half plane defined by the segment of direction $\vec{u}$ and containing A_1. However this may appear as too restrictive and a fuzzy dilation [5] by a structuring element having decreasing membership degrees when going away from the direction orthogonal to $\vec{u}$ [2] is more flexible and better matches the intuitive idea. The projection segment can be defined by dilating the part of A_2 the closest to A_1 (obtained by a distance map computation) conditionally to A_2 and computing the axis of inertia of the result. This approach is illustrated in Figure 3. In terms of visibility, it corresponds to a "myopic" vision, in which the parts of A_2 which are too far from A_1 are not seen.

Further work in this direction aims at achieving a continuity from the case where the objects have similar spatial extensions to the one where one becomes much more elongated.

3.   Illustrations

In this Section, we illustrate some of the proposed definitions on brain structures. Figure 4 presents a few brain structures, on a 2D slice. Usual anatomical descriptions make intensive use of spatial relations to describe these objects (see e.g. http://www.chups.jussieu.fr/ext/neuranat) and such descriptions are very useful for recognizing these structures in medical images [7]. The between relation is involved in several of these descriptions: (i) the internal capsule (IC) is *between* the caudate nucleus (CN) and the lenticular nucleus (LN); (ii) the corpus callosum (CC) is *between* the lateral ventricles (LV) and the cingulate gyrus (CG) ; (iii) the medial frontal gyrus (MFG) is *between* the inferior frontal gyrus (IFG) and the superior frontal gyrus (SFG).

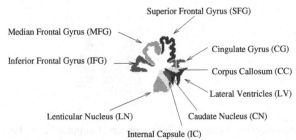

Superior Frontal Gyrus (SFG)

Median Frontal Gyrus (MFG)

Inferior Frontal Gyrus (IFG)

Cingulate Gyrus (CG)

Corpus Callosum (CC)

Lateral Ventricles (LV)

Lenticular Nucleus (LN)   Caudate Nucleus (CN)

Internal Capsule (IC)

Figure 4. A few brain structures (a 2D slice extracted from a 3D atlas).

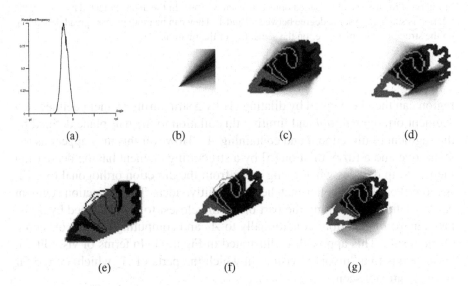

Figure 5. (a) Angle histogram of objects A_1 and A_2 (superior and inferior frontal gyri). (b) Corresponding structuring element ν_1 (ν_2 is its symmetrical with respect to the origin). (c) Definition based on fuzzy dilation (Equation 2). Membership values to $\beta(A_1, A_2)$ vary from 0 (white) to 1 (black). The contours of the median frontal gyrus are superimposed. (d) Definition based on fuzzy dilation, with Equation 3. (e) Convex hull approach. (f) Definition using the admissible segments. (g) Fuzzy visibility approach.

Our definitions were applied to define the region between the aforementioned brain structures. Figure 5 shows (for one example only due to lack of space) the "between" region using the directional dilation (a-d), the convex hull approach (e), the admissible segments (f) and the fuzzy visibility (g). It is clear that the convex hull definition does not deal appropriately with concavities. This problem is solved by the directional dilation and visibility approaches. In (c) and (e), non visible concavities are included in the result, while they have been adequately suppressed in (d), (f) and (g). Also, it should be noted that fuzzy methods (directional dilation and fuzzy visibility) are more appropriate than crisp ones (convex hull and admissible segments): the median frontal gyrus is partly outside the β region defined by crisp approaches while remaining in areas with high membership degrees when using the fuzzy ones. Note that the region where $\beta = 1$ is the same in (f) and (g) and almost the same in (d), therefore these approaches are equivalent for objects completely included in this region.

To evaluate the relation "B is between A_1 and A_2", we computed the normalized intersection of B and β: $\frac{|\beta \cap B|}{|B|}$. A few results are shown in Table 1. They correspond to what is intuitively expected. Higher degrees are obtained with fuzzy methods which again indicates that they deal more appropriately with objects that would not be completely included in the crisp β region. The measures are however quite similar for all approaches, since none of the objects B is located in a concavity.

A_1	A_2	B	$\frac{|\beta \cap B|}{|B|}$ (1)	$\frac{|\beta \cap B|}{|B|}$ (2)	$\frac{|\beta \cap B|}{|B|}$ (3)	$\frac{|\beta \cap B|}{|B|}$ (4)
CN	LN	IC	0.85	0.84	0.84	0.94
LV	CG	CC	1.00	0.93	1.00	1.00
IFG	SFG	MFG	0.78	0.92	0.76	0.95
CG	CN	CC	0.88	0.90	0.88	0.97
CG	CN	LV	0.47	0.63	0.47	0.79
IFG	SFG	IC	0.00	0.02	0.00	0.16
IFG	SFG	LN	0.00	0.00	0.00	0.04

Table 1. A few results obtained with the method of convex hull (1), fuzzy directional dilation (2), admissible segments (3) and with the fuzzy visibility approach (4). The fifth line corresponds to a case where only a part of B is between A_1 and A_2, the relation being thus satisfied with a lower degree than in the previous cases. The last two lines correspond to cases where the relation is not satisfied. Low but non-zero values are obtained with the fuzzy approaches, because of the tolerance on the angles.

4.   Conclusion

We have shown in this paper how a complex spatial relation, "between", can be modeled using simple tools of mathematical morphology and fuzzy mathematical morphology, and addressed the modeling of this relation in cases where objects have similar spatial extensions or very different ones. The proposed definitions of $\beta(A_1, A_2)$ have the following properties: symmetry (by construction); invariance with respect to geometrical operations (translation, rotation). In case the objects have several connected components, the definitions can be easily extended by applying a distributivity principle. For instance if A_1 can be decomposed into connected components as $\cup_i A_1^i$, then we define $\beta(A_1, A_2) = \beta(\cup_i A_1^i, A_2) = \cup_i \beta(A_1^i, A_2)$. Properties of these definitions could be further studied, as well as possible links or transitions from one definition to the other. Future work aims also at defining other measures to assess the degree to which an object B is between A_1 and A_2, in different types of contexts, and at introducing the between relation in structural pattern recognition and spatial reasoning schemes, as done previously for other relationships [7, 8].

Acknowledgements: This work was partially supported by a CAPES / COFECUB grant (number 369/01).

References

[1] M. Aiello and J. van Benthem. A Modal Walk Through Space. *Journal of Applied Non Classical Logics*, 12(3-4):319–364, 2002.

[2] I. Bloch. Fuzzy Relative Position between Objects in Image Processing: a Morphological Approach. *IEEE Trans. on PAMI*, 21(7):657–664, 1999.

[3] I. Bloch. Mathematical Morphology and Spatial Relationships: Quantitative, Semi-Quantitative and Symbolic Settings. In L. Sztandera and P. Matsakis, editors, *Applying Soft Computing in Defining Spatial Relationships*, pages 63–98. Physica Verlag, Springer, 2002.

[4] I. Bloch, O. Colliot, and R. Cesar. Modélisation de la relation spatiale "entre" à partir de notions de convexité et de visibilité floue. In *Rencontres francophones sur la Logique Floue et ses Applications LFA'04*, pages 149–156, Nantes, France, nov 2004.

[5] I. Bloch and H. Maître. Fuzzy Mathematical Morphologies: A Comparative Study. *Pattern Recognition*, 28(9):1341–1387, 1995.

[6] U. Bodenhofer. Fuzzy "Between" Operators in the Framework of Fuzzy Orderings. In B. Bouchon-Meunier, L. Foulloy, and R. R. Yager, editors, *Intelligent Systems for Information Processing: From Representation to Applications*, pages 59–70. Elsevier, 2003.

[7] O. Colliot, O. Camara, R. Dewynter, and I. Bloch. Description of Brain Internal Structures by Means of Spatial Relations for MR Image Segmentation. In *SPIE Medical Imaging*, volume 5370, pages 444–455, San Diego, CA, USA, 2004.

[8] O. Colliot, A. Tuzikov, R. Cesar, and I. Bloch. Approximate Reflectional Symmetries of Fuzzy Objects with an Application in Model-Based Object Recognition. *Fuzzy Sets and Systems*, 147:141–163, 2004.

[9] R. Krishnapuram, J. M. Keller, and Y. Ma. Quantitative Analysis of Properties and Spatial Relations of Fuzzy Image Regions. *IEEE Trans. on Fuzzy Systems*, 1(3):222–233, 1993.

[10] B. Landau and R. Jackendorff. "What" and "Where" in Spatial Language and Spatial Cognition. *Behavioral and Brain Sciences*, 16:217–265, 1993.

[11] Y. Larvor. Notion de méréogéométrie : description qualitative de propriétés géométriques, du mouvement et de la forme d'objets tridimensionnels. PhD thesis, Université Paul Sabatier, Toulouse, 2004.

[12] G. Matheron. *Random Sets and Integral Geometry*. Wiley, New-York, 1975.

[13] Y. Mathet. *Etude de l'expression en langue de l'espace et du déplacement : analyse linguistique, modélisation cognitive, et leur expérimentation informatique*. PhD thesis, Université de Caen, France, December 2000.

[14] K. Miyajima and A. Ralescu. Spatial Organization in 2D Segmented Images: Representation and Recognition of Primitive Spatial Relations. *Fuzzy Sets and Systems*, 65:225–236, 1994.

[15] A. Rosenfeld and R. Klette. Degree of Adjacency or Surroundness. *Pattern Recognition*, 18(2):169–177, 1985.

V

PARTIAL DIFFERENTIAL EQUATIONS
AND EVOLUTIONARY MODELS

SEMIDISCRETE AND DISCRETE WELL-POSEDNESS OF SHOCK FILTERING

Martin Welk[1] and Joachim Weickert[1]

[1]*Mathematical Image Analysis Group*
Faculty of Mathematics and Computer Science, Bldg. 27
Saarland University, 66041 Saarbruecken, Germany
{welk,weickert}@mia.uni-saarland.de

Abstract   While shock filters are popular morphological image enhancement methods, no well-posedness theory is available for their corresponding partial differential equations (PDEs). By analysing the dynamical system of ordinary differential equations that results from a space discretisation of a PDE for 1-D shock filtering, we derive an analytical solution and prove well-posedness. Finally we show that the results carry over to the fully discrete case when an explicit time discretisation is applied.

Keywords: Shock filters, analytical solution, well-posedness, dynamical systems.

1.   Introduction

Shock filters are morphological image enhancement methods where dilation is performed around maxima and erosion around minima. Iterating this process leads to a segmentation with piecewise constant segments that are separated by discontinuities, so-called shocks. This makes shock filtering attractive for a number of applications where edge sharpening and a piecewise constant segmentation is desired.

In 1975 the first shock filters have been formulated by Kramer and Bruckner in a fully discrete manner [6], while first continuous formulations by means of partial differential equations (PDEs) have been developed in 1990 by Osher and Rudin [8]. The relation of these methods to the discrete Kramer–Bruckner filter became clear several years later [4, 12]. PDE-based shock filters have been investigated in a number of papers. Many of them proposed modifications with higher robustness under noise [1, 3, 5, 7, 12], but also coherence-enhancing shock filters [14] and numerical schemes have been studied [11].

Let us consider some continuous d-dimensional initial image $f : \mathbb{R}^d \to \mathbb{R}$. In the simplest case of a PDE-based shock filter [8], one obtains a filtered

C. Ronse et al. (eds.), Mathematical Morphology: 40 Years On, 311–320.
©2005 *Springer. Printed in the Netherlands.*

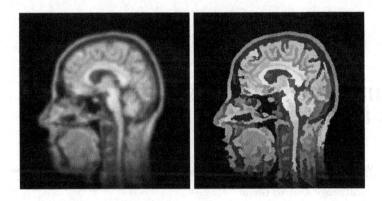

Figure 1. LEFT: Original image. RIGHT: After applying the Osher–Rudin shock filter.

version $u(x, t)$ of $f(x)$ by solving the evolution equation

$$\partial_t u = -\text{sgn}(\Delta u) |\nabla u| \qquad (t \geq 0)$$

with f as initial condition, i. e. $u(0, x) = f(x)$. Experimentally one observes that within finite "evolution time" t, a piecewise constant, segmentation-like result is obtained (see Fig. 1).

Specialising to the one-dimensional case, we obtain

$$\partial_t u = -\text{sgn}(\partial_{xx} u) |\partial_x u| = \begin{cases} |\partial_x u|, & \partial_{xx} u < 0, \\ -|\partial_x u|, & \partial_{xx} u > 0, \\ 0, & \partial_{xx} u = 0. \end{cases} \qquad (1)$$

It is clearly visible that this filter performs dilation $\partial_t u = |\partial_x u|$ in concave segments of u, while in convex parts the erosion process $\partial_t u = -|\partial_x u|$ takes place. The time t specifies the radius of the interval (a 1-D disk) $[-t, t]$ as structuring element. For a derivation of these PDE formulations for classical morphological operations, see e.g. [2].

While there is clear experimental evidence that shock filtering is a useful operation, no analytical solutions and well-posedness results are available for PDE-based shock filters. In general this problem is considered to be too diffi-cult, since shock filters have some connections to classical ill-posed problems such as backward diffusion [8, 7].

The goal of the present paper is to we show that it is possible to establish analytical solutions and well-posedness as soon as we study the *semidiscrete* case with a spatial discretisation and a continuous time parameter t. This case is of great practical relevance, since digital images already induce a natural space discretisation. For the sake of simplicity we restrict ourselves to the 1-D case. We also show that these results carry over to the fully discrete case with an explicit (Euler forward) time discretisation.

Our paper is organised as follows: In Section 2 we present an analytical solution and a well-posedness proof for the semidiscrete case, whereas corresponding fully discrete results are given in Section 3. Conclusions are presented in Section 4.

2. The Semidiscrete Model

Throughout this paper, we are concerned with a spatial discretisation of (1) which we will describe now.

PROBLEM. *Let* $(\ldots, u_0(t), u_1(t), u_2(t), \ldots)$ *be a time-dependent real-valued signal which evolves according to*

$$\dot{u}_i = \begin{cases} \max(u_{i+1} - u_i, u_{i-1} - u_i, 0), & 2u_i > u_{i+1} + u_{i-1}, \\ \min(u_{i+1} - u_i, u_{i-1} - u_i, 0), & 2u_i < u_{i+1} + u_{i-1}, \\ 0, & 2u_i = u_{i+1} + u_{i-1} \end{cases} \tag{2}$$

with the initial conditions

$$u_i(0) = f_i. \tag{3}$$

Assume further that the signal is either of infinite length or finite with reflecting boundary conditions.

Like (1), this filter switches between dilation and erosion depending on the local convexity or concavity of the signal. Dilation and erosion themselves are modeled by upwind-type discretisations [9], and $\dot{u}_i$ denotes the time derivative of $u_i(t)$.

It should be noted that in case $2u_i > u_{i+1} + u_{i-1}$ the two neighbour differences $u_{i+1} - u_i$ and $u_{i-1} - u_i$ cannot be simultaneously positive; with the opposite inequality they can't be simultaneously negative. In fact, always when the maximum or minimum in (2) does not select its third argument, zero, it returns the absolutely smaller of the neighbour differences.

No modification of (2) is needed for finite-length signals with reflecting boundary conditions. In this case, each boundary pixel has one vanishing neighbour difference.

In order to study the solution behaviour of this system, we have to specify the possible solutions, taking into account that the right-hand side of (2) may involve discontinuities. We say that a time-dependent signal $u(t) = (\ldots, u_1(t), u_2(t), u_3(t) \ldots)$ is a *solution* of (2) if

(I) each u_i is a piecewise differentiable function of t,

(II) each u_i satisfies (2) for all times t for which $\dot{u}_i(t)$ exists,

(III) for $t = 0$, the right-sided derivative $\dot{u}_i^+(0)$ equals the right-hand side of (2) if $2u_i(0) \neq u_{i+1}(0) + u_{i-1}(0)$.

We state now our main result.

THEOREM 1 (WELL-POSEDNESS) *For our Problem, assume that the equality $f_{k+1} - 2f_k + f_{k-1} = 0$ does not hold for any pixel f_k which is not a local maximum or minimum of f. Then the following are true:*

(i) **Existence and uniqueness:** *The Problem has a unique solution for all $t \geq 0$.*

(ii) **Maximum–minimum principle:** *If there are real bounds a, b such that $a < f_k < b$ holds for all k, then $a < u_k(t) < b$ holds for all k and all $t \geq 0$.*

(iii) **l_∞-stability:** *There exists a $\delta > 0$ such that for any initial signal $\tilde{f}$ with $\|\tilde{f} - f\|_\infty < \delta$ the corresponding solution $\tilde{u}$ satisfies the estimate*

$$\|\tilde{u}(t) - u(t)\|_\infty < \|\tilde{f} - f\|_\infty$$

for all $t > 0$. The solution therefore depends l_∞-continuously on the initial conditions within a neighbourhood of f.

(iv) **Total variation preservation:** *If the total variation of f is finite, then the total variation of u at any time $t \geq 0$ equals that of f.*

(v) **Steady state:** *For $t \to \infty$, the signal u converges to a piecewise constant signal. The jumps in this signal are located at the steepest slope positions of the original signal.*

All statements of this theorem follow from an explicit analytical solution of the Problem that will be described in the following proposition.

PROPOSITION 2 (ANALYTICAL SOLUTION) *For our standard problem, let the segment $(f_1, \ldots, f_m)$ be strictly decreasing and concave in all pixels. Assume that the leading pixel f_1 is either a local maximum or a neighbour to a convex pixel $f_0 > f_1$. Then the following hold for all $t \geq 0$:*

(i) *If f_1 is a local maximum of f, $u_1(t)$ is a local maximum of $u(t)$.*

(ii) *If f_1 is neighbour to a convex pixel $f_0 > f_1$, then $u_1(t)$ also has a convex neighbour pixel $u_0(t) > u_1(t)$.*

(iii) *The segment $(u_1, \ldots, u_m)$ remains strictly decreasing and concave in all pixels. The grey values of all pixels at time t are given by*

$$u_k(t) = C \cdot \left(1 + (-1)^k e^{-2t} - e^{-t} \sum_{j=0}^{k-2} \frac{t^j}{j!} (1 + (-1)^{k-j}) \right)$$

$$+ e^{-t} \sum_{j=0}^{k-2} \frac{t^j}{j!} f_{k-j} - (-1)^k f_1 e^{-t} \left(e^{-t} - \sum_{j=0}^{k-2} \frac{(-t)^j}{j!} \right) \tag{4}$$

for $k = 1, \ldots, m$, where $C = f_1(0)$ if f_1 is a local maximum of f, and $C = \frac{1}{2}(f_0(0) + f_1(0))$ otherwise.

(iv) At no time $t \geq 0$, the equation $2u_i(t) = u_{i+1}(t) + u_{i-1}(t)$ becomes true for any $i \in \{1, \ldots, m\}$.

Analogous statements hold for increasing concave and for convex signal segments.

In a signal that contains no locally flat pixels (such with $2f_i = f_{i+1} + f_{i-1}$), each pixel belongs to a chain of either concave or convex pixels led by an extremal pixel or an "inflection pair" of a convex and a concave pixel. Therefore Proposition 2 completely describes the dynamics of such a signal. Let us prove this proposition.

PROOF. We show in steps (i)–(iii) that the claimed evolution equations hold as long as the initial monotonicity and convexity properties of the signal segment prevail. Step (iv) then completes the proof by demonstrating that the evolution equations preserve exactly these monotonicity and convexity requirements.

(i) From (2) it is clear that any pixel u_i which is extremal at time t has $\dot{u}_i(t) = 0$ and therefore does not move. Particularly, if f_1 is a local maximum of f, then $u_1(t)$ remains constant as long as it continues to be a maximum.

(ii) If $u_0 > u_1$, u_0 is convex and u_1 concave for $t \in [0, T)$. Then we have for these pixels

$$\begin{aligned} \dot{u}_0 &= u_1 - u_0 \,, \\ \dot{u}_1 &= u_0 - u_1 \end{aligned} \tag{5}$$

which by the substitutions $y := \frac{1}{2}(u_0 + u_1)$ and $v := u_1 - u_0$ becomes

$$\begin{aligned} \dot{y} &= 0 \,, \\ \dot{v} &= -2v \,. \end{aligned}$$

This system of linear ordinary differential equations (ODEs) has the solution $y(t) = y(0) = C$ and $v(t) = v(0) \exp(-2t)$. Backsubstitution gives

$$\begin{aligned} u_0(t) &= C \cdot (1 - e^{-2t}) + f_0 e^{-2t} \,, \\ u_1(t) &= C \cdot (1 - e^{-2t}) + f_1 e^{-2t} \,. \end{aligned} \tag{6}$$

This explicit solution is valid as long as the convexity and monotonicity properties of u_0 and u_1 do not change.

(iii) Assume the monotonicity and convexity conditions required by the proposition for the initial signal hold for $u(t)$ for all $t \in [0, T)$. Then we have in all cases, defining C as in the proposition, the system of ODEs

$$\begin{aligned} \dot{u}_1 &= -2(u_1 - C) \,, \\ \dot{u}_k &= u_{k-1} - u_k \,, \qquad k = 2, \ldots, m \end{aligned} \tag{7}$$

for $t \in [0, T)$. We substitute further $v_k := u_k - C$ for $k = 1, \ldots, m$ as well as $w_1 := v_1$ and $w_k := v_k + (-1)^k v_1$ for $k = 2, \ldots, m$. This leads to the system

$$
\begin{aligned}
\dot{w}_1 &= -2w_1 \,, \\
\dot{w}_2 &= -w_2 \,, \\
\dot{w}_k &= w_{k-1} - w_k \,, \qquad k = 3, \ldots, m \,.
\end{aligned}
\tag{8}
$$

This system of linear ODEs has the unique solution

$$
\begin{aligned}
w_1(t) &= w_1(0)e^{-2t} \,, \\
w_k(t) &= e^{-t} \sum_{j=0}^{k-2} \frac{t^j}{j!} w_{k-j}(0) \,, \qquad k = 2, \ldots, m
\end{aligned}
$$

which after reverse substitution yields (4) for all $t \in [0, T]$.

(iv) Note that (5) and (7) are systems of linear ODEs which have the unique explicit solutions (6) and (4) for all $t > 0$. As long as the initial monotonicity and convexity conditions are satisfied, the solutions of (2) coincide with those of the linear ODE systems.

We prove therefore that the solution (4) fulfils the monotonicity condition

$$
u_k(t) - u_{k-1}(t) < 0 \,, \qquad k = 2, \ldots, m
$$

and the concavity conditions

$$
u_{k+1}(t) - 2u_k(t) + u_{k-1}(t) < 0 \qquad k = 1, \ldots, m
$$

for all $t > 0$ if they are valid for $t = 0$. To see this, we calculate first

$$
\begin{aligned}
u_k(t) - u_{k-1}(t) = e^{-t} \sum_{j=0}^{k-2} \frac{t^j}{j!} (f_{k-j} - f_{k-1-j}) \\
+ 2e^{-t}(-1)^{k-1} \left(e^{-t} - \sum_{j=0}^{k-2} \frac{(-t)^j}{j!} \right) (f_1 - C) \,.
\end{aligned}
$$

By hypothesis, $f_{k-j} - f_{k-1-j}$ and $f_1 - C$ are negative. Further, $\exp(-t) - \sum_{j=0}^{k-2} (-t)^j/j!$ is the error of the (alternating) Taylor series of $\exp(-t)$, thus having the same sign $(-1)^{k-1}$ as the first neglected member. Consequently, the monotonicity is preserved by (4) for all $t > 0$.

Second, we have for $k = 2, \ldots, m - 1$

$$
\begin{aligned}
u_{k+1}(t) - 2u_k(t) + u_{k-1}(t) = e^{-t} \sum_{j=0}^{k-1} \frac{t^j}{j!} (f_{k+1-j} - 2f_{k-j} + f_{k-j-1}) \\
+ 4e^{-t}(-t)^k \left(\sum_{j=0}^{k-1} \frac{(-t)^j}{j!} \right) (f_1 - C)
\end{aligned}
$$

which is seen to be negative by similar reasoning as above.

Concavity at $u_m(t)$ follows in nearly the same way. By extending (4) to $k = m + 1$, one obtains not necessarily the true evolution of u_{m+1} since that pixel is not assumed to be included in the concave segment. However, the true trajectory of u_{m+1} can only lie below or on that predicted by (4).

Third, if f_1 is a maximum of f, then $u_1(t)$ remains one for all $t > 0$ which also ensures concavity at u_1. If f_1 has a convex neighbour pixel $f_0 > f_1$, we have instead

$$u_2(t) - 2u_1(t) + u_0(t) = e^{-t}(f_2 - 2f_1 + f_0) + 4e^{-t}(1 - e^{-t})(f_1 - C) < 0$$

which is again negative for all $t > 0$.

Finally, we remark that the solution (6) ensures $u_0(t) > u_1(t)$ for all $t > 0$ if it holds for $t = 0$. That convexity at u_0 is preserved can be established by analogous reasoning as for the concavity at u_1.

Since the solutions from the linear systems guarantee preservation of all monotonicity and convexity properties which initially hold for the considered segment, these solutions are the solutions of (2) for all $t > 0$. □

We remark that uniqueness fails if the initial signal contains non-extremal locally flat pixels. More details for this case are given in a preprint [15].

3. Explicit Time Discretisation

In the following we discuss an explicit time discretisation of our time-continuous system. We denote the time step by $\tau > 0$. The time discretisation of our Problem then reads as follows:

TIME-DISCRETE PROBLEM. *Let* $(\ldots, u_0^l, u_1^l, u_2^l, \ldots)$, $l = 0, 1, 2, \ldots$ *be a series of real-valued signals which satisfy the equations*

$$\frac{u_i^{l+1} - u_i^l}{\tau} = \begin{cases} \max(u_{i+1} - u_i, u_{i-1} - u_i, 0), & 2u_i > u_{i+1} + u_{i-1}, \\ \min(u_{i+1} - u_i, u_{i-1} - u_i, 0), & 2u_i < u_{i+1} + u_{i-1}, \\ 0, & 2u_i = u_{i+1} + u_{i-1} \end{cases} \quad (9)$$

with the initial conditions

$$u_i^0 = f_i ; \quad (10)$$

assume further that the signal is either of infinite length or finite with reflecting boundary conditions.

THEOREM 3 (TIME-DISCRETE WELL-POSEDNESS) *Assume that in the Time-Discrete Problem the equality* $f_{k+1} - 2f_k + f_{k-1} = 0$ *does not hold for any pixel* f_k *which is not a local maximum or minimum of* f. *Assume further that* $\tau < 1/2$. *Then the statements of Theorem 1 are valid for the solution* *of the Time-Discrete Problem if only $u_k(t)$ for $t > 0$ is replaced everywhere by u_k^l with $l = 0, 1, 2, \ldots$*

The existence and uniqueness of the solution of the Time-Discrete Problem for $l = 0, 1, 2, \ldots$ is obvious. Maximum–minimum principle, l_∞-stability, total variation preservation and the steady state property are immediate consequences of the following proposition. It states that for $\tau < 1/2$ all qualitative properties of the time-continuous solution transfer to the time-discrete case.

PROPOSITION 4 (TIME-DISCRETE SOLUTION) *Let u_i^l be the value of pixel i in time step l of the solution of our Time-Discrete Problem with time step size $\tau < 1/2$. Then the following hold for all $l = 0, 1, 2, \ldots$:*

  (i) *If u_1^l is a local maximum of u^l, then u_1^{l+1} is a local maximum of u^{l+1}.*

  (ii) *If u_1^l is a concave pixel neighbouring to a convex pixel $u_0^l > u_1^l$, then u_1^{l+1} is again concave and has a convex neighbour pixel $u_0^{l+1} > u_1^{l+1}$.*

  (iii) *If the segment $(u_1^l, \ldots, u_m^l)$ is strictly decreasing and concave in all pixels, and u_1^l is either a local maximum of u^l or neighbours to a convex pixel $u_0^l > u_1^l$, then the segment $(u_1^{l+1}, \ldots, u_m^{l+1})$ is strictly decreasing.*

  (iv) *Under the same assumptions as in (iii), the segment $(u_1^{l+1}, \ldots, u_m^{l+1})$ is strictly concave in all pixels.*

  (v) *If $2u_i^l = u_{i+1}^l + u_{i-1}^l$ holds for no pixel i, then $2u_i^{l+1} = u_{i+1}^{l+1} + u_{i-1}^{l+1}$ also holds for no pixel i.*

  (vi) *Under the assumptions of (iii), all pixels in the range $i \in \{1, \ldots, m\}$ have the same limit $\lim_{l \to \infty} u_i^l = C$ with $C := u_1^l$ if u_1^l is a local maximum, or $C := \frac{1}{2}(u_0^l + u_1^l)$ if it neighbours to the convex pixel u_0^l.*

Analogous statements hold for increasing concave and for convex signal segments.

PROOF. Assume first that u_1^l is a local maximum of u^l. From the evolution equation (9) it is clear that $u_j^{l+1} \le u_j^l + \tau(u_1^l - u_j^l)$ for $j = 0, 2$. For $\tau < 1$ this entails $u_j^{l+1} < u_1^l = u_1^{l+1}$, thus (i).

If instead u_i^l is a concave neighbour of a convex pixel $u_0^l > u_1^l$, then we have $u_1^{l+1} = u_1^l + \tau(u_0^l - u_1^l)$ and $u_0^{l+1} = u_0^l + \tau(u_1^l - u_0^l)$. Obviously, $u_0^{l+1} > u_1^{l+1}$ holds if and only if $\tau < 1/2$. For concavity, note that $u_2^{l+1} \le u_2^l + \tau(u_1^l - u_2^l)$ and therefore $u_0^{l+1} - 2u_1^{l+1} + u_2^{l+1} \le (1-\tau)(u_0^l - 2u_1^l + u_2^l) + 2\tau(u_1^l - u_0^l)$. The right-hand side is certainly negative for $\tau \le 1/2$. An analogous argument secures convexity at pixel 0 which completes the proof of (ii).

In both cases we have $u_1^{l+1} \geq u_1^l$. Under the assumptions of (iii), (iv) we then have $u_k^{l+1} = u_k^l + \tau(u_{k-1}^l - u_k^l)$ for $k = 2, \ldots, m$. If $\tau < 1$, it follows that $u_k^l < u_k^{l+1} \leq u_{k-1}^l$ for $k = 2, \ldots, m$ which together with $u_1^{l+1} \geq u_1^l$ implies that $u_{k-1}^{l+1} > u_k^l$ for $k = 2, \ldots, m$ and therefore (iii).

For the concavity condition we compute

$$u_{k-1}^{l+1} - 2u_k^{l+1} + u_{k+1}^{l+1} = (1-\tau)(u_{k-1}^l - 2u_k^l + u_{k+1}^l) + \tau(u_{k-2}^l - 2u_{k-1}^l + u_k^l)$$

for $k = 3, \ldots, m - 1$. The right-hand side is certainly negative for $\tau \leq 1$ which secures concavity in the pixels $k = 3, \ldots, m - 1$. Concavity in pixel m for $\tau \leq 1$ follows from essentially the same argument; however, the equation is now replaced by an inequality since for pixel $m+1$ we know only that $u_{m+1}^{l+1} \leq u_{m+1}^l + \tau(u_m^l - u_{m+1}^l)$. If u_1^l is a local maximum and therefore $u_1^{l+1} = u_1^l$, we find for pixel 2 that $u_1^{l+1} - 2u_2^{l+1} + u_3^{l+1} = (1-\tau)(u_1^l - 2u_2^l + u_3^l) + \tau(u_2^l - u_1^l)$ which again secures concavity for $\tau \leq 1$. As was proven above, concavity in pixel 1 is preserved for $\tau \leq 1/2$ such that (iv) is proven.

Under the hypothesis of (v), the evolution of all pixels in the signal is described by statements (i)–(iv) or their obvious analoga for increasing and convex segments. The claim of (v) then is obvious.

Finally, addition of the equalities $C - u_1^{l+1} = (1 - 2\tau)(C - u_1^l)$ and $u_{i-1}^{l+1} - u_i^{l+1} = (1 - \tau)(u_{i-1}^l - u_i^l)$ for $i = 2, \ldots, m$ implies that

$$C - u_k^{l+1} = (1 - \tau)(C - u_k^l) - \tau(C - u_1^l) < (1 - \tau)(C - u_k^l)$$

for all $k = 1, \ldots, m$. By induction, we have

$$C - u_k^{l+l'} \leq (1 - \tau)^{l'}(C - u_k^l)$$

where the right-hand side tends to zero for $l' \to \infty$. Together with the monotonicity preservation for $\tau < 1/2$, statement (vi) follows. $\square$

4. Conclusions

Theoretical foundations for PDE-based shock filtering has long been considered to be a hopelessly difficult problem. In this paper we have shown that it is possible to obtain both an analytical solution and well-posedness by considering the space-discrete case where the partial differential equation becomes a dynamical system of ordinary differential equations (ODEs). Moreover, corresponding results can also be established in the fully discrete case when an explicit time discretisation is applied to this ODE system.

We are convinced that this basic idea to establish well-posedness results for difficult PDEs in image analysis by considering the semidiscrete case is also useful in a number of other important PDEs. While this has already been demonstrated for nonlinear diffusion filtering [13, 10], we plan to investigate a number of other PDEs in this manner, both in the one- and the higher-dimensional case. This should give important theoretical insights into the dynamics of these experimentally well-performing nonlinear processes.

References

[1] L. Alvarez and L. Mazorra. Signal and image restoration using shock filters and anisotropic diffusion. *SIAM Journal on Numerical Analysis*, 31:590–605, 1994.

[2] R. W. Brockett and P. Maragos. Evolution equations for continuous-scale morphology. In *Proc. IEEE International Conference on Acoustics, Speech and Signal Processing*, volume 3, pages 125–128, San Francisco, CA, March 1992.

[3] G. Gilboa, N. A. Sochen, and Y. Y. Zeevi. Regularized shock filters and complex diffusion. In A. Heyden, G. Sparr, M. Nielsen, and P. Johansen, editors, *Computer Vision – ECCV 2002*, volume 2350 of *Lecture Notes in Computer Science*, pages 399–413. Springer, Berlin, 2002.

[4] F. Guichard and J.-M. Morel. A note on two classical shock filters and their asymptotics. In M. Kerckhove, editor, *Scale-Space and Morphology in Computer Vision*, volume 2106 of *Lecture Notes in Computer Science*, pages 75–84. Springer, Berlin, 2001.

[5] P. Kornprobst, R. Deriche, and G. Aubert. Nonlinear operators in image restoration. In *Proc. 1997 IEEE Computer Society Conference on Computer Vision and Pattern Recognition*, pages 325–330, San Juan, Puerto Rico, June 1997. IEEE Computer Society Press.

[6] H. P. Kramer and J. B. Bruckner. Iterations of a non-linear transformation for enhancement of digital images. *Pattern Recognition*, 7:53–58, 1975.

[7] S. Osher and L. Rudin. Shocks and other nonlinear filtering applied to image processing. In A. G. Tescher, editor, *Applications of Digital Image Processing XIV*, volume 1567 of *Proceedings of SPIE*, pages 414–431. SPIE Press, Bellingham, 1991.

[8] S. Osher and L. I. Rudin. Feature-oriented image enhancement using shock filters. *SIAM Journal on Numerical Analysis*, 27:919–940, 1990.

[9] S. Osher and J. A. Sethian. Fronts propagating with curvature-dependent speed: Algorithms based on Hamilton–Jacobi formulations. *Journal of Computational Physics*, 79:12–49, 1988.

[10] I. Pollak, A. S. Willsky, and H. Krim. Image segmentation and edge enhancement with stabilized inverse diffusion equations. *IEEE Transactions on Image Processing*, 9(2):256–266, February 2000.

[11] L. Remaki and M. Cheriet. Numerical schemes of shock filter models for image enhancement and restoration. *Journal of Mathematical Imaging and Vision*, 18(2):153–160, March 2003.

[12] J. G. M. Schavemaker, M. J. T. Reinders, J. J. Gerbrands, and E. Backer. Image sharpening by morphological filtering. *Pattern Recognition*, 33:997–1012, 2000.

[13] J. Weickert. *Anisotropic Diffusion in Image Processing*. Teubner, Stuttgart, 1998.

[14] J. Weickert. Coherence-enhancing shock filters. In B. Michaelis and G. Krell, editors, *Pattern Recognition*, volume 2781 of *Lecture Notes in Computer Science*, pages 1–8, Berlin, 2003. Springer.

[15] M. Welk, J. Weickert, I. Galić. Theoretical Foundations for 1-D Shock Filtering. Preprint, Saarland University, Saarbruecken, 2005.

A VARIATIONAL FORMULATION OF PDE'S FOR DILATIONS AND LEVELINGS

Petros Maragos[1]

[1] *National Technical University of Athens,*
School of Electrical & Computer Engineering, Athens 15773, Greece.
maragos@cs.ntua.gr

Abstract   Partial differential equations (PDEs) have become very useful modeling and computational tools for many problems in image processing and computer vision related to multiscale analysis and optimization using variational calculus. In previous works, the basic continuous-scale morphological operators have been modeled by nonlinear geometric evolution PDEs. However, these lacked a variational interpretation. In this paper we contribute such a variational formulation and show that the PDEs generating multiscale dilations and erosions can be derived as gradient flows of variational problems with nonlinear constraints. We also extend the variational approach to more advanced object-oriented morphological filters by showing that levelings and the PDE that generates them result from minimizing a mean absolute error functional with local sup-inf constraints.

Keywords: scale-spaces, PDEs, variational methods, morphology.

1. Introduction

Partial differential equations have a become a powerful set of tools in image processing and computer vision for modeling numerous problems that are related to multiscale analysis. They need continuous mahematics such as differential geometry and variational calculus and can benefit from concepts inspired by mathematical physics. The most investigated partial differential equation (PDE) in imaging and vision is the linear isotropic heat diffusion PDE because it can model the Gaussian scale-space, i.e. its solution holds all multiscale linear convolutions of an initial image with Gaussians whose scale parameter is proportional to their variance. In addition, to its scale-space interpretation, the linear heat PDE can also be derived from a variational problem. Specifically, if we attempt to evolve an initial image into a smoother version by minimizing the L_2 norm of the gradient magnitude, then the PDE that results as the gradient descent flow to reach the minimizer is identical to the linear heat PDE. All

C. Ronse et al. (eds.), Mathematical Morphology: 40 Years On, 321–332.
©2005 *Springer. Printed in the Netherlands.*

the above ideas are well-known and can be found in numerous books dealing with classic aspects of PDEs and variational calculus both from the viewpoint of mathematical physics, e.g. [5], as well as from the viewpoint of image analysis, e.g. [12, 6, 14].

In the early 1990s, inspired by the modeling of the Gaussian scale-space via the linear heat diffusion PDE, three teams of researchers (Alvarez, Guichard, Lions & Morel [1], Brockett & Maragos [3, 4], and Boomgaard & Smeulders [19]) independently published nonlinear PDEs that model various morphological scale-spaces. Refinements of the above works for PDEs modeling multiscale morphology followed in [8, 7, 6]. However, in none of the previous works the PDEs modeling morphological scale-spaces were also given a direct variational interpretation. There have been only two indirect exceptions: i) Heijmans & Maragos [7] unified the morphological PDEs using Legendre-Fenchel 'slope' transforms, which are related to Hamilton-Jacobi theory and this in turn is related to variational calculus. ii) Inspired by the level sets methodology [13], it has been shown in [2, 15] that binary image dilations or erosions can be modeled as curve evolution with constant (± 1) normal speed. The PDE of this curve evolution results as the gradient flow for evolving the curve by maximizing or minimizing the rate of change of the enclosed area; e.g. see [17] where volumetric extensions of this idea are also derived. Our work herein is closer to [17].

In this paper we contribute a new formulation and interpretation of the PDEs modeling multiscale dilations and erosions by showing that they result as gradient flows of optimization problems where the volume under the graph of the image is maximized or minimized subject to some nonlinear constraints. Further, we extend this new variational interpretation to more complex morphological filters that are based on global constraints, such as the levelings [10, 11, 9].

2.   Background

Variational Calculus and Scale-Spaces

A standard variational problem is to find a function $u = u(x, y)$ that minimizes the 'energy' functional

$$J[u] = \int \int F(x, y, u, u_x, u_y) dx dy \tag{1}$$

usually subject to natural boundary conditions, where F is a second-order continuously differentiable function. A necessary condition satisfied by an extremal function u is the Euler-Langange PDE $[F]_u = 0$, where $[F]_u$ is the Euler (variational) derivative of F w.r.t. u. In general, to reach the extremal function that minimizes J, we can set up a gradient steepest descent procedure starting from an initial function $u_0(x, y)$ and evolving it into a function $u(x, y, t)$, where t is an artificial marching parameter, that satisfies the evolution PDE

$$\frac{\partial u}{\partial t} = -[F]_u, \quad [F]_u = F_u - \frac{\partial F_{u_x}}{\partial x} - \frac{\partial F_{u_y}}{\partial y} \tag{2}$$

This PDE is called the *gradient flow* corresponding to the original variational problem. In some cases, as $t \to \infty$ the gradient flow will reach the minimizer of J. If we wish to *maximize* J, the corresponding gradient flow is $u_t = [F]_u$. (Short notation for PDEs: $u_t = \partial u / \partial t$, $u_x = \partial u / \partial x$, $u_y = \partial u / \partial y$, $\nabla u = (u_x, u_y)$, $\nabla^2 u = u_{xx} + u_{yy}$.)

In the gradient flow formulation the evolving function $u = u(x, y, t)$ is a family of functions depending on the time parameter t and hence $J[u] = J(t)$. Then [5]

$$\frac{d}{dt} J[u] = \int \int u_t [F]_u dx dy \tag{3}$$

Thus, we can also view the Euler derivative $[F]_u$ as the *gradient of the functional $J[u]$ in function space*. This implies that, in getting from an arbitrary u_0 to the extremal, the PDE (2) of the gradient flow provides us with the fastest possible rate of decreasing J.

In scale-space analysis, we also start from an initial image $u_0(x, y)$ and evolve it into a function $u(x, y, t)$ with $u(x, y, 0) = u_0(x, y)$. The mapping $u_0 \mapsto u$ is generated by some multiscale filtering at scale $t \geq 0$ or by some PDE. The PDEs of several known scale-spaces (e.g. the Gaussian) have a variational interpretation since they can be derived as gradient flows of functional minimization problems where the marching time t coincides with the scale parameter. For example, if $F = (1/2)||\nabla u||^2$, the gradient flow corresponding to minimizing $J = \int \int F$ is the isotropic heat diffusion PDE $u_t = \nabla^2 u$.

PDEs for Dilation/Erosion Scale-Spaces

Let $k : \mathbb{R}^m \to \overline{\mathbb{R}}$, $m = 1, 2, ...,$ be a unit-scale upper-semicontinuous concave structuring function. Let $k_t(x) = tk(x/t)$ be its multiscale version, where both its values and its support have been scaled by a parameter $t \geq 0$. The *multiscale* Minkowski dilation $\oplus$ and erosion $\ominus$ of $f : \mathbb{R}^m \to \mathbb{R}$ by k_t are defined as the scale-space functions $\delta(x, t) = (f \oplus k_t)(x)$ and $\varepsilon(x, t) = (f \ominus k_t)(x)$:

$$\delta(x, t) = \bigvee_{y \in \mathbb{R}^m} f(y) + k_t(x - y), \quad \varepsilon(x, t) = \bigwedge_{y \in \mathbb{R}^m} f(y) - k_t(y - x),$$

where $\bigvee$ and $\bigwedge$ denote supremum and infimum, $\delta(x, 0) = \varepsilon(x, 0) = f(x)$. If $k(x, y)$ is *flat*, i.e. equal to 0 at points $(x, y) \in B$ and $-\infty$ else, where B is a unit disk, the PDEs generating the multiscale *flat* dilations and erosions of 2D images $f(x,y)$ by a disk B are [1, 4, 19]

$$\delta_t = ||\nabla\delta|| = \sqrt{(\delta_x)^2 + (\delta_y)^2}, \quad \varepsilon_t = -||\nabla\varepsilon|| \tag{4}$$

For 1D signals $f(x)$, B becomes the interval $[-1, 1]$ and the above PDEs become [4]

$$\delta_t = |\delta_x|, \quad \varepsilon_t = -|\varepsilon_x| \tag{5}$$

If k is the compact-support spherical function, i.e. $k(x,y) = (1 - x^2 - y^2)^{1/2}$ for $x^2 + y^2 \leq 1$ and $-\infty$ else, the PDE generating these spherical dilations is [4]

$$\delta_t = \sqrt{1 + (\delta_x)^2 + (\delta_y)^2}. \tag{6}$$

3. Variational Approach for Dilation PDEs

Let $u_0(x,y)$ be some smooth initial image over a rectangular image domain R with zero values outside R. Without loss of generality, we can assume that $u_0(x,y) \geq 0$ over R; otherwise, we consider as initial image the function $u_0 - \bigwedge u_0$. Let $u(x,y,t)$ be some scale-space analysis with $u(x,y,0) = u_0(x,y)$ that results from growing u_0 via dilation of the hypograph (umbra) of u_0 by some 3D structuring element $tB = \{tb : b \in B\}$ of radius $t \geq 0$, where $B \subseteq \mathbb{R}^3$ is a unit-radius compact symmetric convex set. From mathematical morphology we know that this 3D propagation of the graph of u_0 corresponds to a function dilation,

$$u(x,y,t) = u_0(x,y) \oplus k_t(x,y), \quad k_t(x,y) = \sup\{v : (x,y,v) \in tB\}, \tag{7}$$

of u_0 by a structuring function k_t that is the upper envelope of tB. We shall study three special cases of B: 1) a vertical line segment B_v, 2) a horizontal disk B_h, and 3) a sphere B_n. From (7), the three corresponding dilation scale-spaces are:

$B =$ v.line: $u(x,y,t) = u_0(x,y) + t$
$B =$ disk: $u(x,y,t) = \bigvee_{||(a,b)||\leq t} u_0(x - a, y - b)$
$B =$ sphere: $u(x,y,t) = \bigvee_{||(a,b)||\leq t} u_0(x - a, y - b) + t\sqrt{1 - (\frac{a}{t})^2 - (\frac{b}{t})^2}$
$$\tag{8}$$

While we know the gererating PDEs for the above scale-spaces (see [4]), in this paper our goal is to provide a variational interpretation for these PDEs and their solutions in (7). Define the multiscale *volume* functional

$$V(t) = \int\int u(x,y,t)dxdy = \int\int_{R(t)} u(x,y,t)dxdy \tag{9}$$

where $R(t)$ is the Minkowski dilation of the initial rectangular domain R with the projection of tB onto the plane. We wish to find the PDE generating u by creating a gradient flow that maximizes the rate of growth of $V(t)$. The classic approach [5] is to consider the time derivative $\dot{V}(t) = dV/dt$ as in (3). However, this is valid only when u is allowed to vary by remaining a function, e.g. $u \to u + \Delta tg$ where g is a perturbation function. Thus, u is allowed to vary in function space along a ray in the 'direction' of g. However, in our problem we have such a case only when $B = B_v$. In the other two cases u evolves as a graph by dilating its surface with a 3D set ΔtB. To proceed, we convert the problem to a more usual variational formulation (i) by modeling the propagation of the graph of u as the evolution of a multiscale parameterized closed surface $\vec{S}(q_1, q_2, t)$, and (ii) by expressing the volume V as a surface integral around this closed surface. A similar approach as in step (i) has also been used in [18] for geometric flows of images embedded as surfaces in higher-dimensional spaces.

We start our discussion from a simpler (but conceptually the same as above) case where $u_0 = u_0(x)$ is a *1D nonnegative image* with nonzero values over an interval R. Let $u(x, t) = u_0(x) \oplus k_t(x)$ be the multiscale dilation of u_0 by a structuring function k_t that is the upper envelope of a 2D set tB where B is a 2D version of the previous 3D unit-radius symmetric convex set; i.e., B is either 1) a vertical line segment B_v, or 2) a horizontal line segment B_h, or 3) a disk B_n. First, we model the propagation of the graph of u as the evolution of a multiscale parameterized curve $\vec{C}(q, t) = (x(q, t), y(q, t))$, whose top part is the graph of u traced when $q \in R(t)^s$ and whose bottom part is the interval $R(t)$ traced when $q \in R(t)$ [where $R^s = \{-q : q \in R\}$]. This implies

$$
\begin{aligned}
y(q, t) = u(x, t), \quad x_q = -1, \ y_q = -u_x, \quad q \in R(t)^s \\
y(q, t) = 0, \qquad\quad x_q = 1, \qquad\qquad\quad q \in R(t)
\end{aligned}
\tag{10}
$$

where subscripts denote partial derivatives. Then, we consider the *area $A(t)$* under u and express it (using Green's theorem) as a line integral around this closed curve:

$$
A(t) = \int u(x, t)dx = \frac{1}{2}\int_{C(t)} (xy_q - yx_q)dq = \frac{1}{2}\int_0^{L_c(t)} < \vec{C}, \vec{N} > ds
\tag{11}
$$

where s is arclength, $< \cdot >$ denotes inner product, $\vec{N}$ is the outward unit normal vector of the curve, $L_c(t) = L(t) + \mathrm{Len}(R(t))$ is the length of the closed curve $C(t)$, and $L(t) = \int_{R(t)} \sqrt{1 + u_x^2}dx$ is the length of the graph of u. Next follows our first main result.

Theorem 1 *Maximization of the area functional $A(t)$ when the graph of $u(x, t)$ is dilated by tB with unit curve speed, where B is any of the following unit-radius 2D summetric convex sets, has a gradient flow governed by the* *following corresponding PDEs:*

$$B = \text{vert.line} \implies u_t = 1 \tag{12}$$

$$B = \text{horiz.line} \implies u_t = |u_x| \tag{13}$$

$$B = \text{disk} \implies u_t = \sqrt{1 + |u_x|^2} \tag{14}$$

with $u(x, 0) = u_0(x)$.

Proof: Since we evolve u toward increasing $A(t)$, the graph curve speed $\vec{C}_t(q, t)$ must point outward for all $q \in R(t)^s$, i.e. $< \vec{C}_t, \vec{N} > \geq 0$. By (10), we can write the area functional as

$$A(t) = \int_{R(t)} u \, dx = \frac{1}{2} \int_0^{L(t)} < \vec{C}, \vec{N} > ds + \text{Len}[R(t)] \tag{15}$$

Differentiating (15) w.r.t. t yields

$$\frac{d}{dt} A(t) = \int_0^{L(t)} < \vec{C}_t, \vec{N} > ds + \text{const} \tag{16}$$

where $\text{const} = d\text{Len}[R(t)]/dt$. When B is the disk, the velocity $\vec{C}_t$ is allowed any direction and hence selecting $\vec{C}_t = \vec{N}$ guarantees that $\dot{A}(t)$ assumes a maximum value (i.e. the flow has a direction in function space in which $A(t)$ is increasing most rapidly). When B is the vertical line, $\vec{C}_t$ must have only a constant vertical component. When B is the horizontal line, $\vec{C}_t$ must have only a horizontal component with value ± 1 according to the sign of u_x. Thus, the three choices for structuring element B induce the following curve velocities:

$$\begin{aligned}
B = \text{vert.line} &\implies \vec{C}_t = (x_t, y_t) = (0, 1) \\
B = \text{horiz.line} &\implies \vec{C}_t = (x_t, y_t) = (\text{sgn}(-u_x), 0) \\
B = \text{disk} &\implies \vec{C}_t = (x_t, y_t) = \vec{N} = \frac{(-u_x, 1)}{\sqrt{1 + u_x^2}}
\end{aligned} \tag{17}$$

In all three cases we shall use the relation

$$u_t = y_t - u_x x_t \tag{18}$$

which follows from $y(q, t) = u(x, t)$. When B is the vertical line, we have $x_t = 0$ and $y_t = 1$. Hence, $u_t = 1$ which proves (12). When B is the horizontal line, $y_t = 0$ and $x_t = \text{sgn}(-u_x)$ which yields $u_t = |u_x|$ and proves (13). When B is the disk, we have $x_t = -u_x/v$ and $y_t = 1/v$ where $v = \sqrt{1 + u_x^2}$. This and (18) yield $u_t = 1/v + u_x^2/v = v$, which proves (14). $\square$

The volumetric extension of the above ideas to the case of a 2D nonnegative image $u_0(x, y)$ whose graph surface is dilated by 3D sets tB to give the graph of a scale-space function $u(x, y, t)$ is conceptually straightforward. First, we model the boundary of the *ordinate set* of u (i.e. the part of the umbra of u lying above the planar domain R) as a multiscale parameterized closed surface $\vec{S}(q_1, q_2, t)$, where (q_1, q_2) parameterize the surface as the two local coordinates and are related to (x, y). The top part of this closed surface is the graph of u and the bottom part is the planar domain $R(t)$ of u. Second, we express the volume $V(t)$ and its derivative as a surface integral around this closed parameterized surface:

$$V(t) = \frac{1}{3} \int < \vec{S}, \vec{N} > d\vec{S}, \quad \frac{d}{dt}V(t) = \int < \vec{S}_t, \vec{N} > d\vec{S} \quad (19)$$

For arbitrary 3D shapes enclosed by a surface, the above formulas were used in [17] to derive volume minimizing flows for shape segmentation.

Theorem 2 *Maximization of the volume functional $V(t)$ when the graph surface of $u(x, y, t)$ is dilated by tB with unit surface speed, where B is any of the following unit-radius 3D symmetric convex sets, has a gradient flow governed by the following corresponding PDEs:*

$$B = \text{vert.line} \implies u_t = 1 \quad (20)$$

$$B = \text{horiz.disk} \implies u_t = ||\nabla u|| = \sqrt{u_x^2 + u_y^2} \quad (21)$$

$$B = \text{sphere} \implies u_t = \sqrt{1 + ||\nabla u||^2} \quad (22)$$

with $u(x, y, 0) = u_0(x, y)$.

Proof: Due to lack of space we sketch the main ideas. Write (19) as

$$V(t) = \frac{1}{3} \int_{S_{top}} \vec{S} \cdot \vec{N} d\vec{S} + \text{Area}(R(t)), \quad \dot{V}(t) = \int_{S_{top}} \vec{S}_t \cdot \vec{N} d\vec{S} + \text{const} \quad (23)$$

where S_{top} is the top part of the surface. Over this part we select the optimum surface velocity vector $\vec{S}_t = (x_t, y_t, z_t)$ that maximizes the volume rate of change. Then we exploit the relationships among x, y and the local surface coordinates q_1, q_2 as well as the relation $z(q_1, q_2, t) = u(x, y, t)$ to express u_t as a function of u_x, u_y, which yields the PDE for u. $\qquad\Box$

So far, we have found a variational interpretation of some well-known multiscale morphological dilations and their corresponding PDEs as area or volume maximization problems. It is straightforward to derive the corresponding multiscale *erosions* and their PDEs by considering the dual problem of area or volume *minimization*. We omit the proofs.

Among the three cases for B, only when B is a vertical line we can also derive the corresponding PDE by using standard variational calculus, as follows.

Proposition 1 *Maximizing the functional* $J[u] = \int \int_R u(x, y, t) dx dy$ *has a gradient flow governed by the PDE* $u_t = 1$.

Proof: By writing $J = \int F$ with $F(u) = u$, the gradient flow will have the general form of (2), i.e. $u_t = [F]_u$. This yields $u_t = 1$, which is the PDE (20). □

In Theorems 1 and 2 we derived the morphological PDEs by maximizing area or volume functionals, either unconstrained if we move in the space of functions (as was the case when B is the vertical line and as explained in Prop.1) or with some geometrical constraints if we move in the space of graphs. Next we interpret our variational results for the multiscale *flat* dilations and erosions as a maximization and minimization, respectively, of the area or volume of the image u but under the constraint that all evolutions u have the same global sup or inf as u_0. This constrained optimization will prove useful for the levelings too.

Theorem 3 *(a) Maximizing the volume functional by keeping invariant the global supremum*

$$\max \int \int_R u \, dx dy \quad \text{s.t.} \quad \bigvee u = \bigvee u_0 \tag{24}$$

has a gradient flow governed by the PDE generating flat dilation by disks:

$$u_t = ||\nabla u||, \quad u(x, y, 0) = u_0(x, y) \tag{25}$$

Similarly, the dual problem of minimizing the volume functional by keeping invariant the global infimum

$$\min \int \int_R u \, dx dy \quad \text{s.t.} \quad \bigwedge u = \bigwedge u_0 \tag{26}$$

has a gradient flow governed by the isotropic flat erosion PDE:

$$u_t = -||\nabla u||, \quad u(x, y, 0) = u_0(x, y) \tag{27}$$

(b) For 1D signals $u(x)$, maximizing (or minimizing) the area functional by keeping invariant the global supremum (or infimum) has a gradient flow governed by the PDE generating flat dilations (or erosions) by intervals $[-t, t]$:

$$\begin{aligned} \max \int_R u \, dx \quad \text{s.t.} \quad \bigvee u = \bigvee u_0 &\implies u_t = |u_x| \\ \min \int_R u \, dx \quad \text{s.t.} \quad \bigwedge u = \bigwedge u_0 &\implies u_t = -|u_x| \end{aligned} \tag{28}$$

with initial condition $u(x, 0) = u_0(x)$.

Proof: Under the sup constraint, the velocity vector for the propagation of the graph of u must have a zero vertical component. Hence, the only directions allowed to propagate the graph of u must be parallel to the image plane. This expansion is done at maximum speed if it corresponds to dilations of the graph (and equivalently of the level sets) of u by horizontal disks in the 2D case and by horizontal line segments in the 1D case. Thus, we have the case of multiscale dilations of the graph of u by horizontal disks or lines for which we use the results of Theorems 1 and 2. Similarly for the erosions. $\square$

4.   Variational Approach for Levelings

Here we consider morphological smoothing filters of the reconstruction type. Imagine creating a type of image simplification like a 'cartoon' by starting from a *reference* image $r(x, y)$ consisting of several parts and a *marker* image $u_0(x, y)$ (initial seed) intersecting some of these parts and by evolving u_0 toward r in a monotone way such that all evolutions $u(x, y, t)$, $t \geq 0$, satisfy the following partial ordering, $\forall x, y \in R$

$$t_1 < t_2 \implies r(x, y) \preceq_r u(x, y, t_2) \preceq_r u(x, y, t_1) \preceq_r u_0(x, y) \qquad (29)$$

The partial order $u \preceq_r f$ means that $r \wedge f \leq r \wedge u$ and $r \vee f \geq r \vee u$. Further, if we partition the following regions R^- and R^+ formed by the zero-crossings of $r - u_0$

$$
\begin{aligned}
R^- &= \{(x, y) : r(x, y) \geq u_0(x, y)\} = \bigsqcup_i R_i^- \\
R^+ &= \{(x, y) : r(x, y) < u_0(x, y)\} = \bigsqcup_i R_i^+
\end{aligned}
\qquad (30)
$$

into connected subregions, then the evolution of u is done by maintaining all local maxima and local minima of u_0 inside these subregions R_i^- and R_i^+, respectively:

$$\bigvee_{R_i^-} u = \bigvee_{R_i^-} u_0 \text{ and } \bigwedge_{R_i^+} u = \bigwedge_{R_i^+} u_0, \quad R = (\bigsqcup_i R_i^-) \sqcup (\bigsqcup_i R_i^+) \qquad (31)$$

where $\bigsqcup$ denotes disjoint union. Since the order constraint $r \preceq_r u \preceq_r u_0$ implies that $|r - u| \leq |r - u_0|$, the above problem is equivalent to the following constrained minimization

$$\min \iint_R |u - r| dx dy \text{ s.t. } \bigvee_{R_i^-} u = \bigvee_{R_i^-} u_0, \quad \bigwedge_{R_i^+} u = \bigwedge_{R_i^+} u_0 \qquad (32)$$

Theorem 4 *A gradient flow for the optimization problem (32) is given by the following PDE*

$$
\begin{aligned}
\partial u(x, y, t) / \partial t &= -\text{sgn}(u - r) \|\nabla u\| \\
u(x, y, 0) &= u_0(x, y)
\end{aligned}
\qquad (33)
$$

Proof: By writing the integral $\iint |u - r|$ as

$$\iint_R |u - r| = \sum_{R_i^-} \iint_{R_i^-} (r - u) + \sum_{R_i^+} \iint_{R_i^+} (u - r) \qquad (34)$$

we can decompose the global problem (32) into local constraint maximization and minimization problems over the regions R_i^- and R_i^+ respectively. Applying Theorem 3 to these local problems yields local evolutions that act as flat dilations when $u < r$ and as erosions when $u > r$. The PDE (33) has a switch that joins these two actions into a single expression. □

The PDE (33) was introduced in [11] and then studied systematically in [9]. For each t, at pixels (x, y) where $u(x, y, t) < r(x, y)$ it acts as a dilation PDE and hence shifts outwards the surface of $u(x, y, t)$ but does not introduce new local maxima. Wherever $u(x, y, t) > r(x, y)$ the PDE acts as a flat erosion PDE and reverses the direction of propagation. In [9] it was proved that this PDE has a steady-state $u_\infty(x) = \lim_{t\to\infty} u(x, t)$ which is a *leveling* of r with respect to u_0, denoted by $u_\infty = \Lambda(u_0|r)$.

Levelings are nonlinear filters with many interesting scale-space properties [11] and have been used for image pre-segmentation [11, 16]. They were defined geometrically in [10, 11] via the property that if p, q are any two close neighbor pixels then the variation of the leveling between these pixels is bracketed by a larger same-sign variation in the reference image r; i.e., if g is a leveling of r, then

$$g(p) > g(q) \implies r(p) \geq g(p) > g(q) \geq r(q) \qquad (35)$$

In [9] they were defined algebraically as fixed points of triphase operators $\lambda(f|r)$ that switch among three phases, an expansion, a contraction, and the reference r. Further, the leveling of r w.r.t. $f = u_0$ can be obtained as the limit of iterations of λ:

$$u_\infty = \Lambda(u_0|r) \triangleq \lim_{n\to\infty} \lambda^n(u_0|r) \preceq_r \cdots \lambda(u_0|r) \preceq_r u_0 \qquad (36)$$

The simplest choise for λ is $\lambda(f|r) = [r \wedge \delta(f)] \vee \varepsilon(f)$, where δ and ε are dilations and erosions by a small disk, but there are many more sophisticated choises [11, 9]. A numerical scheme proposed in [9] to solve the PDE (33) also involves iterating a discrete algorithm that is essentially a discrete triphase operator whose iteration limit yields a discrete leveling.

Levelings have many interesting scale-space properties [11]. Due to (29) and (35), they preserve the coupling and sense of variation in neighbor image values, which is good for edge preservation. Further, due to (31) the levelings do not create any new regional maxima or minima. In practice, they can reconstruct whole image objects with exact preservation of their boundaries and edges. The reference image plays the role of a global constraint.

5. Conclusions

We have developed a new formulation based on functional extremization to derive the PDEs generating well-known multiscale morphological operators, both of the basic type acting locally on the image like dilations and erosions by compact kernels, as well as of the reconstruction type like the levelings which depend on global constraints. The functionals used were the image volume/area for dilations and the L_1 norm of residuals between the simplified image and the reference for the levelings. Maximization or minimization of these functionals was done subject to some nonlinear constraints. This variational approach to multiscale morphology gives a new insightful interpretation to morphological operators and offers useful links with optimization problems.

Acknowledgments

This research work was supported by the Greek GSRT program ΠΕΝΕΔ – 2001, by the Greek Education Ministry's research program 'Pythagoras', and by the European Union under the Network of Excellence MUSCLE.

References

[1] L. Alvarez, F. Guichard, P.L. Lions, and J.M. Morel, "Axioms and Fundamental Equations of Image Processing", *Archiv. Rat. Mech.*, 123 (3):199-257, 1993.

[2] A. Arehart, L. Vincent and B. Kimia, "Mathematical Morphology: The Hamilton-Jacobi Connection," in *Proc. ICCV'93*, pp.215-219, 1993.

[3] R.W. Brockett and P. Maragos, "Evolution Equations for Continuous-Scale Morphology," in *Proc. ICASSP-92*, San Francisco, Mar. 1992.

[4] R.W. Brockett and P. Maragos, "Evolution Equations for Continuous-Scale Morphological Filtering", *IEEE Trans. Signal Process.*, 42: 3377-3386, Dec. 1994.

[5] R. Courant and D. Hilbert, *Methods of Mathematical Physics*, Wiley-Interscience Publ. 1953 (Wiley Classics Edition 1989).

[6] F. Guichard and J.-M. Morel, *Image Analysis and P.D.E.s*, book to appear.

[7] H.J.A.M. Heijmans and P. Maragos, "Lattice Calculus and the Morphological Slope Transform," *Signal Processing*, vol.59, pp.17-42, 1997.

[8] P. Maragos, "Differential Morphology and Image Processing", *IEEE Trans. Image Processing*, vol.5, pp.922-937, June 1996.

[9] P. Maragos, "Algebraic and PDE Approaches for Lattice Scale-Spaces with Global Constraints", *Int. J. Comp. Vision*, 52 (2/3), pp.121-137, May 2003.

[10] F. Meyer, "The Levelings", in *Mathematical Morphology and Its Applications to Image and Signal Processing*, H. Heijmans and J. Roerdink, editors, KluwerAcad. Publ., 1998, p.199-207.

[11] F. Meyer and P. Maragos, "Nonlinear Scale-Space Representation with Morphological Levelings", *J. Visual Commun. & Image Representation*, 11, p.245-265, 2000.

[12] J.-M. Morel and S. Solimini, *Variational Methods in Image Processing*, Birkhauser, 1994.

[13] S. Osher and J. Sethian, "Fronts Propagating with Curvature-Dependent Speed: Algorithms Based on Hamilton-Jacobi Formulations", *J. Comput. Physics*, 79, pp. 12–49, 1988.

[14] G. Sapiro, *Geometric Partial Differential Equations and Image Analysis*, Cambridge University Press, 2001.

[15] G. Sapiro, R. Kimmel, D. Shaked, B. Kimia, and A. Bruckstein, "Implementing Continuous-scale Morphology via Curve Evolution", *Pattern Recognition*, 26(9), pp. 1363–1372, 1993.

[16] J. Serra, "Connections for Sets and Functions", *Fundamentae Informatica 41*, pp.147-186, 2000.

[17] K. Siddiqi, Y. B. Lauziere, A. Tannenbaum and W. Zucker, "Area and Length Minimizing Flows for Shape Segmentation", *IEEE Trans. Image Process.*, 7 (3):433-443, Mar. 1998.

[18] N. Sochen, R. Kimmel and R. Malladi, "A General Framework for Low Level Vision", *IEEE Trans. Image Process.*, 7 (3): 310-318, Mar. 1998.

[19] R. Van den Boomgaard and A. Smeulders, "The Morphological Structure of Images: The Differential Equations of Morphological Scale-Space", *IEEE Trans. Pattern Anal. Mach. Intellig.*, vol.16, pp.1101-1113, Nov. 1994.

STOCHASTIC SHAPE OPTIMISATION

Costin Alin Caciu, Etienne Decencière and Dominique Jeulin
Centre de Morphologie Mathématique, Ecole des Mines de Paris
35, rue Saint Honoré, 77300 Fontainebleau, France
Costin.Caciu@cmm.ensmp.fr

Abstract   We present a constrained shape optimisation problem solved via metaheuristic stochastic techniques. Genetic Algorithms are briefly reviewed and their adaptation to surface topography optimisation is studied. An application to flow optimisation issues is presented.

Keywords: Shape, Topography, Stochastic Optimisation, Genetic Algorithms

Introduction

Geometrical shape or surface topography optimisations are quite numerous among the optimisation problems, since they are encountered and play often an important role in most technical domains. Traditionally, deterministic models or expensive experimental tests are employed to solve these issues.

Computing power has grown considerably in the past years, dramatically promoting the interest in numerical techniques. Numerical simulations and models become issues for problems in many domains. They opened the path, serving as objective functions, to a large number of stochastic optimisation techniques working on discretised search spaces.

We will focus here on *Genetic Algorithms* (GA) [1, 4]. The main advantage of this numerical optimisation approach, compared to deterministic ones, is its extended search space and its robustness. In return, it requires complex and onerous computations, and its convergence rate is particulary slow, which represents its main weakness.

We will first describe (Section 1) the precise problem that led us to this study. Section 2 gives a tutorial introduction to stochastic optimisation and in Section 3 we describe the GAs' adaptation to our specific problem. Section 4 exposes the results of GAs' application for a particular fitness function. A summary and the conclusion are presented in Section 5.

C. Ronse et al. (eds.), Mathematical Morphology: 40 Years On, 333–342.
©2005 *Springer. Printed in the Netherlands.*

1.   Problem

Let us assume we have to optimise a shape S representing an opened volume (Figure 1) in a smooth surface $\mathcal{P}$. The optimisation is controlled by a fitness function f and restricted by several constraints $\mathcal{C}$:

- S is a connected shape.

- S has a constant volume V and a constant opened surface S_s.

- the depth in any point of S may vary between d_{min} and d_{max}.

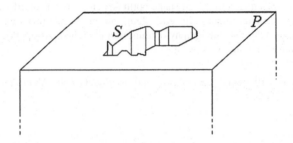

Figure 1.   3D shape to optimise

To build a model as general as possible, the fitness function f may be any measure on S respecting $\mathcal{C}$.

For the problem above, the implementation of a deterministic optimisation is not an option. The shape S may take complex forms, impossible to define analytically; likewise, the fitness function f may be also difficult to estimate analytically; moreover, in many situations, the complexity of the employed fitness function may lead to unpredictable deterministic relations between S and f, impossible to pursuit. In consequence, in order to escape these difficulties and to preserve a wide space in search for improvement, we have to work with discretised shapes and numerical fitness functions, and employ appropriate *stochastic optimisation methods*.

A coarse discretisation of a random surface shape S is presented in Figure 2(a). In this 2D image representation the depth of the opening is coded by the grey levels of pixels (8 bits), the white background being assigned to $\mathcal{P}$ and the grey pixels to the various depths of S. A three-dimensional representation of the previous image is illustrated in Figure 2(b).

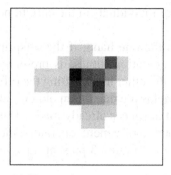

(a) 2D coarse image representation

(b) 3D coarse image representation

Figure 2.   Image representation

2.   Stochastic techniques and Genetic Algorithms

Stochastic optimisation refers to the minimisation (or maximisation) of a function in the presence of randomness in the optimisation process. The randomness may be present either in the search procedure, or as noise in measurements.

The fundamental idea behind stochastic programming is the concept of *recourse*. Recourse is the ability to take corrective action after a random event has taken place. Common methods of stochastic optimisation include *Genetic Algorithms* (GA) and *Simulated Annealing* (SA) [2], well-fitted to our problem; we will pursuit our study here with GAs, the adaptation of SA methods to this case would be rather similar.

A Genetic Algorithm is an implementation of an adaptive plan, employing specific structural modifiers (operators) in the search for optimally performing structures in a particular environment. They are computationally simple, yet very powerful in their search for improvement.

A population of individuals is successively modified from one generation to another. Individuals in the population are known as *chromosomes* and are coded by strings of *genes*. Each gene can have several different forms, known as *alleles*, producing differences in the set of characteristics associated with that gene.

Genetic Algorithms are based on evolution. In evolution, the problem which each species faces is one of searching for beneficial adaptations to a complicated and changing environment. The knowledge that each species has gained is embodied in the makeup of the chromosomes of it members. Future populations can only develop via reproduction of individuals in the current population.

There exists an apparent dilemma. On the one hand, if the offspring are simple duplicates of fit members of the population, fitness is preserved, but there is no provision for improvement. On the other hand, letting the offspring be produced by simple random variation, yields the maximum of new variants, but makes no provision for the retention of advances already made. The solution of the dilemma lies in the combined action of genetic operators, the most commonly used being *mutation* and *crossover*. Figure 3 presents an overview of the components of a simple GA.

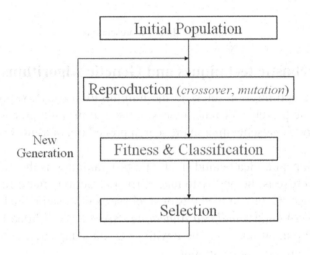

Figure 3.   Schematic GA

Crossover is the key to GAs' power and consists basically in the exchange of genetic material between parents' chromosomes, allowing beneficial material to be combined in their offspring. With crossover, *beneficial mutations* on two parents can be combined immediately when they reproduce and this is more probable when most successful parents reproduce more often.

Mutation is a process wherein one allele of a gene is randomly replaced by another one, to yield a new structure. Though mutation is one of the most familiar of genetic operators, its role in adaptation is frequently misinterpreted. Mutation's primary role is generally associated to generating new structures for trial. Nevertheless, this is a role very efficiently filled by crossover, with one condition: the crossover operator must dispose of the full range of alleles. Once an allele is lost from population, the crossover operator has no way of reintroducing it. To sum up, mutation is a *background* operator, assuring that the crossover operator has full range of alleles, so that the adaptive plan is not trapped on local optima.

For a genetic algorithm to perform its adaptive task, there has to be some means of assigning an observable performance to each individual. A *fitness function* is employed, which attributes some objective measure of how well an individual performs an environment. This fitness function will play the role of selection criterion for the population's individuals.

3. Constrained GA

As shown in Section 1, we deal with a constrained optimisation problem, i.e. the search space is restricted to some subspace of the defining space of the shape to optimise. The *constraints* are equalities and inequalities (C) the solution is required to satisfy. There have been many attempts to solve constrained GA optimisation problems [3, 5]. There are two different approaches, either based on introducing a *penalty function* for constraint violation or building specific *domain dependent* algorithms.

Penalty function GA

The most widely used method to treat constraints [3] is to incorporate them in the objective function, and to use standard methods for unconstrained optimisation.

In order to use standard GAs to our optimisation problem, the individuals in the genetic search space like the one in Figure 2(a) have to be mapped to a finite length string, over some alphabet. As a rule, when coding a problem for genetic search, one should select the smallest possible alphabet which permits a natural expression of the problem, this being known as the principle of minimal alphabets [1]. This leads automatically to code each grey-level pixel of the individual using 8 binary digits, like illustrated in Figure 4. Choosing an order to scan the image, coding is done by simply mapping the pixel binary code to its appropriate position in the portion of the chromosome. It is an approach quite similar to morphological filters coding, an example which is clearly treated in [6, 7].

Once the coding process is finished, the fitness function is modified and becomes some weighted sum of the original and some penalty for every constraint violation. The GA is ready for use. There is no general solution for the choice of the weight coefficients, this being rather problem-dependent.

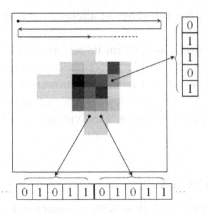

Figure 4.   Standard coding

Domain specific GA

The best results obtained by GAs on constrained problems use however domain-dependent methods, where the genetic operators, the fitness function and the search space itself are tailored to take the constraints into account. When possible, this approach, which uses as much domain-specific knowledge as possible, is probably the best way to tackle constraints [5].

The GA employed to solve our problem will keep the generic form presented in Section 2. Nevertheless, important modifications will be brought to the coding and the genetic operators. Considering the severe constraints C (Section 1) the optimum must satisfy, the most appropriate coding for the individuals (Figure 1) will be that of Figure 2(a) and not the one using the smallest possible alphabet; indeed, working with individuals represented by strings of binaries would induce laborious constraint verifications. Hence, for complexity and computation time reasons, the grey-level image representation is chosen to code the individuals, being the most suited to use for the control of the connectivity, volume and surface constraints; an image represents a chromosome and the genes correspond to the pixels of the image.

The constraints' integration into the algorithm is done at the *random individual generation* level. Each time new individuals are introduced in the population, the random generation process is controlled by the constraints. In consequence, every new individual satisfies the constraints. Since the population is made of (before applying the genetic reproducing operators) new randomly generated individuals and also of old performing individuals selected by the fitness function, the genetic operators automatically preserve the constraints' verification.

From a practical point of view, a *crossover under constraints* operator working on the defined individuals is difficult to realise. For complexity and computation time reasons again, this operator will be substituted. Instead, we will define three *mutation* operators that will serve, from a *functional* point of view, as *mutation* and *crossover* at the same time (Figure 5).

The first one (M_1) occurs with the probability mp_1 and consists in swapping genes corresponding to two different positions in S as shows Figure 5(a).

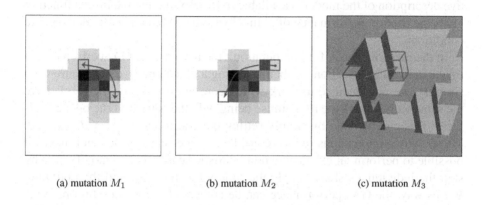

(a) mutation M_1 (b) mutation M_2 (c) mutation M_3

Figure 5. Mutation operators

M_2 is the second mutation operator which modifies the shape of the surface opened by S, moving a border pixel to a different near-border position. It occurs with the mp_2 probability and is illustrated in Figure 5(b).

The third mutation operator (M_3) is represented in Figure 5(c). It consists in subtracting a small volume at a random position in S and adding it at another position, or genetically speaking, substituting one allele of a gene with another one, chosen randomly, and also modifying a second different gene in order to keep the constraints satisfied. From a genetic functional point of view this is a pure mutation operator as it permits to introduce new or lost alleles in the population.

Despite the fact that there is no conventional crossover operator, the role of the *beneficial mutations* (Section 2) will be played by M_1 and M_2, which will hence fulfill the *crossover* functionality, while M_3 and the percentage of randomly generated new individuals in each population will ensure the availability of a full range of alleles, i.e. the *mutation* functionality.

4.   Application

In order to prove whether or not stochastic methods are fitted to solve the shape optimisation problem described in Section 1, the domain specific GA (Section 3) was tested.

The fitness function used for optimisation is a numerical model simulating a hydrodynamic contact between a smooth surface and S; the model outputs an estimation of the hydrodynamic contact friction. For more details an exhaustive description of the model is available in [8]. To sum up, the fitness function gives the friction performance of S in a hydrodynamic contact, under specific conditions.

In the performed test, for computation time reasons, the GA works with 2D image individuals (Section 3) obtained using a rather coarse discretisation. To simplify, we also decided to work with a constant depth, i.e. the pixels have the same value, the shape to optimise being only the pattern of the opened surface. This can be done by simply setting the constraint $d_{min} = d_{max} = d$. These two simplifications reduce considerably the search space and makes it possible to perform an exhaustive non-heuristic search, i.e. to modify step by step the individual's shape, apply the fitness function and find the optimum. In this way, the GAs performance can be compared to a successful optimisation. Figure 6 contains the tests' results. From left to right are present the best individuals of the population at different optimisation stages given by the iteration number. In Figure 6(a) is illustrated the non-heuristic optimisation. The first individual is a thin stripe and the last one contains the optimal shape. Figures 6(b) and 6(c) present two different genetic evolutions of the same algorithm, the first reaching near-optimal configuration after 1506 iterations and the second after 663 iterations. The starting population is randomly generated.

Although the computation time required for convergence is generally essential for GAs, it is a parameter highly related to the complexity of the fitness function and is hence problem-dependent; therefore, the computation time needed by our application would be rather irrelevant for further convergence investigations. The algorithm's parameters are:

- the individuals are coded on 2D images of size 32×32.

- $S_s = 32$ pixels, $d = 60$ grey levels and $V = S_s \cdot d$.

- $mp_1 = mp_2 = mp_3 = 1/3$ and the population size is 128.

- percentage of selected performing individuals: 50%.

It is interesting to note that the algorithm has correct evolutions reaching near-optimal configurations relatively quickly considering the complexity of the fitness function used; it should be noticed also that the second GA evolution is only one mutation far from optimum.

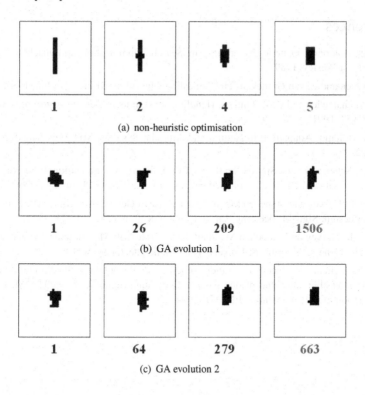

1   2   4   5

(a) non-heuristic optimisation

1   26   209   1506

(b) GA evolution 1

1   64   279   663

(c) GA evolution 2

Figure 6.   GA application

5. Conclusion

In this paper, a generic approach has been presented for stochastic shape optimisation, using Genetic Algorithms. An application of GAs to a real elementary topography optimisation, in fluid dynamics domain, was presented; the tests' results were encouraging, illustrating the algorithm's ability to find simple optimal shapes. As an extension to the present work, another area of investigation for future research may be the use of the algorithms developed to perform texture or more complex shape optimisation.

References

[1] D.E. Goldberg, Genetic Algorithms in Search, Optimization and machine Learning, Addison-Wesley, 1989.

[2] R. Otten and L. van Ginneken, The Annealing Algorithm, Kluwer, Boston, 1989.

[3] Z. Michalewicz and C. Z. Janikov, Handling constraints in Genetic Algorithms, ICGA, p. 151-157, 1991.

[4] J.H. Holland, Adaptation in Natural and Artificial Systems, MIT Press/Bradford Books, 1992.

[5] Marc Schoenauer and Spyros Xanthakis, Constrained GA optimization, Proceedings of the 5th International Conference on Genetic Algorithms, Urbana Champaign, 1993.

[6] Neal R. Harvey and Stepehen Mmarshall, The use of Genetic Algorithms in morphological filter design, Signal Processing: Image Communication, n. 8, p. 55-71, 1996.

[7] Neal R. Harvey and Stepehen Marshall, Grey-Scale Soft Morphological Filter Optimization by Genetic Algorithms, University of Strathclyde, Glasgow, 1997.

[8] C. A. Caciu and E. Decencière, Etude numérique de l'écoulement 3D d'un fluide visqueux incompressible entre une plaque lisse et une plaque rugueuse, N-03/04/MM, Centre de Morphologie Mathematique, ENSMP, 2004.

ON THE LOCAL CONNECTIVITY NUMBER OF STATIONARY RANDOM CLOSED SETS

Evgueni Spodarev[1] and Volker Schmidt[1]

[1]*Universität Ulm*
Abteilung Stochastik
D-89069 Ulm, Germany
spodarev@mathematik.uni-ulm.de, schmidt@mathematik.uni-ulm.de

Abstract   Random closed sets (RACS) in the d–dimensional Euclidean space are considered, whose realizations belong to the extended convex ring. A family of nonparametric estimators is investigated for the simultaneous estimation of the vector of all specific Minkowski functionals (or, equivalently, the specific intrinsic volumes) of stationary RACS. The construction of these estimators is based on a representation formula for the expected local connectivity number of stationary RACS intersected with spheres, whose radii are small in comparison with the size of the whole sampling window. Asymptotic properties of the estimators are given for unboundedly increasing sampling windows. Numerical results are provided as well.

Keywords: Mathematical morphology; random closed sets; stationarity; Minkowski functionals; intrinsic volumes; nonparametric estimation; local Euler–Poincaré characteristic; principal kinematic formula; Boolean model

Introduction

The theory of random closed sets (RACS) and its morphological aspects with emphasis on applications to image analysis have been developed in the second half of the 20th century. This scientific process has been significantly influenced by the pioneering monographs of G. Matheron [6] and J. Serra [15, 16]. It turned out that *Minkowski functionals* or, equivalently, *intrinsic volumes* are important characteristics in order to describe binary images, since they provide useful information about the morphological structure of the underlying RACS. In particular, the so–called *specific intrinsic volumes* of stationary RACS have been intensively studied for various models from stochastic geometry.

There exist several approaches to the construction of statistical estimators for particular specific intrinsic volumes of stationary RACS in two and three

343

C. Ronse et al. (eds.), Mathematical Morphology: 40 Years On, 343–354.
©2005 *Springer. Printed in the Netherlands.*

dimensions. However, in many cases, only little is known about goodness properties of these estimators, like unbiasedness, consistency, or distributional properties. Furthermore, an extra algorithm has to be designed for the estimation of each specific intrinsic volume separately.

In contrast to this situation, the method of moments proposed in the present paper provides a unified theoretical and algorithmic framework for simultaneous nonparametric estimation of all specific intrinsic volumes, in an arbitrary dimension $d \geqslant 2$. The construction principle of these estimators, which is similar to the approach considered in [11], is based on a representation formula for the (expected) local connectivity number of stationary RACS intersected with spheres, whose radii are small in comparison with the size of the whole sampling window. It can be considered as a statistical counterpart to a method for the simultaneous computation of all intrinsic volumes of a deterministic polyconvex set based on the principal kinematic formula.

Our estimators are unbiased by definition. Moreover, under suitable integrability and mixing conditions, they are mean–square consistent and asymptotically normal distributed. This can be used in order to establish asymptotic tests for the vector of specific intrinsic volumes.

Notice that the method of moments (which is also called the method of intensities by some authors) has been used in the analysis of various further statistical aspects of models from stochastic geometry, for example, in order to estimate the intensity of germs and other characteristics of the Boolean model; see e.g. [7], and Sections 5.3–5.4 in [13].

The present paper is organized as follows. Some necessary preliminaries on Minkowski functionals and intrinsic volumes, respectively, are given in Section 1. In Section 2, the computation of intrinsic volumes of deterministic polyconvex sets is briefly discussed. The above–mentioned representation formula for the (expected) local connectivity number of stationary RACS is stated in Section 3; see Proposition 3.1. We give an alternative proof of this representation formula which makes use of an explicit extension of Steiner's formula for convex bodies to the convex ring. The result of Proposition 3.1 is then used in Section 4 in order to construct a family of nonparametric estimators for all $d + 1$ specific intrinsic volumes simultaneously. The construction principle of these estimators is described and their asymptotic properties are discussed. A related family of least–squares estimators is also provided in Section 4. In Section 5, some aspects of variance reduction using *kriging of the mean* are touched upon. Finally, in Section 6 numerical results are given for the planar Boolean model with spherical primary grains. They are compared with those obtained by another method described in [10] for the computation of specific intrinsic volumes.

1. Minkowski functionals and intrinsic volumes

Let $d \geqslant 2$ be an arbitrary fixed integer and let $\mathcal{K}$ be the family of all *convex bodies*, i.e., compact convex sets in $\mathbb{R}^d$. The *convex ring* $\mathcal{R}$ in $\mathbb{R}^d$ is the family of all finite unions $\bigcup_{i=1}^{m} K_i$ of convex bodies $K_1, \ldots, K_m \in \mathcal{K}$. The elements of $\mathcal{R}$ are called *polyconvex sets*. Furthermore, the *extended convex ring* $\S$ is the family of all subsets $A \subset \mathbb{R}^d$ such that $A \cap K \in \mathcal{R}$ for any $K \in \mathcal{K}$. For $A, B \subset \mathbb{R}^d$, the *Minkowski sum* $A \oplus B$ and the *Minkowski difference* $A \ominus B$ are defined by $A \oplus B = \{x + y : x \in A, y \in B\}$ and $A \ominus B = \{x \in \mathbb{R}^d : B + x \subset A\}$, respectively. For any Borel set $B \subset \mathbb{R}^d$, denote by $V_d(B)$ its Lebesgue measure. It is well known that there exist nonnegative functionals $V_j : \mathcal{K} \to [0, \infty), j = 0, \ldots, d$ such that for each $r > 0$ the volume $V_d(K \oplus B_r(o))$ of the so-called *parallel body* $K \oplus B_r(o)$ of any $K \in \mathcal{K}$ is given by *Steiner's formula*

$$V_d(K \oplus B_r(o)) = \sum_{j=0}^{d} r^{d-j} \kappa_{d-j} V_j(K), \tag{1}$$

where $o \in \mathbb{R}^d$ denotes the origin, $B_r(x) = \{y \in \mathbb{R}^d : |y - x| \leqslant r\}$ is the closed ball with midpoint $x \in \mathbb{R}^d$ and radius r, and κ_j is the volume of the unit ball in $\mathbb{R}^j$. The functionals V_j are called *intrinsic volumes*. They are closely related to the widely known *quermassintegrals* or *Minkowski functionals* W_j given by $W_j(K) = V_{d-j}(K) \kappa_j / \binom{d}{j}$, $K \in \mathcal{K}$. There exists a unique additive extension of the functionals V_j to the convex ring $\mathcal{R}$ given by the *inclusion–exclusion formula*

$$V_j(K_1 \cup \ldots \cup K_m) = \sum_{k=1}^{m} (-1)^{k-1} \sum_{i_1 < \ldots < i_k} V_j(K_{i_1} \cap \ldots \cap K_{i_k}) \tag{2}$$

for any $K_1, \ldots, K_m \in \mathcal{K}$. The intrinsic volumes $V_j(K), j = 0, \ldots, d$ provide information about the morphological structure of the polyconvex set $K \in \mathcal{R}$. For example, $V_d(K)$ is the usual volume, $2V_{d-1}(K)$ is the surface area, and the *Euler–Poincaré characteristic* $V_0(K)$ reflects the connectivity properties of K. Notice that in the planar case, that is $d = 2$, $V_0(K)$ is equal to the number of "clumps" minus the number of "holes" of $K \in \mathcal{R}$, i.e., the number of connected outer boundary components of K minus the number of its inner boundary components. In particular, $V_0(K) = 1$ for any convex body $K \neq \emptyset$. Furthermore, for any $K \in \mathcal{R}$ and $q, x \in \mathbb{R}^d$, $q \neq x$, the so-called *index* $J(K, q, x)$ of K is given by

$$J(K, q, x) = 1 - \lim_{\delta \to +0} \lim_{\varepsilon \to +0} V_0\left(K \cap B_{|x-q|-\varepsilon}(x) \cap B_\delta(q)\right) \tag{3}$$

for $q \in K$. For all $q \notin K$, we put $J(K, q, x) = 0$. In particular, $J(\emptyset, q, x) = 0$ for arbitrary $q, x \in \mathbb{R}^d$.

2.   Computation of intrinsic volumes of a polyconvex set

Given a polyconvex set $K \subset \mathbb{R}^d$, apply the principal kinematic formula of integral geometry (cf. formula (4.5.3) in [12]) to the Euler–Poincaré characteristic of the intersection of K with an arbitrary translation of the ball $B_r(o)$. This yields

$$\int_{K \oplus B_r(o)} V_0(K \cap B_r(x)) \, dx = \sum_{j=0}^{d} r^{d-j} \kappa_{d-j} V_j(K), \qquad (4)$$

where the integration domain is $K \oplus B_r(o)$ since $V_0(K \cap B_r(x)) = 0$ for $x \notin K \oplus B_r(o)$. Introduce the notation $R_r = \int_{K \oplus B_r(o)} V_0(K \cap B_r(x)) \, dx$.

Writing equation (4) for $d+1$ pairwise different radii $r_0, \ldots, r_d$ yields the following system of $d+1$ linear equations:

$$A_{r_0 \ldots r_d} V = R, \qquad (5)$$

where $V = \left(V_0(K), \ldots, V_d(K) \right)^{\top}$, $R = (R_{r_0}, \ldots, R_{r_d})^{\top}$ and

$$A_{r_0 \ldots r_d} = \begin{pmatrix} r_0^d \kappa_d & r_0^{d-1} \kappa_{d-1} & \cdots & r_0^2 \kappa_2 & r_0 \kappa_1 & 1 \\ r_1^d \kappa_d & r_1^{d-1} \kappa_{d-1} & \cdots & r_1^2 \kappa_2 & r_1 \kappa_1 & 1 \\ \cdots & \cdots & \cdots & \cdots & \cdots & \cdots \\ r_d^d \kappa_d & r_d^{d-1} \kappa_{d-1} & \cdots & r_d^2 \kappa_2 & r_d \kappa_1 & 1 \end{pmatrix} \qquad (6)$$

is a regular matrix. Then, $V = A_{r_0 \ldots r_d}^{-1} R$ is the unique solution of (5). The integrals R_{r_i} can be approximated by

$$\widehat{R}_{r_i} = \Delta^d \sum_{k=1}^{m} V_0(K \cap B_r(x_k)), \qquad (7)$$

where the points $x_1, \ldots, x_m$ belong to a d–dimensional cubic lattice with mesh size Δ. Thus, the vector V can be computed numerically as

$$V \approx A_{r_0 \ldots r_d}^{-1} \widehat{R}, \qquad (8)$$

where $\widehat{R}$ is the vector $(\widehat{R}_{r_0}, \ldots, \widehat{R}_{r_d})^{\top}$ of approximations given in (7). This numerical solution heavily depends on the choice of radii $r_0, \ldots, r_d$. To reduce this dependence, a least–squares method can be used; see also [5]. Instead of (5), consider the (overdetermined) system of linear equations $\widehat{R} = A_{r_0 \ldots r_{k-1}} x$ for $k > d+1$ pairwise different radii $r_0, \ldots, r_{k-1}$ where $x = (x_0, \ldots, x_d)^{\top} \in \mathbb{R}^{d+1}$. It is well known that the vector

$$v^* = \left(A_{r_0 \ldots r_{k-1}}^{\top} A_{r_0 \ldots r_{k-1}} \right)^{-1} A_{r_0 \ldots r_{k-1}}^{\top} \widehat{R} \qquad (9)$$

is the unique solution of the *least–squares minimization* problem

$$| \widehat{R} - A_{r_0...r_{k-1}} v^* | = \min_{x \in \mathbb{R}^{d+1}} | \widehat{R} - A_{r_0...r_{k-1}} x |$$

and, therefore, can be regarded as an approximation to the vector V of intrinsic volumes of K. For a discussion of the practical choice of radii $r_0, \ldots, r_{k-1}$, see [4, 5].

In general, the numerical solutions (8) and (9) of (5) do not necessarily preserve the positivity property of the volume $V_d(K)$ and the surface area $2V_{d-1}(K)$. Practically one can cope with this problem by changing the values and the number of radii r_i as well as distances between them. For a detailed discussion, see [4].

3. Stationary random closed sets

Let Ξ be a stationary random closed set (RACS) in $\mathbb{R}^d$ whose realizations belong to the extended convex ring $\S$ with probability 1. Recall that stationarity of Ξ means the invariance of its distribution with respect to arbitrary translations in $\mathbb{R}^d$. More details on stationary RACS can be found in many books; see e.g. [6, 7, 13, 15, 16, 18].

Specific intrinsic volumes

For any $K \in \mathcal{R}$, let $N(K) = \min\{m \in \mathbb{N} : K = \bigcup_{i=1}^m K_i, \; K_i \in \mathcal{K}\}$ denote the minimal number of convex components of the set K, where we put $N(K) = 0$ if $K = \emptyset$. Assume that

$$E \, 2^{N(\Xi \cap [0,1]^d)} < \infty. \tag{10}$$

Then, for any sequence $\{W_n\}$ of compact and convex observation windows $W_n = nW$ with $W \in \mathcal{K}$ such that $V_d(W) > 0$ and $o \in \mathrm{int}(W)$, the limit $\overline{V}_j(\Xi) = \lim_{n \to \infty} E V_j(\Xi \cap W_n)/V_d(W_n)$ exists for each $j = 0, \ldots, d$ (see [13], Theorem 5.1.3) and is called the *j*th *specific intrinsic volume* of Ξ.

Local Euler–Poincaré characteristic

The expectation $E V_0 (\Xi \cap B_r(x))$ is called *local Euler–Poincaré characteristic* or, equivalently, *local connectivity number* of Ξ, where $r > 0$ is an arbitrary fixed number. For $r = 1$, the following representation formula for $E V_0 (\Xi \cap B_1(x))$ can be found e.g. in [13], Corollary 5.3.2, where its proof is based on the principal kinematic formula. In the present paper, we give an alternative proof for any $r > 0$, which makes use of an explicit extension of Steiner's formula (1) to the convex ring; see [12].

PROPOSITION 3.1 *For any $r \geqslant 0$ and $x \in \mathbb{R}^d$, it holds*

$$E\, V_0\left(\Xi \cap B_r(x)\right) = \sum_{j=0}^{d} r^{d-j} \kappa_{d-j} \overline{V}_j(\Xi)\,. \tag{11}$$

Proof. Consider the stationary random field $\{Z_r(x),\ x \in \mathbb{R}^d\}$, where

$$Z_r(x) = \sum_{q \in \partial \Xi \cap B_r(x),\ q \neq x} J(\Xi \cap B_r(x), q, x)\,.$$

and $J(\Xi \cap B_r(x), q, x)$ is given by (3). In [11], we showed that $E\, Z_r(x) = \sum_{j=0}^{d-1} r^{d-j} \kappa_{d-j} \overline{V}_j(\Xi)$ holds for any $x \in \mathbb{R}^d$. Thus, it suffices to prove that $E\, V_0\left(\Xi \cap B_r(x)\right) = E\, Z_r(x) + \overline{V}_d(\Xi)$. Notice that the function $f(r) = E\, Z_r(x)$ is continuously differentiable as a polynomial in r, where $f(r) = \int_0^r f^{(1)}(s)\, ds$ since $f(0) = 0$. Furthermore, for any $s > 0$, we have

$$f^{(1)}(s) = \frac{d}{ds} E\, V_0(\Xi \cap B_s(x))\,, \tag{12}$$

where the derivative on the right–hand side does not depend on x by the stationarity of Ξ. In order to show (12), let A_o be a sufficiently small open cube with diagonals crossing at the origin o such that $A_o \subset \text{int}(B_s(o))$. Then, for any $\Delta s > 0$, we have

$$
\begin{aligned}
Z_{s+\Delta s}(o) - Z_s(o) &= \sum_{q \in \partial \Xi \cap (B_{s+\Delta s}(o) \setminus B_s(o))} J\left(\Xi \cap B_{s+\Delta s}(o), q, o\right) \\
&= \sum_{q \in \partial \Xi \cap (B_{s+\Delta s}(o) \setminus B_s(o))} J\left((\Xi \setminus A_o) \cap B_{s+\Delta s}(o), q, o\right) \\
&= V_0\left((\Xi \setminus A_o) \cap B_{s+\Delta s}(o)\right) - V_0\left((\Xi \setminus A_o) \cap B_s(o)\right) \\
&= V_0\left(\Xi \cap B_{s+\Delta s}(o)\right) - V_0\left(\Xi \cap B_s(o)\right),
\end{aligned}
$$

where the third equality follows from the fact that

$$\sum_{q \in \partial A \cap B_r(o)} J\left((A \setminus A_o) \cap B_r(o), q, o\right) = V_0\left((A \setminus A_o) \cap B_r(o)\right)$$

for each $A \in \S$ and for any $r > 0$ such that $A_o \subset \text{int}(B_r(o))$. This gives

$$
\begin{aligned}
f^{(1)}(s) &= \lim_{\Delta s \downarrow 0} E\, \frac{Z_{s+\Delta s}(o) - Z_s(o)}{\Delta s} \\
&= \lim_{\Delta s \downarrow 0} E\, \frac{V_0(\Xi \cap B_{s+\Delta s}(o)) - V_0(\Xi \cap B_s(o))}{\Delta s} = \frac{d}{ds} E\, V_0(\Xi \cap B_s(o))\,.
\end{aligned}
$$

Now, using (12), $f(r)$ can be rewritten as

$$
\begin{aligned}
f(r) &= \int_0^r \frac{d}{ds} E\, V_0(\Xi \cap B_s(o))\, ds = E\, V_0(\Xi \cap B_r(o)) - E\, V_0(\Xi \cap \{o\}) \\
&= E\, V_0(\Xi \cap B_r(o)) - E\, 1_\Xi(o) = E\, V_0(\Xi \cap B_r(o)) - \overline{V}_d(\Xi),
\end{aligned}
$$

where 1_Ξ denotes the indicator of Ξ. $\qquad\square$

It is well known that the Minkowski functionals of polyconvex sets can be defined through the Euler–Poincaré characteristics of their lower dimensional sections by means of Crofton's formula; see e.g. [13, 18]. Proposition 3.1 immediately implies that

$$
\overline{V}_j(\Xi) = \frac{1}{(d-j)!\,\kappa_{d-j}} \cdot \frac{d^{(d-j)} E\, V_0(\Xi \cap B_r(x))}{dr^{(d-j)}} \Bigg|_{r=0}
$$

for any $j = 0, \ldots, d-1$ and $x \in \mathbb{R}^d$. Thus, similarly to Crofton's formula, the specific intrinsic volumes of stationary RACS can be expressed by their local Euler–Poincaré characteristics.

4.   Estimation of specific intrinsic volumes

In this section, similar to the approach considered in [11], the method of moments is used to construct joint nonparametric estimators for the specific intrinsic volumes $\overline{V}_j(\Xi)$, $j = 0, \ldots, d$.

Indirect estimation via local Euler–Poincaré characteristics

For any $d+1$ positive pairwise different radii r_j, Proposition 3.1 yields the following system of $d+1$ linear equations with respect to the variables $\overline{V}_j(\Xi)$, $j = 0, \ldots, d$:

$$
A_{r_0 \ldots r_d}\, v = c, \tag{13}
$$

where $A_{r_0 \ldots r_d}$ is the matrix introduced in (6), $v = (\overline{V}_0(\Xi), \ldots, \overline{V}_d(\Xi))^\top$ and $c = (E\, V_0(\Xi \cap B_{r_0}(o)), \ldots, E\, V_0(\Xi \cap B_{r_d}(o)))^\top$. Similar to the deterministic case of Section 2, choose an appropriate estimator $\widehat{c}$ for c and define the estimator $\widehat{v}$ for v by

$$
\widehat{v} = A_{r_0 \ldots r_d}^{-1} \widehat{c} \tag{14}
$$

in order to estimate the vector v of specific intrinsic volumes from a single realization of Ξ observed in a certain window $W \in \mathcal{K}$. For any $r > 0$ such that $V_d(W \ominus B_r(o)) > 0$, consider the stationary random field $\{Y_r(x),\ x \in \mathbb{R}^d\}$ with $Y_r(x) = V_0(\Xi \cap B_r(x))$. An unbiased estimator for $y_r = E\, Y_r(o)$ is given by $\widehat{y}_r = \int_{W \ominus B_r(o)} Y_r(x)\, \mu(dx)$, where μ is an arbitrary probability measure concentrated on $W \ominus B_r(o) \subset \mathbb{R}^d$. For instance, μ can be the normalized Lebesgue measure $\mu(\cdot) = V_d(\cdot \cap W \ominus B_r(o))/V_d(W \ominus B_r(o))$ on $W \ominus B_r(o)$, or a discrete measure $\mu(\cdot) = \sum_{i=1}^{m} w_i \delta_{x_i}(\cdot)$ with measurements at locations $x_1, \ldots, x_m \in W \ominus B_r(o)$ and weights $w_i > 0$ such that $w_1 + \ldots + w_m = 1$. Notice that integration is performed over the reduced window $W \ominus B_r(o)$ to avoid edge effects, since the computation of $V_0(\Xi \cap B_r(x))$ for $x \in W$ requires the knowledge of Ξ in the r–neighborhood of x while Ξ is observed only within W. Thus, assuming that $V_d(W \ominus B_{r_j}(o)) > 0$ for each $j = 0, \ldots, d$, an unbiased estimator $\widehat{c}$ for c is given by

$$\widehat{c} = \left(\int_{W \ominus B_{r_0}(o)} Y_{r_0}(x)\,\mu(dx), \ldots, \int_{W \ominus B_{r_d}(o)} Y_{r_d}(x)\,\mu(dx) \right)^{\top}.$$

Mean–square consistency and asymptotic normality

For $\mu(\cdot) = V_d(\cdot \cap W \ominus B_r(o))/V_d(W \ominus B_r(o))$, the integral

$$\widehat{y}_r = \int_{W \ominus B_r(o)} Y_r(x)\,\mu(dx)$$

is the least–squares estimator for y_r, which is mean–square consistent as $W \uparrow \mathbb{R}^d$ provided that some integrability conditions are satisfied; see e.g. [3], p. 131. This means that for a sequence $\{W_n\}$ of unboundedly increasing sampling windows with $W_n = nW$, we have $E\left(\widehat{y}_{r,n} - y_r\right)^2 \to 0$ as $n \to \infty$, where $\widehat{y}_{r,n} = \int_{W_n \ominus B_r(o)} Y_r(x)\,\mu_n(dx)$ and

$$\mu_n(\cdot) = V_d(\cdot \cap W_n \ominus B_r(o))/V_d(W_n \ominus B_r(o));$$

see also [11]. Assuming that $E\,4^{N(\Xi \cap [0,1]^d)} < \infty$, it can be shown that the *covariance functions* $C_{r_i r_j}(x) = E\left(Y_{r_i}(o)Y_{r_j}(x)\right) - y_{r_i}y_{r_j}$ are well defined; $i, j = 0, \ldots, d$. Furthermore, under suitable mixing conditions on Ξ and assuming that $\int_{\mathbb{R}^d} |C_{r_i r_j}(x)|\,dx < \infty$, the random vector

$$\sqrt{V_d(W_n \ominus B_r(o))}\left(\widehat{y}_{r_0,n} - y_{r_0}, \ldots, \widehat{y}_{r_d,n} - y_{r_d}\right)$$

is asymptotically normal distributed, where the asymptotic covariance matrix is given by $\left(\int_{\mathbb{R}^d} C_{r_i r_j}(x)\,dx\right)_{i,j=0}^{d}$ and can be consistently estimated; see [3], Section 3.1, and [11]. Notice that the integrability and mixing conditions mentioned above are fulfilled, for example, for rapidly mixing germ–grain models including the well–known *Boolean model*; see e.g. [6, 7, 14]. We also remark that the estimator $\widehat{v}_n = A_{r_0 \ldots r_d}^{-1}\left(\widehat{y}_{r_0,n}, \ldots, \widehat{y}_{r_d,n}\right)^{\top}$ for v is mean–square consistent and asymptotically normal distributed, provided that the estimator $\left(\widehat{y}_{r_0,n}, \ldots, \widehat{y}_{r_d,n}\right)^{\top}$ for c possesses these properties.

Least–squares estimator

The least-squares approach of Section 2 also applies (with minor changes) to the case of stationary RACS. For $k > d + 1$ pairwise different radii

$r_0, \ldots, r_{k-1}$ such that $V_d(W \ominus B_{r_j}(o)) > 0$, $j = 0, \ldots, k$, the corresponding solution of the least squares minimization problem is $v^* = (A_{r_0 \ldots r_{k-1}}^\top A_{r_0 \ldots r_{k-1}})^{-1} A_{r_0 \ldots r_{k-1}}^\top \widehat{y}$, where $\widehat{y} = (\widehat{y}_{r_0}, \ldots, \widehat{y}_{r_{k-1}})^\top$ with $\widehat{y}_{r_j} = \int_{W \ominus B_{r_j}(o)} Y_{r_j}(x) \mu(dx)$. Notice that the estimator $v^* = (v_0^*, \ldots, v_d^*)^\top$ for the vector $v = (\overline{V}_0(\Xi), \ldots, \overline{V}_d(\Xi))^\top$ of specific intrinsic volumes of Ξ is much more robust with respect to the choice of radii $r_0, \ldots, r_{k-1}$ than the estimator $\widehat{v}$ given in (14).

5. Estimation variance and spatial sampling designs

Besides unbiasedness, another important criterion for goodness of the estimator $\widehat{v}$ given in (14) is related to its variance properties, where the radii $r_0, \ldots, r_d$ and the averaging probability measure μ should be chosen in such a way that the *estimation variance* $\sigma^2 = Var(\widehat{v}) = E|\widehat{v} - v|^2$ is possibly small.

Bound on the estimation variance

Unfortunately, it seems to be impossible to determine the estimation variance $\sigma^2 = Var(\widehat{v}) = E|A_{r_0 \ldots r_d}^{-1}(\widehat{c} - c)|^2$ explicitly. However, it is easy to get an upper bound for σ^2. Namely, (14) implies that

$$\sigma^2 \leq \|A_{r_0 \ldots r_d}^{-1}\|^2 E|\widehat{c} - c|^2 = \|A_{r_0 \ldots r_d}^{-1}\|^2 \sum_{j=0}^{d} Var(\widehat{y}_{r_j}), \qquad (15)$$

where

$$\|A_{r_0 \ldots r_d}^{-1}\|^2 = \max_{i=0,\ldots,d} \lambda_i\left((A_{r_0 \ldots r_d} A_{r_0 \ldots r_d}^\top)^{-1}\right) = \frac{1}{\min_{i=0,\ldots,d} \lambda_i(A_{r_0 \ldots r_d} A_{r_0 \ldots r_d}^\top)}$$

is the squared matrix norm of $A_{r_0 \ldots r_d}^{-1}$ and $\lambda_i(A)$ is the ith eigenvalue of the matrix A. Notice that $\|A_{r_0 \ldots r_d}^{-1}\|$ is finite. Thus, it is reasonable to choose $r_0, \ldots, r_d$ and μ such that the bound in (15) becomes small. Consider the variance $Var(\widehat{y}_{r_j}) = E(\widehat{y}_{r_j} - y_{r_j})^2$ appearing in (15). For any fixed $r > 0$, let $\mathcal{P}$ denote the family of all probability measures on $W \ominus B_r(o)$ and let the function $L : \mathcal{P} \to (0, \infty)$ be defined by $L(\mu) = E(\widehat{y}_r - y_r)^2$ for each $\mu \in \mathcal{P}$. By Fubini's theorem, we can write

$$E(\widehat{y}_r - y_r)^2 = \int_{W \ominus B_r(o)} \int_{W \ominus B_r(o)} C_{rr}(x - x') \, \mu(dx)\mu(dx'). \qquad (16)$$

Suppose that $L(\mu_0) = \min_{\mu \in \mathcal{P}} L(\mu)$ holds for some $\mu_0 \in \mathcal{P}$. Then, using the methods of variational analysis developed e.g. in [8] (see also [17]), it can be shown that the function $g(x) = \int_{W \ominus B_r(o)} C_{rr}(x - h) \mu_0(dh)$ necessarily has the following properties:

$$g(x) = L(\mu_0) \quad \mu_0\text{-a.e.} \qquad \text{and} \qquad g(x) \geq L(\mu_0) \quad \text{for all } x \in \mathbb{R}^d.$$

Discrete sampling designs

Suppose now that $L(\mu_0) = \min_{\mu \in \mathcal{P}} L(\mu)$ holds for some discrete probability measure $\mu_0 \in \mathcal{P}$ such that $\mu_0(\cdot) = \sum_{i=1}^{m} w_i \delta_{x_i}(\cdot)$ for some integer $m \geqslant 1$, where $x_1, \ldots, x_m \in W \ominus B_r(o)$ and $w_1, \ldots, w_m > 0$ with $w_1 + \ldots + w_m = 1$. Then, it can be shown that $L(\mu_0) = (e^\top Q_r^{-1} e)^{-1} = \left(\sum_{i,j=1}^{n} q_{ij}^{-1} \right)^{-1}$ holds provided that the number of atoms m and the atoms $x_1, \ldots, x_m$ themselves satisfy the condition

$$q_r^\top(x) Q_r^{-1} e \geqslant 1 \qquad \text{for all } x \in \mathbb{R}^d \qquad (17)$$

and the covariance matrix $Q_r = (C_{rr}(x_i - x_j))_{i,j=1}^{m}$ is regular, where $e = (1, \ldots, 1)^\top$, $Q_r^{-1} = \left(q_{ij}^{-1} \right)_{i,j=1}^{m}$ denotes the inverse matrix of Q_r and $q_r(x) = (C_{rr}(x - x_1), \ldots, C_{rr}(x - x_n))^\top$ for any $x \in \mathbb{R}^d$. Moreover, in this case, the vector of weights $w = (w_1, \ldots, w_m)^\top$ is given by

$$w = L(\mu_0) Q_r^{-1} e. \qquad (18)$$

Notice that, for fixed sampling points $x_1, \ldots, x_m$, formula (18) coincides with the *kriging of the mean*; see [19]. In this case, the estimator $\hat{y}_r$ with weights given by (18) is also known as the generalized least–squares estimator of the trend; see [9], p. 11. On the other hand, the locations $x_1, \ldots, x_n$ can be chosen iteratively using gradient algorithms described e.g. in [9].

Anyhow, the choice of an appropriate number m of sampling points, locations $x_1, \ldots, x_m$ and weights $w_1, \ldots, w_m$ depends on the covariance function $C_{rr} : \mathbb{R}^d \to \mathbb{R}$ which is unknown in general. Therefore, $C_{rr}(h)$, $h \in \mathbb{R}^d$ has to be estimated from data. Sometimes it is preferable to consider the *variogram function* $\gamma_r : \mathbb{R}^d \to \mathbb{R}$ with $\gamma_r(h) = \frac{1}{2} E (Y_r(x) - Y_r(x + h))^2$, $h, x \in \mathbb{R}^d$, instead of C_{rr} since it can be estimated more easily. For corresponding estimation techniques and algorithms, see e.g. [1, 2, 19]. Since $\gamma_r(h) = C_{rr}(o) - C_{rr}(h)$ holds for any $h \in \mathbb{R}^d$, (17) and (18) can be rewritten as

$$p_r^\top(x) \Gamma_r^{-1} e \leqslant 1 \qquad \text{for all } x \in \mathbb{R}^d \qquad (19)$$

and

$$w = \gamma_0 \Gamma_r^{-1} e, \qquad (20)$$

respectively, where $\Gamma_r = (\gamma_r(x_i - x_j))_{i,j=1}^{m}$, $\gamma_0 = (e^\top \Gamma_r^{-1} e)^{-1}$ and $p_r(x) = (\gamma_r(x - x_1), \ldots, \gamma_r(x - x_m))^\top$.

6.   Numerical results

To test the performance of the above estimation method, 200 realizations of a planar Boolean model Ξ $(d = 2)$ with circular grains were generated in the observation window $W = [0, 1000]^2$. Let λ be the intensity of the stationary

Poisson point process $X = \{X_i\}$ of germs and let the grains Ξ_i be independent circles with radii that are uniformly distributed on $[20, 40]$. Then, Ξ is given by $\Xi = \bigcup_{i=1}^{\infty}(\Xi_i + X_i)$. The intensity λ was chosen to fit the volume fractions $\overline{V}_2(\Xi) = 0.2, 0.5, 0.8$, respectively. For each realization, the vector v of specific intrinsic volumes of Ξ was estimated using the radii $r_0 = 10$, $r_{i+1} = r_i + 1.3$, $i = 0, \ldots, 49$ in the least–squares method. In the estimation, sampling was performed on the regular square lattice of points $x_1, \ldots, x_m$ with mesh size $\Delta = 5$. Finally, vector $\overline{v}^* = (\overline{v}_0^*, \overline{v}_1^*, \overline{v}_2^*)$ was built being the arithmetic mean over the results of 200 realizations. Its values are compared with the theoretical counterparts $v = (\overline{V}_0(\Xi), \overline{V}_1(\Xi), \overline{V}_2(\Xi))$ in Table 1. Additionally, the specific intrinsic volumes were estimated by the method described in [10] from the same 200 realizations of Ξ. The resulting arithmetic means $\tilde{v}_0, \tilde{v}_1, \tilde{v}_2$ are also presented in Table 1. To compare the precision of both algorithms, the relative error $\delta_{A,B} = \frac{B-A}{A} \cdot 100\%$ of an estimated quantity B with respect to the theoretical value A is given. It is clear from Table 1 that the performance of

Table 1. Theoretical and estimated values of specific intrinsic volumes

$\overline{V}_2(\Xi)$	0.2	0.5	0.8
$\overline{v}_2^*$	0.194299	0.490611	0.793328
$\tilde{v}_2$	0.199362	0.498217	0.798085
$\delta_{\overline{V}_2, \overline{v}_2^*}, \%$	-2.85	-1.88	-0.83
$\delta_{\overline{V}_2, \tilde{v}_2}, \%$	-0.32	-0.36	-0.24
$2\overline{V}_1(\Xi)$	0.011476	0.02228	0.020693
$2\overline{v}_1^*$	0.012123	0.023402	0.021547
$2\tilde{v}_1$	0.011361	0.021947	0.02022
$\delta_{\overline{V}_1, \overline{v}_1^*}, \%$	5.64	5.04	4.13
$\delta_{\overline{V}_1, \tilde{v}_1}, \%$	-1.0	-1.5	-2.29
$\overline{V}_0(\Xi) \times 10^4$	0.4778163	0.3919529	-0.6059316
$\overline{v}_0^* \times 10^4$	0.4348681	0.1555031	-0.10798772
$\tilde{v}_0 \times 10^4$	0.4312496	0.1595565	-0.10672334
$\delta_{\overline{V}_0, \overline{v}_0^*}, \%$	-8.99	-60.33	78.22
$\delta_{\overline{V}_0, \tilde{v}_0}, \%$	-9.75	-59.29	76.13

our algorithm is comparable to that of the method described in [10]. Hovever, the above results can be improved by taking e.g. $\Delta = 1$. In fact, the precision of our computations can be controlled by changing the sampling design as well as the number and values of dilation radii. The increase of the numbers of radii and sampling points results in a higher precision. This implies longer run times. Hence, the parameters of the algorithm should be tuned in accordance with the needs of specific applications; see [4] for an extensive discussion.

Acknowledgments

The authors are very grateful to Simone Klenk and Johannes Mayer for the implementation of the algorithm in Java and numerical experiments. Furthermore, they would like to thank the referees for their comments that helped to improve the paper.

References

[1] J.-P. Chilès and P. Delfiner. *Geostatistics: Modelling Spatial Uncertainty*. J. Wiley & Sons, New York, 1999.

[2] N. A. C. Cressie. *Statistics for Spatial Data*. J. Wiley & Sons, New York, 2nd edition, 1993.

[3] A. V. Ivanov and N. N. Leonenko. *Statistical Analysis of Random Fields*. Kluwer, Dordrecht, 1989.

[4] S. Klenk, J. Mayer, V. Schmidt, and E. Spodarev. Algorithms for the computation of Minkowski functionals of deterministic and random polyconvex sets. Preprint, 2005.

[5] S. Klenk, V. Schmidt, and E. Spodarev. A new algorithmic approach for the computation of Minkowski functionals of polyconvex sets. Preprint, 2004. Submitted.

[6] G. Matheron. *Random Sets and Integral Geometry*. J. Wiley & Sons, New York, 1975.

[7] I. S. Molchanov. *Statistics of the Boolean Model for Practitioners and Mathematicians*. J. Wiley & Sons, Chichester, 1997.

[8] I. S. Molchanov and S. A. Zuyev. Variational analysis of functionals of a Poisson process. *Mathematics of Operations Research*, 25:485–508, 2000.

[9] W. G. Müller. *Collecting Spatial Data*. Physica–Verlag, Heidelberg, 2001.

[10] J. Ohser and F. Mücklich. *Statistical Analysis of Microstructures in Materials Science*. J. Wiley & Sons, Chichester, 2000.

[11] V. Schmidt and E. Spodarev. Joint estimators for the specific intrinsic volumes of stationary random sets. Preprint, 2004. To appear in Stochastic Processes and their Applications.

[12] R. Schneider. *Convex Bodies. The Brunn–Minkowski Theory*. Cambridge University Press, Cambridge, 1993.

[13] R. Schneider and W. Weil. *Stochastische Geometrie*. Teubner Skripten zur Mathematischen Stochastik. Teubner, Stuttgart, 2000.

[14] J. Serra. The Boolean model and random sets. *Computer Graphics and Image Processing*, 12:99–126, 1980.

[15] J. Serra. *Image Analysis and Mathematical Morphology*. Academic Press, London, 1982.

[16] J. Serra. *Image Analysis and Mathematical Morphology: Theoretical Advances*, volume 2. Academic Press, London, 1988.

[17] E. Spodarev. Isoperimetric problems and roses of neighborhood for stationary flat processes. *Mathematische Nachrichten*, 251(4):88–100, 2003.

[18] D. Stoyan, W. S. Kendall, and J. Mecke. *Stochastic Geometry and its Applications*. J. Wiley & Sons, Chichester, 2nd edition, 1995.

[19] H. Wackernagel. *Multivariate Geostatistics*. Springer, Berlin, 2nd edition, 1998.

VI

TEXTURE, COLOUR
AND MULTIVALUED IMAGES

INTERSIZE CORRELATION OF GRAIN OCCURRENCES IN TEXTURES AND ITS APPLICATION TO TEXTURE REGENERATION

Akira Asano,[1] Yasushi Kobayashi[2] and Chie Muraki[3]

[1]*Faculty of Integrated Arts and Sciences*
Hiroshima University, Japan
asano@mis.hiroshima-u.ac.jp

[2]*Graduate School of Engineering*
Hiroshima University, Japan

[3]*Research Institute for Radiation Biology and Medicine*
Hiroshima University, Japan
chiem@mikeneko.jp

Abstract   A novel method of texture characterization, called *intersize correlation* of grain occurrences, is proposed. This idea is based on a model of texture description, called "Primitive, Grain and Point Configuration (PGPC)" texture model. This model assumes that a texture is composed by arranging *grains,* which are locally extended objects appearing actually in a texture. The grains in the PGPC model are regarded to be derived from one *primitive* by the homothetic magnification, and the size of grain is defined as the degree of magnification. The intersize correlation is the correlation between the occurrences of grains of different sizes located closely to each other. This is introduced since homothetic grains of different sizes often appear repetitively and the appearance of smaller grains depends on that of larger grains. Estimation methods of the primitive and grain arrangement of a texture are presented. A method of estimating the intersize correlation and its application to texture regeneration are presented with experimental results. The regenerated texture has the same intersize correlation as the original while the global arrangement of large-size grains are completely different. Although the appearance of the resultant texture is globally different from the original, the semi-local appearance in the neighborhood of each large-size grain is preserved.

Keywords: texture, granulometry, skeleton, size distribution

C. Ronse et al. (eds.), Mathematical Morphology: 40 Years On, 357–366.
©2005 *Springer. Printed in the Netherlands.*

1.   Introduction

Texture in the context of image processing is an image structure whose characteristics are given by the size, shape, and arrangement of its parts. Various methods of texture analysis, for example the cooccurrence matrix method and the spatial frequency method, have been proposed [1, 2]. These methods measure global or statistical characteristics of a texture based on its randomness.

Our aim of texture analysis is description, rather than measurement. Texture description assumes a model of texture generation, and estimates the model parameters of a texture and regenerates a new texture whose visual impression is related to the original [3–7]. These methods are based on the random nature of textures.

We have proposed a model of texture description, called "Primitive, Grain and Point Configuration (PGPC)" texture model, and parameter estimation methods based on this model [8]. The PGPC texture model is based on the following observation of the texture suggested by Gestalt psychology: A repetitive appearance of similar objects of a moderate size is organized to be a meaningful structure by the human cognitive process. This observation suggests that a texture is neither completely deterministic nor completely random, but is often locally deterministic and globally random or regular, and that an appropriate texture description model has to be locally deterministic as well as globally deterministic or stochastic. Our model assumes that a texture is composed by arranging *grains* regularly or randomly on an image, and a grain is defined as a locally extended object actually appearing in a texture. The grains in the PGPC model are regarded to be derived from one *primitive* by some shape modifications, since the texture is regarded to be composed by the arrangement of similar small object, as explained in the above observation. The primitive is a model parameter estimated from a texture, and its shape determines local deterministic characteristics of the texture. The grain arrangement determines global, and often stochastic, characteristics of the texture. The primitive and grain arrangement can be estimated using an optimization process based on the granulometry [9, 10] and skeletonization [9, 11].

We propose a novel texture characterization based on the PGPC texture model, called *intersize correlation* of grain occurrences, in this paper. If the modification on the primitive is limited to a homothetic magnification, the *size* of grain is defined as the degree of magnifications. The intersize correlation is the correlation between the occurrences of grains of different sizes located closely to each other. The motivation of introducing the intersize correlation is the observation that the appearance of smaller grains depends on that of larger grains; For example, a large grain is surrounded anisotropically by several smaller grains and the group of these grains appears repetitively. This kind of characteristics is suitable to be expressed by the PGPC model, since it assumes a primitive and the derivation of homothetic grains.

In this paper, we show an estimation method of the primitive and grain arrangements of a texture, and we present an experimental estimation method of the intersize correlation of a texture. We also present an application of the intersize correlation to texture regeneration with experimental results. We generate a new texture where its intersize correlation is the same as that of the original and the global arrangement of large-size grains are completely different from that of the original. This is achieved by generating a different arrangement of locations for large-size grains, generating a random arrangement of small-size grains around each location of the large-size grain following the same intersize correlation as the original texture, and locating the grains at corresponding locations. Although the appearance of the resultant texture is globally different from the original, the visual impression of the semi-local appearance, which is the appearance of small-size grains around a large-size grain, for example the anisotropy of the locations of small-size grains, is preserved.

A similar idea of image modification is found in [12], which derives the skeleton of the target image and regenerates another image by locating a different structuring element on the skeleton, for an application to the modification of the impression of calligraphic characters. This method employs another structuring element that is not related to the original image, although our texture regeneration method employs the primitive and the intersize correlation estimated from the original texture.

The PGPC model is closely related to the well-known Boolean set model [13]. It describes a random figure by a germ process generating grains to be arranged and the Poisson point process determining where to locate the grains. The Boolean set model, however, restricts the point process of grain arrangement to the Poisson point process, and does not introduce the concept of grain size. Thus the intersize correlation cannot be considered by the Boolean set model, and this is an advantage of the PGPC model.

2. Granulometry and skeleton

Granulometry and size distribution function

Opening of an image X with respect to a structuring element B means the residue of X obtained by removing smaller structures than B. It indicates that the opening works as a filter to distinguish object structures by their sizes. Let $2B, 3B, \ldots$, be homothetic magnifications of the basic structuring element B. The r-times magnification of B, denoted rB, is usually defined in the context of mathematical morphology as follows:

$$rB = \begin{cases} B \oplus B \oplus \cdots \oplus B \quad ((r-1) \text{ times of} \oplus) & r > 0, \\ \{0\} & r = 0, \end{cases} \tag{1}$$

where $\{0\}$ denotes a single dot at the origin and $\oplus$ denotes the Minkowski set addition. In the case of gray scale images and structuring elements, the umbra of the result is derived by the above operation applied for their umbrae.

However, the difference between the extent of r-times magnification and that of $(r + 1)$-times magnification is often too large in the definition of Eq. (1). Thus we redefine the magnification as follows:

$$rB = \begin{cases} B \oplus C \oplus \cdots \oplus C & ((r-1) \text{ times of} \oplus) \quad r > 0, \\ \{0\} & r = 0, \end{cases} \tag{2}$$

where C is another small structuring element.

We then perform opening of X with respect to the homothetic structuring elements, and obtain the image sequence $X_B, X_{2B}, X_{3B}, \ldots$. In this sequence, X_B is obtained by removing the regions smaller than B, X_{2B} is obtained by removing the regions smaller than X_{2B}, X_{3B} is obtained by removing the regions smaller than $3B$, ... If it holds that $X \subseteq X_B \subseteq X_{2B} \subseteq X_{3B} \subseteq \ldots$, this sequence of openings is called *granulometry* [9].

We then calculate the ratio of the area (for binary case) or the sum of pixel values (for gray scale case) of X_{rB} to that of the original X at each r. The area of an image is defined by the area occupied by an image object, i. e. the number of pixels composing an image object in the case of discrete images. The function from a size r to the corresponding ratio is monotonically decreasing, and unity when the size is zero. This function is called *size distribution function*. The size distribution function of size r indicates the area ratio of the regions whose sizes are greater than or equal to r.

Skeleton

The morphological skeleton $SK(X, B)$ is characterized as follows [14, 15]:

$$SK(X, B) \quad = \quad \bigcup_r SK_r(X, B), \tag{3}$$

$$SK_r(X, B) \quad = \quad (X \ominus r\check{B}) - (X \ominus r\check{B})_B, \tag{4}$$

for each r such that $SK_r(X, B) \neq \emptyset$. $SK_r(X, B)$ is often referred as medial axis transform. In the above equation, $\ominus$ denotes the Minkowski set subtraction, $\check{B}$ denotes the reflection of B with respect to the origin. We mainly employ the medial axis transform in this paper, and this is similarly obtained for gray scale images by applying gray scale morphological operations.

3.   PGPC texture model

The Primitive, Grain and Point Configuration (PGPC) texture model represents a texture image X as follows:

$$X = \bigcup_r B_r \oplus \Phi_r \tag{5}$$

for nonempty Φ_r, where B_r denotes a grain, and Φ_r is a set indicating pixel positions to locate the grain rB. In the case of binary image, Φ_r is a subset of $\mathbb{Z}^2$. In the case of gray scale image, Φ_r is a subset of $\mathbb{Z}^3$ if we employ the umbra expression to the grains rB.

The PGPC texture model regards a texture as an image composed by a regular or irregular arrangement of objects that are much smaller than the image itself and resemble each other. The objects arranged in a texture are called *grains,* and the grains are regarded to be derived from one or a few typical objects called *primitives.*

We assume here that $\{0B, 1B, \ldots, rB, \ldots\}$ are homothetic magnifications of a small object B as defined in Eq. (2), and that B_r in Eq. (5) is equivalent to rB for each r. In this case, B is regarded as the primitive and r is referred as *size* of the magnification, and X_{rB} is regarded as the texture image composed by the arrangement of rB only. It follows that $rB - (r+1)B$ indicates the region included in the arrangement of rB but not included in that of $(r + 1)B$. Consequently, $X_{rB} - X_{(r+1)B}$ is the region where r-size grains are arranged if X is expressed by employing an arrangement of grains which are preferably large magnifications of the primitive. The sequence $X - X_B$, $X_B - X_{2B}, \ldots, X_{rB} - X_{(r+1)B}, \ldots$ is the decomposition of the target texture to the arrangement of the grains of each size.

Estimation of primitive and grain arrangement

The sequence of the texture decomposition, mentioned in the previous subsection, can be derived using any structuring element. Thus it is necessary to estimate the appropriate primitive that is a really typical representative of the grains. We employ an idea that the structuring element yielding the simplest grain arrangement is the best estimate of the primitive, similarly to the principle of minimum description length (MDL). The simple arrangement locates a few number of large magnifications for the expression of a large part of the texture image, contrarily to the arrangement of a large number of small-size magnifications. We derive the estimate by finding the structuring element minimizing the integral of $1 - F(r)$, where $F(r)$ is the size distribution function with respect to size r. The function $1 - F(r)$ is 0 for $r = 0$ and monotonically increasing, and 1 for the maximum size required to compose the texture by the magnification of this size. Consequently, if the integral of $1 - F(r)$ is minimized, the sizes of employed magnifications concentrate to relatively large sizes, and the structuring element in this case expresses the texture using preferably large magnifications. We regard this structuring element as the estimate of primitive.

We estimate the gray scale structuring element in two steps: the shape of structuring element is estimated by the above method in the first step, and the gray scale value at each pixel in the primitive estimated in the first step is then estimated. However, if the above method is applied to the gray scale estimation, the estimate often has a small number of high-value pixel and other pixels whose values are almost zero. This is because the umbra of any object can be composed by arranging a one-pixel structuring element. This is absolutely not a desired estimate. Thus we minimize $1 - F(1)$, i. e. the residual area of X_B instead of the method in the first step. Since the residual region cannot be composed of even the smallest magnification, the composition by this structuring element and its magnification is the most admissible when the residual area is the minimum. The exploration of the structuring element can be performed by the simulated annealing, which iterates a modification of the structuring element and finds the best estimate minimizing the evaluation function described in the above [8].

Once the primitive B is estimated, an estimate of Φ_r is obtained by the morphological skeletonization employing B as the structuring element. This is because the grains B_i are assumed to be homothetic magnifications of the primitive B, and the skeleton is defined as the locus of the origin of homothetically magnified structuring elements exactly covering the original image objects with preferably large magnifications at each position on the locus.

Note that the arrangement derived by the skeletonization is redundant. This is the reason why we do not employ the estimation of the primitive by finding the simplest grain arrangement, i .e. minimizing the numbers of points in Φ_r, although such minimization is a direct application of the MDL principle.

4.   Intersize correlation of grain occurrences

Once the primitives and grain arrangements are estimated using the PGPC model, it is observed in the grain arrangements of textures that a large-size grain is often surrounded by smaller-size grains. This indicates that there exists a correlation between the occurrences of the grain locations corresponding to size r and those to size $r - 1$, and it expresses visual impressions of semi-local grain appearances. We call it *intersize correlation* of grain occurrences between size r and $r - 1$.

Let x_1 and x_2 be grain location points. The probability that a grain of size r is located at x_1 is denoted as $\Pr\{x_1 \in \Phi_r\}$. The conditional probability that a grain of size $(r - 1)$ is located at x_2 under the condition that $x_1 \in \Phi_r$ is denoted as $\Pr\{x_2 \in \Phi_{r-1}|x_1 \in \Phi_r\}$. We assume that

$$\Pr\{x_2 \in \Phi_{r-1}|x_1 \in \Phi_r\} = f(x_2 - x_1) \qquad (6)$$

for all x_1, where f is a function of the relative position of x_2 to x_1 only, i. e. translation-invariant. This is because a texture often has a globally repetitive structure. We suppose that the function f decreases when the distance $|x_2 - x_1|$ increases, and depends on the direction of $x_2 - x_1$. We consider that the function f characterizes the intersize correlation.

Estimation of intersize correlation and application to texture regeneration

We consider a random rearrangement of grain locations while preserving the intersize correlation of the original texture, and regenerate a modified texture having a different global grain arrangement while the visual impression of semi-local grain appearances in the original textures are preserved.

We estimate the intersize correlation of the original texture for characterizing mainly its anisotropy of the occurrence of small-size grains around each large-size grain in our experiment. It is assumed that an 11×11-pixel window is centered at each point in Φ_r, as illustrated in Fig. 1(a). The window is divided into four subwindows as shown in Fig. 1(b), and the number of points in Φ_{r-1} in each subwindow is counted. If the windows overlap and share a point in Φ_{r-1} at several points in Φ_r, the point is counted as if the number of the point were not unity but $\frac{\alpha}{m} + \epsilon$, where m is the number of overlapping windows at this point and α and ϵ are tuning parameters. This value indicates the dependency of occurrences of points in Φ_{r-1} around closely located points in Φ_r. If these points occur independently, $\alpha = m$ and $\epsilon = 0$, i. e. the number of points in Φ_{r-1} are evaluated as it is. Although we have introduced the above idea in our experiments, the proper estimate of these parameters is still an open problem. The ratio of the sum of the number of points in Φ_{r-1} in the subwindow of a direction over the whole image to the total number of points in Φ_{r-1} in the windows over the whole image is calculated. The ratio corresponding to each subwindow is regarded as the estimated occurrence probability of the points corresponding to size $r - 1$ around the pixel to size r in the direction of this subwindow.

In the case of gray scale textures, the pixel value range from 0 to 255 is divided into several subranges of the same width, and the same calculation as the above is carried out for each subrange. The estimated occurrence probability for each direction is obtained for each subrange of pixel values.

Texture generation by random rearrangement of large-size grains

We generate a new texture from the original by distributing randomly the grain location points corresponding to size r, and generating the points corresponding to size $(i - 1)$ randomly independently at each position in the above

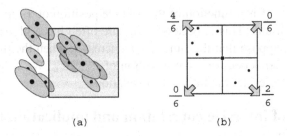

(a)   (b)

Figure 1.   Schematic illustration of the evaluation of intersize correlation. (a) An example of grain location. Each ellipse indicates a grain and the dot in each ellipse indicates the located point. The gray square indicates the window centered at a point in Φ_r. (b) the number of points in Φ_{r-1} is counted in each subwindow.

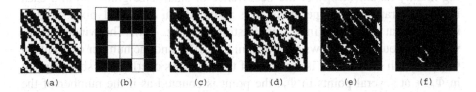

(a)   (b)   (c)   (d)   (e)   (f)

Figure 2.   An example of binary texture regeneration preserving the intersize correlation. (a) Original texture. (b) Estimated primitive. (c) Restored texture by locating the grains on the original Φ_2 and Φ_1 without any modifications. (d) Regenerated texture based on a random arrangement of the grains of size 2 and the arrangement of the grains of size 1 following the estimate intersize correlation. (e) Skeleton Φ_1. (f) Skeleton Φ_2.

four subwindows around the points corresponding to size r with the occurrence probabilities estimated as the above. Figure 2 shows an experimental example of a binary texture. Figure 2(a) shows the original binary texture, and (b) shows the estimated primitive. Each small square in (b) corresponds to one pixel in the primitive, and the shape is expressed by the arrangement of white squares. The primitive is explored from figures of nine connected pixels within 5×5-pixel square. The small structuring element C employed for magnification in Eq. (2) is the 2×2-pixel square. Figure 2(c) shows the restored texture by locating the grains on the original Φ_2 and Φ_1 without any modifications. It indicates that the estimated primitive and grain arrangements express the original texture properly. Figure 2(d) shows the result of texture regeneration by locating the grains of size 2 randomly by a Poisson point process and arranging the grains of size 1 following the estimated occurrence probability in each direction. Figures 2(e) and (f) show Φ_1 and Φ_2, i. e. the skeletons, respectively, estimated from the original image (a).

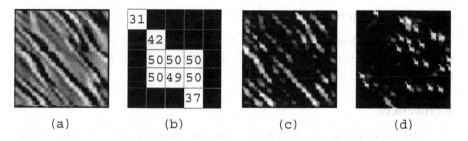

(a)	(b)	(c)	(d)

Figure 3.   An example of gray scale texture regeneration. (a) – (d) are similar to Fig. 2.

Figures 3(a)–(d) are the results in the case of a gray scale image. The gray scale value in the estimated primitive (b) is explored by setting the initial pixel value to 50 and modifying the value in the range of 0 to 100. The small structuring element C for magnification in Eq. (2) is defined as the 2×2-pixel square with pixel value 1 for all the four pixels.

These results show that the regenerated textures preserve local characteristics of the original texture by the primitive as well as semi-local characteristics around the large-size grains by the preservation of intersize correlation, although they look globally different from the original.

5.   Conclusions

We have proposed a novel method of texture characterization, called intersize correlation of grain occurrences. This idea is based on a model of texture description, called "Primitive, Grain and Point Configuration (PGPC) texture model." This model assumes that a texture is composed by arranging grains derived from one primitive by the homothetic magnification, and defines the size of grain. The intersize correlation is the occurrence correlation between grains of different sizes located closely to each other.

We have presented estimation methods of the primitive, grain arrangements, and the intersize correlation of a texture. We have also shown an application of the intersize correlation to texture regeneration, which generates a globally different texture preserving the visual impression of semi-local grain appearances in the original texture, by experiments for both binary and gray scale textures.

The granulometry used in this paper is based on openings, which suggests that a texture composed of brighter grains on the darker background is assumed. Although the brighter grains are extracted in our experiment for the gray scale image, the extraction of the darker grains by the granulometry based on closings may be useful for more precise characterizations. The correlation estimation method employed in this paper is just an experimental, and its characterizations by theoretical investigations and comparison with other methods are also our future problem.

Acknowledgments

This research has been supported by the Grant-in-Aid for Scientific Research of the Ministry of Education, Culture, Sports, Science and Technology of Japan, No. 14750297.

References

[1] Mahamadou I., and Marc A. (2002). "Texture classification using Gabor filters," Pattern Recognition Letters, Vol 23, 1095–1102.

[2] Liu, F., and Picard, R. W. (1996). "Periodicity, Directionality, and Randomness: Wold features for image modeling and retrieval," IEEE Trans. Pattern Anal. Machine Intell., Vol. 18, 722–733.

[3] Tomita, T., Shirai, Y., and Tsuji. Y. (1982). "Description of Textures by a Structural Analysis," IEEE Trans. Pattern Anal. Machine Intell., Vol. PAMI-4, 183–191.

[4] Sand, F., and Dougherty, E. R. (1998). "Asymptotic granulometric mixing theorem: morphological estimation of sizing parameters and mixture proportions," Pattern Recognition, Vol. 31, no. 1, 53–61.

[5] Balagurunathan, Y., and Dougherty, E. R. (2003). "Granulometric parametric estimation for the random Boolean model using optimal linear filters and optimal structuring elements," Pattern Recognition Letters, Vol. 24., 283–293.

[6] Gimel'farb, G. (2001). "Characteristic interation structures in Gibbs texture modelling," in *Imaging and Vision Systems: Theory, Assesment and Applications* J. Blanc-Talon and D. Popescu, Ed. Huntington: Nova Science Publishers, 71–90.

[7] Aubert, A. and Jeulin, D. (2000). "Estimation of the influence of second and third order moments on random sets reconstruction," Pattern Recognition, Vol. 33, no. 6, 1083–1104.

[8] Asano, A., Ohkubo, T., Muneyasu, M., and Hinamoto T. (2003). "Primitive and Point Configuration Texture Model and Primitive Estimation using Mathematical Morphology," Proc. 13th Scandinavian Conference on Image Analysis, Göteborg, Sweden; Springer LNCS Vol. 2749, 178–185.

[9] Heijmans, H. J. A. M. (1994). *Morphological Image Operators,* San Diego: Academic Press.

[10] Maragos P. (1989). "Pattern Spectrum and Multiscale Shape Representation," IEEE Trans. Pattern Anal. Machine Intell. Vol. 11, 701–716.

[11] Soille, P. (2003). *Morphological Image Analysis,* 2nd Ed. Berlin: Springer.

[12] Li, W., Hagiwara, I., Yasui, T., and Chen, H. (2003). "A method of generating scratched look calligraphy characters using mathematical morphology," Journal of Computational and Applied Mathematics, Vol. 159, 85–90.

[13] Serra, J. Ed. (1988). *Image Analysis and Mathematical Morphology Volume 2. Technical Advances,* Academic Press, London, Chaps. 15 and 18.

[14] Lantuejoul, Ch. (1980). "Skeletonization in quantitative metallography," *Issues in Digital Image Processing,* R. M. Haralick and J. C. Simon, Eds., Sijthoof and Noordoff.

[15] Serra. J. (1982). *Image analysis and mathematical morphology.* London: Academic Press.

[16] Kobayashi. Y., and Asano. A. (2003). "Modification of spatial distribution in Primitive and Point Configuration texture model," Proc. 13th Scandinavian Conference on Image Analysis, Göteborg, Sweden; Springer LNCS Vol. 2749, 877–884.

TEXTURE SEGMENTATION USING AREA MORPHOLOGY LOCAL GRANULOMETRIES

Neil D. Fletcher * and Adrian N. Evans

Department of Electronic and Electrical Engineering
University of Bath, BA2 7AY, UK

N.D.Fletcher@bath.ac.uk A.N.Evans@bath.ac.uk

Abstract Texture segmentation based on local morphological pattern spectra provides an attractive alternative to linear scale spaces as the latter suffer from blurring and do not preserve the shape of image features. However, for successful segmentation, pattern spectra derived using a number of structuring elements, often at different orientations, are required. This paper addresses this problem by using area morphology to generate a single pattern spectrum, consisting of a local granulometry and anti-granulometry, at each pixel position. As only one spectrum is produced, segmentation is performed by directly using the spectrum as the feature vector instead of taking pattern spectrum moments. Segmentation results for a simulated image of Brodatz textures and test images from the Outex texture database show the potential of the new approach.

Keywords: Texture analysis, granulometries, area morphology

Introduction

Automated texture classification and segmentation remains a challenging task for computer vision algorithms. Texture refers to the variation of intensity in a local neighbourhood and therefore cannot be defined by a single pixel [8]. In supervised texture classification schemes a feature vector, consisting of a number of textural features, is evaluated against a selected library of feature vectors for particular textures. Alternatively, if an a priori library is not available clustering techniques can be used to classify the feature vectors into an appropriate number of classes [10]. As texture can be said to consist of a distribution of image features at different scales, multiscale texture analysis schemes are an attractive alternative to the traditional fixed scale approach. In

*Neil Fletcher supported by an EPSRC CASE award in conjunction with QinetiQ.

367

C. Ronse et al. (eds.), Mathematical Morphology: 40 Years On, 367–376.
©2005 *Springer. Printed in the Netherlands.*

practice, multiscale feature vectors can be found by decomposing the signal into elements of different resolutions with a linear filterbank and then extracting the structural information in each sub-band. Popular methods include Gabor filtering [9] and, more recently, wavelet analysis [17]. However, despite its popularity, linear image analysis suffers from a number of drawbacks including the shifting and blurring of image features, a scale parameter that is not related to a size-based definition of scale and the fact that it produces filtered images that do not correspond to the shape of image features [11].

An alternative method of decomposition uses mathematical morphology operators to produce an improved representation both in terms of scale and the position of sharp edged objects. In practice, this can be achieved by differencing a series of openings and closings by increasing scale structuring elements to produce a granulometric size distribution for greyscale images, termed a pattern spectrum [11], that has its roots in the binary granulometries of Matheron and Serra [12, 16]. Feature vectors are formed by taking moments of the local pattern spectra produced by a number of different structuring elements, where the term "local" refers to a window centred on the pixel of interest [7]. In terms of texture classification, morphological scale spaces obtained using standard open and close filters suffer from two disadvantages. Firstly, the shape of the structuring element produces edge movement such as corner rounding and, secondly, a large number of structuring elements may be required to capture textual features. The latter problem is further exacerbated when linear features are present in the texture at different orientations. Consequently, many pattern spectra are required at each image point and this explains in part why only moments of the spectra are used to form the feature vectors.

Area morphology operators [18, 15] address both these problems. The scale spaces resulting from increasing area open-close (AOC) and area close-open (ACO) operations exhibit the property of strong causality thus guaranteeing the preservation of edge positions through scale [1]. In addition, as area operations can be considered as the maximum (resp. minimum) of openings (resp. closings) with all possible connected structuring elements with a given number of elements [5], the need for a set of differently orientated structuring elements is removed. An area morphology scale space classification scheme using a feature vector whose elements are the intensities of a particular pixel at a given set of scales is described in [1]. Here, a new local granulometric texture analysis technique is proposed that combines the advantageous properties of area morphology scale spaces with a local pattern spectrum approach.

This paper is arranged as follows. Section 1 discusses local granulometric texture analysis in more detail and the new area morphology local granulometric method is described in section 2. Experimental results for a compound image consisting of a number of Brodatz textures and images from the Outex texture database are presented in section 3. Finally, discussion and conclusions are given in section 4.

1. Morphological Texture Analysis

The granulometric size distributions first proposed by Matheron [12] provide a morphological description of the granularity or texture in an image (see also chapter 8 of [6]). Granulometries are obtained by the repeated application of multiscale non-linear filters. For example, consider the opening of a binary image image I with a set of structuring elements $\{B_i\}$. As the opening operator is increasing, when B_i is a subimage of B_{i+1} and $B_{i+1} \circ B_i = B_{i+1}$, then $I \circ B_{i+1}$ is a subimage of $I \circ B_i$. Therefore as $B_1, B_2, B_3, \ldots B_k$ is an increasing sequence of structuring elements, the filtered images will form the decreasing sequence

$$I \circ B_1 \supset I \circ B_2 \supset I \circ B_3 \supset \ldots \supset I \circ B_k. \tag{1}$$

The total number of pixels remaining after each successive opening results in the decreasing function $\Psi(k)$ where, for a given value of K, $\Psi(k) = 0$ for $k \geq K$. Various textural information can be extracted from $\Psi(k)$ depending on the shape of the structuring element. The sequence of images I_k given by

$$I_k = \{X \circ B_k : 0 \leq k \leq K - 1\} \tag{2}$$

defines the granulometry and the resulting function $\Psi(k)$ is known as the size distribution [7]. Since $\Psi(k)$ is a decreasing size distribution function its normalisation yields the distribution function

$$\Phi(k) = 1 - \frac{\Psi(k)}{\Psi(1)}. \tag{3}$$

The discrete derivative of (3) produces the discrete density function

$$\delta\Phi(k) = \{\Phi(k) - \Phi(k-1) : 1 \leq k \leq K - 1\} \tag{4}$$

that defines the discrete pattern spectrum. The pattern spectrum for greyscale images, PS_I, can be described by the decreasing sequence

$$PS_I(B_{+k}) = A[I \circ B_k - I \circ B_{k+1}], \tag{5}$$

where $\circ$ is now a greyscale opening, k is the size parameter, the structuring element B defines the shape parameter and $A[I] = \sum_{xy} I(x, y)$ [11]. Keeping the shape of B fixed produces a decreasing sequence of greyscale openings reminiscent of (1) and the result is a size histogram of I relative to B. The pattern spectrum can be extended to contain negative values by replacing the greyscale opening in (5) with a greyscale closing giving

$$PS_I(B_{-k}) = A[I \bullet B_k - I \bullet B_{k-1}]. \tag{6}$$

Normalising the local pattern spectra of (5) and (6) by the total energy in the original image defines the size density [11],

$$s_I(k) = \frac{1}{A[I]} \begin{cases} A[I \circ B_k - I \circ B_{k+1}], & \text{for } k \geq 0 \\ A[I \bullet B_k - I \bullet B_{k-1}], & \text{for } k \leq -1. \end{cases} \quad (7)$$

This density is equivalent to the pattern spectrum for binary images of (4) with the discrete differentiation being replaced by a point-wise difference of functions.

To obtain the full shape-size classification for $I(x, y)$ using either the binary or the greyscale approach an appropriate set of the structuring elements is required and, for each, the size parameter k is varied over its range, producing a discrete pattern spectrum for each structuring element.

Ganulometric-Based Texture Segmentation

Granulometric size distributions can be used as features for texture-based classification schemes. However, if the overall aim is the segmentation of images containing more than one texture, the classification needs to be performed at pixel-level and the features should reflect the local size distribution. This is the notion behind the local pattern spectrum methods of [7, 6] that compute local size distributions centred on a pixel of interest.

For binary images, local granulometric size distributions can be generated by placing a window W_x at each image pixel position x and, after each opening operation, counting the number of remaining pixels within W_x. This results in a local size distribution, Ψ_x, that can be normalised in the same manner as the global granulometry to give the local distribution $\Phi_x(k)$. Differentiating gives the density $\delta\Phi_x$ that yields the local pattern spectrum at pixel x, providing a probability density which contains textural information local to each pixel position.

As acknowledged in [7], the reason for calculating local pattern spectra on binary images is computational complexity. The analysis of grey-scale images is preferable as more image information is available for the classification algorithm. This can be achieved by replacing the density $\delta\Phi_x$ with a local greyscale size density calculated by applying (7) within a local window W_x. Textural feature extraction can be achieved by employing several structuring elements to generate a number of local granulometries, each providing different texture qualities, to produce a more robust segmentation.

Supervised classification schemes compare the observed pattern spectra moments at each pixel position to a database of texture moments [7]. For every pixel the distance between the local size density and every entry in the database of codevectors is calculated. The texture class with the minimum distance is then assigned to the pixel position, resulting in a texture class map.

Feature vectors that possess common attributes will form clusters in the feature space. The distances between points inside a cluster are smaller than those between points which are members of different clusters and therefore the clusters can be regarded as texture classes. In this manner, popular clustering algorithms such as the k-means [10] can be employed to provide an unsupervised texture segmentation scheme. This was the approach used to cluster granulometric moments for segmenting grey-scale mammogram images [2].

2.   Area Morphology Local Granulometries

Morphological area openings and closings [18, 15] remove connected components smaller than a given area and can be used to generate a scale space in which area provides a scale parameter equivalent to structuring element size in standard morphology. Area openings and closings can be defined by operations on a threshold decomposition or, alternatively, by

$$\gamma_\lambda^a(I) = \bigvee_{B \in A_\lambda} (I \circ B) \tag{8}$$

and

$$\varphi_\lambda^a(I) = \bigwedge_{B \in A_\lambda} (I \bullet B) \tag{9}$$

respectively, where A_λ is the set of connected subsets with area $\geq \lambda$ [5]. This alternative definition illustrates the fact that area operators select the most appropriately shaped structuring at each pixel, the only constraint being its area. Therefore, the operations adapt to the underlying image structure and eliminate any shape bias and artificial patterns associated with fixed structuring elements.

In terms of texture classification, area morphology scale spaces are attractive as they can capture all the structures contained in an image. This is advantageous as textured images generally contain far more structures that can be described by a family of fixed structuring elements. In addition, area scale spaces have the property of strong causality which ensures that edge positions are preserved through scale [1, 3]. Computational complexity issues have also been greatly reduced with the development of fast area opening and pattern spectra algorithms [13].

Successful feature classification was demonstrated by Acton and Mukherjee using a scale space vector containing the intensity of each pixel at a selected set of scales [1]. To extend the use of area morphology operators to texture segmentation a new approach based on local area morphology granulometries is proposed. The area operators are applied to the image and then local pattern spectra, containing information of all shapes at each scale, are calculated over

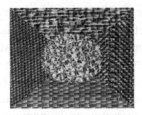

Figure 1 Brodatz test image.

a square window W_x by

$$s_I(\lambda) = \frac{1}{A_x[I]} \left\{ \begin{array}{ll} A_x[\gamma_\lambda^a(I) - \gamma_{\lambda+1}^a(I)], & \text{for } \lambda \geq 1 \\ A_x[\varphi_{-\lambda}^a(I) - \varphi_{-\lambda+1}^a(I)], & \text{for } \lambda \leq -1. \end{array} \right. \qquad (10)$$

The size of the local window is kept relatively small to ensure it captures the texture at a particular pixel. However, the scale space vector for each window can, in theory, have as many elements as there are pixels. To reduce the size of the vector the granulometry can be truncated at an area above which there is considered to be no textural information. This approach is computationally simpler than the scale space sampling of [1]. In addition, as the resulting pattern spectrum contains a limited number of elements it can be used directly to provide a feature vector for classification purposes, instead of indirectly through pattern spectra moments [7].

3.   Experimental Results

A simulated image created from five natural textures from the Brodatz album [4] (see figure 1) was used to quantify the performance of the new local area granulometry texture segmentation technique. To provide a set of comparative results the test image was first segmented using binary and greyscale versions of the local granulometry segmentation scheme described in [7, 6]. The binary test image was created by thresholding figure 1 so that equal numbers of pixels were assigned to each value. For both the binary and greyscale methods five structuring elements, four linear (vertical, horizontal, positive diagonal and negative diagonal) and one circular, were used to create local size distributions using a 33×33 window. The first 3 central moments of the local pattern spectra were used as textural features, augmented by two additional features, $MaxLin$ and $Linearity$, giving a 17 element feature vector. The $MaxLin$ feature finds the maximum linear dimension regardless of direction while the $Linearity$ feature differentiates between regions of large linear components and circular components with similar diameters.

The database of codevectors used for the supervised classifier were obtained by generating the 17 feature images for homogeneous texture and then calculating the mean value of each feature image. The supervised segmentation

True	Classified as:				
class:	D77	D84	D55	D17	D24
D77	61.91%	24.73%	0.13%	5.34%	7.90%
D84	0.00%	97.44%	1.20%	0.00%	1.36%
D55	0.00%	18.86%	80.74%	0.00%	0.40%
D17	0.00%	26.18%	1.40%	59.30%	13.12%
D24	0.00%	2.67%	10.83%	0.00%	86.50%

Figure 2. Supervised segmentation result for Brodatz image using binary pattern spectra.

results achieved by the binary and greyscale schemes using a minimum distance classifier are shown in figures 2 and 3 respectively. Comparison of these results shows the improved performance of greyscale spectra. The overall correct classification, given by the mean value of the diagonal for each confusion matrix, rises from 77.18% (binary) to 83.15% (greyscale). However, many intra-region classification errors are still present resulting, in part, from the introduction of new features with increasing scale [1].

The classification result for the supervised area morphology local granulometry scheme is given in figure 4. Here, the maximum area size was set to 78 pixels, a value equal to the area of the largest structuring element in the previous results, and the local window size was 33×33 as before. The granulometry and anti-granulometry were found using (10) resulting in a 144 element local size density. The results are an improvement on the greyscale structuring element approach with an average correct classification of 89.95%.

In texture segmentation, adjacent pixels are highly probable to belong to the same texture class and the incorporation of spatial information in the feature space has been shown to reduce misclassifications in regions of homogeneous texture and at texture boundaries [9]. In practice, this can be achieved by including the spatial co-ordinates of each pixel as two extra features. As no a priori information about the co-ordinates of each texture class is available this approach is incompatible with a supervised segmentation scheme. Instead,

True	Classified as:				
class:	D77	D84	D55	D17	D24
D77	82.98%	3.94%	1.25%	11.26%	0.56%
D84	0.00%	92.06%	0.04%	6.25%	1.64%
D55	0.24%	37.66%	60.27%	1.32%	0.51%
D17	0.00%	12.12%	0.01%	87.27%	0.61%
D24	0.00%	5.07%	1.59%	0.17%	93.17%

Figure 3. Supervised segmentation result for Brodatz image using greyscale pattern spectra.

True	Classified as:				
class:	D77	D84	D55	D17	D24
D77	96.56%	0.65%	0.00%	0.79%	2.00%
D84	0.32%	92.37%	3.37%	0.99%	2.95%
D55	0.69%	9.03%	79.05%	6.77%	4.46%
D17	6.84%	3.62%	2.18%	86.15%	1.20%
D24	1.10%	0.50%	2.79%	0.00%	95.62%

Figure 4.   Supervised segmentation result for Brodatz image using local area pattern spectra.

a k-means clustering algorithm can be used to provide an unsupervised segmentation. The result achieved by this approach is presented in figure 5 and achieves an overall correct classification of 93.26%. Inspection of the figure shows that intra-region classification errors are virtually eliminated with the only misclassifications occurring at region boundaries.

To provide a more rigourous test the local area morphology granulometry segmentation method was applied to a number of images from the Outex textural database [14]. Each entry in the database is a 512×512 image of 5 compound textures. An example of one such image and the ground truth class map used by all images in the database are shown in figure 6. As these test images possess irregular texture boundaries they provide a challenging segmentation problem. Figure 7 shows a typical segmentation for one of the Outex images and an averaged confusion matrix from 3 segmentation results, which has an overall classification accuracy of 93.78%. Further examination of the confusion matrix reveals that the majority of errors occur from the misclassification of pixels in the central texture. For comparison, an unsupervised greyscale segmentation scheme employing structuring elements only classified 69.34% of pixels correctly, while the unsupervised Gabor filterbank technique of [9] achieved an average classification accuracy of 95.82%.

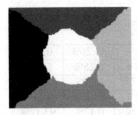

True	Classified as:				
class:	D77	D84	D55	D17	D24
D77	99.94%	0.06%	0.00%	0.00%	0.00%
D84	5.45%	88.34%	0.00%	2.49%	3.73%
D55	5.81%	0.00%	89.35%	3.20%	1.64%
D17	0.00%	0.03%	2.03%	97.94%	0.00%
D24	4.97%	0.71%	1.38%	2.20%	90.74%

Figure 5.   Unsupervised segmentation result for Brodatz image using local area pattern spectra.

Figure 6. Outex test image and ground truth of 5 classes shown with increasing greyscale.

4. Discussion and Conclusions

A new approach to morphological texture segmentation using area morphology local granulometries has been presented that combines the advantages of area morphology with a local granulometry approach. Pixel-wise classification is achieved using a feature vector consisting of a local area morphology size density. Segmentation results on a simulated image of Brodatz textures show the supervised scheme to out-perform supervised segmentation schemes based on fixed structuring elements.

An unsupervised area granulometry texture segmentation scheme has also been presented that produces improved results by virtue of incorporating the spatial co-ordinates as additional feature vector elements. Results show that this unsupervised technique produces a better performance than the other morphological segmentation methods in terms of overall correct classifications and also eliminates the training stage required by supervised schemes.

Application of the unsupervised scheme to images from the Outex texture database produced an overall classification performance that significantly improves on equivalent schemes employing structuring elements and is comparably with that of a Gabor segmentation scheme. One difference between the local area granulometries approach and those based on Gabor filtering and wavelets [17] is that the former currently calculates its distributions within

True	Classified as class:				
class:	1	2	3	4	5
1	98.57%	0.65%	0.72%	0.00%	0.05%
2	1.58%	98.09%	0.00%	0.23%	0.10%
3	2.83%	0.00%	95.21%	1.50%	0.45%
4	0.00%	3.10%	3.55%	93.01%	0.35%
5	4.91%	3.73%	4.15%	3.19%	84.02%

Figure 7. Unsupervised segmentation result for Outex images using local area pattern spectra. Example segmentation (left) and average confusion matrix for 3 images (right).

a fixed-sized local window, while the others increase their window size with scale. This may explain why the results achieved are only comparable with existing linear methods despite the advantages of the morphological approach. The incorporation of variable window size within our scheme is an area on ongoing research, as is the automatic selection of scales for the feature vector.

References

[1] S.T. Acton and D.P. Mukherjee. Scale space classification using area morphology. *IEEE Transactions on Image Processing*, 9(4):623–635, April 2000.

[2] S. Baeg, A.T. Popov, V.G. Kamat, S. Batman, K. Sivakumar, N. Kehtarnavaz, E.R. Dougherty, and R.B. Shah. Segmentation of mammograms into distinct morphological texture regions. In *IEEE Symp. Computer Based Medical Systems*, pages 20–25, 1998.

[3] J.A. Bangham, R. Harvey, P.D. Ling, and R.V. Aldridge. Morphological scale-space preserving transforms in many dimensions. *Journal of Electronic Imaging*, 5(3):283–299, July 1996.

[4] P. Brodatz. *A Photographic Album for Artists and Designers*. Dover, New York, 1966.

[5] F. Cheng and A.N. Venetsanopoulos. An adaptive morphological filter for image processing. *IEEE Transactions on Image Processing*, 1(4):533–539, 1992.

[6] E.R. Dougherty and R.A. Lotufo. *Hands-on Morphological Image Processing*. SPIE Press, 2003.

[7] E.R. Dougherty, J.T. Newell, and J.B. Pelz. Morphological texture-based maximum likelihood pixel classification based on local granulometric moments. *Pattern Recognition*, 25(10):1181–1198, 1992.

[8] R.M. Haralick. Statistical and structural approaches to texture. *Proceedings of the IEEE*, 67(5):786–804, May 1979.

[9] A.K. Jain and F. Farrokhnia. Unsupervised texture segmentation using Gabor filters. *Pattern Recognition*, 24(12):1167–1186, 1991.

[10] T. Kohonen. The self-organizing map. *Proceedings of the IEEE*, 78(9):1464–1480, 1990.

[11] P. Maragos. Pattern spectrum and multiscale shape representation. *IEEE Transactions on Pattern Analysis and Machine Intelligence*, 11(7):701–716, 1989.

[12] G. Matheron. *Randon Sets and Integral Geometry*. Wiley, 1975.

[13] A. Meijster and M.H.F. Wilkinson. A comparison of algorithms for connected set openings and closings. *IEEE Transactions on Pattern Analysis and Machine Intelligence*, 24(4):484–494, April 2002.

[14] T. Ojala, T. Mäenpää, M. Pietikäinen, J. Viertola, J. Kyllonen, and S. Huovinen. Outex - new framework for empirical evaluation of texture analysis algorithms. In *Int. Conf. on Pattern Recognition*, volume 1, pages 701–706, Canada, 2002.

[15] P. Salembier and J. Serra. Flat zones filtering, connected operators, and filters by reconstruction. *IEEE Transactions on Image Processing*, 4(8):1153–1160, August 1995.

[16] J. Serra. *Image Analysis and Mathematical Morphology*. Academic Press, 1982.

[17] M. Unser. Texture classification and segmentation using wavelet frames. *IEEE Transactions on Image Processing*, 4(11):1549–1560, 1995.

[18] L. Vincent. Morphological area openings and closings for grey-scale images. In *Shape in Picture: Mathematical Description of Shape in Grey-level Images*, pages 196–208, 1993.

ILLUMINATION-INVARIANT MORPHOLOGICAL TEXTURE CLASSIFICATION

Allan Hanbury,[1] Umasankar Kandaswamy[2] and Donald A. Adjeroh[2]

[1]*PRIP group, Institute of Computer-Aided Automation,*
Vienna University of Technology, Favoritenstraße 9/1832, A-1040 Vienna, Austria
hanbury@prip.tuwien.ac.at

[2]*Lane Department of Computer Science and Electrical Engineering*
West Virginia University, Morgantown, WV 26505, USA
(umasank, adjeroh)@csee.wvu.edu

Abstract   We investigate the use of the standard morphological texture characterisation methods, the granulometry and the variogram, in the task of texture classification. These methods are applied to both colour and greyscale texture images. We also introduce a method for minimising the effect of different illumination conditions and show that its use leads to improved classification. The classification experiments are performed on the publically available Outex 14 texture database. We show that using the illumination invariant variogram features leads to a significant improvement in classification performance compared to the best results reported for this database.

Keywords: Mathematical morphology, texture, variogram, granulometry, illumination invariance

1.   Introduction

The principal tools in the morphological texture analysis toolbox are the variogram, which is a generalisation of the covariance, and the granulometry [19, 20]. These have been used successfully in a number of applications [20]. It is nevertheless desirable to place these tools in the context of current research on texture analysis methods. To this end, we first discuss how they fit into the framework of structural and perceptual properties of texture. Then we compare their performance to that of the best reported method using a standard benchmark. Additionally, an approach to solving the problem of computing illumination-invariant texture features is presented. By illumination invariance we mean that the feature vector describing a texture should be independent of the illumination conditions in which the texture image is captured.

C. Ronse et al. (eds.), Mathematical Morphology: 40 Years On, 377–386.

There has recently been much effort at comparing the performance of texture feature calculation methods on standard publically-available databases, such as the Outex databases [13]. This is done for tasks such as texture classification and texture segmentation. Classification results exist, for example, for the Local Binary Pattern (LBP) [14] and the Gabor filter approaches [12]. In this paper, we compare the performance of the standard morphological texture description methods for the task of texture classification.

Texture analysis tools have mostly been applied to greyscale images. Colour textures have however received much attention recently, with many greyscale texture analysis methods being extended to colour images. There are three main approaches to the analysis of a colour texture [15]:

Parallel approach: Colour and texture information is processed separately. The global colour is characterised, usually by means of a colour histogram. The intensity is used with greyscale texture descriptors to characterise the texture.

Sequential approach: Colour information is processed first to create an image labelled by scalars. Greyscale texture algorithms are then applied to this labelled image.

Integrative approach: This can be divided into single- and multi-channel strategies. Single-channel strategies apply greyscale texture analysis algorithms to each colour channel separately, while multi-channel strategies handle two or more channels simultaneously.

Many greyscale texture description techniques have been recast in the integrative framework: cooccurrence matrices [1, 15], run length [5] and Gabor filtering [16]. There is however no agreement yet as to whether the integrative approach functions better [15] or worse [12] than the parallel approach.

We begin with a brief overview of the morphological texture description methods (Section 2), and relate these to the perceptual properties of texture in Section 3. Our proposed transformation allowing illumination invariant classification of textures is presented in Section 4. The experimental setup, texture features used and results are presented in Sections 5, 6 and 7 respectively.

2.   Morphological texture processing

We briefly summarise the variogram and granulometry as well as their extensions to colour textures.

Variogram

The variogram is a notion which generalises the covariance [19]. We make use of it here as it is easier to generalise to colour images than the covariance.

To calculate a variogram of an image $f(\mathbf{x})$, a direction α and a unit displacement vector $\hat{\mathbf{h}}$ in this direction must be chosen. For various multiples of vector $\hat{\mathbf{h}}$, written $q\hat{\mathbf{h}}$, the following value

$$V(q, \alpha) = \frac{1}{2}\mathcal{E}\left[f(\mathbf{x}) - f_\alpha\left(\mathbf{x} + q\hat{\mathbf{h}}\right)\right]^2 \qquad (1)$$

is plotted against q, where $f_\alpha\left(\mathbf{x} + q\hat{\mathbf{h}}\right)$ is the displacement of image f in direction α by distance q. The expectation value $\mathcal{E}$ of the greyscale differences squared is calculated only in the region in which the original and displaced images overlap.

The generalisation of the variogram to colour images was suggested by Lafon et al. [10]. It is an integrative multi-channel strategy in which the difference in Equation 1 is replaced by the Euclidean distance in the CIELAB colour space. The CIELAB space was designed such that this distance corresponds to the perceptual difference between two colours expressed in CIELAB coordinates. We also use the Euclidean distance in the RGB space.

Granulometry

The granulometry in materials science, which is used to characterise granular materials by passing them through sieves of increasing mesh size while measuring the mass retained by each sieve, is transposed to image data by opening the image with a family of openings γ_λ of increasing size λ [20]. The mass is replaced by the image volume Vol, i.e. the sum of the pixel values. The normalised granulometric curve of an image f is a plot of $\mathrm{Vol}\left[\gamma_\lambda(f)\right]/\mathrm{Vol}(f)$ versus λ. The most useful structuring elements are discs and lines. Negative values of λ are interpreted as a closing with a structuring element of size λ.

Due to the extremely large number of ways of applying an opening to a colour image [7], we decided to use an integrative single-channel strategy and apply the granulometry to each channel of the image in the RGB and CIELAB spaces.

3. Structural properties of texture

Texture has been characterised in terms of two sets of properties: spatial relation properties and perceptual properties. Rao [17] developed a taxonomy of textures based on their spatial relations. He defined the following four texture classes, examples of which are shown in Figure 1:

Strongly ordered: Textures made up of a specific placement of primitive elements, or of a distribution of a class of elements.

Weakly ordered: Textures exhibiting a certain level of specificity of orientation at each position.

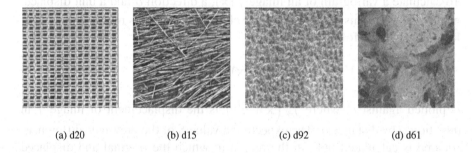

(a) d20   (b) d15   (c) d92   (d) d61

Figure 1.   Examples of the four texture classes defined by Rao: (a) Strongly ordered. (b) Weakly ordered. (c) Disordered. (d) Compositional. The reference number of each texture in the Brodatz album [2] is shown below each image.

Disordered: These textures are not oriented or repetitive, and could be described based on their roughness.

Compositional: Textures which do not fit completely into one of the above three texture classes, but can be described as a combination of them.

Perceptual properties are those which humans take into account when looking at texture. Rao and Lohse [18] performed an experiment designed to find the high-level texture characteristics which are important in the perception of texture by humans. They asked 20 people to each classify 30 greyscale textures from the Brodatz album [2] into an unspecified number of classes, and then to group similar classes so as to create a tree of texture similarity. Using an analysis by multidimensional scaling, the authors determined that the most important characteristic is regularity, followed by the degree of orientation specificity and lastly descriptional complexity. Mojsilović et al. [11] did a similar experiment in which 28 people were asked to specify numerically the similarity between every combination of 20 colour textures obtained from a textile catalogue. Using multidimensional scaling, they identified five important perceptual characteristics, of which two are specific to colour textures and three correspond to those of Rao and Lohse. They are, in order of decreasing importance: overall colour, directionality and orientation, regularity and placement, colour purity and complexity and heaviness. Chetverikov [3] relates orientation specificity to anisotropy and points out that regularity and anisotropy are hierarchical — a periodic texture is always anisotropic as characteristic orientations are defined by the directions of the periodicity.

The morphological texture descriptors described above are both well adapted to describing regularity and orientation specificity. The variogram of a regular texture should have a visible periodicity. An orientation specificity will be characterised by different variograms for different directions. The granulometry by discs is well suited to identifying regularity — regular structures should have the same size and/or be separated by structures having the same size. Orientation specificity can be detected by computing granulometries using linear structuring elements in a number of directions. The granulometry should also be able to extract useful information from textures which are not periodic and not orientation specific, such as those made up of a random distribution of differently sized grains.

4. Illumination invariance

Even though colour is the most appealing feature [11], it is also the most vulnerable image indexing feature when the illumination under which image content is captured is varied. The problem of illumination variance has been addressed for the last few years, but there is no solid model for textures that provides illumination invariant features. It is obvious that a change in illumination does not change the texture content and hence should not change the texture features. However, the majority of texture analysing algorithms are sensitive to illumination changes. Even in conjunction with popular colour constancy normalization schemes (such as those proposed by Funt, Chatterjee and Finlayson [4]), colour histogram based classification methods and integrative single channel texture analyzing strategies fail to perform well for the Outex 14 database [13]. Overall, the average classification score under illumination varying conditions drops by 25% compared to constant illumination conditions [12].

In this work we propose to use Minvariance Model [9], in which the change in the pixel values is modelled as a function of relative change in two interdependent variables. In general using Lambert's Law, one can determine the resultant colour coefficients using the following equation:

$$\vec{\rho} = \int_w S(\lambda)E(\lambda)\mathcal{F}(\lambda)d\lambda \qquad (2)$$

where λ is wavelength, $\vec{\rho}$ is the sensor responses (the resultant *RGB* pixel values), $\mathcal{F}$ is the visual apparatus response function, E is the incident illumination and S is the surface reflectance at location x on the texture surface. Under the assumption that local surface reflectance is a constant at location x, the *RGB* pixel values can be expressed as a product function of sensor response function and illumination. We rewrite Equation 2 as a continuous function Ψ of two functions Ω and χ that are implicitly connected by the third variable λ,

$$\vec{\rho} = \Psi(\Omega, \chi) \qquad (3)$$

$$\Omega = \mathcal{F}(\lambda), \chi = E(\lambda) \qquad (4)$$

Figure 2.   An example of each of the 68 Outex 14 textures (from [12]).

Using the chain rule of differentiation, we can write the total differential coefficient of $\vec{\rho}$ as

$$d\vec{\rho} = \frac{\partial \vec{\rho}}{\partial \Omega} \cdot d\Omega + \frac{\partial \vec{\rho}}{\partial \chi} \cdot d\chi \tag{5}$$

given the condition $\triangle\lambda \to 0$, $\triangle\Omega \to 0$, and $\triangle\chi \to 0$, where $\triangle$ indicates the change in the corresponding parameter. We obtain the following condition for illumination invariant pixel values by equating the above equation to zero. That is

$$d\Omega = \mathcal{T}(\vec{\rho})^{-1} \cdot d\chi \tag{6}$$

where $\mathcal{T}$ denotes the ratio between partial derivaties, i.e. the ratio between the partial change in pixel values due to a change in illumination and the partial change in pixel values due to a change in sensor sensitivity. In practice, a histogram stretching technique [6] is used to achieve the above condition on colour textures. A detailed description of the Minvariance Model can be found in [9]. A similar approach to illumination invariance using histogram equalisation has been suggested in [8]. We have derived a similar technique, but using a different theoretical model.

5.   Experimental setup

Classification experiments were performed on the Outex 14 texture database [13], which contains 68 textures. An example of each texture is shown in

Figure 2. The training set was obtained by acquiring a 100dpi image of size 746×538 pixels of each texture illuminated by a 2856K incandescent CIE A light source. Each image was then divided into 20 non-overlapping sub-images, each of size 128×128 pixels, producing 1360 training images. The test set is made up of differently illuminated samples of the same textures, once again with 20 sub-images per texture. The illumination sources used are a 2300K horizon sunlight and a 4000K fluorescent TL84. For each illumination source, 1360 images are available, making a total of 2720 test images. The textures have the same rotations under the three light sources.

6. Texture features and classification

Granulometry and Variogram texture feature vectors were calculated on three images: colour images represented in the RGB (RGB) and CIELAB (Lab) colour spaces, and a greyscale image containing luminance values (L), the latter being the L^* values in the CIELAB space. To transform the RGB images into the CIELAB space, the first step was done using the RGB to XYZ transformation matrix calibrated to the CIE A white point given in [12].

Four variograms are calculated for each image in directions $\alpha = 0°, 45°, 90°$ and $135°$ for values of $q = 1, 2, 3, \ldots, 50$ using Equation 1. These four variograms are then concatenated to form the variogram feature vector. For the greyscale image L, Equation 1 is used directly. For the colour images RGB and Lab, the subtraction in Equation 1 is replaced by the Euclidean distance. The feature vectors for both greyscale and colour images therefore contain 200 features.

The granulometry feature vector is the granulometry curve using linear structuring elements in four directions, as for the variograms. We use the convention that a linear structuring element of size ℓ has a length of $2\ell + 1$ pixels. For a granulometric curve in a single direction, the structuring elements range in size from -25 (i.e. a closing with a linear structuring element of length 49 pixels) to size 25 in steps of size 2. For the greyscale image L, the feature vector therefore contains 104 features. For the colour images RGB and Lab, the granulometry curves for each channel were concatenated to form a feature vector containing 312 features.

Computing a variogram feature vector on a 128×128 image requires on average 0.68 seconds for a greyscale image and 1.16 seconds for a colour image. A granulometry feature vector requires on average 10.7438 seconds for a colour image. The experiments were done in a MATLAB 6.1 environment, using a personal computer with an AMD $Athlon^{[TM]}$ MP 1800$^+$ 1.5GHz processor and 1GB RAM.

Textures were classified using a kNN classifier, in which the distance between feature vectors was calculated using the Kullback-Leibler distance,

Table 1. Classification scores for Outex 14. The methods are the granulometry $\mathcal{G}$ and Variogram $\mathcal{V}$. The subscript indicates the colour space used.

Methods	No minvariance			With minvariance		
	TL84	horizon	Average	TL84	horizon	Average
$\mathcal{G}_{RGB}$	41.54	47.13	**44.34**	60.44	65.59	**63.02**
$\mathcal{G}_{Lab}$	37.43	56.10	**46.77**	69.04	72.13	**70.56**
$\mathcal{G}_L$	16.40	22.72	**19.56**	24.41	19.41	**21.91**
$\mathcal{V}_{RGB}$	73.46	65.66	**69.56**	65.59	60.44	**63.02**
$\mathcal{V}_{Lab}$	65.76	73.75	**69.76**	74.12	55.22	**64.67**
$\mathcal{V}_L$	70.07	73.01	**71.54**	77.35	78.82	**78.09**
Best result from [12]			**69.5**			

which is well suited to comparing probability distributions. The Kullback-Leibler distance between two vectors **p** and **q** having components p_k and q_k is:

$$d = \sum_k p_k \log_2 \frac{p_k}{q_k} \qquad (7)$$

We used a value of $k = 3$. For the results in [12], with which we are comparing our results, a value of $k = 1$ was used. However, for the experiments in [12], the classifier was trained and tested using only half of the images in the benchmark database. As we are using the full database, we use $k = 3$ to make the comparison more fair. The classification performance is measured as the percentage of test set images classified into the correct texture class.

7. Results

The results of the texture classification experiments are shown in Table 1. The left part of the table shows the results without illumination invariance and the right part with illumination invariance. For each feature extraction method, the classification performance on the two halves of the test database corresponding to different illuminants (TL84 and horizon) are shown. Finally, the average of these two values is shown in bold. For comparison, the best result of classification on the Outex 14 database, reported in [12], is given. This value of 69.5% was obtained using the Local Binary Pattern (LBP) texture features.

The granulometry features for the colour images gave consistently better classification rates than those for the greyscale images. Furthermore, the use of the illumination invariant features with the granulometry improved the classification significantly for the colour images, but little for the greyscale images.

The variogram applied to the non-illumination invariant images produced superior results to the granulometry. These results all have the same magnitude as the best result by the LBP method [12]. The use of the illumination invariant features had the undesired effect of lowering the classification rate for the colour images, but raising it significantly for the greyscale images. It can be assumed that the transformation used to produce the illumination invariance distorts the distances in the RGB or CIELAB spaces which are used in the colour version of the variogram, thereby leading to lower classification rates, but this remains to be investigated further. The use of illumination invariant features for the greyscale images leads to a significant improvement. The best result of 78.09% is obtained by applying the variogram to the illumination invariant luminance images. This is an improvement of 8.6% with respect to the best LBP result.

8. Conclusion

We have discussed the standard morphological texture characterisation tools, the variogram and granulometry, in terms of the structural and perceptual properties of texture, and compared their performance for texture classification. In general, it is can be seen that classification using feature vectors based on the variogram performs better than that using granulometric curves. On average, the use of our proposed illumination invariant features improves the classification for all features except for the colour variogram features. The illumination invariant variogram features applied to the luminance image results in a significant improvement in classification performance compared to the best performance reported in the literature.

Due to the theme of the conference, we have concentrated on comparing classification results using the standard morphological approaches, and have shown that the use of illumination invariant features can lead to a significant improvement in classification. It still remains, of course, to investigate the improvement in the classification results when using the LBP with our proposed illumination invariant features.

The fact that the best classification performance is obtained for greyscale images supports the assertion of Mäenpää et al. [12] that colour information usually only improves texture classification marginally. However, the performance of the texture classification using the colour granulometry features is better than that using the greyscale features, which counters this assertion. We therefore also plan to investigate if the application of the variogram in an integrative single-channel way (i.e. concatenating the variogram of each colour channel to produce the feature vector, as done for the granulometry features) leads to an improvement over the integrative multi-channel strategy used.

Acknowledgments

This work was supported by the Austrian Science Foundation (FWF) under grant P17189-N04, and the European Union Network of Excellence MUSCLE (FP6-507752).

References

[1] V. Arvis, C. Debain, M. Berducat, and A. Benassi. Generalisation of the cooccurrence matrix for colour images: Application to colour texture classification. *Image Analysis and Stereology*, 23(1):63–72, 2004.

[2] P. Brodatz. *Textures: a photographic album for artists and designers*. Dover, 1966.

[3] D. Chetverikov. Fundamental structural features in the visual world. In *Proceedings of the International Workshop on Fundamental Structural Properties in Image and Pattern Analysis*, pages 47–58, 1999.

[4] G. Finlayson, S. Chatterjee, and B. Funt. Color angular indexing. *The Fourth European Conference on Computer Vision, European Vision Society*, II:16–25, 1996.

[5] I. Foucherot, P. Gouton, J. C. Devaux, and F. Truchetet. New methods for analysing colour texture based on the Karhunen-Loeve transform and quantification. *Pattern Recognition*, 37:1661–1674, 2004.

[6] R. Gonzalez and R. Woods. *Digital Image Processing*. Peason Education, Inc, 2002.

[7] A. Hanbury. Mathematical morphology applied to circular data. In P. Hawkes, editor, *Advances in Imaging and Electron Physics*, volume 128, pages 123–204. Academic Press, 2003.

[8] S. D. Hordley, G. D. Finlayson, G. Schaefer, and G. Y. Tian. Illuminant and device invriant colour using histogram equalisation. Technical Report SYS-C02-16, University of East Anglia, 2002.

[9] U. Kandaswamy, A. Hanbury, and D. Adjeroh. Illumination minvariant color texture descriptors. Manuscript in preparation.

[10] D. Lafon and T. Ramananantoandro. Color images. *Image Analysis and Stereology*, 21(Suppl 1):S61–S74, 2002.

[11] A. Mojsilović, J. Kovačević, D. Kall, R. J. Safranek, and S. K. Ganapathy. The vocabulary and grammar of color patterns. *IEEE Trans. on Image Processing*, 9(3):417–431, 2000.

[12] T. Mäenpää and M. Pietikäinen. Classification with color and texture: jointly or separately? *Pattern Recognition*, 37:1629–1640, 2004.

[13] T. Ojala, T. Mäenpää, M. Pietikäinen, J. Viertola, J. Kyllönen, and S. Huovinen. Outex – new framework for empirical evaluation of texture analysis algorithms. In *Proceedings of the 16th ICPR*, volume 1, pages 701–706, 2002.

[14] T. Ojala, M. Pietikäinen, and T. Mäenpää. Multiresolution gray-scale and rotation invariant texture classification with local binary patterns. *IEEE Trans. on Pattern Analysis and Machine Intelligence*, 24(7):971–987, 2002.

[15] C. Palm. Color texture classification by integrative co-occurrence matrices. *Pattern Recognition*, 37:965–976, 2004.

[16] C. Palm and T. M. Lehmann. Classification of color textures by gabor filtering. *Machine Graphics and Vision*, 11(2/3):195–219, 2002.

[17] A. R. Rao. *A Taxonomy for Texture Description and Identification*. Springer-Verlag, 1990.

[18] A. R. Rao and G. L. Lohse. Identifying high level features of texture perception. *CVGIP: Graphical Models and Image Processing*, 55(3):218–233, 1993.

[19] J. Serra. *Image Analysis and Mathematical Morphology*. Academic Press, London, 1982.

[20] P. Soille. *Morphological Image Analysis*. Springer, second edition, 2002.

UNIFIED MORPHOLOGICAL COLOR PROCESSING FRAMEWORK IN A LUM/SAT/HUE REPRESENTATION

Jesús Angulo

Centre de Morphologie Mathématique - Ecole des Mines de Paris
35 rue Saint Honoré - 77300 Fontainebleau - France
angulo@cmm.ensmp.fr

Abstract   The extension of lattice based operators to color images is still a challenging task in mathematical morphology. The first choice of a well-defined color space is crucial and we propose to work on a lum/sat/hue representation in norm L_1. We then introduce an unified framework to consider different ways of defining morphological color operators either using the classical formulation with total orderings by means of lexicographic cascades or developing new transformations which takes advantage of an adaptive combination of the chromatic and the achromatic (or the spectral and the spatio-geometric) components. More precisely, we prove that the presented saturation-controlled operators cope satisfactorily with the complexity of color images. Experimental results illustrate the performance and the potential applications of the new algorithms.

Keywords: color mathematical morphology, luminance/saturation/hue, lexicographic orderings, reconstruction, gradient, top-hat, leveling, segmentation

1.   Introduction

Mathematical morphology is the application of lattice theory to spatial structures [16] (i.e. the definition of morphological operators needs a totally ordered complete lattice structure). Therefore the extension of mathematical morphology to color images is difficult due to the vectorial nature of the color data. Fundamental references to works which have formalized the vector morphology theory are [17] and [8].

Here we propose here a unified framework to consider different ways of defining morphological color operators in a luminance, saturation and hue color representation. This paper is a summary of the Ph. D. Thesis of the author [1] done under the supervision of Prof. Jean Serra (full details of the algorithms and many other examples can be found in [1]).

C. Ronse et al. (eds.), Mathematical Morphology: 40 Years On, 387–396.
©2005 *Springer. Printed in the Netherlands.*

2.   Luminance/Saturation/Hue color in norm L_1

The primary question to deal with color images involves choosing a suitable color space representation for morphological processing. The RGB color representation has some drawbacks: components are strongly correlated, lack of human interpretation, non uniformity, etc. A polar representation with the variables luminance, saturation et hue (lum/sat/hue) allows us to solve these problems. The HLS system is the most popular lum/sat/hue triplet. In spite of its popularity, the HLS representation often yields unsatisfactory results, for quantitative processing at least, because its luminance and saturation expressions are not norms, so average values, or distances, are falsified. In addition, these two components are not independent, which is not appropriate for a vector decomposition. The reader can find a comprehensive analysis of this question by Serra [20]. The drawbacks of the HLS system can be overcome by various alternative representations, according to different norms used to define the luminance and the saturation. The L_1 norm system has already been introduced in [18] as follows:

$$
\begin{cases}
l = \frac{1}{3}\left(max + med + min\right) \\
s = \begin{cases} \frac{3}{2}\left(max - l\right) & if \ l \geq med \\ \frac{3}{2}\left(l - min\right) & if \ l \leq med \end{cases} \\
h = k\left[\lambda + \frac{1}{2} - (-1)^\lambda \left(\frac{max+min-2med}{2s}\right)\right]
\end{cases}
\tag{1}
$$

where max, med and min refer the maximum, the median and the minimum of the RGB color point (r, g, b), k is the angle unit ($\pi/3$ for radians and 42 to work on 256 grey levels) and $\lambda = 0$, if $r > g \geq b$; 1, if $g \geq r > b$; 2, if $g > b \geq r$; 3, if $b \geq g > r$; 4, if $b > r \geq g$; 5, if $r \geq b > b$ allows to change to the color sector. In all processing that follows, the l, s and h components are always those of the system (1), named LSH representation.

3.   Morphological color operators from LSH

For detailed exposition on complete lattice theory refer to [7]. Let E, T be nonempty set. We denote by $\mathcal{F}(E, T)$ the power set T^E, i.e., the functions from E onto T. If T is a complete lattice, then $\mathcal{F}(E, T)$ is a complete lattice too. Let f be a grey level image, $f : E \rightarrow T$, in this case $T = \{t_{min}, t_{min} + 1, \cdots, t_{max}\}$ is an ordered set of grey-levels. Given the three sets T^l, T^s, T^h, we denote by $\mathcal{F}(E, [T^l \otimes T^s \otimes T^h])$ or $\mathcal{F}(E, T^{lsh})$ all color images in a LSH representation (T^{lsh} is the product of T^l, T^s, T^h, i.e., $c_i \in T^{lsh} \Leftrightarrow c_i = \{(l_i, s_i, h_i); l_i \in T^l, s_i \in T^s, h_i \in T^h\}$). We denote the elements of $\mathcal{F}(E, T^{lsh})$ by $\mathbf{f}$, where $\mathbf{f} = (f_L, f_S, f_H)$ are the color component functions. Using this representation, the value of $\mathbf{f}$ at a point $x \in E$, which lies in T^{lsh}, is denoted by $\mathbf{f}(x) = (f_L(x), f_S(x), f_H(x))$. Note that the sets T^l, T^s corresponding to the luminance and the saturation are complete totally ordered lattices. The hue component is an angular function defined on the unit circle, $\mathcal{T}^h = \mathcal{C}$, which has no partial ordering. Let $a : E \to \mathcal{C}$ be an angular function, the angular difference [15, 9] is defined as $a_i \div a_j =| a_i - a_j |$ if $| a_i - a_j | \leq 180°$ or $a_i \div a_j = 360° - | a_i - a_j |$ if $| a_i - a_j | > 180°$. It is possible to fix an origin on $\mathcal{C}$, denoted by h_0. We can now define a h_0-centered hue function by computing $f_H(x) \div h_0$. The function $(f_H \div h_0)(x)$ is an ordered set and therefore leads to a total complete ordered lattice denoted by $\mathcal{T}^{h \div h_0}$. We propose to distinguish two main classes of morphological color operators, the vector-to-vector operators or VV-operators and the vector-to-scalar operators or VS-operators. Let $\mathbf{f}, \mathbf{g} \in \mathcal{F}(E, \mathcal{T}^{lsh})$ be two color images in LSH color space and $h \in \mathcal{F}(E, \mathcal{T})$ a grey level image. An operator Ψ is called a VV-operator if $\Psi : \mathcal{T}^{lsh} \to \mathcal{T}^{lsh}$; $\mathbf{g} = \Psi(\mathbf{f})$. An operator Φ is called a VS-operator if $\Phi : \mathcal{T}^{lsh} \to \mathcal{T}$; $h = \Phi(\mathbf{f})$. In addition, a connective criterion $\sigma : \mathcal{F} \otimes \mathcal{P}(E) \to [0,1]$ can be applied to a color image $\mathbf{f}$ for segmenting and obtaining a partition D_σ (see Serra's segmentation theory [19]). For the sake of simplicity, we consider that a segmentation operator based on a connective criterion is a VS-operator. Different ordering relationships between vectors have been studied [5]. The marginal ordering or M-ordering is a partial ordering, based on the usual pointwise ordering (i.e., component by component). Another more interesting ordering is called conditional ordering or C-ordering, where the vectors are ordered by means of some marginal components sequentially selected according to different conditions. This is commonly named as lexicographic ordering which is a total ordering. Using a M-ordering for the elements of $\mathcal{F}(E, \mathcal{T}^{lsh})$ we can introduce color vector values in the transformed image that are not present in the input image (the problem of "false colors"). The application of a C-ordering in $\mathcal{F}(E, \mathcal{T}^{lsh})$ preserves the input color vectors. When dealing with operators for color images $\mathcal{F}(E, \mathcal{T}^{lsh})$ C-orderings are indicated to build VV-operators ($\mathbf{g} = \Psi(\mathbf{f})$), introducing no new colors [21], but can be also used for VS-operators ($h = \Phi(\mathbf{f})$). An inconvenient of the C-orderings (vectorial approach) is the computational complexity of the algorithms which leads to slow implementations. However, in practice, for many applications (e.g. segmentation and feature extraction) involving VS-operators, total orderings are not required as well as increment based operators (e.g. gradients and top-hats) can be defined in the unit circle $\mathcal{T}$ without fixing an origin on the hue component [9]. We consider therefore that M-orderings can be interesting for developing color operators. Let ψ_i be the mapping $\psi_i : \mathcal{T} \to \mathcal{T}$ an operator for grey level images (marginal operator). In general, a separable marginal operator is formalized by $h = \Xi(\Psi_1(f_L), \Psi_2(f_S), \Psi_3(f_H))$, where Ξ is a merging function (linear or non-linear) to combine the components. Obviously, although less useful, M-operators can be also applied to VV-operators (i.e., $\mathbf{g} = (\Psi_1(f_L), \Psi_2(f_S), \Psi_3(f_H))$).

4.   Total orderings using lexicographical cascades

Let $c_i = (u_i, v_i, w_i)$ and $c_j = (u_j, v_j, w_j)$ be two arbitrary color points, i.e., $c_i, c_j \in T^{lsh}$, where the generic components (u_k, v_k, w_k) are $f_L(x)$, $f_L(x)$ the and negative of the h_0-centered hue $(f_H(x) \div h_0)$ (the closest value $f_H(x)$ to h_0 must be the supremum) of the color image f at point x. The Ω-*lexicographical ordering* or $<_\Omega$ is defined as

$$c_i <_\Omega c_j \; if \; \begin{cases} u_i < u_j & or \\ u_i = u_j & and \; v_i < v_j \quad or \\ u_i = u_j & and \; v_i = v_j \quad and \; w_i < w_j \end{cases}$$

We denote the lexicographical cascade by Ω_{uvw}. In this case the priority is given to the component u, then to v and finally to w. Obviously, it is possible to define other orderings for imposing a dominant role to any other of the vector components. The drawback of this kind of orderings is that most of vector pairs are sorted by the chosen first component. There is a simple way in order to make Ω-ordering more flexible which involves the linear reduction of the dynamic margin of the first component, applying a division by a constant and rounding off, i.e., changing u by $\lceil u/\alpha \rceil$. It is named an $\alpha-modulus$ Ω-*lexicographical ordering*. The value for α controls the influence degree of the first component with regard to the others (above all the second one, since the cascade almost never reaches the third row).

We then define three main families of lexicographical orderings from the representation LSH: luminance-based $\Omega_{l|_\alpha s(h \div h_0)}$, saturation-based $\Omega_{s|_\alpha l(h \div h_0)}$ and hue-based $\Omega_{(h \div h_0)sl}$. The value of h_0 yields an important degree of freedom which allows us to act on a specific hue. A disadvantage of the hue-based ordering is its instability for the low saturation points (different solutions can be used which are based on a weighting of the hue by the saturation [1]). The C-ordering color morphology has been widely studied in the framework of lum/sat/hue representations such as we propose here (e.g. by Hanbury and Serra [10], by Ortiz *et al.* [14]), but most of works are very preliminary studies, being limited to the basic operators

Once these orders have been defined, the morphological color VV-operators are defined in the standard way. We limit our developments to the flat operators. The *color erosion* of an image f at pixel x by the structuring element B of size n is $\varepsilon_{\Omega,nB}(f)(x) = \{f(y) : f(y) = \inf_\Omega[f(z)], z \in n(B_x)\}$, where $\inf_\Omega$ is the infimum according to the lexicographical ordering Ω. The corresponding *color dilation* $\delta_{\Omega,nB}$ is obtained by replacing the $\inf_\Omega$ by the $\sup_\Omega$. A *color opening* $\gamma_{\Omega,nB}$ is an erosion followed by a dilation, and a *color closing* $\varphi_{\Omega,nB}$ is a dilation followed by an erosion. Once the color opening and closing are defined it is obvious how to extend other classical operators like the *alternate sequential filters* or the *granulometries*. Moreover, using a vectorial distance to calculate the difference point-by-point of two images $d(f, g)(x)$,

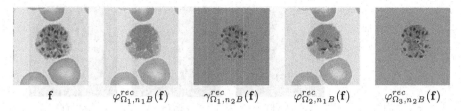

$$\mathbf{f} \qquad \varphi^{rec}_{\Omega_1, n_1 B}(\mathbf{f}) \qquad \gamma^{rec}_{\Omega_1, n_2 B}(\mathbf{f}) \qquad \varphi^{rec}_{\Omega_2, n_1 B}(\mathbf{f}) \qquad \varphi^{rec}_{\Omega_3, n_2 B}(\mathbf{f})$$

Figure 1. Detection of inclusions in erythrocytes (malaria diagnosis) using color openings/closings by reconstruction. The LSH lexicographical orderings are $\Omega_1 = \Omega_{ls(h \div h_0)}$, $\Omega_2 = \Omega_{(h \div h_0)sl}$ with $h_0 = 90$ (green-yellow, opposite to blue-purple), $\Omega_3 = \Omega_{(h \div h_0)sl}$ with $h_0 = 270$ (blue-purple) ; and where $n_1 = 15$, $n_2 = 200$, B is an unit square SE.

$d : \mathcal{T}^{lsh} \times \mathcal{T}^{lsh} \to \mathcal{T}$, $x \in E$, we can easily define the two most classical VS-operators: the *morphological gradient*, i.e., $\varrho_\Omega(\mathbf{f}) = d(\delta_{\Omega, B}(\mathbf{f}), \varepsilon_{\Omega, B}(\mathbf{f}))$, and the *top-hat transformation*, i.e., $\rho^+_{\Omega, nB}(\mathbf{f}) = d(\mathbf{f}, \gamma_{\Omega, nB}(\mathbf{f}))$. In addition, we propose also the extension of the operators "by reconstruction" implemented using the *color geodesic dilation* which is based on restricting the iterative dilation of a function marker $\mathbf{m}$ by B to a function reference $\mathbf{f}$ [22], i.e., $\delta^n_\Omega(\mathbf{m}, \mathbf{f}) = \delta^1_\Omega \delta^{n-1}_\Omega(\mathbf{m}, \mathbf{f})$, where $\delta^1_\Omega(\mathbf{m}, \mathbf{f}) = \delta_{\Omega, B}(\mathbf{m}) \wedge_\Omega \mathbf{f}$. The *color reconstruction* by dilation is defined by $\gamma^{rec}_\Omega(\mathbf{m}, \mathbf{f}) = \delta^i_\Omega(\mathbf{m}, \mathbf{f})$, such that $\delta^i_\Omega(\mathbf{m}, \mathbf{f}) = \delta^{i+1}_\Omega(\mathbf{m}, \mathbf{f})$ (idempotence). In a similar way the *color leveling* $\lambda_\Omega(\mathbf{m}, \mathbf{f})$ is computed by means of an iterative algorithm with geodesic dilations and geodesic erosions until idempotence [12].

In figure 1 an example of application for the detection of inclusions in red blood cells (parasites of the malaria *Plasmodium Vivax*) is given. The inclusions are two types of dark structures: blue-purple ones and brown others, which can be detected separately. Using openings/closings by reconstruction on $\Omega_{ls(h \div h_0)}$ ordering, all inclusions are removed/enhanced together whereas choosing the adequate h_0 angle, the hue-based ordering allows a more specific selection of the blue-purples ones. Note also that working on the hue-based $\Omega_{(h \div h_0)sl}$ ordering, it is possible to use a color closing to remove or to enhance the structures according to the hue origin (h_0 is the color of the structure or h_0 is the opposite on $\mathcal{C}$). Another example of color filtering is shown in figure 4: a color leveling using a luminance-based α-modulus ordering to simplify the texture/contours of the image [2] (the value of $\alpha = 10$ has shown to achieve robust and pleasant levelings) where the marker is a median filter of size 11×11.

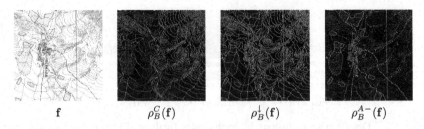

$$\mathbf{f} \qquad \rho_B^C(\mathbf{f}) \qquad \rho_B^{\downarrow}(\mathbf{f}) \qquad \rho_B^{A-}(\mathbf{f})$$

Figure 2.   Color top-hat's for detail extraction in cartographic image.

5.   Marginal orderings and merging by saturation-controlled operators

The saturation s is associated to the intensity of the hue h and has the intrinsic role of discrimination of the color points as chromatic (high s value) or achromatic (low s value). In this section, we discuss how to define marginal separable saturation-controlled VS-operators which cope satisfactorily with the complexity of color images. We propose also a hybrid VS- and VV-operator to filter adaptively color images. We suppose here f_S is normalized between 0 and 1.

Color top-hats for feature extraction

In the sense of Meyer [11], there are two versions of the top-hat for numerical functions ($f : E \rightarrow \mathcal{T}$). The *white top-hat* is the residue of the initial image f and an opening $\gamma_B(f)$, i.e. $\rho_B^+(f) = f - \gamma_B(f)$ (extracting bright structures) and the *black top-hat* is the residue of a closing $\varphi_B(f)$ and f, i.e. $\rho_B^-(f) = \varphi_B(f) - f$ (extracting dark structures). This numerical residue involves increments and hence can be defined to circular functions as the hue component. The circular centered top-hat [9] of an angular function is defined by $\rho_B^\circ(a(x)) = -\sup\{\inf[a(y) \div a(x), y \in B(x)]\}$ (extracting fast angular variations). Starting from these grey-level transformations, let us propose a series of definitions for the top-hat of a color image $\mathbf{f}$ from a LSH representation. The *chromatic top-hat* is given by $\rho_B^C(\mathbf{f}) = [f_S \times \rho_B^\circ(f_H)] \vee \rho_B^+(f_S)$. This operator extracts the fast variations of color regions on a saturated color background (i.e. saturated color peaks on uniform color regions) and the fast variations of saturated color regions on an achromatic (unsaturated) background (i.e. saturated color peaks on achromatic regions). The *white-achromatic top-hat* is the difference between the chromatic top-hat and the global bright top-hat, $\rho_B^{A+} = \rho_B^C - \rho_B^{\uparrow}$, where the *global bright top-hat* is calculated by $\rho_B^{\uparrow}(\mathbf{f}) = \rho_B^+(f_L) \vee \rho_B^-(f_S)$. It characterises the fast variations of bright regions (i.e. positive peaks of luminance) and the fast variations of achromatic regions

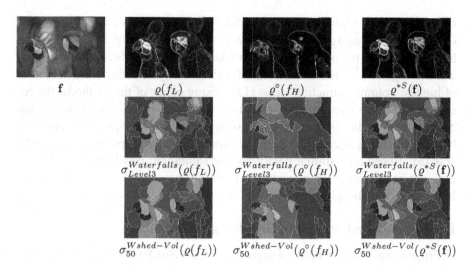

$$\mathbf{f} \qquad \varrho(f_L) \qquad \varrho^\circ(f_H) \qquad \varrho^{*S}(\mathbf{f})$$

$$\sigma^{Waterfalls}_{Level3}(\varrho(f_L)) \qquad \sigma^{Waterfalls}_{Level3}(\varrho^\circ(f_H)) \qquad \sigma^{Waterfalls}_{Level3}(\varrho^{*S}(\mathbf{f}))$$

$$\sigma^{Wshed-Vol}_{50}(\varrho(f_L)) \qquad \sigma^{Wshed-Vol}_{50}(\varrho^\circ(f_H)) \qquad \sigma^{Wshed-Vol}_{50}(\varrho^{*S}(\mathbf{f}))$$

Figure 3. Color gradients and segmentation by watershed transformation.

on a saturated background (i.e. unsaturated peaks: black, white and grey on color regions). The *black-achromatic top-hat* is the difference $\rho_B^{A-} = \rho_B^C - \rho_B^\downarrow$, where the *global dark top-hat* is obtained by $\rho_B^\downarrow(\mathbf{f}) = \rho_B^-(f_L) \vee \rho_B^-(f_S)$. Dually, it copes with the fast variations of dark regions (i.e. negative peaks of luminance) and the fast variations of achromatic regions on a saturated background. The term $\rho_B^-(f_S)$ appears in both $\rho_B^\uparrow$ and $\rho_B^\downarrow$ to achieve symmetrical definitions. Figure 2 shows the color top-hats of a cartographic image [4]. The extracted objects are different and certain kinds of details are better defined on one top-hat than on the other. Their contributions are consequently complementary.

Color gradient for segmentation

The *morphological gradient* by Beucher [6] is the numerical residue of a dilation and an erosion, i.e., $\varrho(f) = \delta_B(f) - \varepsilon_B(f)$ (where B is an unitary disk). In a similar way as for the top-hat, a version has been defined for the angular functions. The *circular centered gradient* is given by [9] $\varrho^\circ(a(x)) = \vee[a(x) \div a(y), y \in B(x)] - \wedge[a(x) \div a(y), y \in B(x)]$.

We introduce the *color gradient* from a LSH representation by means of the following barycentric merging function: $\varrho^{*S}(\mathbf{f}) = f_S \times \varrho^\circ(f_H) + (1 - f_S) \times \varrho(f_L)$. The *watershed transformation*, a pathwise connection, is one of the most powerful tools for segmenting images. Typically, the function to flood is a gradient function which yields the transitions between the regions. The color gradient $\varrho^{*S}(\mathbf{f})$ may therefore be used for segmenting color images.

Figure 3 depicts a comparative example of the partitions obtained by watershed algorithms using different gradients. The approach $\sigma_{Level3}^{Waterfalls}$ is the level 3 of a non-parametric pyramid of watershed (waterfalls algorithm [6]) and the $\sigma_{50}^{Wshed-Vol}$ is a marker-based watershed by selecting the 50 minima of highest volume extinction value [13]. Using either of the methods, the results obtained by means of the saturation weighting-based color gradient are better than working only on the luminance or on the hue gradient and even better than taking as color gradient the supremum of the three marginal gradients [3]. Based on a similar paradigm, the saturation, considered as a binary key, can be also used for merging the partitions associated to the hue and the luminance (see also [3]).

Regional-based color leveling for simplification

Finally, we shall introduce a regional-based color leveling algorithm. The rationale behind this technique is to work on two steps. First, to obtain a partition of the image $D_\sigma(\mathbf{f}) = \{R_i\}_{i=1}^n$ using the precedent color segmentation algorithm. Now, according the mean value of the saturation in each region R_i, the region is classified as chromatic or achromatic, and then, in the second step, each color image region $R_i(\mathbf{f})$ is independently leveled with $\lambda_{\Omega_{(h \div h_0)sl}}$ or $\lambda_{\Omega_{ls(h \div h_0)}}$ respectively (the marker is the median color of $R_i(\mathbf{f})$). In fact, this technique is an example of combination of a M-ordering operator (color gradient) followed by a C-ordering (leveling) adapted to the nature of the region. The results obtained by this filtering approach (see example in figure 4) yields very strong simplifications (in terms of color flat zones reduction) but keeping enough visual information. Consequently, it can be useful for region-based coding applications.

Figure 4 Morphological simplification of a color image. Comparison of a color levelling and a regional-based color levelling.

$\mathbf{f}$ $\lambda_\Omega(VMF_n(\mathbf{f}), \mathbf{f})$ $\lambda_{D_\sigma}(\mathbf{f})$

6.   Conclusions and perspectives

The extension of mathematical morphology operators to multi-valued functions is neither direct nor general, above all if the aim is to obtain useful transformations. We have focused on color images in order to develop specific well-adapted morphological color operators. To achieve that, we have proceeded in three steps. ▷1 Use of a color representation (system LSH in norm L_1) which yields: (i) a correct formalization from a mathematical viewpoint,

(ii) an intuitive interpretation of effects (as it is usual in mathematical morphology); ▷2 Explore the direct extension of morphological operators by using lexicographic orderings on the LSH system, proving the improvement for filtering applications when compared to the use of luminance only; ▷3 Introduce new marginal operators which take advantage of an adaptive combination of the chromatic and the achromatic (or the spectral and the spatio-geometric) components. Moreover, these separable mechanisms allow the application of classical grey level implementations with simple complexity elements to be added.

We can conclude that the dichotomy C-ordering vs. M-ordering for color operators can be integrated in an unified framework providing a wide range of operators. We have demonstrated by means of different applications on real images (biomedical microscopy, cartography, segmentation and coding in multimedia, etc.) the advantages of our new algorithms. We believe that the proposed methodology opens new possibilities for the application of mathematical morphology to color. Especially, we are working on three issues: (i) geodesic color reconstruction for specific object extraction, (ii) skeletons and thinnings of color objects, (iii) color granulometries.

References

[1] Angulo, J. (2003) *Morphologie mathématique et indexation d'images couleur. Application à la microscopie en biomédecine.* Ph.D. Thesis, Ecole des Mines, Paris, December 2003.

[2] Angulo, J. and Serra, J. (2003) "Morphological coding of color images by vector connected filters," in *Proc. of IEEE 7th International Symposium on Signal Processing and Its Applications (ISSPA'03)*, Vol. I, 69–72.

[3] Angulo, J. and Serra, J. (2003) "Color segmentation by ordered mergings," in *Proc. of IEEE International Conference on Image Processing (ICIP'03)*, Vol. 2, 125–128.

[4] Angulo, J. and Serra J. (2003) "Mathematical Morphology in Color Spaces Applied to the Analysis of Cartographic Images," in *Proc. of International Workshop Semantic Processing of Spatial Data (GEOPRO'03)*, p. 59–66, IPN-Mexico Ed.

[5] Barnett, V. (1976) "The ordering of multivariate data," *J. Statist. Soc. America A*, 139(3):318–354.

[6] Beucher, S. and Meyer, F. (1992) "The Morphological Approach to Segmentation: The Watershed Transformation," In *(Dougherty, Ed.) Mathematical Morphology in Image Processing*, Marcel-Dekker, 433–481.

[7] Birkhoff, G. (1967) *Lattice theory*. American Mathematical Society, Providence.

[8] Goutsias, J., Heijmans, H.J.A.M. and Sivakumar, K. (1995) "Morphological Operators for Image Sequences," *Computer Vision and Image Understanding*, 62(3): 326–346.

[9] Hanbury, A. and Serra, J. (2001) "Morphological operators on the unit circle," *IEEE Transactions on Image Processing*, 10: 1842–1850.

[10] Hanbury, A., Serra, J. (2001) "Mathematical morphology in the HLS colour space," *Proc. 12th British Machine Vision Conference (BMV'01)*, II-451–460.

[11] Meyer, F. (1977) "Constrast features extraction," In (Chermant Ed.) *Quantitative Analysis of Microstructures in Materials Science, Biology and Medecine*, Riederer Verlag, 374–380.

[12] Meyer, F. (1998) "The levelings," In (Heijmans and Roerdink Eds.), *Mathematical Morphology and its Applications to Image and Signal Processing,* Kluwer, 199–206.

[13] Meyer, F. (2001) "An Overview of Morphological Segmentation," *International Journal of Pattern Recognition and Artificial Intelligence,* 7:1089–1118.

[14] Ortiz, F., Torres, F., Angulo, J., Puente, S. (2001) "Comparative study of vectorial morphological operations in different color spaces," In *Intelligent Robots and Computer Vision XX,* Vol. SPIE 4572, 259–268.

[15] Petters II, R.A. (1997) "Mathematical morphology for angle-valued images," In *Non-Linear Image Processing VIII,* Vol. SPIE 3026, 84–94.

[16] Serra, J. (1982, 1988) *Image Analysis and Mathematical Morphology. Vol I,* and *Image Analysis and Mathematical Morphology. Vol II: Theoretical Advances,* Academic Press, London.

[17] Serra, J. (1992) "Anamorphoses and Function Lattices (Multivalued Morphology)," In *(Dougherty, Ed.) Mathematical Morphology in Image Processing,* Marcel-Dekker, 483–523.

[18] Serra, J. (2002) "Espaces couleur et traitement d'images," *CMM-Ecole des Mines de Paris,* Internal Note N-34/02/MM, October 2002, 13 p.

[19] Serra, J. (2005) "A Lattice Approach to Image Segmentation," To appear in *Journal of Mathematical Imaging and Vision.*

[20] Serra, J. (2005) "Morphological Segmentation of Colour Images by Merging of Partitions," Proc. of ISMM '05.

[21] Talbot, H., Evans, C. et Jones, R. (1998) "Complete ordering and multivariate mathematical morphology: Algorithms and applications," In *Proc. of the International Symposium on Mathematical Morphology (ISMM'98),* Kluwer, 27–34.

[22] Vincent, L. (1993) "Morphological Grayscale Reconstruction in Image Analysis: Applications and Efficient Algorithms," *IEEE Transactions on Image Processing,* 2(2):176–201.

ITERATIVE AREA SEEDED REGION GROWING FOR MULTICHANNEL IMAGE SIMPLIFICATION

Dominik Brunner[1]* and Pierre Soille[2]†

[1]*Institute for the Protection and Security of the Citizen*

[2]*Institute for Environment and Sustainability*

Joint Research Centre of the European Commission, I-21020 Ispra, Italy

Abstract   Motivated by the unsuitability of the image extrema paradigm for processing multiphase or multichannel images, we propose a solution in the context of image simplification based on a combination of the flat zone and seeded region growing paradigms. Concepts and results are illustrated on satellite images.

Keywords: lambda flat zone, mathematical morphology, area filter.

Introduction

While the image extrema paradigm is a valid assumption for numerous applications, it does not apply to images containing more than two phases such as satellite images with various crop fields. In this situation, objects correspond to regions of homogeneous grey tone rather than simply maxima and minima. It follows that filters acting solely on the image extrema may not produce the desired filtering effect. For example, Fig. 1 displays the flat zones of a satellite image processed by the self-dual alternating filter based on 8-connected area opening and closing up to an area of 25 pixels. Although this filter ensures that all extrema of the filtered image are larger or equal to the size of the filter, flat zones belonging to non-extrema regions can be arbitrarily small in the filtered image. Indeed, transition regions and non-extrema plateaus may be preserved by this filter even if their extent is smaller than the selected area parameter [8]. Likewise, the extrema paradigm does not apply to multichannel images owing to the lack of total ordering between vectors of more than one dimension. It

*Now with SAP, Walldorf, Germany, dominik@d-brunner.de
†Corresponding author, Pierre.Soille@jrc.it

C. Ronse et al. (eds.), Mathematical Morphology: 40 Years On, 3397–406.
©2005 *Springer. Printed in the Netherlands.*

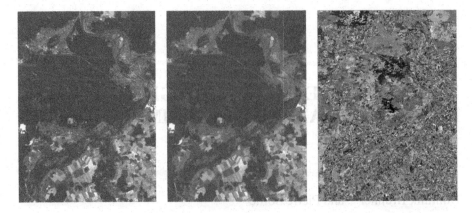

Figure 1. *Left:* Subset of the 3rd channel of a Landsat image with Fontainebleau at the upper left, 481×641 pixels (25 m resolution). *Middle:* Self-dual alternating sequential 8-connected area opening and closing up to 25 pixels applied to the left image. *Right:* Partition of the middle image (labelled 8-connected flat zones). Note the presence of many flat zones with an area less than 25 pixels. In this example, the number of 8-connected flat regions drops from 151,715 to 83,808.

follows that objects of multichannel images are often considered as regions whose pixels have similar vector values.

In this paper, we propose to tackle these problems in the context of multi-channel images by combining the flat zone [7] with the seeded region growing [1] paradigms while considering an iterative scheme. The proposed methodology is developed in Sec. 1. Related works are briefly discussed in Sec. 2. We then conclude and present some ideas for future research in Sec. 3.

1.   Methodology

The proposed methodology proceeds iteratively until a given area threshold value is reached. Denoting the current area value by i, which is initialised to 2 pixels, it can be summarised as follows: (i) select all flat zones whose area is greater than or equal to i, (ii) grow the selected flat zones so as to obtain a new partition of the image definition domain into flat zones, and (iii) stop if i has reached the threshold value, otherwise increment i by 1 and go to (i). The two main steps of our procedure are detailed hereafter as well as post processing steps to further improve the simplification results with the aim to vectorise the output image.

Quasi-flat zone filtering

We extend the definition of quasi-flat zones [5] to multichannel images as follows. Two pixels x and y belong to the same quasi-flat zone of a multichannel image $\mathbf{f} = (f_1, \ldots, f_n)$ if and only if for there exists a discrete path $\mathcal{P} = (p_1, \ldots, p_m)$ such that $\mathbf{x} = p_1$ and $\mathbf{y} = p_m$ and, for all $i \in \{1, \ldots, m-1\}$, p_i and p_{i+1} satisfy the following symmetrical relationship: $|f_j(p_i) - f_j(p_{i+1})| \leq \lambda_j$, for all $j \in \{1, \ldots, n\}$. The maximum differences in each channel can be combined in a difference vector $\boldsymbol{\lambda} = (\lambda_1, \ldots, \lambda_n)$. In the sequel, we call the resulting flat zones, the $\boldsymbol{\lambda}$-*flat zones*. Note that lambda flat zone an alternative definition based on a multidimensional distance measurement is proposed in [10]. The fast breadth first stack algorithm described in [9, p. 38] for grey level flat zones can be extended to λ-flat zones as follows:

```
1.  for all pixel p of the image f
2.    if lbl(p) = NOTLABELLED
3.      offset ← p;
4.      lval ← lval + 1;
5.      lbl(p) ← lval;
6.      for all neighbours p' of p
7.        hasSameValue ← true;
8.        for all channels c of image f
9.          if f_c(p') is not in
               [f_c(offset) − λ_c, f_c(offset) + λ_c]
10.           hasSameValue ← false;
11.       if hasSameValue = true
12.         push(p');
13.         lbl(p') ← lval;
14.   while (isStackEmpty()=false)
15.     q ← pop();
16.     offset ← q;
17.     for all neighbours q' of q
18.       hasSameValue ← true;
19.       for all channels c of image f
20.         if f_c(q') is not in
               [f_c(offset) − λ_c, f_c(offset) + λ_c]
21.           hasSameValue ← false;
22.       if hasSameValue = true
23.         push(q');
24.         lbl(q') ← lval;
```

Those labelled regions whose area are equal to or exceed the current threshold value are then selected as seeds. Similarly to alternating sequential filters, it is essential to consider an iterative procedure, starting with the smallest possible size. Indeed, the number of flat regions versus the area of these flat regions is a monotonically decreasing function with usually only few regions remaining for sizes larger than 2 pixels, even when increasing the value of λ (in all experiments of this paper we use the same λ value for all channels). This is illustrated in Fig. 2. When processing multichannel images, a sufficient number of seeds can only be obtained by considering λ-flat zones because the probability to find neighbour pixels with the same vector values decreases with the number of channels.

Iterative seeded region growing

The iterative seeded region growing is based on the seeded region growing (SRG) algorithm described in [1] and enhanced in [4] to achieve order independence (ISRG). We propose to extend the ISRG algorithm to multicomponent images and call it the MCISRG algorithm. The mean value $\bar{r}$ of a region R can be seen as a point in an n-dimensional space, i.e. $\bar{r} \in \mathbb{R}^n$, with its coordinates

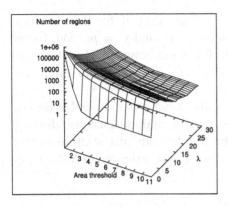

Figure 2. Number of λ-flat zones (log scale) versus region area and λ value where $\lambda = (\lambda, \lambda, \lambda)$. The input image is a subset of the 6 channel Landsat scene whose third band is shown in Fig. 1.

at $x_1, \ldots, x_n$. Each coordinate x_i of this point is defined as follows:

$$x_i = \sum_{\mathbf{x} \in R} f_i(\mathbf{x}) / \mathrm{card}(R), \tag{1}$$

where $\mathrm{card}(R)$ is the number of pixels in this region R and $f_i(\mathbf{x})$ is the value of the pixel $\mathbf{x}$ in the ith channel. The values of a pixel p, which is a neighbour of R, is also a point in the n-dimensional space. It follows that the new priority $\Delta(p, \bar{r})$ of p for the region R can be defined by the Euclidean distance between p and $\bar{r}$:

$$\Delta(p, \bar{r}) = \left(\sum_{i=1}^{n} [f_i(p) - f_i(\bar{r})]^2 \right)^{0.5}. \tag{2}$$

Alternatively, the distances can be measured from the initial seeds rather than the grown seed regions. The MCISRG algorithm can be implemented using an ascending priority queue (PQ), which orders its elements from the smallest to the largest one according to the priority value. This value is in this case the distance $\Delta(p, \bar{r})$ from the point p to the mean value of the region R. A new element can be inserted unrestricted, but only the element with the smallest Δ value can be removed [2]. An element of the PQ is a pqDatum storing the position, the priority of the pixel, and the region to which the distance of the pixel was calculated which is equal to the region to which the pixel would be assigned to if the Δ value is smallest. The function pqInsert(pqDatum) inserts one element in the PQ, pqRemove() returns the pqDatum with the smallest Δ value and deletes it from the queue, while isPQEmpty() returns false if there are still elements in the queue and true otherwise. The function setValuesOfPQDatum(pqDatum, distance, regionLabel, pixel) stores the attributes distance, regionLabel and pixel in the pqDatum, while getPQDatumDistance(pqDatum) returns the distance, getPQDatumPixel(pqDatum) the pixel, and getPQDatumRegion(pqDatum) the region of the attribute pqDatum.

The data structure `meanDatum` is needed in order to handle the mean values of the regions (one per channel). The variable `allMeans` holds for each region, which is represented by the information of the pixels in the seed image `seedIm`, one `meanDatum`. The function `getMean(allMeans, regionLabel)` returns the `meanDatum` of the corresponding region that is specified with the `regionLabel` attribute. The method `addValueToMean(meanDatum, f, p)` is used to add the values of the pixel `p` in the multispectral image `f` to the `meanDatum`, while `getDistanceToMean(meanDatum, f, p)` returns the Euclidean distance Δ, which is calculated with the equation 2, from the pixel `p` of the image `f` to the mean value of the region represented with the `meanDatum`. Finally, the function `getChannelMean(meanDatum, c)` returns the mean value of the channel `c` of the `meanDatum`.

The stacks *neighbour holding queue* `nq` and *holding queue* `hq` are used to store the pixels, which are neighbours of seeds and the pixels which were processed and need some postprocessing. For the stacks, the methods `isStackEmpty(queue)`, `push(queue,pixel)`, and `pop(queue)` are used for storing pixels, loading pixels, and checking whether the stack is empty or not. Several labels represent the different states of a pixel: `FIRSTSEED` is the smallest region label in the seed image, `NOSEED` means that the pixel does not belong to a region yet, `IN_PQ` indicates that the pixel is already in the priority queue, and `IN_NHQ` expresses that the pixel is in the neighbour holding queue. The constant `MAXDISTANCE` sets the `minDistance` variable to its maximum value so that the first comparison between the current `distance` value and the `minDistance` is always true. The pseudo code for the MCISRG algorithm is:

```
1. for all pixel p of the image seedIm
2.     if seedIm(p) >= FIRSTSEED
3.         meanDatum ← getMean(allMeans, seedIm(p));
4.         addValueToMean(meanDatum, f, p);
5.     else if seedIm(p) = NOSEED
6.         for all p' which are neighbours from p
7.             if seedIm(p') >= FIRSTSEED
8.                 push(nhq, p);
9.                 seedIm(p) ← IN_NHQ;
10.                break;
11. while (isStackEmpty(nhq) = false) AND (isPQEmpty() = false)
12.    while (isStackEmpty(nhq))
13.        minDistance ← MAXDISTANCE;
14.        q ← pop(nhq);
15.        for all q' which are neighbours from q
16.            if seedIm(q') >= FIRSTSEED
17.                meanDatum ← getMean(allMeans, seedIm(q'));
18.                distance ← getDistanceToMean(meanDatum, f, q);
```

```
19.                if distance < minDistance
20.                      minDistance ← distance;
21.                      regionLabel ← seedIm(q');
22.              setValuesOfPQDatum(pddatum, mindistance, regionLabel, q);
23.              pqInsert(pqdatum);
24.          seedIm(q) ← IN_PQ;
25.    if isPQEmpty() = false
26.         pqDatum ← pqRemove();
27.         distance ← getPQDatumDistance(pqDatum);
28.         p ← getPQDatumPixel(pqdatum);
29.         if seedIm(p) = IN_PQ
30.             seedIm(p) ← getPQDatumRegion(pqDatum);
31.             push(hq, p);
32.         while (isPQDatum = false)
33.             pqdatum ← pqRemove();
34.             if distance != getPQDatumDistance(pqDatum)
35.                 pqInsert(pqDatum);
36.                 break;
37.             p ← getPQDatumPixel(pqDatum);
38.             if seedIm(p) = IN_PQ
39.                 seedIm(p) ← getPQDatumRegion(pqDatum);
40.             push(hq, p);
41.     while isStackEmpty(hq) = false
42.         p ← pop(hq);
43.         meanDatum ← getMean(allMeans, seedIm(p));
44.         addValueToMean(meanDatum, f, p);
45.         for all p' which are neighbours from p
46.             if seedIm(p') = NOSEED OR seedIm(p') = IN_PQ
47.                 push(nhq, p');
48.                 seedIm(p') ← IN_NHQ;
49. for all pixel p of the image seedIm
50.     for all channels c of the image f
51.         meanDatum ← getMean(allMeans, seedIm(p));
52.         f_c(p) ← getChannelMean(meanDatum, c);
```

For example, Fig. 3a shows the output of the proposed filtering by iterating it up to an area threshold of 25 and 100 pixels while using 8-connectivity. Contrary to the partition produced by the self-dual alternating sequential area opening/closing filter, each flat zone of the iterative area filter of the flat zones up to an area of n pixels contains at least n pixels (compare right image of Fig. 3 with that of Fig. 1). Furthermore, the number of λ-flat zones decreases drastically: from 103,973 3-flat zones to 2,972.

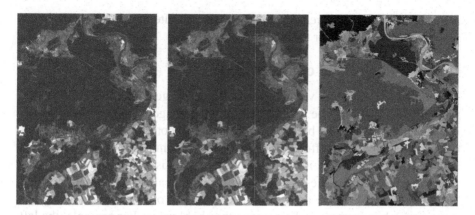

Figure 3. Iterative area seeded region growing of 6 channels of a satellite image subset whose third channel is shown in Fig. 1a. The left image shows the output of the third channel for area of 25 pixels (2,972 regions), the middle one for an area of 100 pixels (829 regions), while the the right image displays the flat zones of the left image (i.e., threshold value of 25 pixels). Note that contrary to the right image of Fig. 1, each flat zone contains at least 25 pixels. In addition, the resulting partition is the same for all channels.

Merging of regions with similar contrast

The proposed iterative seeded region growing requires the selection of a minimum area as input parameter. Therefore, an object of a given area (e.g., a field) may consist of more than one region in the output image if the selected minimum area is smaller than or equal to half the area of this object. However, half this area may already be too large if there is a need to preserve objects (e.g., houses) having such an area. To get around this problem, we initially choose a size parameter leading to a detailed enough image for the considered application. Then, a contrast coefficient measuring the dissimilarity between each region and its neighbouring pixels is calculated. We propose to define this coefficient as follows:

$$\kappa_R = \sum_{p \in \rho^+(R)} \Delta(p, \bar{r}) / \mathrm{card}(\rho^+(R)), \qquad (3)$$

where $\rho^+(R)$ denotes the external boundary of the region R.

The merging of regions whose contrast coefficient is smaller than the threshold value is a complex issue especially if we want to achieve order independence. We propose therefore an alternative and simpler approach based on the MCISRG algorithm. The segmented image is first labelled, the contrast coefficient using the equation 3 is then calculated for each region, and finally each region in the label image which has a smaller contrast value than the threshold value is deleted. The remaining regions are then used as seeds for MCISRG algorithm. The sole difference with the area based filtering lies therefore in the way seeds are selected for the MCISRG algorithm.

Calculating optimal contrast value

The optimal threshold value depends on the image content. It follows that the optimal value must be determined for each image. One possibility is the specification of the value by the user. This is time consuming for the user, because he/she has to try different threshold values, compare the results by hand, and decide which image is best. In practice, this is not feasible and therefore we developed a process to determine automatically the optimal threshold value. It is based on the analysis of the evolution of the number of regions remaining when increasing the contrast threshold value. For example, the left diagram of Fig. 4 shows this relationship for the whole Landsat image subset shown in Fig. 1a. The optimal value is then simply defined as the contrast value corresponding to the inflection point of this curve, i.e., maximum derivative, see right diagram of Fig. 4. Tests performed on a series of images have shown

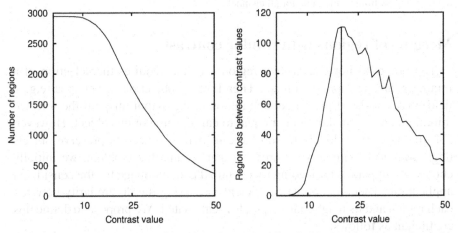

Figure 4. The graph on the left side plots the number of regions versus the contrast threshold value for the 6-channel input image whose 3rd channel is shown in Fig. 1a. The graph on the right side displays the corresponding gradient magnitude. The vertical line in the right graph highlights the maximum of the gradient which is selected as optimal contrast threshold value.

that the automatically selected values are in accordance with values determined manually. The contrast values at the maximum points of the graphs are within the range of threshold values, which were identified manually. In consequence, this value can be seen as the *optimal contrast threshold value*. Figure 5 illustrates the output of the MCISRG merging process using the automatically derived optimal contrast threshold value for the panchromatic channel of the Landsat scene of Fontainebleau. When the input image is getting smaller, the derivative of the curve may display numerous local maxima with two or more

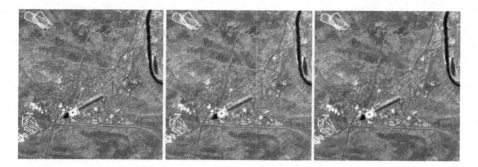

Figure 5.   *Left:* Landsat panchromatic image of Fontainebleau (12.5m resolution), 481 × 481 pixels (122,582 0-flat regions). *Middle:* Output of iterative seeded region growing up to a size of 25 pixels (4,071 0-flat regions). *Right:* After MCISRG merging using the optimum contrast threshold value of 4 grey levels (2,399 0-flat regions).

maxima having a magnitude close or even equal to that of the absolute maximum. In this situation, we consider the contrast value corresponding to the first maximum rather than the absolute maximum (conservative approach).

2.   Link with other approaches

Salembier *et al.* [6] also propose a filter suppressing all flat regions whose area is less than a given threshold. It is based on the processing of the region adjacency graph of the flat zones using an area merging criterion and setting the grey level of the merged region to the median value of the largest region (or the arithmetic mean of the two merged regions if they have both the same size) while considering an ad hoc merging order. Note that contrary to our approach, this type of process defines a connected operator. That is, when a flat zone is below the threshold level, it cannot be shared by two different flat zones. The same comment applies to the method proposed by Crespo *et al.* [3]. Furthermore, we advocate the use of an iterative procedure to ensure a smooth degradation whatever the targeted minimum area value.

3.   Concluding remarks and perspectives

The proposed algorithm is useful not only for simplifying an image but also for contrast enhancement when using just a size of two pixels (no iteration). Indeed, the largest loss of regions occur at the first step. For example, for the sample image used in this paper, we go from 430,150 to 1,840 0-flat regions when considering an area of 2 pixels. Similarly to the distances which can be measured either from the seed or the grown seed regions, the values allocated to the latter regions could be set to those of the seeds rather than to the mean values of both regions. By doing so, the result image should be even more contrasted. For simplification purposes, we have shown that larger sizes and an iterative process is required, similarly to that introduced for alternating sequential filters. Note that although it is essential to consider an iterative procedure, once the smallest size (i.e, 2 pixels) has been performed, similar results may be obtained by incrementing the size by more than just one pixel. This does not improve the quality of the final result (decreasing quality with increasing step size) but speeds up the processing (trade-off between quality and speed). Note also that the proposed MCISRG algorithm can be adapted so as to produce a connected operator. This will be detailed in a subsequent paper.

Before vectorising the simplified image as required by vector based Geographical Information systems (GIS), there is still a need to further simplify the boundary of the resulting partition. To achieve this goal, we are currently investigating the use of morphological operators taking into account the shape rather than just the area of the segments.

References

[1] R. Adams and L. Bischof. Seeded region growing. *IEEE Transactions on Pattern Analysis and Machine Intelligence*, 16(6):641–647, 1994.

[2] E. Breen and D. Monro. An evaluation of priority queues for mathematical morphology. In J. Serra and P. Soille, editors, *Mathematical Morphology and its Applications to Image Processing*, pages 249–256. Kluwer Academic Publishers, 1994.

[3] J. Crespo, R. Schafer, J. Serra, C. Gratin, and F. Meyer. The flat zone approach: a general low-level region merging segmentation method. *Signal Processing*, 62(1):37–60, 1997.

[4] A. Mehnert and P. Jackway. An improved seeded region growing. *Pattern Recognition Letters*, 18:1065–1071, 1997.

[5] F. Meyer and P. Maragos. Nonlinear scale-space representation with morphological levelings. *J. of Visual Communication and Image Representation*, 11(3):245–265, 2000.

[6] P. Salembier, L. Garrido, and D. Garcia. Auto-dual connected operators based on iterative merging algorithms. In H. Heijmans and J. Roerdink, editors, *Mathematical Morphology and its Applications to Image and Signal Processing*, volume 12 of *Computational Imaging and Vision*, pages 183–190, Dordrecht, 1998. Kluwer Academic Publishers.

[7] P. Salembier and J. Serra. Flat zones filtering, connected operators, and filters by reconstruction. *IEEE Transactions on Image Processing*, 4(8):1153–1160, August 1995.

[8] P. Soille. On the morphological processing of objects with varying local contrast *Lecture Notes in Computer Science*, 2886

[9] P. Soille. *Morphological Image Analysis: Principles and Applications*. Springer-Verlag, Berlin Heidelberg New York, 2nd edition, 2003 [Reprinted with corrections in 2004]. See also http://ams.jrc.it/soille/book2ndprint

[10] F. Zanoguera and F. Meyer. On the implementation of non-separable vector levelings. In H. Talbot and R. Beare, editors, *Proceedings of VIth International Symposium on Mathematical Morphology*, pages 369–377, Sydney, Australia, 2002. Commonwealth Scientific and Industrial Research Organisation.

MORPHOLOGY FOR HIGHER-DIMENSIONAL TENSOR DATA VIA LOEWNER ORDERING

Bernhard Burgeth,[1] Nils Papenberg, [1] Andres Bruhn, [1] Martin Welk, [1] Christian Feddern, [1] and Joachim Weickert [1]

[1] *Saarland University*
Faculty of Mathematics and Computer Science
Bld. 27.1, P.O. Box 15 11 50, 66041 Saarbruecken
Germany

{burgeth,papenberg,bruhn,welk,feddern,weickert} @mia.uni-saarland.de

Abstract   The operators of greyscale morphology rely on the notions of maximum and minimum which regrettably are not directly available for tensor-valued data since the straightforward component-wise approach fails.

This paper aims at the extension of the maximum and minimum operations to the tensor-valued setting by employing the Loewner ordering for symmetric matrices. This prepares the ground for matrix-valued analogs of the basic morphological operations. The novel definitions of maximal/minimal matrices are rotationally invariant and preserve positive semidefiniteness of matrix fields as they are encountered in DT-MRI data. Furthermore, they depend continuously on the input data which makes them viable for the design of morphological derivatives such as the Beucher gradient or a morphological Laplacian. Experiments on DT-MRI images illustrate the properties and performance of our morphological operators.

Keywords: Mathematical morphology, dilation, erosion, matrix-valued images, diffusion tensor MRI, Loewner ordering

Introduction

A fruitful and extensive development of morphological operators has been started with the path-breaking work of Serra and Matheron [11, 12] almost four decades ago. It is well documented in monographs [8, 13–15] and conference proceedings [7, 16] that morphological techniques have been successfully used to perform shape analysis, edge detection and noise suppression in numerous applications. Nowadays the notion of image also encompasses tensor-valued data, and as in the scalar case one has to detect shapes, edges and eliminate noise. This creates a need for morphological tools for matrix-valued data.

407

C. Ronse et al. (eds.), Mathematical Morphology: 40 Years On, 407–416.
©2005 *Springer. Printed in the Netherlands.*

Matrix-valued concepts, that truly take advantage of the interaction of the different matrix-channels have been developed for median filtering [20], for active contour models and mean curvature motion [5], and for nonlinear regularisation methods and related diffusion filters [17, 19]. In [4] the basic operations dilation and erosion as well as opening and closing are transfered to the matrix-valued setting at least for 2×2 matrices. However, the proposed approaches lack the continuous dependence on the input matrices which poses an insurmountable obstacle for the design of morphological derivatives.

The goal of this article is to present an alternative and more general approach to morphological operators for tensor-valued images based on the Loewner ordering. The morphological operations to be defined should work on the set $\text{Sym}(n)$ of symmetric $n \times n$ matrices and have to satisfy conditions such as:

(i) Continuous dependence of the basic morphological operations on the matrices used as input for the aforementioned reasons,

(iii) preservation of the positive semidefiniteness of the matrix field since DT-MRI data sets posses this property,

(iii) rotational invariance.

It is shown in [4] that the requirement of rotational invariance already rules out the straightforward component-wise approach. In this paper we will introduce a novel notion of the minimum/maximum of a finite set of symmetric matrices which will exhibit the above mentioned properties.

The article has the following structure: The subsequent section gives a brief account of the morphological operations we aim to extend to the matrix-valued setting. In section 3 we present the crucial maximum and minimum operations for matrix-valued data based on the Loewner ordering. We report the results of our experiments with various morphological operators applied to real DT-MRI images in section 4. The last section 5 provides concluding remarks .

1.   Morphological Operators

Standard morphological operations utilise the so-called *structuring element* to work on images represented by scalar functions $f(x, y)$ with $(x, y) \in \mathbb{R}^2$. Greyscale *dilation* $\oplus$, resp., *erosion* $\ominus$ w.r.t. B is defined by

$$(f \oplus B)(x, y) \quad := \quad \sup \{f(x-x', y-y') \mid (x', y') \in B\},$$
$$(f \ominus B)(x, y) \quad := \quad \inf \{f(x-x', y-y') \mid (x', y') \in B\}.$$

The combination of dilation and erosion gives rise to various other morphological operators such as *opening* and *closing*,

$$f \circ B := (f \ominus B) \oplus B, \qquad f \bullet B := (f \oplus B) \ominus B,$$

the *white top-hat* and its dual, the *black top-hat*

$$\text{WTH}(f) := f - (f \circ B), \qquad \text{BTH}(f) := (f \bullet B) - f,$$

finally, the *self-dual top-hat*, $\text{SDTH}(f) := (f \bullet B) - (f \circ B)$.

The boundaries of objects are the loci of high greyvalue variations in an image which can be detected by gradient operators. The so-called *Beucher gradient*

$$\varrho_B(f) := (f \oplus B) - (f \ominus B),$$

as well as the *internal* and *external gradient*,

$$\varrho_B^-(f) := f - (f \ominus B), \qquad \varrho_B^+(f) := (f \oplus B) - f$$

are analogs to the norm of the gradient $\|\nabla f\|$ if the image f is considered as a differentiable function.

The application of shock filtering to matrix-valued data calls for an equivalent of the Laplace operator $\Delta f = \partial_{xx} f + \partial_{yy} f$ appropriate for this type of data. A *morphological Laplacian* has been introduced in [18]. However, we use a variant given by

$$\Delta_m f := \varrho_B^+(f) - \varrho_B^-(f) = (f \oplus B) - 2 \cdot f + (f \ominus B).$$

This form of a Laplacian acts as the second derivative $\partial_{\eta\eta} f$ where η stands for the direction of the steepest slope. Therefore it allows us to distinguish between influence zones of minima and maxima of the image f, a property essential for the design of shock filters.

The idea underlying *shock filtering* is applying either a dilation or an erosion to an image, depending on whether the pixel is located within the influence zone of a minimum or a maximum [10]:

$$\delta_B(f) := \begin{cases} f \oplus B & \text{if} \quad \text{trace}(\Delta_m f) \leq 0, \\ f \ominus B & \text{otherwise.} \end{cases}$$

2. Maximal and Minimal Matrices with Respect to Loewner Ordering

In this section we describe how to obtain the suitable maximal (minimal) matrix that majorises (minorises) a given finite set of symmetric matrices. We start with a very brief account of some notions from convex analysis necessary for the following.

A subset C of a vector space V is named *cone*, if it is stable under addition and multiplication with a positive scalar. A subset B of a cone C is called *base* if every $y \in C, y \neq 0$ admits a unique representation as $y = r \cdot x$ with $x \in B$ and $r > 0$. We will only consider a cone with a convex and compact base.

The most important points of a closed convex set are its *extreme points* characterised as follows: A point x is an extreme point of a convex subset $S \subset V$ of a vector space V if and only if $S \setminus \{x\}$ is still convex. The set of all extreme points of S is denoted ext(S). All extreme points are necessarily boundary points, ext(S) $\subset$ bd(S). Each convex compact set S in a space of finite dimension can be reconstructed as the set of all convex combinations of its extreme points [1, 9]: $S = $ convexhull(ext(S)).

The Cone of the Loewner Ordering

Let Sym(n) denote the vector space of symmetric $n \times n$-matrices with real entries. It is endowed with the scalar product $\langle A, B \rangle := \sqrt{\text{trace}(A^\top B)}$. The corresponding norm is the Frobenius norm for matrices: $\|A\| = \sum_{i,j=1}^{n} a_{ij}$.

There is also a natural partial ordering on Sym(n), the so-called *Loewner ordering* defined via the cone of positive semidefinite matrices Sym$^+(n)$ by

$$A, B \in \text{Sym}(n) : \quad A \geq B :\Leftrightarrow A - B \in \text{Sym}^+(n),$$

i.e. if and only if $A - B$ is positive semidefinite.

This partial ordering is *not* a lattice ordering, that is to say, the notion of a unique supremum and infimum with respect to this ordering does not exist [3]. The (topological) interior of Sym$^+(n)$ is the cone of positive definite matrices, while its boundary consists of all matrices in Sym(n) with a rank strictly smaller than n. It is easy to see that, for example, the set $\{M \in \text{Sym}^+(n) : \text{trace}(M) = 1\}$ is a convex and compact base of the cone Sym$^+$(n). It is known [1] that the matrices $v v^\top$ with unit vectors $v \in \mathbb{R}^n$, $\|v\| = 1$ are the extreme points of the set $\{M \in \text{Sym}^+(n) : \text{trace}(M) = 1\}$ [1]. They have by construction rank 1 and for any unit vector v we find $v v^\top v = v \cdot \|v\|^2 = v$ which implies that 1 is the only non-zero eigenvalue. Hence trace($v v^\top$)= 1. Because of this extremal property the matrices $v v^\top$ with $\|v\| = 1$ carry the complete information about the base of Loewner ordering cone: convexhull($\{v v^\top : v \in \mathbb{R}^n, \|v\| = 1\}$) is a base for the Loewner ordering cone.

The *penumbra* $P(M)$ of a matrix $M \in$ Sym(n) is the set of matrices N that are smaller than M w.r.t. the Loewner ordering:

$$P_0(M) := \{N \in \text{Sym(n)} : N \leq M\} = M - \text{Sym}^+(n),$$

where we used the customary notation $a + r S := \{a + r \cdot s : s \in S\}$ for a point $a \in V$, a scalar r and a subset $S \subset V$.

Using this geometric description the problem of finding the maximum of a set of matrices $\{A_1, \dots, A_m\}$ amounts to determining the minimal penumbra covering their penumbras $P_0(A_1), \dots, P_0(A_m)$. Its vertex represents the wanted maximal matrix $\overline{A}$ that dominates all A_i w.r.t the Loewner ordering.

However, the cone itself is too complicated a structure to be handled directly. Instead we associate with each matrix $M \in$ Sym(n) a *ball* in the subspace $\{A : \text{trace}(A) = 0\}$ of all matrices with zero trace as a *completely descriptive set*. We will assume for the sake of simplicity that trace$(M) \geq 0$. This ball is constructed in two steps: First, from the statements above we infer that the set $\{M - \text{trace}(M) \cdot \text{convexhull}\{v\, v^\top : v \in \mathbb{R}, \|v\| = 1\}\}$ is a base for $P_0(M)$ contained in the subspace $\{A : \text{trace}(A) = 0\}$. We observe that the identity matrix E is perpendicular to the matrices A from this subspace, $\langle A, E \rangle = \sqrt{\text{trace}(A)} = 0$, and hence the orthogonal projection of M onto $\{A : \text{trace}(A) = 0\}$ is given by

$$m := M - \frac{\text{trace}(M)}{n} E.$$

Second, the extreme points of the base of $P_0(M)$ are lying on a sphere with center m and radius

$$r := \|M - \text{trace}(M)v\, v^\top - m\| = \text{trace}(M)\sqrt{1 - \frac{1}{n}}.$$

Consequently, if the center m and radius r of a sphere in $\{A \in \text{Sym}(n) : \text{trace}(A) = 0\}$ are given the vertex M of the associated penumbra $P_0(M)$ is obtained by

$$M = m + \frac{r}{n} \frac{1}{\sqrt{1 - \frac{1}{n}}} E.$$

With this information at our disposal, we can reformulate the task of finding a suitable maximal matrix $\overline{A}$ dominating the matrices $\{A_1, \ldots, A_m\}$: The *smallest* sphere enclosing the spheres associated with $\{A_1, \ldots, A_m\}$ determines the matrix $\overline{A}$ that dominates the A_i. It is minimal in the sense, that there is no smaller one w.r.t. the Loewner ordering which has this "covering property" of its penumbra.

This is a non trivial problem of computational geometry and we tackle it by using a sophisticated algorithm implemented by B. Gaertner [6]. Given a set of points in $\mathbb{R}^d$ it is capable of finding the smallest ball enclosing these points. Hence for each $i = 1, \ldots, m$ we sample within the set of extreme points $\{A_i - \text{trace}(A_i)v\, v^\top\}$ of the base of $P_0(A_i)$ by expressing v in 3d-spherical coordinates, $v = (\sin\phi\cos\psi, \sin\phi\sin\psi, \cos\phi)$ with $\phi \in [0, 2\pi[, \psi \in [0, \pi[$.

The case $n = 2$ can be visualised by embedding Sym(2) in $\mathbb{R}^3$ via $A = (a_{ij})_{i,j=1,2} \longleftrightarrow (a_{11}, a_{22}, a_{12})$ as it is indicated in Figure 1. The penumbras of the matrices $\{A_1, \ldots, A_m\}$ are covered with the minimal penumbral cone whose vertex is the desired maximal matrix $\overline{A}$. For presentational purposes an additional orthogonal transformation has been applied such that the x-y-plane coincides with $\{A \in \text{Sym}(2) : \text{trace}(A) = 0\}$. The minimal element $\underline{A}$ is

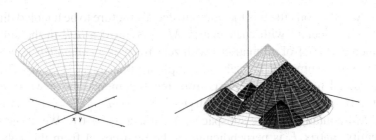

Figure 1.   (a) **Left:** Image of the Loewner cone Sym$^+$(2). (b) **Right:** Cone covering four penumbras of other matrices. The tip of each cone represents a symmetric 2×2 matrix in $I\!R^3$. Each of the cones (and hence its generating matrix) is uniquely determined by its circular base. The minimal disc covering the smaller discs belongs to the selected maximal matrix $\overline{A}$

obtained through the formula

$$\underline{A} = \big(\max(A_1^{-1}, \ldots, A_m^{-1})\big)^{-1}$$

inspired by its well-known counterpart for real numbers. The construction of maximal and minimal elements ensures their rotational invariance, their positive semidefiniteness and continuity. These properties are passed on to the above mentioned morphological operations.

3.   Experimental Results

In our numerical experiments we use positive definite data. A 128×128 layer of 3-D tensors which has been extracted from a 3-D DT-MRI data set of a human head. For detailed information about the acquisition of this data type the reader is referred to [2] and the literature cited there. The data are represented as ellipsoids via the level sets of the quadratic form $\{x^\top A x : x \in I\!R^3\}$ associated with a matrix $A \in$ Sym$^+$(3). The color coding of the ellipses reflects the direction of their principle axes.

Due to quantisation effects and measurement imprecisions our DT-MRI data set of a human head contains not only positive definite matrices but also singular matrices and even matrices with negative eigenvalues, though the negative values are of very small absolute value. While such values do not constitute a problem in the dilation process, the erosion, relying on inverses of positive definite matrices, has to be regularised. Instead of the exact inverse A^{-1} of a given matrix A we use $(A + \varepsilon I)^{-1}$ with a small positive ε.

Due to the complexity of the not yet fully optimised procedures the running time to obtain dilation and erosion is about two orders of magnitude longer than in the case of comparable calculations with grey value data.

Figure 2 displays the original head image and the effect of dilation and erosion with a ball-shaped structuring element of radius $\sqrt{5}$. For the sake of brevity we will denote in the sequel this element by BSE($\sqrt{5}$). As it is ex-

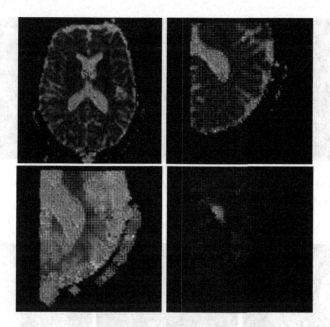

Figure 2.   **(a) Top left:** 2-D tensor field extracted from a DT-MRI data set of a human head. **(b) Top right:** enlarged section of left image. **(c) Bottom left:** dilation with BSE($\sqrt{5}$). **(d) Bottom right:** erosion with BSE($\sqrt{5}$).

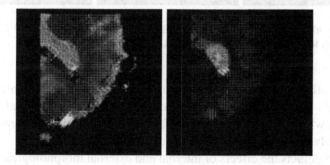

Figure 3.   **(a) Left:** closing with BSE($\sqrt{5}$). **(b) Right:** opening with BSE($\sqrt{5}$).

pected from scalar-valued morphology, the shape of details in the dilated and eroded images mirrors the shape of the structuring element. In Figure 3 the results of opening and closing operations are shown. In good analogy to their scalar-valued counterparts, both operations restitute the coarse shape and size of structures. The output of top hat filters can be seen in Figure 4. As in the scalar-valued case, the white top hat is sensitive for small-scale details formed by matrices with large eigenvalues, while the black top hat responds with high values to small-scale details stemming from matrices with small eigenvalues.

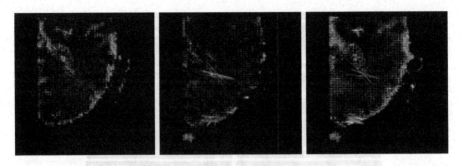

Figure 4.   (a) **Left:** white top hat with BSE($\sqrt{5}$). (b) **Middle:** black top hat with BSE($\sqrt{5}$). (c) **Right:** self-dual top hat with BSE($\sqrt{5}$).

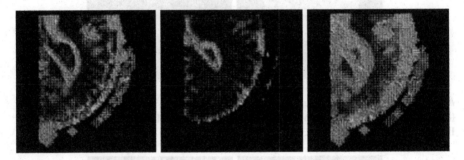

Figure 5.   (a) **Left:** external gradient with BSE($\sqrt{5}$). (b) **Middle:** internal gradient with BSE($\sqrt{5}$). (c) **Right:** Beucher gradient with BSE($\sqrt{5}$).

Very long ellipses also seen in the yellow spot in Figure 3, are partially arte-facts caused by the tool for graphical representation. The self-dual top hat as the sum of white and black top hat results in homogeneously extreme matrices rather evenly distributed in the image.

Figure 5 depicts the effects of internal and external morphological gradients and their sum, the Beucher gradient for our sample matrix field. The action of the Laplacian Δ_m and its use for steering a shock filter can be seen in Figure 6: While applying dilation in pixels where the trace of the Laplacian is negative, the shock filter acts as an erosion wherever the trace of the Laplacian is posi-tive. The output is an image where regions with larger and smaller eigenvalues are separated more clearly than in the original image.

4.   Conclusion

In this paper we determined suitable maximal and minimal elements $\overline{A}$, $\underline{A}$ in the space of symmetric matrices Sym(3) with respect to the Loewner order-ing. Thus we have been able to transfer fundamental concepts of mathematical

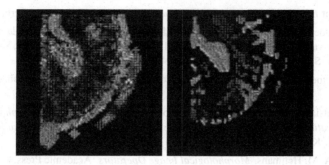

Figure 6.   (a) **Left:** morphological Laplacian with BSE($\sqrt{5}$). (b) **Right:** result of shock filtering with BSE($\sqrt{5}$).

morphology to matrix-valued data. The technique developed for this purpose is considerably more general and sustainable than former approaches for the case of 2×2-matrices. The present approach has potential to cope successfully even with 5×5-matrix fields. We obtained appropriate analogs with desirable continuity properties for the notion of maximum and minimum, the corner stones of mathematical morphology. Therefore we succeeded in designing morphological derivatives and shock filters for tensor fields, aside from the standard morphological operations. The practicability of various morphological operations on positive definite matrix-fields is confirmed by several experiments. Future work will focus on faster performance and the development of more sophisticated morphological operators for matrix-valued data.

Acknowledgments

We are grateful to Anna Vilanova i Bartroli (Eindhoven Institute of Technology) and Carola van Pul (Maxima Medical Center, Eindhoven) for providing us with the DT-MRI data set and for discussing data conversion issues. We also thank Prof. Hans Hagen (University of Kaiserslautern) for very helpful discussions.

References

[1] A. Barvinok. *A Course in Convexity*, volume 54 of *Graduate Studies in Mathematics*. American Mathematical Society, Providence, 2002.

[2] P. J. Basser. Inferring microstructural features and the physical state of tissues from diffusion-weighted images. *Nuclear Magnetic Resonance in Biomedicine*, 8:333–344. Wiley, New York, 1995.

[3] J. M. Borwein and A. S. Lewis. *Convex Analysis and Nonlinear Optimization*. Springer, New York, 1999.

[4] B. Burgeth, M. Welk, C. Feddern, and J. Weickert. Morphological operations on matrix-valued images. In T. Pajdla and J. Matas, editors, *Computer Vision – ECCV 2004*, volume

3024 of *Lecture Notes in Computer Science*, pages 155–167. Springer, Berlin, 2004.

[5] C. Feddern, J. Weickert, B. Burgeth, and M. Welk. Curvature-driven pde methods for matrix-valued images. Technical Report 104, Department of Mathematics, Saarland University, Saarbrücken, Germany, May 2004.

[6] B. Gärtner. http://www2.inf.ethz.ch/personal/gaertner/. WebPage last visited: July 2nd, 2004.

[7] J. Goutsias, L. Vincent, and D. S. Bloomberg, editors. *Mathematical Morphology and its Applications to Image and Signal Processing*, volume 18 of *Computational Imaging and Vision*. Kluwer, Dordrecht, 2000.

[8] H. J. A. M. Heijmans. *Morphological Image Operators*. Academic Press, Boston, 1994.

[9] J.-B. Hiriart-Urruty and C. Lemarechal. *Fundamentals of Convex Analysis*. Springer, Heidelberg, 2001.

[10] H. P. Kramer and J. B. Bruckner. Iterations of a non-linear transformation for enhancement of digital images. *Pattern Recognition*, 7:53–58, 1975.

[11] G. Matheron. *Eléments pour une théorie des milieux poreux*. Masson, Paris, 1967.

[12] J. Serra. *Echantillonnage et estimation des phénomènes de transition minier*. PhD thesis, University of Nancy, France, 1967.

[13] J. Serra. *Image Analysis and Mathematical Morphology*, volume 1. Academic Press, London, 1982.

[14] J. Serra. *Image Analysis and Mathematical Morphology*, volume 2. Academic Press, London, 1988.

[15] P. Soille. *Morphological Image Analysis*. Springer, Berlin, 1999.

[16] H. Talbot and R. Beare, editors. *Proc. Sixth International Symposium on Mathematical Morphology and its Applications*. Sydney, Australia, April 2002. http://www.cmis.csiro.au/ismm2002/proceedings/.

[17] D. Tschumperlé and R. Deriche. Diffusion tensor regularization with contraints preservation. In *Proc. 2001 IEEE Computer Society Conference on Computer Vision and Pattern Recognition*, volume 1, pages 948–953, Kauai, HI, December 2001. IEEE Computer Society Press.

[18] L. J. van Vliet, I. T. Young, and A. L. D. Beckers. A nonlinear Laplace operator as edge detector in noisy images. *Computer Vision, Graphics and Image Processing*, 45(2):167–195, 1989.

[19] J. Weickert and T. Brox. Diffusion and regularization of vector- and matrix-valued images. In M. Z. Nashed and O. Scherzer, editors, *Inverse Problems, Image Analysis, and Medical Imaging*, volume 313 of *Contemporary Mathematics*, pages 251–268. AMS, Providence, 2002.

[20] M. Welk, C. Feddern, B. Burgeth, and J. Weickert. Median filtering of tensor-valued images. In B. Michaelis and G. Krell, editors, *Pattern Recognition*, volume 2781 of *Lecture Notes in Computer Science*, pages 17–24, Berlin, 2003. Springer.

VII

APPLICATIONS IN IMAGING SCIENCES

USING WATERSHED AND MULTIMODAL DATA FOR VESSEL SEGMENTATION: APPLICATION TO THE SUPERIOR SAGITTAL SINUS

N. Passat[1,2], C. Ronse[1], J. Baruthio[2], J.-P. Armspach[2] and J. Foucher[3]

[1]*LSIIT, UMR 7005 CNRS-ULP, Strasbourg I University, France*

[2]*IPB, UMR 7004 CNRS-ULP, Strasbourg I University, France*

[3]*U405 INSERM Strasbourg I University, France*

Abstract   Magnetic resonance angiography (MRA) provides 3-dimensional data of vascular structures by finding the flowing blood signal. Classically, algorithms dedicated to vessel segmentation detect the cerebral vascular tree by only seeking the high intensity blood signal in MRA. We propose here to use both cerebral MRA and MRI and to integrate a priori anatomical knowledge to guide the segmentation process. The algorithm presented here uses mathematical morphology tools (watershed segmentation and grey-level operators) to carry out a simultaneous segmentation of both blood signal in MRA and blood and wall signal in MRI. It is dedicated to the superior sagittal sinus segmentation but similar strategies could be considered for segmentation of other vascular structures. The method has been performed on 6 cases composed of both MRA and MRI. The results have been validated and compared to other results obtained with a region growing algorithm. They tend to prove that this method is reliable even when the vascular signal is inhomogeneous or contains artefacts.

Keywords: vessel segmentation, watershed segmentation, a priori knowledge, MRA, MRI

1.   Introduction

Magnetic resonance angiography (MRA) is a technique [5] frequently used to provide 3D images of cerebral vascular structures. The availability of precise information about brain vascular networks is fundamental for planning and performing neurosurgical procedures, but also for detecting pathologies such as aneurysms and stenoses. Since all classical image processing tools have been applied more or less successfully to the case of vessel segmentation, it might be interesting to explore new kinds of algorithms involving a priori knowledge. In a previous paper [8] we proposed a first attempt to use anatomical knowledge as a way to guide a segmentation algorithm. A major

C. Ronse et al. (eds.), Mathematical Morphology: 40 Years On, 419–428.
©2005 *Springer. Printed in the Netherlands.*

breakthrough of this work was the creation of an atlas dividing the head into different areas presenting homogeneous vessel properties. The use of this atlas enables to store a priori knowledge concerning the vessels located in each area and then to propose ad hoc segmentation algorithms. In this paper, we propose an algorithm dedicated to one of these areas, containing a main vessel of the venous tree: the superior sagittal sinus (SSS). This algorithm is based on mathematical morphology tools (watershed segmentation and grey-level operators). It also integrates a priori anatomical knowledge and uses both MRA and MRI data in order to take advantage of both acquisition techniques. It uses a multi-resolution slice by slice process, simultaneously segmenting the flowing blood signal in MRA and the blood and vessel wall in MRI. This paper is organized as follows. In Section 2, we review previous approaches concerning vessel segmentation. In Section 3, we describe the way to use anatomical knowlege. In Section 4, the proposed algorithm is described. In Section 5, technical details concerning the method and the database used for validation are provided. In Section 6, the method is tested and compared to a region growing algorithm. Discussion and projects are presented in Section 7.

2.   Related work

The vessel segmentation methods can be divided into several categories, corresponding to the main strategies used to carry out the segmentation. The first proposed strategies were based on filtering [4]. Method based on mathematical morphology (hysteresis thresholding in [7], grey level erosions and dilations in [3] or grey-scale skeletonization in [10]) and region growing [11] have also been proposed. More recently, methods based on vessel tracking [6], and crest line detection [1] have also been proposed.

It has to be noticed that very few vessel segmentation methods have been designed to process multimodal data. A method proposed in [9] for cerebral vascular structures visualization, uses both 3D MRA and 2D X-ray images. A method has been proposed by us in [8], where angiographic and non angiographic data are involved in an atlas-based region growing algorithm. Nevertheless, the simultaneous use of images from different modalities is quite unusual. The algorithm presented here, based on watershed [2] segmentation and mathematical morphology operators, proposes to uses both MRA and MRI to take advantage of anatomical knowledge concerning the brain superficial venous structures.

3.   A priori knowledge integration

The SSS presents many invariant properties (i.e. properties being identical for every subjects) which can be useful for guiding a segmentation process. These properties and a way to use them are described as follows.

Trajectory properties: A way to guide normal planes computation. The regular trajectory of the SSS and its position relatively to the cerebral median plane and the surface of the head theoretically enable to compute successive planes being perpendicular to the sinus axis. Indeed, if the surface of the head and the sagittal median plane of the brain can be found, then their intersection provides a curve. A normal plane computation using the points of that curve finally gives planes being normal to the sinus axis too. Using that strategy, it becomes possible to perform a segmentation of the sinus slice by slice.

Structures intensity and relative positions: A way to guide watershed segmentation. Observing MRA slices (Figure 1, right lower pictures), one can see that only the flowing blood is generally quite visible as it presents the highest intensity. In MRI, more structures can be observed in an easier way (Figure 1, right upper pictures). Although the flowing blood does not present a very high intensity, it can be observed surrounded by the dura mater. The brain hemispheres present a nearly identical intensity, such as a part of the skull. These four structures are separated by areas of low intensity and their relative positions are globally invariant.

MRA and MRI intensity properties can then be used to perform watershed segmentation on slices of the sinus region. Indeed, a gradient computation on the MRA should correctly delineate the blood from the remaining structures. A watershed segmentation could also be directly used on the MRI images to segment the different structures (considering low intensity regions as the frontier between them). Since the main problem of watershed segmentation remains oversegmentation, it is important to choose correct markers to initialize it. This could be done here by sharing information between MRA and MRI segmentation (the segmentation of blood in MRA could be used to find a marker for the dura mater in MRI, and vice versa) or by sharing information between successive MRI or MRA segmentations.

Structure homogeneity along the sinus trajectory: Justifying an iterative slice by slice strategy. The sinus and its neighboring structures present quite invariant position properties. By observing slices at different points on the SSS trajectory, we can also observe that their size and distance from each other are different (Figure 1 right pictures) but vary smoothly. This property could be efficiently used to start from one slice and successively generate markers to initialize segmentation of the neighboring slices. This is generally done in vessel tracking algorithms. A weakness of the vessel tracking approach is that a segmentation error in one slice will generally have consequences on all the following ones. In order to avoid such problems, an alternative could be to propose an iterative approach. For each slice, it consists in starting, from an average image of the current slice and its neighbors. A segmentation of this

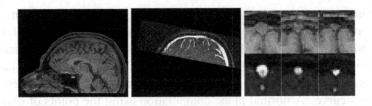

Figure 1.   Left: T1 MRI sagittal slice of the whole head. Middle: MRA sagittal slice of the top of the head. Right: slices of MRA and MRI data at various positions on the SSS.

average slice can then be used to generate markers for a new segmentation of a new average slice, closer from the current slice. This process can then be iterated until segmenting the real slice.

4.   Method

Input and output.   The method takes as input a classical MRI and a MRA of the same patient. They must contain at least the top of head and have to be correctly superimposed. If they are not, they can be superimposed by performing a rigid registration (using translations and rotations). Figure 1 illustrates an example of such data. The method provides two resulting images: a segmentation of the flowing blood detected in the MRA and a segmentation of the dura mater surrounding the blood and then forming the sinus wall, from the MRI.

Preprocessing.   The segmentation process is not carried out on the global images but on slices that must be normal to the sinus axis. A first step then consists in computing these slices. The sinus axis is parallel to a curve obtained by intersecting the cerebral sagittal median plane and the head surface. The surface of the head can be easily found by thresholding, while the sagittal median plane of the brain can be found by an histogram analysis of the MRA image. The intersection of that plane and the surface of the head then provides a discrete curve. A normal vectors computation on the points of the curve finally enables to compute planes being normal to the curve and also to the sinus axis.

That step then provides two sets of slices of the SSS. The first set is composed of MRI slices while the second contains MRA slices, the n-th slice of the first set corresponding to the n-th slice of the second. It has been experimentally observed that sets of 256 slices were sufficient to carry out the segmentation. Moreover, we sample the slices to keep 23×26 voxel-slices located 11 mm away from the head surface, assuming that small slices centered on the sinus and containing neighboring structures enable to obtain correct results with a lower computation time.

Definitions and notations.   In the following, a slice s of $n \times m$ voxels will be considered as a function $[0, n-1] \times [0, m-1] \rightarrow \mathbb{N}$. We will always assume that $\alpha \in \{mra, mri\}$. Let $\{s_i^{mra}\}_{i=0}^t$ and $\{s_i^{mri}\}_{i=0}^t$ be sequences of MRA and MRI slices. It has to be noticed that for any $a, b, c \in [0, t]$ such that $a < b < c$ the slice s_b^α is physically located between s_a^α and s_c^α. For each $i \in [0, t]$ let $\{I_k^i\}_{k=0}^\omega$ (in our case $\omega = 50$) be a sequence of intervals around i, decreasing from $[0, t]$ to $\{i\}$:

$$I_0^i = [0, t],$$

$$\forall x, y \in [0, \omega], x < y \Rightarrow I_y^i \subset I_x^i,$$

$$I_\omega^i = [i, i].$$

The lenght of an interval I will be denoted by $|I|$. For each $i \in [0, t]$ let $\{s_{i,k}^{mra}\}_{k=0}^\omega$ and $\{s_{i,k}^{mri}\}_{k=0}^\omega$ be the averaged sequences over I_k^i:

$$s_{i,k}^\alpha = \frac{1}{|I_k^i|} \sum_{m \in I_k^i} s_m^\alpha,$$

In the following, for $i \in [0, t]$ the result of the segmentation of s_i^α will be denoted b_i^α. Similarly, for $i \in [0, t]$ and $k \in [0, \omega]$, the result of the segmentation of $s_{i,k}^\alpha$ will be denoted $b_{i,k}^\alpha$. Then we will obtain:

$$b_{i,\omega}^\alpha = b_i^\alpha.$$

General description.   For each slice s_i^{mra}, we have defined two sequences $\{s_{i,k}^{mra}\}_{k=0}^\omega$ and $\{s_{i,k}^{mri}\}_{i=0}^\omega$. These sequences start respectively from an average image of all the MRA and MRI slices and finally come respectively to the i-th MRA and MRI slice. Assuming that both sequences will smoothly converge from an average image to the current slice, we propose the following segmentation strategy based on an iterative process:

1 initial segmentation of $s_{i,0}^{mra}$ and $s_{i,0}^{mri}$;

2 for $k = 1$ to ω:

    (a) segmentation of $s_{i,k}^{mri}$, using $b_{i,k-1}^{mra}$ and $b_{i,k-1}^{mri}$;

    (b) segmentation of $s_{i,k}^{mra}$, using $b_{i,k-1}^{mra}$ and $b_{i,k}^{mri}$.

The process starts from average images and iteratively uses previous segmentations of both modalities to carry out the current segmentation. This current segmentation is carried out by using a watershed algorithm while previous segmentations are used for creation of markers dedicated to watershed initialization. In the following paragraphs, the segmentation process is more precisely described. The gradient computation and template creation steps are explained in specific paragraphs.

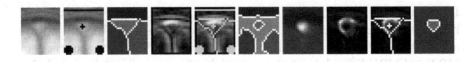

Figure 2.   Initialization process. From left to right: MRI average slice ($s_{i,0}^{mri}$); MRI with four markers; first segmentation; MRI gradient; MRI gradient with five markers; final MRI segmentation ($b_{i,0}^{mri}$); MRA average slice ($s_{i,0}^{mra}$); MRA gradient; MRA gradient with two markers; MRA segmentation ($b_{i,0}^{mra}$).

Initialization.   The first step consists in performing the segmentation of $s_{i,0}^{mra}$ and $s_{i,0}^{mri}$ for $i \in [0, \omega]$. By definition, for all $i, j \in [0, \omega]$, $s_{i,0}^{\alpha} = s_{j,0}^{\alpha}$. Thus the initialization step only requires to segment two average slices, one for the MRA and the MRI. The initialization is organized as follows (Figure 2):

1   grey-level opening of the MRA slice with a flat structuring element (3×3 cross): the points of maximum value become the markers for the flowing blood in the MRA and MRI slices;

2   successive grey-level openings of the MRI slice with three flat structuring elements (a one voxel width line and two 5×5 circular elements): the points of maximum value become the markers for the brain and the bone in the MRI slices;

3   watershed segmentation of the MRI slice using the four markers;

4   gradient computation of the MRI slice and watershed segmentation of the gradient MRI slice using the four markers plus a new marker provided by the frontier between the four regions of the previous segmentation;

5   gradient computation of the MRA slice and watershed segmentation of the gradient MRA slice using one marker plus a new marker provided by the frontier between the four regions of the first MRI segmentation.

Standard shape and size of the different structuring elements have been chosen in order to fit the different structures to find. That step finally provides $b_{i,0}^{mri}$ and $b_{i,0}^{mra}$ then enabling to initialize the iterative process for each $i \in [0, t]$.

Iterative process.   For each $i \in [0, t]$, at the step k, the iterative process consists in first segmenting the current MRI slice ($s_{i,k}^{mri}$) by using the MRA and MRI segmentation of the previous step ($b_{i,k-1}^{mra}$ and $b_{i,k-1}^{mri}$). Then the current MRA slice can be segmented using the current MRI segmentation ($b_{i,k}^{mri}$) and the previous MRA segmentation ($b_{i,k-1}^{mra}$). A step calculating $b_{i,k}^{mra}$ and $b_{i,k}^{mri}$ can be decomposed as follows (Figure 3):

Figure 3. A step of the iterative process for one slice. From left to right: MRI average slice ($s_{i,k}^{mri}$); MRI with four markers; first segmentation; MRI gradient; MRI gradient with five markers; final MRI segmentation ($b_{i,k}^{mri}$); MRA average slice ($s_{i,k}^{mra}$); MRA gradient; MRA gradient with two markers; MRA segmentation ($b_{i,k}^{mra}$).

1 creation of four markers to initialize $s_{i,k}^{mri}$ segmentation;

2 watershed segmentation of $s_{i,k}^{mri}$, using the four markers;

3 gradient computation of $s_{i,k}^{mri}$ and watershed segmentation of $s_{i,k}^{mri}$ gradient, using the four markers plus one marker provided by the frontier between the four regions found by the previous watershed segmentation;

4 creation of one marker to initialize $s_{i,k}^{mra}$ segmentation;

5 gradient computation of $s_{i,k}^{mra}$ and watershed segmentation of $s_{i,k}^{mra}$ gradient, using the marker plus one marker provided by the frontier between the four regions found by the first watershed segmentation of $s_{i,k}^{mri}$;

Gradient computation. For each step, the segmentation of both MRA and MRI slices requires the computation of gradient images. Concerning the MRI slices, the gradient is computed by choosing the maximum intensity variation in the four principal directions, then correctly delineating the four main structures from the low intensity regions. This gradient calculation gives correct results as the four regions of interest have homogeneous intensity levels. This is not the case of the flowing blood in the MRA slices. Indeed, the blood present a very high but heterogeneous level. Since computing a simple gradient does not allow to obtain well defined frontiers, a solution consists in dividing the gradient value calculated at a pixel by the pixel value. This normalized gradient will present low values in homogeneous high intensity regions and high values for the background points located at the frontier with the flowing blood.

Markers generation. At each step of the iterative process and for each slice, it is necessary to generate markers to initialize the watershed segmentations. The segmentation of the MRI slice requires four markers: one for the dura mater, one for both hemispheres and one for the skull. For any MRI slice $s_{i,k}^{mri}$, the markers are generated as follows. First, three templates, for hemispheres and skull are created from $s_{i,k-1}^{mri}$. They are obtained by performing a grey-level erosion with a flat structuring element (3×3 square) on $s_{i,k-1}^{mri}$. For each template, $b_{i,k-1}^{mri}$ is used as a mask to indicate what are the regions where the erosion has to be performed. A fourth template, for the dura mater, is obtained by performing a grey-level erosion with a flat structuring element (3×3 cross) on $s_{i,k-1}^{mri}$. Both $b_{i,k-1}^{mri}$ and $b_{i,k-1}^{mra}$ are used as masks to indicate what are the regions where the erosion has to be performed. Then, for each template a grey-level erosion of $s_{i,k}^{mri}$ using the current template as a grey-level structuring element is performed. After the erosion, a dilation using the template as a binary element is carried out at the maximum point of the eroded image, then providing a marker for watershed initialization.

The segmentation of the MRA slice requires one marker indicating the position of the flowing blood. For any MRA slice $s_{i,k}^{mra}$, the marker is choosen as being the pixel of maximal intensity in the region of $s_{i,k}^{mra}$ delimited by the vascular region segmented in $b_{i,k-1}^{mra}$.

Postprocessing. The segmentation provides two sets of slices. The first gives the segmentation of the flowing blood in the MRA set while the second gives the segmentation of the sinus blood and wall in the MRI set. Then, these slices have to be put back to their correct position in the initial images. During this step, it might happen that small gaps appear between successives slices. Since these gaps are quite small (their thickness is never larger than one voxel), they can then be filled by using a binary closing with a linear structuring element composed of 3 voxels and oriented according to the direction of the axis. As a very last step, the user can also choose to apply a binary opening to smooth the image (we propose here to use a $3 \times 3 \times 3$ cross).

5.   Experimental section

Data acquisition.   A database of 6 patients has been used to validate the efficiency of the proposed algorithm. For each patient, two MR data have been provided (Figure 1): a T1 MRI of the whole head ($280 \times 240 \times 256$ voxels) and a MRA of the top of the head ($192 \times 70 \times 256$ voxels). Voxels are cubes of 1 mm edge.

Complexity and computation time.   The proposed algorithm has a complexity of $O(\omega.t.x.y.m)$, where ω is the number of iterations, t is the number of slices, x and y are the slice dimensions and m is the maximum area of all the used structuring elements ($m \ll x.y$). The images have been segmented with a computer using a 2.4 GHz Pentium IV processor with 2 GB of memory. The average computation time is then 6 minutes. It has to be noticed that the proposed algorithm runs in an entirely automatic fashion.

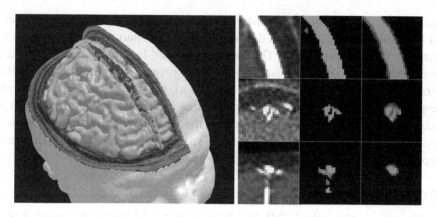

Figure 4. Left: flowing blood surrounded by the SSS wall. Right: results provided by the proposed method and a region growing method. First column: MRA data. Second column: region growing segmentation. Third column: proposed segmentation.

6. Results and discussion

The results obtained with the proposed method have been compared to those provided by a region growing algorithm proposed in [8]. All the validations have been carried out by a human specialist who qualitatively tested both algorithms on each case of the previously described database.

During the validation, it has been observed that the method could segment the flowing blood even when MRA signal was heterogeneous or low. The right pictures of Figure 4 illustrate the main observations of the validations. For highly homogeneous intensity regions, it has been observed that both algorithms provide correct segmentations. Nevertheless, the region growing algorithm is sensitive to aliasing artefacts while the proposed algorithm is more robust, also segmenting the low intensity flowing blood in the middle of the vessel. The proposed algorithm is also able to segment only the sinus while the region growing algorithm also segments connected veins.

7. Conclusion

This paper presents a novel method, based on watershed segmentation and mathematical morphology operators guided by anatomical knowledge. This method is dedicated to SSS segmentation from brain multimodal MR data. It has been tested on 6 cases, providing more precise results than a previously proposed region growing algorithm, even in case of strong inhomogeneity of signal in MRA data. The main originality of this work consists in integrating high level anatomical knowledge, and using both MRA and MRI data in order to guide mathematical morphology tools. A first attempt to integrate anatomical knowledge in a vessel segmentation process had already been proposed in [8], where an atlas was used to divide the brain into areas having homogeneous vascular properties. The method proposed here can be considered as being dedicated to one of these areas (the SSS area), then proposing a reliable strategy for the vessels it contains. This work makes part of a new kind of segmentation strategy consisting in processing each part of a vascular tree in a adapted fashion, instead of processing all the vessels in a global way. Further work will now consist in using this method as a first step for segmentation and topology recovery of the whole cerebral superficial venous tree.

Acknowledgements

The authors thank the EPML #9 CNRS-STIC for its financial support. They also thank C. Maillot for his precious knowledge on human anatomy.

References

[1] S.R. Aylward and E. Bullit. Initialization, noise, singularities, and scale in height ridge traversal for tubular object centerline extraction. *IEEE Transactions on Medical Imaging*, 21:61–75, 2002.

[2] S. Beucher and F. Meyer. *The morphological approach to segmentation: the watershed transformation*, chapter 12, pages 433–481. E.R. Dougherty, Ed. Marcel Dekker, 1993.

[3] H.E. Cline, D.R. Thedens, C.H. Meyer, D.G. Nishimura, T.K. Foo, and S. Ludke. Combined connectivity and a grey-level morphological filter in magnetic resonance coronary angiography. *Magnetic Resonance in Medicine*, 43:892–995, 2000.

[4] Y.P. Du and D.L. Parker. Vessel enhancement filtering in three-dimensional angiograms using long-range signal correlation. *Journal of Magnetic Resonance Imaging*, 7:447–450, 1997.

[5] C.L. Dumoulin and H.R. Hart. Magnetic resonance angiography. *Radiology*, 161:717–720, 1986.

[6] N. Flasque, M. Desvignes, J.M. Constans, and M. Revenu. Acquisition, segmentation and tracking of the cerebral vascular tree on 3D magnetic resonance angiography images. *Medical Image Analysis*, 5:173–183, 2001.

[7] G. Gerig, T. Koller, G Székely, C. Brechbühler, and O. Kübler. Symbolic description of 3-D structures applied to cerebral vessel tree obtained from MR angiography volume data. In *IPMI'93*, volume 687 of *LNCS*, pages 94–111, 1993.

[8] N. Passat, C. Ronse, J. Baruthio, J.-P. Armspach, C. Maillot, and C. Jahn. Atlas-based method for segmentation of cerebral vascular trees from phase-contrast magnetic resonance angiography. In *SPIE Image Processing 2004*, volume 5370, pages 420–431, 2004.

[9] A.R. Sanderson, D.L. Parker, and T.C. Henderson. Simultaneous segmentation of MR and X-ray angiograms for visualization of cerebral vascular anatomy. In *VIP'93*, 1993.

[10] P.J. Yim, P.L. Choyke, and R.M. Summers. Gray-scale skeletonization of small vessels in magnetic resonance angiography. *IEEE Transactions on Medical Imaging*, 19:568–576, 2000.

[11] C. Zahlten, H. Jürgens, C.J.G. Evertsz, R. Leppek, H.-O. Peitgen, and K.J. Klose. Portal vein reconstruction based on topology. *European Journal of Radiology*, 19:96–100, 1995.

USING GREY SCALE HIT-OR-MISS TRANSFORM FOR SEGMENTING THE PORTAL NETWORK OF THE LIVER

Benoît Naegel[1,2], Christian Ronse [1] and Luc Soler [2]

[1]*LSIIT UMR 7005 CNRS-ULP*
Pôle API, Boulevard Sébastien Brant, B.P. 10413, 67412, Illkirch CEDEX, France

[2]*IRCAD, Virtuals*
1, place de l'Hôpital, 67091 Strasbourg Cedex, France

Abstract   In this paper we propose an original method of segmentation of the portal network in the liver. For this, we combine two applications of the grey scale hit-or-miss transform. The automatic segmentation is performed in two steps. In the first step, we detect the shape of the entrance of the portal vein in the liver by application of a grey scale hit-or-miss transform. This gives the seed or starting point of the region-growing algorithm. In a second step, we apply a region-growing algorithm by using a criterion still based on a hit-or-miss. Our method performs better than a previous method based on region-growing algorithm with a single threshold criterion.

Keywords: Vessel segmentation, grey scale hit-or-miss transform, shape detection, CT-scan.

1.   Introduction

In liver surgery, non-healthy segments are removed to prevent tumoral proliferation. The liver, indeed, is an organ composed of eight functional segments. Liver's functional segmentation is based on one of its vessel systems: the portal network. Hence, precise segmentation of this network is highly desirable since it improves the preoperative planning.

In this work, we propose to detect the portal network from 3D CT-scan images of the abdomen. For this, we use a priori knowledge about the intensity and the shape of the starting point of the portal network: the extra-hepatic part of the portal vein. In a first step, we achieve a shape detection of this structure by using a definition of the grey scale hit-or-miss transform operator. In a second step, we use a region-growing algorithm with a hit-or-miss criterion to detect the points belonging to the portal network.

C. Ronse et al. (eds.), Mathematical Morphology: 40 Years On, 429–440.

A method for the segmentation of the portal network has been given by Selle et al. in [15–17] and Zahlten et al. in [26, 27]. A seed voxel of the portal vein close to its entrance into the liver is selected interactively. Then the algorithm iteratively accumulates all 26-connected neighbors whose grey values exceed a given threshold, by using a classical region-growing algorithm. In order to reconstruct the greater part of the portal network without including other structures, the authors compute an optimal threshold for the region-growing algorithm.

In [5, 6], Dokládal proposes two approaches to preserve the topology of the structure: one which adds only simple points during a region growing process and one which homotopically reduces a simply connected superset of the object to segment.

A method based on histogram thresholding and topological refinement is described by Soler in [22].

We can also cite the work of Krissian et al. [8] which describes a template-based approach to detect tubular structures in 3D images.

It should be mentioned that some of the previous methods perform only on the liver mask, which reduces greatly the problem.

The method proposed in this paper tries to combine the advantages of previous methods. Our method performs the segmentation of the portal network from a whole CT-scan image. Moreover, it is fully automatic: it does not require user interaction.

2.   Shape detection: the hit-or-miss transform

A morphological operator that can be used to perform shape detection is the hit-or-miss transform. The binary hit-or-miss transform is widely used, for example in document analysis [2, 3]. Hardware implementations with optical correlators have been studied in [7, 9, 10, 14, 24, 25].

Binary definition

Definition of hit-or-miss transform in the binary case is very common:

$$X \otimes (A, B) = \{x \mid A_x \subseteq X, B_x \subseteq X^c\} \qquad (1)$$

where A and B are two structuring elements. Of course, A and B must be disjoint sets: $A \cap B = \emptyset$.

Grey scale hit-or-miss transform

Few works address the grey scale hit-or-miss transform. Principal definitions are the following:

1 Ronse's definition appears in [12]. Given two structuring functions A, B and a function F, the grey scale hit-or-miss transform is defined on a point by:

$$[F \otimes_1 (A, B)](p) = \begin{cases} (F \ominus A)(p) & \text{if } (F \ominus A)(p) \geq (F \oplus B^*)(p) \\ \bot & \text{otherwise} \end{cases}$$

where $B^*(p) = -B(-p)$ and $\bot$ is the minimum value of function F.

2 Soille's definition appears in [19] and is also described in [21]. In fact two definitions are described: the unconstrained hit-or-miss transform (UHMT) and the constrained hit-or-miss transform (CHMT). UHMT leads to the following definition, applied on a function F, given two structuring sets A and B:

$$[F \otimes_2 (A, B)](p) = \begin{cases} (F \ominus A)(p) - (F \oplus \check{B})(p) & \text{if } (F \ominus A)(p) > \\ & (F \oplus \check{B})(p) \\ 0 & \text{otherwise} \end{cases}$$

where $\check{B} = \{-b \mid b \in B\}$. Obviously this definition can be extended with structuring functions.

Discussion

Grey scale hit-or-miss transform is very interesting since it permits the detection of a structure of a given shape directly in a grey scale image. The two definitions above are useful; however for our application we would like to keep the intensity of the points belonging to the hit-or-miss transform in order to threshold image afterwards.

For our application, we propose to use a slightly modified definition of $\otimes_1$:

$$[F \otimes_3 (A, B)](p) = \begin{cases} F(p) & \text{if } (F \ominus A)(p) > (F \oplus B^*)(p) \\ \bot & \text{otherwise} \end{cases}$$

where A and B are structuring functions.

The three definitions of the grey scale hit-or-miss are illustrated on Figure 1.

3.   Method

Data

We are working on 3D CT-scans of the abdomen. The protocol includes an injection of a contrast medium in order to enhance the healthy parts of the liver. Image is acquired at portal time, the time needed for the contrast medium to reach the portal vein. Thus the portal network is highlighted: its intensity is

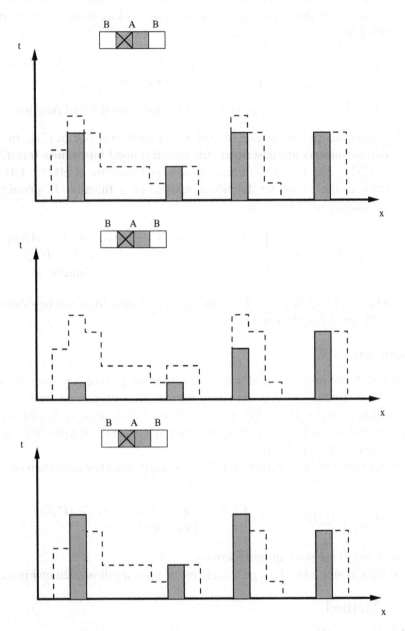

Figure 1.   Grey scale hit-or-miss transform. Up: Ronse's definition. Middle: Soille's definition. Down: proposed definition. Dashed: original function. In grey: result of the hit-or-miss
transform with structuring elements *A* and *B* (*A* and *B* are functions in the first and third figure,
and sets in the second).

greater than the liver's one. Typical size of such 3D images is $512 \times 512 \times 100$, but the number of slices can vary. Images are anisotropic: the size of the voxels, which is not constant between images, is typically 0.5mm $\times$ 0.5mm $\times$ 2mm.

Preprocessing of source image. First, images are reduced to make them isotropic, with a voxel size of two millimeters in all directions. This permits the use of structuring elements of identical shapes in all images. Moreover, the computation of morphological operators is greatly accelerated. Original CT-scan images contain usually random noise. We use a median filter, which is efficient to remove such noise. The neighborhood used is a sphere of 6 millimeter radius. In the sequel, we call I the function defining the preprocessed original image.

First step: detection of the starting point

The entrance of the portal vein in the liver can be considered as the root of the vascular portal network of the liver. The first step of our method consists in the automatic detection of this point. In order to do this, we use robust a priori knowledge of this structure.

A priori knowledge.

- Photometric attributes: due to the acquisition protocol, the portal vein has a greater intensity than the one of the liver. Since the peak of the liver is highly visible on the image histogram, we use this information to retain only the points having a higher intensity than the liver. We use also the fact that the portal vein intensity is lower than the bone intensity.

- Shape attributes: the extra-hepatic part of the portal vein (the starting point of the portal network) has a very characteristic shape. Its origin is constituted by the ending of the superior mesenteric vein (in brief SMV); it makes a change of direction before entering the liver (see Figure 2). Anatomically, it has been observed that the extra-hepatic part of the portal vein presents very few variations between patients.

Principle of the method. The method used to detect the entrance of the portal vein in the liver is based on the fact that portal vein is connected to the SMV which is a vertical cylinder. We perform two grey scale hit-or-miss transforms (by using the third definition): one to detect the quasi horizontal part of the entrance of the portal vein (in brief EPV) in the liver and one for the detection of the SMV. We threshold each image by the upper bound of the intensity of the liver and the lower bound of the intensity of the bones. Then we

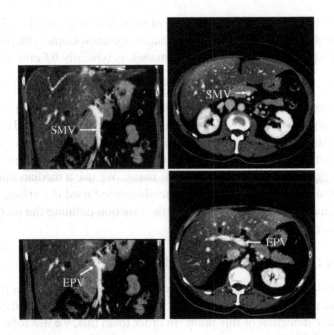

Figure 2.   SMV: superior mesenteric vein. EPV: entrance of the portal vein in the liver (extra-hepatic part of the portal vein). Upper and lower left: Frontal view. Upper and lower right: Axial view. (Original images)

keep only the structures of the first image that are connected to the structures of the second image.

Detection of the superior mesenteric vein (SMV).   A stable referential used in anatomy is the vertebral column. We use the hypothesis that the SMV is located in front of the vertebral column to compute a region of interest (ROI) (see Figure 3, left). Since the SMV is a vertical cylindrical trunk, we perform in this ROI a grey scale hit-or-miss transform with two structuring elements: a discrete vertical cylinder C orthogonal to the axial plane of 12mm length and 2mm radius, and a discrete hollow cylinder H parallel to the latter of 2mm length, 12mm radius and 4mm thickness. By using this operator, we keep only vertical cylindrical structures surrounded by a darker neighborhood. We threshold this result with the highest intensity of the liver $liver_{high}$ and the lowest intensity of the bones $bones_{low}$. After these steps, we obtain an image constituted of the SMV, and other structures. Typically, we also obtain aorta and inferior veina cava.

We call I_1 the resulting image:

$$I_1 = \text{TH}(I \otimes_3 (C, H), liver_{high}, bones_{low})$$

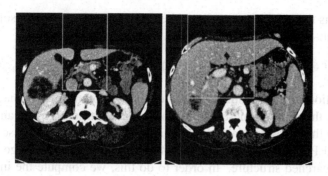

Figure 3. Left: 3D region of interest for the detection of the SMV. We keep only the region located in front of the vertebral column and delimited by its left and right edges. Right: 3D region of interest for the detection of the EPV. Example on one slice (axial view).

Figure 4. Mean shape of the EPV on 18 cases. Left: Sagittal view. Middle: Frontal view. Right: Axial view.

$$\text{where TH}(I, s_1, s_2)(p) = \begin{cases} I(p) & \text{if } s_1 \leq I(p) \leq s_2 \\ \bot & \text{otherwise} \end{cases}$$

Detection of the entrance of the portal vein in the liver (EPV). We use another region of interest for the detection of EPV: since the portal vein makes a change of direction before entering the liver, the ROI should be slightly larger than for the SMV. We add to the region of interest of the SMV another zone of same width in the liver direction (see Figure 3, right). The entrance of the portal vein (EPV) in the liver can be described as a quasi horizontal cylinder. This structure has very few variations between patients. We can see on Figure 4 the mean shape of EPV on 18 cases. However, size of EPV can vary between patients.

We perform multiple grey scale hit-or-miss transforms to detect EPV. To deal with the size variability, we use a set of structuring elements of different sizes. The first structuring element C remains constant, while the second structuring elements H (hollow cylinders) have a variable radius. Structuring elements used are: $\text{SE} = \{(C, H_1), (C, H_2), (C, H_3)\}$, where C is an elementary discrete horizontal cylinder orthogonal to the sagittal plane, of 6mm length and 2mm radius; the H_i's are hollow discrete cylinders parallel to C, of same length, with radius of $4 + 2 * i$ mm, and 4mm thickness. We compute the image

I_2, which is the pointwise maximum between the three images resulting from the three hit-or-miss transforms:

$$I_2 = \text{TH}(I \otimes_3 (C, H_1) \vee I \otimes_3 (C, H_2) \vee I \otimes_3 (C, H_3), liver_{high}, bones_{low})$$

Combination of both results.   We now have two images: an image I_1 constituted by the superior mesenteric vein and other structures, and an image I_2 containing the entrance of the portal vein in the liver and some false positives. Since the EPV is connected to the SMV, we combine both images to retain only the searched structure. In order to do this, we compute the intersection image of I_1 and I_2: INTER $= I_1 \wedge I_2$. To keep only the entrance of the portal vein in the liver, we perform a geodesical reconstruction by dilation of INTER in I_2: $I_3 = \text{Rec}_\oplus(I_2, \text{INTER})$. Finally, image I_3 contains the EPV. Some false positives can also exist, but the combination of images I_1 and I_2 has drastically reduced their number. To keep only the EPV, we keep the biggest connected component of the image.

Second step: propagation using a hit-or-miss criterion

We have now detected the starting point of the portal network. To segment the vascular network, we use these assumptions:

1 Portal network is connected and can be seen as a tree having as root the EPV.

2 Branches of the portal network are bright tubular structures surrounded by dark areas.

According to the first assumption, we use a region-growing algorithm having for seed the EPV. According to the second assumption, we propose to use a criterion of propagation based on a local grey scale hit-or-miss transform. The criterion, that can also be seen as a contrast criterion, is the following:

$$C(p) = \begin{cases} \text{true} & \text{if at least one of} \\ & [I \otimes_3 (o, R_1)](p), [I \otimes_3 (o, R_2)](p), [I \otimes_3 (o, R_3)](p) \\ & \text{is} >\perp \\ \text{false} & \text{otherwise} \end{cases}$$

where o is a structuring element composed of the origin. The R_i's are structuring elements used to constrain the point p to belong to a tubular structure. Formally, structuring elements R_i are three orthogonal discrete rings parallel to the three planes XY, XZ, YZ. They are defined by a radius and a thickness (see Figure 5). These parameters must be chosen according to the thickness and the density of the vascular

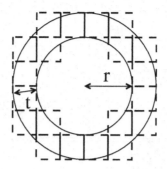

Figure 5.   Discrete ring (dashed) superimposed on a continuous ring of radius r and thickness t

network. In order to detect branches of radius r_{branch}, we should use rings which have a minimal radius $r = r_{branch}$. However, if the network is dense, r should not be too high. For our application, these parameters are chosen empirically.

To perform the region-growing, we use only the C criterion. Starting from the seed point (the EPV), we accumulate iteratively all 26-connected neighbors for which the criterion is *true*.

Discussion.   This step of region-growing is based on a hit-or-miss criterion, which permits the detection of bright tubular structures. We use only three structuring elements (the three orthogonal rings) in order not to be too constraining in the detection. This permits, for example, the detection of forks which are essential components of a network.

4.   Results

We can see on Figure 6 (left) the result of our method obtained with rings of radius 10mm and 2mm thickness. This result is compared to the method of Selle et al. (Figure 6, center and right), which use also a region-growing algorithm but with a single threshold criterion. The optimal threshold is computed by analyzing the histogram of the propagation, i.e. for each threshold the number of segmented voxels is computed. The drawback of a region-growing algorithm based only on a threshold is that for a low threshold, the propagation extends outwards of the vascular network (Figure 6, right). With a higher threshold (optimal threshold on Figure 6, center), few branches are included.

Actually, a true optimal threshold, i.e. a threshold that makes a correct separation between the voxels belonging to the vascular network and those belonging to other structures (in our application, the liver's voxels) does not exist in our case.

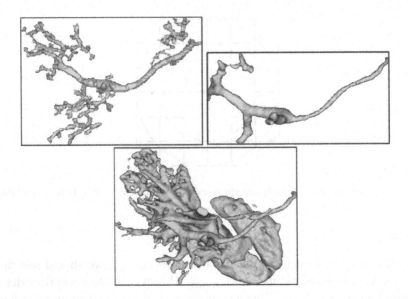

Figure 6.   Comparison between our method (upper left) and Selle's and al. method (upper right: with the optimal threshold; down: with a low threshold).

Visually, our method includes more branches than the Selle's algorithm, without the drawback of segmenting non-vessel structures. This is due to the fact that we include in the segmentation only points for which local neighborhood indicates that the point is likely included in a tubular structure.

It should be noted that, using both methods, not only the portal network is segmented. We can see on the right the splenic vein and part of the spleen network. This is not an issue since we can afterwards keep only the network located on the left of the seed point (the EPV).

We have successfully tested this method on 16 cases. The first step of automatic detection of the extra-hepatic segment of the portal vein performs well in all cases. The second step permits to obtain a segmentation of the portal network of the liver. The number of branches that are segmented and the global quality of the segmentation is variable between the cases, depending on the quality of the CT-scan acquisition.

5.   Conclusion

In this paper we have described an original method for the segmentation of the portal network of the liver. This method is based on robust a priori knowledge and shape detection. The second step is based on a region-growing algorithm with a modified criterion that uses a hit-or-miss transform.

This algorithm performs better than the Selle's et al. algorithm, since more branches are included in the segmentation while it does not extend outwards of the vascular network.

However our method should be evaluated quantitatively, but in the field of vessel segmentation ground-truth is seldom available. Moreover manual segmentation by an expert is very time consuming and prone to errors due to the high inter-slice distance.

Future works will include a quantitative validation of our results by comparison with manually delineated networks. There is still a lot of research to be done towards using the full potential of the grey scale hit-or-miss transform in pattern recognition. One of the subject that should be investigated further concerns the choice of the structuring elements.

Acknowledgments

The authors wish to thank S. Thery and J. Lamy for providing a mesh computing software and a 3D visualization tool.

References

[1] D. Bloomberg. Multiresolution morphological approach to document image analysis. In *International Conference on Document Analysis and Recognition, Saint-Malo, France*, pages 963–971, 1991.

[2] D. Bloomberg and P. Maragos. Generalized hit-miss operations. In *SPIE Conference Image analysis and morphological image processing.*, volume 1350, pages 116–128, July 1990.

[3] D. Bloomberg and L. Vincent. Pattern matching using the blur hit-miss transform. *Journal of Electronic Imaging*, 9(2):140–150, April 2000.

[4] D. Casasent and E. Botha. Optical symbolic substitution for morphological transformations. *Applied Optics*, 27(18):3806–3810, 1988.

[5] P. Dokládal, C. Lohou, Perroto L., and G. Bertrand. A new thinning algorithm and its application to extraction of blood vessels. In *BioMedSim'99*, pages 32–37, April 1999.

[6] P. Dokládal, C. Lohou, L. Perroton, and G. Bertrand. Liver blood vessels extraction by a 3-D topological approach. In *MICCAI'99*, pages 98–105, September 1999.

[7] J.B. Fasquel. *Une méthode opto-informatique de détection et de reconnaissance d'objets d'intérêt: Application à la détection des lésions cancéreuses du foie et à la vérification en temps-réel de signatures manuscrites*. Thèse de Doctorat, Université Louis Pasteur, Strasbourg I, 2002.

[8] K. Krissian, G. Malandain, N. Ayache, R. Vaillant, and Y. Trousset. Model-based detection of tubular structures in 3D images. *Computer Vision and Image Understanding*, 80(2):130–171, 2000.

[9] H. Liu, M. Wu, G. Jin, G. Cheng, and Q. He. An automatic human face recognition system. *Optics and Lasers in Engineering*, 30:305–314, 1998.

[10] H. Liu, M. Wu, G. Jin, Y. Yan, and G. Cheng. Optical implementation of the morphological hit-miss transform using extensive complementary encoding. *Optics Communications*, 156:245–251, 1998.

[11] C. Ronse. Extraction of narrow peaks and ridges in images by combination of local low rank and max filters: implementation and applications to clinical angiography. Working document, Philips Research Laboratory Brussels, October 1988.

[12] C. Ronse. A lattice-theoretical morphological view on template extraction in images. *Journal of Visual Communication and Image Representation*, 7(3):273–295, 1996.

[13] C. Ronse. Removing and extracting features in images using mathematical morphology. *CWI Quarterly*, 11(4):439–457, 1998.

[14] R. Schaefer and D. Casasent. Nonlinear optical hit-miss transform for detection. *Applied Optics*, 34(20):3869–3882, 1995.

[15] D. Selle, B. Preim, A. Schenk, and H.-O. Peitgen. Analysis of vasculature for liver surgery plannning. *IEEE Transactions on Medical Imaging*, 21(11):1344–1357, 2002.

[16] D. Selle, T. Schindewolf, C. Evertsz, and H. Peitgen. Quantitative analysis of CT liver images. In *Proceedings of the First International Workshop on Computer-Aided Diagnosis*, pages 435–444. Elsevier, 1998.

[17] D. Selle, W. Spindler, B. Preim, and H.-O. Peitgen. *Mathematical Methods in Medical Image Processing: Analysis of vascular structures for Preoperative Planning in Liver Surgery*, pages 1039–1059. Springer, 2000.

[18] D. Sinha and P. Laplante. A rough set-based approach to handling spatial uncertainty in binary images. *Engineering Applications of Artifical Intelligence*, 17:97–110, 2004.

[19] P. Soille. Advances in the analysis of topographic features on discrete images. In *DGCI: 10th International Conference*, volume 2301 of *Lecture Notes in Computer Science*, pages 175–186. Springer, March 2002.

[20] P. Soille. On morphological operators based on rank filters. *Pattern Recognition*, 35:527–535, 2002.

[21] P. Soille. *Morphological Image Analysis: Principles and Applications*. Springer-Verlag, Berlin, Heidelberg, 2nd edition, 2003.

[22] L. Soler. *Une nouvelle méthode de segmentation des structures anatomiques et pathologiques: application aux angioscanners 3D du foie pour la planification chirurgicale*. Thèse de Doctorat, Université d'Orsay, 1999.

[23] M. Van Droogenbroeck and H.. Talbot. Fast computation of morphological operations with arbitrary structuring elements. *Pattern Recognition Letters*, 17(14):1451–1460, 1996.

[24] F.T.S Yu and S. Jutamulia. *Optical Pattern Recognition*. Cambridge University Press, first edition, 1998.

[25] S. Yuan, M. Wu, G. Jin, X. Zhang, and L. Chen. Programmable optical hit-miss transformation using an incoherent optical correlator. *Optics and Lasers in Engineering*, 24:289–299, 1996.

[26] C. Zahlten, H. Jürgens, C. Evertsz, R. Leppek, H. Peitgen, and K. Klose. Portal vein reconstruction based on topology. *European Journal of Radiology*, 19(2):96–100, 1995.

[27] C. Zahlten, H. Jürgens, and H. Peitgen. Reconstruction of branching blood vessels from CT-data. In R. Scateni, J. van Wijk, and P. Zanarini, editors, *Workshop of Visualization in Scientific Computing*, pages 41–52. Springer-Verlag, 1995.

BLOOD CELL SEGMENTATION USING MINIMUM AREA WATERSHED AND CIRCLE RADON TRANSFORMATIONS

F. Boray Tek, Andrew G. Dempster and Izzet Kale

University of Westminster
Applied DSP and VLSI Research Group
[f.b.tek,dempsta,kalei]@westminster.ac.uk

Abstract   In this study, a segmentation method is presented for the images of microscopic peripheral blood which mainly contain red blood cells, some of which contain parasites, and some white blood cells. The method uses several operators based on mathematical morphology. The cell area information which is estimated using the area granulometry (area pattern spectrum) is used for several steps in the method. A modified version of the original watershed algorithm [31] called minimum area watershed transform is developed and employed as an initial segmentation operator. The circle Radon transform is applied to the labelled regions to locate the cell centers (markers). The final result is produced by applying the original marker controlled watershed transform to the Radon transform output with its markers obtained from the regional maxima. The proposed method can be applied to similar blob object segmentation problems by adapting red blood cell characteristics for the new blob objects. The method has been tested on a benchmark set and scored a successful correct segmentation rate of 95.40%.

Keywords: Blood cell, watershed, area granulometry, minimum area watershed transform (MAWT), circle Radon Transform (CRT)

Introduction

The first step in a computerized microscopic blood cell image analysis system is segmentation. The blood cells have to be segmented into separate regions for further analysis. The term "blood cell segmentation" has been used to refer to different notions in the literature. In white blood cell (WBC) segmentation papers [32], [15], [21], "blood cell segmentation" tends to refer to the localization of white blood cells and segmentation of the white blood cell body into the structures such as cytoplasm and nucleus. As opposed to this, here "the blood cell segmentation" refers to the segmentation of an image of

441

C. Ronse et al. (eds.), Mathematical Morphology: 40 Years On, 441–454.
©2005 *Springer. Printed in the Netherlands.*

mainly red blood cells (RBC) into separate labelled regions each representing an individual blood cell for further analysis.

Some methods exist which are directly related to the red blood cell segmentation as described in this paper [7], [6], [24], [22], [4], [2]. Several techniques have been proposed for segmentation and preventing under-segmentation: granulometries and regional extrema analysis [7], [6] distance transform and area tophats [24], [22], Bayesian colour segmentation and watershed segmentation improved by discrete chamfer distance [4]. However, most of the studies do not provide evaluation of the segmentation performance nor do they solve the under-segmentation problem completely. Hence, the under-segmentation remains a problem in blood cell segmentation. In blob segmentation, a common technique to reduce under-segmented regions is to utilize the distance transform on the binary image [24]. However, we observed that this approach also produces over-segmented regions when dividing the under-segmented regions.

Here, we develop a blood cell segmentation method using the watershed transformation [31] and the Radon transformation [28]. The method employs the watershed transformation in two different stages. First, we introduce the minimum area watershed transform (MAWT) to obtain an initial segmentation. After extracting markers from this segmentation using the circle Radon transformation, a marker-controlled watershed transformation [26] is applied to the Radon transformed image to obtain a new final segmentation. Most of the steps in the proposed method use blood cell radius (or area) as an attribute. The radius and the area are estimated using area granulometry (area pattern spectrum) [23]. Furthermore, a benchmark set has been prepared manually for evaluating segmentation performance. The proposed algorithm and an earlier algorithm proposed by Rao *et.al.* [24], [22] are compared by testing on the same benchmark set.

In section 2 the method will be explained in details. In section 3 the experimental method will be explained and a comparison table given for the evaluated segmentation performances. Concluding remarks and discussions are given in the last section.

1.   Method

The proposed method will be explained in three major steps. The first step is the estimation of the cell area and radius. Next, the input image will be roughly segmented by applying a constrained watershed transform on the morphological gradient [25] using this area information. The output of this rough segmentation does not eliminate background regions which do not contain cells and includes under-segmented regions which contain more than one cell. The background regions will be eliminated by a morphological thresholding tech-

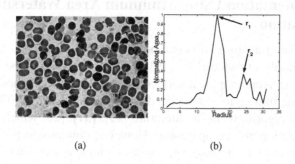

Figure 1. (a) Input image (I), (b) area granulometry (Ag) and peak radii r_1, r_2.

nique using the image histogram information. The final step utilizes the circle Radon transform to obtain a circularity map which is used for locating the cell centers (markers) and reapplying segmentation.

Area and Radius Estimation

The granulometry is a useful tool to obtain a priori information about the size of the objects in the image before processing.[30], [16]. In [3], Breen and Jones give generalized definitions of granulometry in relation to the attribute openings and also present an efficient algorithm to implement attribute granulometries. A fast implementation for the area granulometry (area pattern spectrum) is presented in [18]. Area granulometry was used in [23] to estimate blood cell radius.

The blood cells appear in the image (I) as darker, curved and convex regions (Figure1(a)). Calculating area granulometry (area pattern spectrum) on the negative of I gives a good estimate of the blood cell area (A) and the pseudo-radius ($r = \sqrt{\frac{A}{\pi}}$) [23]. Figure 1(b) shows size (area) distribution of the negative of the image in Figure 1(a). It can be seen that the differential volume plotted against r shows a peak at a particular radius index. We empirically observed that this radius index r_1 (and area A_1) is due to RBCs. This is because most of the image area is covered by the RBCs and the areas of these cells are very close to each other. There is also a larger radius r_2 where a smaller peak is caused by the white blood cells, enlarged RBCs due to the parasites, touching RBCs with very weak boundaries. These two radii will be used in the succeeding steps.

Initial Segmentation Using Minimum Area Watershed Transformation

The watershed transformation is a powerful morphological image segmentation tool initially introduced by [19]. In the watershed transformation the image is seen as a topographical surface. In the output, watershed lines divide the image plane into regions associated with regional minima. There are several watershed algorithms in the literature [19], [31], [20] which differ in the flooding realization and computation. However, common to all is that every regional minimum will be associated with a unique region (label) in the output. Applying a watershed transform on the image directly is generally useless unless the objects are flat (grey level) regions. Hence, a marker controlled transform is proposed which basically replaces the regional minima with the externally supplied markers. This technique transforms the watershed segmentation problem to marker extraction [19]. Some marker extraction techniques can be found in [26], [14], [27].

Our discussion on the minimum area watershed transformation is based on the watershed algorithm presented in [26], [31]. The minimum area watershed transformation (MAWT) is a modification to the original watershed transformation, which ensures the area of the labelled regions are above a given threshold. Same concept is studied in [1] with a progressive merging strategy called an "attribute-based absorptions" (ABA) method. In [21] an area constrained watershed is briefly mentioned as a modification for overcoming the over-segmentation problem. Here we will detail the modification, introduce the algorithm and explain the relation with the morphological area closing operator.

Definition.   It is easy to represent the minimum area watershed by the watershed definition in terms of flooding simulations as presented in [26]. Here, it is necessary to include Soille's definition:

$$
\begin{aligned}
X_{h_{min}} &= RMIN_{h_{min}}(f) \\
\forall h &\in [h_{min}, h_{max} - 1], \\
X_{h+1} &= RMIN_{h+1}(f) \cup IZ_{T_{t \le h+1(f)}}(X_h)
\end{aligned}
\tag{1}
$$

where f is a grey scale image, h is the grey level value, the T_h is threshold operator at level h, X_h is the set of the catchment basins up to level h, $RMIN_h(f)$ is the set of *all* regional minima at the level h, and $IZ_{T_{t \le h+1(f)}}(X_h)$ denotes the geodesic influence zones of X_h at the new threshold level $h + 1$. The set of all catchment basins $X_{h_{max}}$ are formulated recursively. Beginning from the minimum grey level h_{min}, in successive threshold levels, catchment basins X_{h+1} are updated according to the union of the geodesic influence zones of basins and newly emerged minima. The minimum area modification substitutes *every* regional minimum ($rmin$) belonging to minima at level h ($RMIN_h(f)$) with the $rmin^a$:

$$rmin^a = \left\{ \begin{array}{cc} rmin & if \quad AREA(rmin) \geq a \\ \{\} & else \end{array} \right\} \tag{2}$$

Algorithm. The implementation is realized from Soille's watershed transformation algorithm [26] by modifying the last section about the new minima.

fi: (INPUT IMAGE),fo: (OUTPUT IMAGE), $A\dot{}th$: (AREA THRESHOLD)
h: (GREY LEVEL VALUE), $A\dot{}counter$: (AREA COUNTER)
$fifo\dot{}add$: (ADD PIXEL TO TOP OF FIFO QUEUE), $fifo\dot{}empty$: (RETURN 1 IF FIFO EMPTY),
$fifo\dot{}retrieve$: (RETRIEVE LAST PIXEL AND DELETE FROM QUEUE)
$AQ\dot{}clean$: (EMPTY AUXILIARY QUEUE), $AQ\dot{}add$: (ADD PIXEL TO AUXILIARY QUEUE), $AQ\dot{}remask$: (RELABEL ALL PIXELS IN AUXILIARY QUEUE AS *MASK*)

```
...
{check for new minima}
∀ pixel p such that fᵢ(p) = h {
   if fₒ(p) = MASK {
      current_label+ = 1;
      fifo_add(p); fₒ(p)←current_label;
      A_counter←1; AQ_clean();
      while fifo_empty() = false {
         p'←fifo_retrieve();
         ∀ pixel p''∈N_G(p') {
            if fₒ(p'') = MASK {
               fifo_add(p''); fₒ(p'')←current_label;
               AQ_add(p''); A_counter+ = 1;
      }}}}
   {now checking for area}
   if A_counter < A_th AQ_remask()
}
```

Algorithm 1: MAWT algorithm in pseudocode

The three boxed lines in Algorithm 1 show the modification for the minimum area watersheds. In the first line a simple counter ($A\_counter$) is initialized for measuring area and an auxiliary queue is initialized by cleaning previous entries (AQ_clean). In the second boxed line, the pixel added to the FIFO is also added to the auxiliary queue for later use and the area counter is increased. When the process reaches the third boxed line, the pixels belonging to the current label are all registered in the auxiliary queue and the area is stored in the counter. Comparing the counter with the threshold (A_th), the algorithm resets the labels to the *MASK* if the area is not sufficient. Reseting pixel labels to the *MASK* passes pixels to the next grey level $(h + 1)$. These pixels will be regarded as being the same as pixels of level $h + 1$. This process is iteratively repeated at successive grey levels until A_th is met.

To summarize, we have added an area measuring step and control statement to the Soille's algorithm. If the area of the newly emerged minimum is greater than the specified threshold we label it, else we leave it unlabelled and test again at the next grey level. The output of the original watershed algorithm will be the same if the threshold is chosen as 1. The process reduces the number of calculated regions as the threshold is increased.

Relation to Area Closing.   As proved in [29], the relation between area opening and regional maxima can be stated as:
$$\gamma_\lambda^a(f) \rightarrow AREA(RMAX(f)) \geq \lambda$$
and by duality the relation between area closing and regional minimum is such:
$$\phi_\lambda^a(f) \rightarrow AREA(RMIN(f)) \geq \lambda$$
The area closing (ϕ_λ^a) operation on a grey level image (f) ensures the area of every regional minimum in $RMIN(f)$ is greater than or equal the closing parameter λ. Therefore every regional minimum $rmin$ will be equal to $rmin^a$ in 2. This suggests ϕ_λ^a followed by the watershed transform is equivalent to the MAWT with area threshold (A_th) equal to λ. This is valid for all values including $\lambda = 1$.

As stated earlier, MAWT checks the area of every regional minimum against the parameter λ and passes them to next grey level $(h+1)$ if the area is smaller. Passing to next grey level means this pixel values will be processed as they are $h + 1$. Which shows that the MAWT algorithm performs as an embedded area closing ϕ_λ^a operation in the watershed transformation.

The connection between the area closing and the MAWT leads us to think about more attributes which are already used in attribute closings or openings [3]. The attributes such as depth, volume or moments can be useful [1], and can be implemented by applying small modifications to the Algorithm 1.

Application.   An initial segmentation (S_1) is going to be calculated on the morphological image gradient using the MAWT algorithm. Since we have calculated the area A_1 from the area granulometry already, it is possible to use it as a threshold for the regional minima area. However, the blood cells are not flat, detached regions. Instead of using A_1, using a fraction $(\frac{A_1}{2})$ of it can be sufficient for the initial segmentation. In Figure 2(a) the morphological gradient of the input image is shown. The labelled output image is presented in Figure 2(b), 2(c), and 2(d), respectively for the original, MAWT with $A\_th = 10$, and $A\_th = \frac{A_1}{2}$. Figure 2(d) represents the labelled initial segmentation image (S_1) which we will be used in next steps of the method. The S_1 is not a

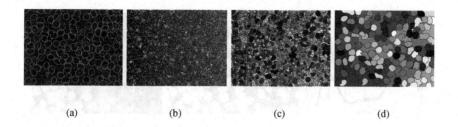

Figure 2. (a) Image gradient (b) watershed labels with the original algorithm and (c)watershed labels with the MAWT algorithm $A\_th = 10$ applied to gradient, (d) watershed labels (S_1) with the MAWT algorithm $A\_th = \frac{A_1}{2}$ applied to gradient. Note both foreground and background regions are labelled.

binary or a grey level image: all pixels on the watershed lines have label zero and all other regions have unique labels greater than zero.

Although MAWT reduces the number of regions and calculates useful boundaries, it is not sufficient to complete RBC segmentation. The output also contains background regions and under-segmented regions. These problems have to be solved with further processing.

Cleaning Background. The initial segmentation (S_1) obtained from the MAWT contains background and under-segmented regions as shown in Figure 2(d). We eliminate the background regions by employing the double threshold method (DBLT) described in [26]. In this method, the input image is thresholded twice to produce two binary images (wide and narrow threshold images). Then the wide threshold binary image is reconstructed from the narrow image used as a marker image.

$$DBLT[_{t_1 \leq t_2 \leq t_3 \leq t_4}](I) = R^\delta_{T_{[t_1,t_4]}(I)}\left[T_{[t_2,t_3]}(I)\right], \qquad (3)$$

where T is the threshold, R^δ is the morphological reconstruction operator, and $[t_1, t_4]$, $[t_2, t_3]$ are the wide and narrow threshold intervals respectively.

In our application, both lower bound t_1, t_2 values are set to 1, and t_3, t_4 lower bound values are obtained from the gray level histogram peaks. To obtain two peaks, the image histogram is iteratively smoothed by an averaging filter until only two global peaks remain (Figure 3(a)). The smaller and larger indices of these peaks are taken to be p_1 and p_2, respectively. The t_3 and t_4 are calculated using these levels and their middle level (Figure 3(a)). Narrow and wide threshold images are obtained using t_3 and t_4 threshold levels (Figures 3(b) and 3(c)) respectively.

The binary foreground mask is shown in Figure 3(b). The S_1 is then multiplied by this mask and relabelled (Figure 3(c)) after closing small holes by

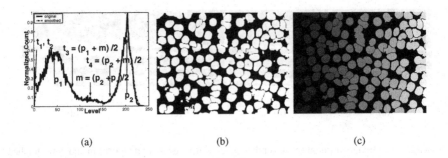

(a)   (b)   (c)

Figure 3.   (a) Original and smoothed histogram of the image showing peaks and calculation of threshold levels,(b) The foreground mask, (c) masked and relabelled S_1 after closing holes

reconstruction. Now, as the background labels (segments) have been removed, the remaining problem is the under-segmented regions.

Marker Extraction Using Radon Transform

Definition.   Generally, parameterized shape detection studies utilize the mathematical formalism of the Hough transform instead of the Radon transform. However, a recent study has shown that the two methods are equivalent [28]. The difference is in computational interpretations. When data which is to be transformed is not sparse, the Hough transform is computationally more expensive than the Radon transform. However, there are several applications and realizations of the Hough transformation, in the literature [17], [5], [9], [12], [8] while far fewer for Radon [13],[10],[11]. We have preferred to use the contour integral description that exists in the Radon transformation which enables the implementation via convolution and can be realized even faster when implemented in Fourier domain [28]:

$$R_{c(\mathbf{p})}\{I\}(\mathbf{p}) = \int_{c(\mathbf{p})} I(\mathbf{x})\,d\mathbf{x} = \int_{\Re^D} \delta(C(\mathbf{x};\mathbf{p}))\,I(\mathbf{x})d\mathbf{x}, \qquad (4)$$

where $R_{c(\mathbf{p})}\{I\}(\mathbf{p})$: Radon transform, $\mathbf{x}$: spatial coordinates, $\mathbf{p}$: N-dimensional shape parameter vector, $c(\mathbf{p})$:contour of the shape having parameter $\mathbf{p}$, $I(x)$: D dimensional image, $C(\mathbf{x};\mathbf{p})$: parameterized shape, δ: delta-dirac function.

The $C(\mathbf{x};\mathbf{p})$ can be any generalized function with parameter set $\mathbf{p}$. In case of a circle $\mathbf{p}$ can be written in $((x_0, y_0), r)$ form, where (x_0, y_0) is the centre and the r is the radius of the circular contour (i.e. 1 on circle 0 off circle). Thus for circle and an image of $I(x, y)$,

$$R_{c((x_0,y_0),r)}\{I\} = \int_{\Re^2} \delta((x-x_0)^2 + (y-y_0)^2, r^2)\, I(x,y)dx\, dy \qquad (5)$$

If the transform is going to produce a response on every pixel (every possible (x_0, y_0)), the above continuous integral can be transformed to a convolution with a shift-invariant circle kernel having radius r. Excluding (x_0, y_0) from our parameter vector, **p** reduces to variable r. Thus, we can write our transform formula as:

$$R(r)\{I\} = I * K_r, \qquad (6)$$

where $K_r = \delta(x^2 + y^2, r^2)$.

Generally the image I in 6 is an edge representation of the original image. The edge image (E_I) can be extracted by calculating the difference between S_1 and the erosion of S_1 with structuring element B (i.e the erosion gradient).

$$E_I = S_1 - \epsilon_B(S_1) \qquad (7)$$

Since S_1 is labelled, the edge image E_I will provide the region contours with unique labels. Label information will be useful if we incorporate it into our formula. However, there is no simple substitution between E_I and I for equation 6 such that: $R(r)\{I\} = E_I * K_r$, Instead, each labels' own R should be calculated separately or by a label control operation embedded into the convolution:

$$L = \{l_1, l_2, l_3, ..., l_N\}, \quad R^L(r) = \bigcup_{l=1}^{N} \left\{ E_I^{(l)} * K_r \right\} \qquad (8)$$

where l_i denotes the ith label and $E_I^{(l)}$ connected edge pixels belonging to the ith label in the edge image.

Application. To summarize, we utilized a labelled circle Radon transformation on the S_1 boundary lines (E_I). Instead of searching for resolvable circles and detecting the correct radius, we are interested in finding centroids of circles with radii closer to values r_1, r_2. Hence, we can calculate a cumulative circularity map in labels by summing R^L calculations among two radii r_1, r_2, then find peaks by searching for the regional maxima.

$$R^L_{r_1, r_2} = \sum_{r=r_1, r_2} R^L(r)$$

Because of the summation along the radius above, the circularity map calculation ($R^L_{r_1, r_2}$) reduces different r calculations to a cumulative concern. Since

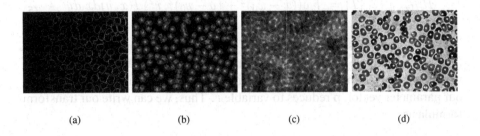

(a)   (b)   (c)   (d)

Figure 4. (a) The labelled edge image (E_I) for S_1 in Figure 3(c) (b) Labelled cumulative Radon transform $R^L_{r_1,r_2}$ on E_I, (c) Unlabelled Radon Transform R for thresholded E_I (i.e $E_I > 0$) for r_1, r_2, (d) Markers extracted from $R^L_{r_1,r_2}$ shown on gray level image.

we calculated the radii r_1 and r_2 from the area granulometry, with the summation the transform provides a 2-dimensional mapping $R^L_{r_1,r_2}$ of the (x, y) coordinates resulting in peaks at the possible centroids.

Figure 4(a) shows the labelled edge image E_I computed according to equation 7. The circularity map ($R^L_{r_1,r_2}$) calculated for r_1, r_2 (equation 3) on the E_I in Figure 4(b). An unlabelled calculation of the transformation (R) (by direct convolution with thresholded E_I>0) is presented for comparison 4(c). The latter is more noisy due to the effect of the adjacent region boundaries.

After $R^L_{r_1,r_2}$ is calculated, markers (M) can be extracted by searching for the regional maxima. Regional maxima are calculated with a structuring element defined as a disk of radius $\frac{r_1}{4}$, then dilated to unite very close points (Figure 4(d)). The radius of the disk is set big enough to reduce the number over-segmentations (too many markers) while allowing some under-segmentations, which will be solved on a second pass described in the next section.

Finalizing Segmentation. To finalize segmentation, we apply the marker controlled watershed transformation using the markers calculated in the previous step M. For the input, we employed the $R^L_{r_1,r_2}$ negative obtained in the previous section.

In Figure 5(a) the markers and the watershed lines corresponding to these markers are shown. However, the marker finding algorithm in the marker extraction step depends on the regional maxima, and the size of the structuring element used in calculation. Hence, it can still result in under-segmented or over-segmented regions. To detect under-segmentations we apply area opening by size $1.5 * \pi r_1^2$. To separate these regions, another regional maximum extraction process is applied by a disk structuring element of $r = 0.7 * r_1$. Then the marker controlled watershed is employed again to find the correct regions. In a similar way, the over-segmented regions are detected by area opening by size $\frac{A_1}{2}$, and recovered by dilation. Figure 5(b) shows the under-segmented and

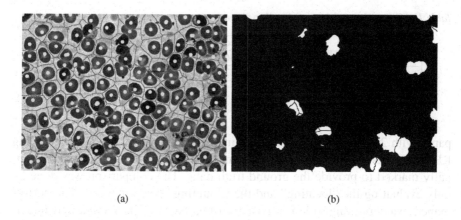

(a)   (b)

Figure 5.   (a) Watershed lines corresponding to markers, (b) over- and under-segmented regions

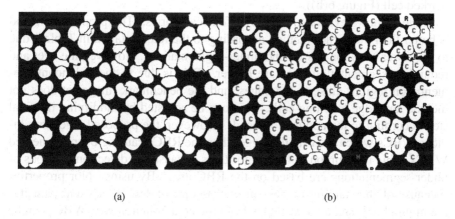

(a)   (b)

Figure 6.   (a) Final segmentation, (b)evaluation marks: correct (pink 'C'), under (green 'U'), missed(red 'M'), over and redundant(blue 'R') segmentations.

over-segmented regions. Finally, there are some cells connected to the edges of the image frame. These are reduced by the reconstruction method described in [26]. The final result can be seen in Figure 6(a).

2.   Experimental Results

The proposed method has been tested on images containing RBCs, malaria parasites, and some WBCs. The results were found to be satisfactory in most of the cases. Furthermore, the algorithm is compared with our previous best algorithm employing marker (area top hats) controlled watersheds and distance transform in combination [24], [22]. A benchmark image set was pre-

Table 1.   Comparison of the segmentation algorithms

Method	correct %	missed %	under %	over&redundant %
Proposed Method	95.40	1.46	3.12	8.14
Rao *et.al.* [22]	91.27	3.12	5.60	4.22

pared with 20 microscopic blood cell images containing 2177 cells including RBCs, WBCs and infected RBCs by malaria parasites. The cells were manually marked to provide the ground truth data. The comparison was done by only evaluating the "locating" and the "counting" performances. The performance evaluation algorithm is based on heuristics. The missed and under-segmentation rates were calculated directly by comparing labelled cell locations to manual markers. The over and redundant segmentation rates were calculated from all labelled regions which did not coincide with a manually marked cell (Figure 6(b))

Table 1 shows evaluated results in four categories. The proposed method is better than the earlier method according to the rates detailed in Table 1 due to the lower rates for missed and under-segmentations. However, the over and redundant segmentation rates are higher. This is due to the redundant segmentation regions close to the edges, and the benchmark image set does not include any edge touching cells. This can be adjusted by modifying the final step according to the application needs. However, the proposed algorithm divides WBCs which are usually bigger than the RBCs causing over-segmented WBCs. This is common to both algorithms because the detection methods of under-segmentations are based on the RBC area. By using color properties (because of the staining process in microscopic slides, WBCs and parasites are highlighted) and area attributes from the area granulometry, WBCs can be detected early and excluded from this processing.

3.   Conclusion and Discussions

We have introduced and explained a novel morphological blob segmentation algorithm comprised of area granulometry, minimum area watershed transformation, and circle Radon transformation. The proposed algorithm is tailored to and tested on microscopic images which mainly contain red blood cells. It gives better results than the algorithm that uses the distance transform and area tophats.

We have introduced the minimum area watershed transformation (MAWT) which embeds an area attribute in the original watershed algorithm. If employed with area granulometry, the MAWT becomes a powerful segmentation tool. However, depending on the nature of the problem, the transform is not completely capable of detecting or preventing under or over-segmentation in the objects. However, the MAWT algorithm only uses area, but it can be extended to use different attributes separately or in combination if needed.

We have proposed the labelled circle Radon transformation for marker extraction. We applied the Radon transformation to the labelled edge (contour) image which produces more efficient responses than the commonly used binary edge image representation. Although the blood cells are not perfect circular objects, the Radon transform with the help of the labeled edge representation, was quite successful in locating cell centers.

References

[1] M.C. De Andrade, G. Bertrand, and A.A. Araujo. Segmentation of microscopic images by flooding simulation: A catchment basins merging algorithm. In *Proc. of the Society for Imaging and PhotoOptical Engineering*. SPIE, 1997.

[2] J. Angulo and G. Flandrin. Automated detection of working area of peripheral blood smears using mathematical morphology. *A. Cellular Pathology*, 25(1):37–49, 2003.

[3] E.J. Breen and R. Jones. Attribute openings, thinnings, and granulometries. *CVIU*, 64:377–389, 1996.

[4] A. Cosio, F. M. Flores, J. A. P. Castaneda, M. A. Solano, and S. Tato. Automatic counting of immunocytochemically stained cells. In *Proc. 25th Annual Int. Conf. IEEE, Engineering in Medicine and Biology Society*, pages 790–793, 2003.

[5] E.R. Davies. Finding ellipses using the generalised hough transform. *PRL*, 9:87–96, 1989.

[6] C. di Ruberto, A. Dempster, S. Khan, and B. Jarra. Segmentation of blood images using morphological operators. In *ICPR00*, pages Vol III: 397–400, 2000.

[7] C. di Ruberto, A. Dempster, S. Khan, and B. Jarra. Analysis of infected blood cell images using morphological operators. *IVC*, 20(2):133–146, February 2002.

[8] R. O. Duda and P. E. Hart. Use of the hough transformation to detect lines & curves in pictures. *Comm. ACM*, 15(1), 1972.

[9] N. Guil and E. L. Zapata. Lower order circle and ellipse hough transform. *Pattern Recognition*, 30:1729–1744, 1997.

[10] K. V. Hansen and P. A. Toft. Fast curve estimation using preconditioned generalized radon transform. *IEEE Trans. on Image Processing*, 5(12):1651–1661, 1996.

[11] C. L. Hendriks, M. van Ginkel, P.W. Verbeek, and L.J. van Vliet. The generalised radon transform: Sampling and memory considerations. In *Proc. of CAIP2003, Groningen, Netherlands*. Springer Verlag, 2003.

[12] J. Illingworth and J. Kittler. Survey, a survey of the hough transform. *Computer Vis. Graph. and Img. Processing*, 44:87–116, 1988.

[13] V. F. Leavers. Use of the two-dimensional radon transform to generate a taxonomy of shape for the characterization of abrasive powder particles. *IEEE Trans. on PAMI*, 22(12):1411–1423, 2000.

[14] O. Lezoray and H. Cardot. Bayesian marker extraction for color watershed in segmenting microscopic images. In *Proc. Pattern Recognition*, volume 1, pages 739–742, 2002.

[15] Q. Liao and Y. Deng. An accurate segmentation method for white blood cell images. In *Proc. IEEE Int. Symp. on Biomedical Imaging*, pages 245–248, 2002.

[16] P. Maragos. Pattern spectrum and multiscale shape respresentation. *IEEE Trans. PAMI.*, 2:701–716, 1989.

[17] R.A. Mclaughlin. Randomized hough transform: Improved ellipse detection with comparison. *Pattern Recognition Letters*, 19(3-4):299–305, March 1998.

[18] A. Meijster and M.H.F. Wilkinson. Fast computation of morphological area pattern spectra. In *ICIP01*, pages III: 668–671, 2001.

[19] F. Meyer and S. Beucher. Morphological segmentation. *J. Vis. Comms. Image Represent.*, 1:21–46, 1990.

[20] L. Najman and M. Schmitt. Geodesic saliency of watershed contours and hierarchical segmentation. *IEEE Trans. PAMI*, 18:1163–1173, 1996.

[21] J. Park and J. M. Keller. Snakes on the watershed. *IEEE Trans. PAMI.*, 23(10):1201–1205, 2001.

[22] KNR M. Rao. *Application of Mathematical Morphology to Biomedical Image Processing*. PhD thesis, U. Westminster, 2004.

[23] KNR M. Rao and A. Dempster. Area-granulometry: an improved estimator of size distribution of image objects. *Electronics Letters*, 37:50–951, 2001.

[24] KNR M. Rao and A. Dempster. Modification on distance transform to avoid over-segmentation and under-segmentation. In *Proc. VIPromCom-2002, Croatia*, pages 295–301, 2002.

[25] J.F. Rivest, P. Soille, and S. Beucher. Morphological gradients. *Journal of Electronic Imaging*, 2(4), 1993.

[26] P. Soille. *Morphological Image Analysis*. Springer-Verlag, Heidelberg, 2003.

[27] C. Vachier and F. Meyer. Extinction value: a new measurement of persistence. In *IEEE Workshop on Nonlinear Signal and Image Processing*, pages 254–257, 1995.

[28] M. van Ginkel, C.L. Luengo Hendriks, and L.J. van Vliet. A short introduction to the radon and hough transforms and how they relate to each other. Technical report, Quantitative Imaging Group, Delft University of Technology, 1997.

[29] L. Vincent. Morphological area openings and closings for greyscale images. In *Proc. NATO Shape in Picture Workshop*, pages 197–208. Springer Verlag, 1992.

[30] L. Vincent. Granulometries and opening trees. *Fundamenta Informaticae*, 41:57–90, 2000.

[31] L. Vincent and P. Soille. Watersheds in digital spaces: An efficient algorithm based on immersion simulations. *IEEE Trans. PAMI., vol. 13, pp. 583-598, June 1991*, 13:583–598, 1991.

[32] S.E. Volkova, N.Yu Ilyasova, A.V Ustinov, and A.G Khramov. Methods for analysing the images of blood preparations. *Optics & Laser Technology*, 27(2):255–261, 1995.

QUANTIFYING MEAN SHAPE AND VARIABILITY OF FOOTPRINTS USING MEAN SETS

J. Domingo[1], B. Nacher[2], E. de Ves[3], E. Alcantara[2], E. Diaz[3], G. Ayala[4] and A. Page[2]

[1]*Instituto de Robotica, Universidad de Valencia.*
Poligono de la Coma, s/n. Aptdo. 2085, 46071 Valencia (SPAIN)
Juan.Domingo@uv.es

[2]*Instituto de Biomecanica, Universidad Politécnica de Valencia.*
Edificio 9C. Camino de Vera s/n, 46022 Valencia (SPAIN)
beanacfe@ibv.upv.es,ealcanta@ibv.upv.es,afpage@ibv.upv.es

[3]*Departamento de Informatica, Universidad de Valencia.*
Avda. Vte. Andres Estelles,s/n, 46100 Burjasot (SPAIN)
Esther.deVes@informat.uv.es,Elena.Diaz@informat.uv.es

[4]*Departamento de Estadistica e IO, Universidad de Valencia.*
Avda. Vte. Andres Estelles,s/n, 46100 Burjasot (SPAIN)
Guillermo.Ayala@uv.es

Abstract   This paper[1] presents an application of several definitions of a mean set for use in footwear design. For a given size, footprint pressure images corresponding to different individuals constitute our raw data. Appropriate footwear design needs to have knowledge of some kind of typical footprint. Former methods based on contour relevant points are highly sensitive to contour noise; moreover, they lack repeatability because of the need for the intervention of human designers. The method proposed in this paper is based on using mean sets on the thresholded images of the pressure footprints. Three definitions are used, two of them from Vorob'ev and Baddeley-Molchanov and one morphological mean proposed by the authors. Results show that the use of mean sets improves previous methodologies in terms of robustness and repeatability.

Keywords: Mean set, morphological mean, footprint, footwear design

Introduction

Footprints have been used as a source of data for many different applications such as medical diagnosis, anthropometric studies and footwear design.

C. Ronse et al. (eds.), Mathematical Morphology: 40 Years On, 455–464.
©2005 *Springer. Printed in the Netherlands.*

Although foot printing techniques may be inadequate to determine the maximum values for many length and breadth foot dimensions [21], they allow the calculation of a variety of parameters of interest for orthotic applications such as the footprint angle, the foot contact area and the arch index [3]. On the other hand, footprint and foot outline shape analysis has been used in anthropometric studies to establish foot-type classifications [2], to obtain average plantar curves [7] and to detect sex or ethnic morphological differences [8]. Finally, a systematic analysis of plantar foot-shape variation is necessary to identify the critical parameters of the insole design, in order to ensure an adequate fitting between insole and plantar foot shapes [20].

Detailed knowledge of footprint shape and dimensions implies the analysis of averaged values and forms as well as the quantification of size and shape variability. Both aspects of the quantitative description of footprint form, mean values and variability, are crucial in design applications. Two main approaches have been used for this purpose.

The easiest and most-used method is based on the measurement of metric distances, areas, angles or ratios and the calculation of descriptive statistical parameters such as mean, standard deviation or coefficient of variation. This approach provides good information about the distribution of the univariate anthropometric but it has several limitations. Firstly, it needs the selection of a set of landmarks and the use of a particular measurement protocol. The unsuitability of landmark set selection or differences between measurement protocols can generate results which are not comparable ([12], [13]). Moreover, this approach does not reflect the multivariate nature of strongly correlated variables such as foot length, foot width and contact area, for instance [20]. An adequate description of the whole footprint shape variability cannot be made by quantifying the variability of a discrete set of variables, because distorted percentiles are obtained when they are estimated from each individual component ([4], [15]). In short, traditional methods originating from classical descriptive statistics neither allow an efficient way to make shape analysis nor provide a comprehensible description of human foot variability.

An alternative strategy is based on the mathematical analysis and quantification of the whole footprint shape, although these methods are insufficient for extracting and quantifying shape characteristics.

A first approach is based on dimensions or foot outline coordinates using multivariate analysis techniques. The use of multivariate analysis techniques in anthropometric foot description includes factor analysis and principal component analysis [9],[5], multivariate discriminant analysis [16] or clustering techniques [2]. A completely different approach uses Fourier descriptors to represent outlines [10]. This technique allows a quantitative description of footprint shape and its classification independently of size [17]. Finally, some attempts to define averaged plantagram curves or foot outlines have been developed [6]. Despite the interest of these methods they are mainly focused on grouping homogeneous clusters of population and on describing some averaged characteristics, but no attempts to make a multivariate quantification of human variability have been reported.

Thus, alternative techniques for providing realistic statistical shapes are needed for footprint application in many fields like footwear design. In this paper a method is presented that uses techniques from image analysis, and particularly from mathematical morphology, that provide a coherent statistical analysis framework including the determination of mean shapes and, in the future, the identification of extreme shapes and confidence regions.

1. Mean shapes

Different definitions of the mean set of a random compact set can be found in the literature, some of which are particularly important: the Aumann mean, the Vorob'ev mean, the Baddeley-Molchanov mean and morphological means. In this section, a brief review of their definitions is given. However, a more detailed explanation appears in [19] for the Aumann and Vorob'ev means, [1] for the Baddeley-Molchanov mean and [18] for morphological means.

Let Φ be a random compact set on $\Re^n$ i.e. a random closed set taking values in the space K of all compact subsets of $\Re^n$. In the first place let us introduce Aumann's definition of a mean set, based on the concept of *support function*. Given a fixed set X, their *support function* is defined as

$$h(X, u) = \sup\{u * x : x \in X\} \tag{1}$$

where $u * x$ denotes the inner product of u with x and u runs over the unit circumference centred at the origin. The Aumann mean of Φ, $E_a\Phi$ is defined as the convex set with support function given by the expected value of the random variable $h(\Phi, u)$ i.e.

$$h(E_a\Phi, u) = Eh(\Phi, u). \tag{2}$$

The Vorob'ev mean set is based on the thresholding of the *coverage function* of Φ. The coverage function at the point x is defined as the probability of this point belonging to the random set Φ, $P(x \in \Phi)$. Let

$$S_p = \{x \in \Phi : P(x \in \Phi) \geq p\}. \tag{3}$$

The Vorob'ev mean set $E_v\Phi$ is defined as S_{p_0} where p_0 is chosen in such a way that the area of S_{p_0} is equal to the mean area of Φ.

Baddeley and Molchanov proposed a new concept of mean set based on their distance function representation. A good and detailed presentation with many examples can be found in the original reference [1]. We have used the modified version proposed in [11]. Let $\mathcal{F}'$ be the space of nonempty closed sets with hit-miss topology (see [14]) and let $d : \Re^n \times \mathcal{F}' \to \Re$ defined as $d(x, F) = min \{ \| x - y \| : y \in F \}$ where $\| x - y \|$ is the Euclidean distance between x and y. Let $m_W(F_1, F_2) = \left(\int_W (d(x, F_1) - d(x, F_2))^2 dx \right)^{\frac{1}{2}}$, where W is a certain compact set (window). Suppose that $d(x, \Phi)$ is integrable for all x and define the mean distance function $d^*(x) = Ed(x, \Phi)$. Let $\Phi(\epsilon) = \{ x \in W : d^*(x) \leq \epsilon \}$ with $\epsilon \in \Re$. The Baddeley-Molchanov mean of Φ, $E_{bm}\Phi$, is the set $\Phi(\epsilon^*)$ where ϵ^* is chosen such that $\nu(\Phi(\epsilon^*)) = E\nu(\Phi)$.

Finally, morphological mean sets are defined by using basic set transformations taken from Mathematical Morphology (erosion, dilation, opening and closing) to associate a stochastic process to Φ.

Let Φ be a random compact set and let W (the window) be a compact set such that $\Phi \subset W$. Let T be a compact structuring element containing the origin. From this set, the family of homothetic sets to T will be considered. A complete description of W can be obtained from all the erosions of W with the family of λT's. A similar comment applies when erosion is replaced by dilation, opening or closing.

Different random processes will be associated to W, corresponding to each transformation. The first example will be based on erosion. Let Λ_Φ^e be $\Lambda_\Phi^e(x) \equiv -\sup\{\lambda : x \in \epsilon_{\lambda T}(\Phi)$ i.e., at each point x of W, the maximum λ such that x belongs to the smaller (and smaller) $\epsilon_{\lambda T}(\Phi)$. An analogous function is defined by using the dilation as $\Lambda_\Phi^d(x) \equiv \inf\{\lambda : x \in \delta_{\lambda T}(\Phi)$. By replacing erosion and dilation by opening and closing respectively, the next two functions will be defined: $\Lambda_\Phi^o(x) \equiv -\sup\{\lambda : x \in \gamma_{\lambda T}(\Phi)$, and $\Lambda_\Phi^{cl}(x) \equiv \inf\{\lambda : x \in \psi_{\lambda T}(\Phi)$.

Let $\Lambda_\Phi(x)$ be any of the previous random processes, from now on simply Λ-*function*. It will be assumed that $\Lambda_\Phi(x)$ is integrable for all $x \in W$ and let

$$\Lambda_\Phi^*(x) = E\Lambda_\Phi(x). \qquad (4)$$

From the function $\Lambda_\Phi^*(x)$, we can define

$$\Phi_\epsilon = \{ x : \Lambda_\Phi^*(x) \leq \epsilon \}. \qquad (5)$$

Let Φ be a random compact set and Λ_Φ a random process associated to Φ. Let $E_1(\Phi)$ be the set Φ_ϵ (Equation 5) where ϵ is chosen in such a way that

$$\epsilon = \min\{\epsilon' > 0 : \nu(\Phi_{\epsilon'}) = E\nu(\Phi)\} \qquad (6)$$

where $\nu(X)$ stands for the area of a set X. The set $E_1(\Phi)$ will be called the mean set associated with Λ_Φ.

In the next section these definitions of means sets will be applied to the binary images resulting from the thresholding of the pedigraphies of several individuals with the same footwear size so as to summarise each size as a typical shape that can be used for footwear design.

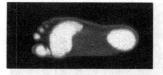

Figure 1. Original grey images

The estimation of the Aumann and Vorob'ev means is simple from the definition. For the other means we need to estimate $E\nu(\Phi)$. From a sample $\phi_q, \ldots, \phi_n$, we take $\hat{E\nu}(\Phi) = \sum_{i=1}^{n} \frac{\nu(\phi_i)}{n}$ and the threshold level is estimated iteratively.

2. Mean sets as summarised descriptions of footprints

A pedigraphy is a real function whose value at each point represents the pressure exerted by the plantar area on that point when an individual stands on his/her two feet. It is delivered by the capture device as a digital image with a resolution of 450×200 and 256 pressure levels. Two typical images are shown in Figure 1. For representation purposes, gray levels are chosen so that brighter represents more pressure. These gray level images clearly have two different elements of interest: the lightest represents the foot part which puts the higher pressures the floor, and the other, of a medium grey level, represents the external shape of the foot. An important difference between these two typical images is worth noting: the inner part of the right-hand image is divided into two unconnected shapes whereas the right-hand one is not divided. All the images in our database are similar to one of these two types of footprints.

A procedure for the automatic extraction of the inner and outer shapes of the footprint using image processing techniques has been devised and programmed, whose brief description follows.

Three main intervals for gray levels appear in the image: the darker one (background), a brighter one (inner shape) and the brightest (outer shape) which appear as peaks in the gray level histograms. A clear three-mode histogram suggests that two thresholds can be found to separate the regions; these thresholds are taken as the gray levels which are the deepest minimum between each pair of maxima, after the histogram has been convolved with a uniform mask of size 10. A consistency check is done to ensure that the thresholds are within the expected gray level range. Thresholding returns two binary images with the inner and outer shape. To smooth small irregularities that still appear in the contour of both shapes due to noise, and to eliminate small inner holes, two morphological operations are applied: first an opening with a binary square of 3×3 pixel structuring element and then a hole closing. Results for a typical case are shown in Figure 2. Visual inspection suggests that the ex-

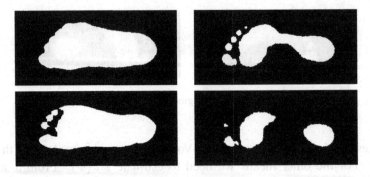

Figure 2. Segmented images obtained from those in figure 1 : outer (left) and inner (right) shapes

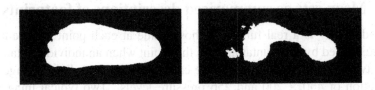

Figure 3. Vorob'ev mean of footprints: outer (left) and inner (right) shape

tracted shapes match the real shapes quite well, but apart from the validation of this procedure by a human expert, its value lies in its repeatability and uniformity for many images, since no free parameters are used, which is a desirable characteristic to assure a consistent comparison.

Once the binary images have been extracted the shape comparison can only be made with aligned images. The alignment process is simply done by moving and rotating the outer shape so that its centre coincides with a given point (the centre of the image) and its main symmetry axes (calculated as the main inertia axes) are respectively parallel to the image borders. Although simple, this alignment procedure has proved to be sufficient for our purposes; a more complicated method based on the detection and matching of relevant points in the contour is being studied for future evaluation.

As we have explained earlier, different definitions of mean set have been proposed in the literature. Among these different definitions we have selected the Vorob'ev mean, the Baddeley-Molchanov mean and the one based on morphological operators to be applied to our samples. The computed mean Vorob'ev sets (outer and inner shape) for the group of images in our database appear in Figure 3. The accuracy of the results of the outer mean set contrasts with the noisy characteristic of the inner mean set. The second definition of mean set applied is the Baddeley-Molchanov mean. The computed mean shapes are shown in Figure 4. If these results are compared with the previous

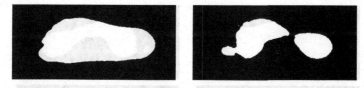

Figure 4.   Baddeley-Molchanov mean of footprints: outer (left) and inner (right) shapes

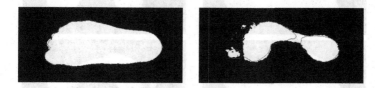

Figure 5.   Comparison between mean sets: Vorob'ev (binary shape) and Baddeley-Molchanov (red overlapped contour)

ones, it seems that the outer contours are very similar to each other. However, the inner ones are quite different. The Baddeley-Molchanov mean set is made up of two connected components whereas the Vorob'ev is made of just one. Besides, the Baddeley-Molchanov mean set is much smoother, as can be seen in Figure 5. Finally, we have applied the mean set definition based on morphological operations. In particular, the erosion and opening operations have been chosen for this experiment. The mean sets obtained with this definition are shown in Figure 6. All these results, the inner and the outer mean sets, are quite smooth, in contrast to the Vorob'ev mean set which seems a noisy version of the inner contour. In order to compare the morphological mean sets when different basic operations are used, the results have been overlapped over the same image (see the third row of figure 6). Figure 7 shows all the computed mean sets overlapped over one of the original images. As this figure illustrates, these shapes are an accurate representation of the set of shapes we are dealing with, mainly of the outer contour, even though there are slight differences among the distinct mean sets. With respect to the inner shapes: some of them have a smoother contour (Baddeley-Molchanov and Morphological mean set), most of them have just one connected component whereas the Baddeley-Molchanov mean set has two connected components. Small contour details are not captured by any of the mean sets, as expected.

A numerical evaluation of these results is difficult, since there is no real or correct result with which the means can be compared, and nor can it be generated from synthetic examples. Nevertheless, it is sensible to enquire which of the three means is most similar to the shapes from which it has been obtained. By analogy with the coefficient of variation in classical statistics (ratio between standard deviation and mean) a parameter σ has been calculated as

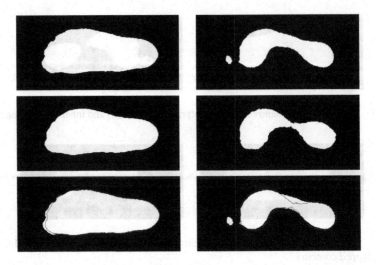

Figure 6.   Morphological means using the erosion and opening operations respectively for inner and outer shapes and overlapped contour (third row)

	Vorob'ev	Baddeley-Molchanov	Morf-erosion	Morf-opening
σ for outer shape	0.0269	0.0271	0.0288	0.0259
σ for inner shape	0.1318	0.1292	0.1393	0.1165

follows: first the symmetric difference between the mean shape and each sample has been taken, whose area will be called a_i. The parameter σ given in Table 6 is the ratio between the mean of the a_i and the area of the mean of the samples. Notice that for all the mean definitions we have used the area of the mean shape is equal to the mean of the areas of the samples.

As can be seen from the table, the results are quite similar, which shows that no definition of mean is clearly superior in terms of closeness to the original samples. Therefore, subjective evaluations have had to be made by experts in the area, who found the provided means much better than former numerical procedures in terms of repeatability and adequacy for the purpose of footwear design. In particular, it overcomes the unresolved issue of how to go from numerical descriptors (mean area, mean perimeter and so on) to a real shape that can be used as a footwear footprint. From amongst all the means, specialists gave the highest rank to the morphological mean by erosion for the outer contour and to the same, but with dilation, for the inner contour.

Future work includes the comparison of numerical descriptors obtained from the mean shapes with those obtained by doing classical statistics on the values obtained from each individual using a large database of pedigraphies that is currently being acquired.

Vorob'ev mean set

Baddeley-Molchanov mean set

Morphological Mean set (erode)

Morphological mean set (opening)

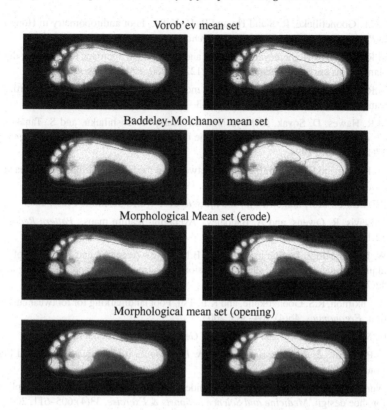

Figure 7. Morphological means using the erode and opening operation respectively for the internal shape

Notes

1. This work has been supported by projects TIC2002-03494 from the Spanish Ministry of Science and Technology, CTIDIA-2002-133, RGY 40/2003 and GV04B-032 and GV04A-177 from the Generalitat Valenciana.

References

[1] A.J. Baddeley and I.S. Molchanov. Averaging of random sets based on their distance functions. *Journal of Mathematical Imaging and Vision*, 8:79–92, 1998.

[2] A. Bataller, E. Alcántara, J.C. González, A.C. Garcia, and S. Alemany. Morphological grouping of Spanish feet using clustering techniques. In E. Hennig and H. Stacoff, A.and Gerber, editors, *Proceedings of Fifth Symposium on Footwear Biomechanics*, pages 12–13, 2001.

[3] P.R. Cavanagh and M.M. Rodgers. The arch index: a useful measure from footprints. *Journal of Biomechanics*, 20(5):547–551, 1987.

[4] G.S. Daniels. The average man? Technical Note WCRD 53-7, Wright Air Development Center, Wrigth-Patterson Air Force Base, Dayton, Ohio, 1952.

[5] E.C.F. Goonetilleke, R. S.and Ho and R. H. Y. So. Foot anthropometry in Hong Kong. In *Proceedings of the ASEAN 97 Conference*, pages 81–98, 1997.

[6] M.R. Hawes, R. Heinemeyer, D. Sovak, and B. Tory. An approach to averaging digitized plantagrams curves. *Ergonomics*, 37(7):1227–1230, 1994.

[7] M.R. Hawes and D. Sovak. Quantitative morphology of the human foot in a North American population. *Ergonomics*, 37(7):1213–1226, 1993.

[8] M.R. Hawes, D. Sovak, M Miyashita, S.J. Kang, Y. Yoshihuku, and S. Tanaka. Ethnic differences in forefoot shape and the determination of shoe comfort. *Ergonomics*, 37(1):187–193, 1994.

[9] M. Kouchi and E. Tsutsumi. Relation between the medial axis of the foot outlina and 3-d foot shape. *Ergonomics*, 39(6):853–861, 1996.

[10] P.E. Lestrel. *Morphometrics for the life sciences*. World Scientific Press, 2000.

[11] T. Lewis, R. Owens, and A. Baddeley. Averaging feature maps. *Pattern Recognition*, 32:1615–1630, 1999.

[12] W. Liu, J. Miller, D. Stefanyshyn, and B.N. Nigg. Accuracy and reliability of a technique for quantifying foot shape, dimensions and structural characteristics. *Ergonomics*, 42(2):346–358, 1999.

[13] A. Luximon, R.S. Goonetikelle, and K.L. Tsu. Foot landmarking for footwear customization. *Ergonomics*, 46(4):364–383, 2003.

[14] G. Matheron. *Random sets and Integral Geometry*. Wiley, London, 1975.

[15] S. Pheasant. *Bodyspace. Anthropometry, Ergonomics and Design*. Taylor and Francis, London, 1986.

[16] Wunderlich R.E. and Cavanagh P.R. Gender differences in adult foot shape: Implications for shoe design. *Medicine and Science in Sports & Exercise*, 33(4):605–611, 2000.

[17] C. Sforza, G. Michielon, N. Frangnito, and V.F Ferrario. Foot asymmetry in healthy adults: Elliptic fourier analysis of standardized footprints. *Journal of Orthopaedic Research*, 16(6):758–765, 1998.

[18] A. Simó, E. de Ves, G. Ayala, and J. Domingo. Resuming shapes with applications. *Journal of Mathematical Imaging and Vision*, 20:209–222, 2004.

[19] D. Stoyan and H. Stoyan. *Fractals, Random Shapes and Point Fields. Methods of Geometrical Statistics*. Wiley, 1994.

[20] B.Y.S. Tsung, M. Zang, Y.B. Fan, and D.A. Boone. Quantitative comparison of platar foot shapes under different wheight-bearing conditions. *Journal of Rehabilitation Research and Development*, 40(6):517–526, 2003.

[21] R.S. Urry and S.C. Wearing. A comparison of footprint indexes calculated form ink and electronic footprints. *Journal of American Podiatric Medical Association*, 91(4):203–209, 2001.

EXPLOITING AND EVOLVING
R^N MATHEMATICAL MORPHOLOGY
FEATURE SPACES

Vitorino Ramos and Pedro Pina
CVRM / Centro de Geo-Sistemas, Instituto Superior Técnico
Av. Rovisco Pais, 1049-001 Lisboa, PORTUGAL
vitorino.ramos@alfa.ist.utl.pt, ppina@alfa.ist.utl.pt

Abstract   A multidisciplinary methodology that goes from the extraction of features till the classification of a set of different granites is presented in this paper. The set of tools to extract the features that characterise the polished surfaces of granites is mainly based on mathematical morphology. The classification methodology is based on a genetic algorithm capable of searching for the input feature space used by the nearest neighbour rule classifier. Results show that is adequate to perform feature reduction and simultaneously improve the recognition rate. Moreover, the present methodology represents a robust strategy to understand the proper nature of the textures studied and their discriminant features.

Keywords: Granite textures, feature extraction, size-intensity diagram, feature reduction, genetic algorithms, nearest neighbour rule classifiers.

1.   Introduction

Natural ornamental stones are quantitatively characterised in many ways, mostly physical, namely through geological-petrographical and mineralogical composition, or by mechanical strength. However, the properties of such products differ not only in terms of type but also in terms of origin, and their variability can also be significant within a same deposit or quarry. Though useful, these methods do not fully solve the problem of classifying a product whose end-use makes appearance so critically important. Traditionally, the industrial selection process is based on visual inspection, giving a subjective characterisation of the appearance of the materials. Thus, one suitable tool to characterise the appearance of these natural stones is digital image analysis. If the identification of the features (colour, size/shape, texture) that characterise a type of material may seem easier to list, the definition of a set of parameters to

C. Ronse et al. (eds.), Mathematical Morphology: 40 Years On, 465–474.
©2005 *Springer. Printed in the Netherlands.*

quantify those features becomes less evident. Those parameters have to clearly characterise each feature for each type of material and should not be redundant.

Mathematical morphology disposes of a set of operators that can handle, in an individual or in a combined mode, with the extraction of the textural features of images in general [3, 12, 16], but also with ornamental stones in particular, namely related to their size/shape, dispersion/orientation and connectivity/neighbourhood. In the present case study, several morphological operators were tested, but there is one that provides significant discrimination between the different textures: it consists of the size-intensity diagram introduced by Lotufo and Trettel [6], that combines both size and intensity information into a single parameter.

In what concerns classification, if the features are significant and synthesize correctly the real texture, a simple approach can be used. Thus, the nearest neighbour rule (k-NNR) was considered to be used, since it is a simple but powerful classification technique [4], even for a small number of training prototypes. In that context, a method was envisaged to perform nearest neighbour optimisation by genetic algorithms (GA) (*i.e.* via feature reduction). In the basic nearest neighbour rule classifier, each sample (described by their features) is used as a prototype and a test sample is assigned to the class of the closest prototype [2]. It has been shown that in some cases a small number of optimised prototypes can even provide a higher classification rate than all training samples used as prototypes. Another approach consists in using genetic algorithms based methods for searching the feature space to apply in nearest neighbour rule prototypes, which is the case presented in this paper. For instance, Brill *et al.* [1] used a k-NNR classifier to evaluate feature sets for counter propagation networks training. Some other authors used the same approach [14, 15] for another kind of classifiers.

2.　　Textures studied: Feature selection and extraction

A collection of 14 Portuguese grey granites with several samples per type was constituted [10]. Although this commercial label includes the real grey types, it also includes other similar colourless types (bluish, whitish and yellowish, for instance). The samples of these types of granites used for the development and testing of our research are 30 cm x 30 cm polished tiles. The digital images were acquired (total set of 237 images) using a colour scanner with a predefined regulation set for the brightness and the contrast parameters [10] having a spatial resolution of 150 dpi and a spectral resolution of 256^3 colours. Sample images of each granite texture and respective description are presented in figure 1 and table 1.

The extraction of features was initially envisaged to be implemented in a two-stage approach: globally and locally. It consisted on the extraction of

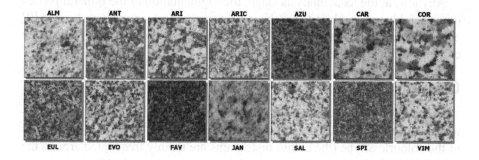

Figure 1. Collection of Portuguese granites (each image corresponds to a 4 cm x 4 cm region of the original polished surface)

Table 1. Types of granites and number of samples available.

Code	Acronym	Name	total	training	test
1	ALM	Branco Almeida	20	16	4
2	ANT	Cinzento de Antas	20	16	4
3	ARI	Branco Ariz	8	6	2
4	ARIC	Cinza Ariz	4	3	1
5	AZU	Azulália	20	16	4
6	CAR	Branco Caravela	20	16	4
7	COR	Branco Coral	20	16	4
8	EUL	Cinzento Sta Eulália	20	16	4
9	EVO	Cinzento de Évora	20	16	4
10	FAV	Favaco	15	11	4
11	JAN	Jané	20	16	4
12	SAL	Pedras Salgadas	10	7	3
13	SPI	SPI	20	16	4
14	VIM	Branco Vimieiro	20	16	4
TOTAL			237	187	50

features before and after the mineral phase segmentation, which corresponds to global and local analysis, respectively.

The global analysis stage consists on the first attempt to characterise the textures, and it is applied to the colour images before the segmentation or phase classification procedure. The main idea, which was also the main expectation, was that the features extracted could be sufficient to discriminate several types of materials, being not needed to proceed to a local analysis stage. In fact, that is what happened, since the classification rates obtained were excellent using only these global features, like is shown in the following sections.

Anyhow, at the local analysis stage, the segmentation of the mineral phases (quartz, feldspars and biotite are the main occurrences) was also performed [9] but the related parameters were not necessary to incorporate in the current classification procedure.

Colour features

Consists of computing the Red (R), Green (G) and Blue (B) individual histograms of the images. Due to the fact that this set of 14 types of granites is of the colourless type (the colour histograms are mainly frequented around the main diagonal of the RGB cube), a conversion to the Hue (H), Intensity (I) and Saturation (S) colour system was performed in order to enhance unnoticed details. These individual histograms (H, I, S) are this way the colour features to be used. Since the total number of features may pose some computational problems a reduction of the number of features was envisaged without filtering their global characteristics. This way, the initial 256 classes per band were reduced to 8 classes of equal frequency (8 octils), $i.e.$, the number of observations that occur in each interval or class is the same, but their length is variable. The minimum value of each band was also retained. Thus, the colour features consist of 27 values: 3*(8+1).

Size features

In what concerns the size of the mineralogical components of the granites, a size distribution can be directly achieved through the grey level images. The size distribution by grey level granulometry is computed through morphological openings or closings of increasing size acting simultaneously over all phases present in the image. The opening and the closing have granulometric properties [7] once are increasing, extensive (closing) and anti-extensive (opening) and idempotent [12] and reflect the size distributions of the lighter and darker minerals, respectively, present in the samples.

Besides the information related to size given by the classical grey level granulometry (measure of the volume), another measure can be achieved by associating it to the distribution of the grey levels. This measure was originally introduced by Lotufo and Trettel [6] and consists on the computation of the grey level distribution for opened or closed images of increasing size. The resulting diagram is of bidimensional type and incorporates this way both size and intensity or grey level information. This diagram is built using a family of openings or closings of increasing size using the cylindrical structuring element $B(r, k)$ of radius r and height k.

The size-intensity diagram by openings γ of a function f, $SI_f^\gamma(r, k)$, is defined as [6]:

$$SI_f^\gamma(r, k) = A(\gamma^{B(r,k)}(f)) \qquad (1)$$

where $A(f)$ is a measuring function of the number of non-zero pixels of f:

$$A(f) = Meas\{x \in Z^2 : f(x) \neq 0\} \tag{2}$$

Similarly, this diagram is also defined for closings φ of increasing size [6]:

$$SI_f^{\varphi}(r,k) = A(\varphi^{B(r,k)}(f)) \tag{3}$$

The size-intensity diagram has granulometrical properties because it satisfies Matheron axioms extended to two parameters. For the opening/closing of $r = 0$, the column $SI_f(0,k)$ of the size-intensity diagram corresponds to the grey level distribution of the initial image. It can also be shown that each row (fixed k, varying r) of the size-intensity diagram gives the granulometry of the binary image thresholded at level k [6].

The different evolution of each granite texture by application of openings or closings of increasing size is clearly marked and results in distinct size-intensity diagrams. The evolution of the image textures by application of openings and closings is illustrated on the Intensity channel (I) of three granite types (ALM, EVO and FAV) in figure 2. The respective diagrams are plotted in figures 3 (for openings) and 4 (for closings) and show distinct behaviours.

The initial features selected were again reduced. For each opened or closed image of the Intensity channel, we considered only 8 octils (8 classes of equal frequency) and the respective minimum value. This way, it were considered 5 opened and 5 closed images per sample using structuring elements of sizes 3, 6, 9, 12 and 15. Thus the number of features in this case is 90: 2*5*(8+1).

To sum up, the total number of features per image to be used by the classifier is 117: 27 are related to colour and 90 are related to size.

In the following, from the features extracted in each granite image, one training and one testing sets were build, whose exact number of samples per type is presented in table 1. On the overall, the test set is composed by 50 random chosen and independent images representing the 14 different types of granites while the training set is constituted by the remaining 187 samples. This way, the training set is represented by a matrix of 187 lines (187 images representing 14 different types of granites) and 117 columns (117 features).

3. Classification with the k-Nearest Neighbour Rule

Nearest neighbour rule (NNR) methods are among the most popular for classification. They represent the earliest general (non-parametric) methods proposed for this problem and were deeply investigated in the fields of statistics and pattern recognition. Recently renewed interest on them emerged in the connectionist literature ("memory" methods) and also in machine learning ("instance-based" methods). Despite their basic simplicity and the fact that many more sophisticated alternative techniques have been developed since

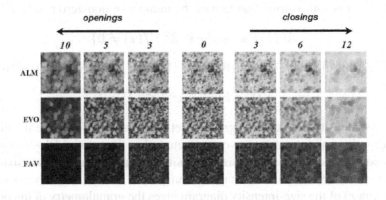

Figure 2.   Application of openings/closings of increasing size.

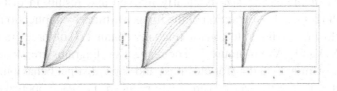

Figure 3.   Size-intensity by openings for granites ALM (left), EVO (centre) and FAV (right).

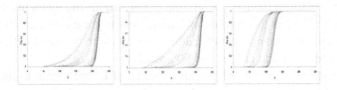

Figure 4.   Size-intensity by closings for granites ALM (left), EVO (centre) and FAV (right).

their introduction, nearest neighbour methods still remain among the most successful for many classification problems.

The k-NNR assigns an object of unknown class to the plurality class among the k labelled "training" objects that are closer to it. Closeness is usually defined in terms of a metric distance on the Euclidean space with the input measurement variables as axes. Nearest neighbour methods can easily be expressed mathematically. Let x be the feature vector for the unknown input, and let m_1, m_2, ..., m_c, be the templates (*i.e.*, perfect, noise-free feature vectors) for the c classes. Then the error in matching x against m_i, is given by the Euclidean norm $\| x - m_i \|$. A minimum-error classifier computes the norm for $i = 1, c$ and chooses the class for which this error is minimum. Since $\| x - m_i \|$ is also the distance from x to m_i, this a minimum-distance classifier. Naturally, the n dimension Euclidean distance d of x (n dimensional feature vector) to a training sample m, is:

$$d = \sqrt{\sum_{i=1}^{N}(x_i - m)^2} \qquad (4)$$

The previous approach may be extended to the k-nearest neighbour rule (k-NNR), where we examine the labels on the k-nearest samples in the input space and then classify them by using a voting scheme. Often in $c = 2$ class problems, k is chosen to be an odd number, to avoid ties. Other significant concerns and possible extensions include the use of a rejection option in instances where there is no clear winner, and the finite sample size performance of the NNR. Given a vector x and a training set H, whose cardinality may be large, the computational expense of finding the nearest neighbour of x may be significant. For this reason, frequent attention has been given to efficient algorithms. The computational savings are typically achieved by a pre-ordering of samples in H, combined with efficient (often hierarchical) search strategies (for extended analysis see [17]).

In order to compare results of our case study, 1-NNR and 3-NNR were implemented using the constructed training set for granite classification (*i.e.*, with all 117 features). The best results in terms of successful recognition were of 98% and gives respect to 1-NNR. In global terms, recognition errors occur between VIM and SAL samples.

4.   Feature space reduction via genetic algorithms

In order to reduce the input feature space, and hypothetically improve the recognition rate, Genetic Algorithms (GA) were implemented. The idea is to reduce the number of features necessary to obtain at least the same recognition rates. GA [5, 8] are search procedures based on the mechanics of natural selection and natural genetics. The GA were developed by J. Holland in the 1960's to allow computers to evolve solutions to difficult search and combinatorial problems, such as function optimisation and machine learning. The basic operation of a GA is conceptually simple (canonical GA): (i) To maintain a population of solutions to a problem: (ii) To select the better solutions for recombination with each other and (iii) To use their offspring to replace poorer solutions.

The combination of selection pressure and innovation (through crossover and mutation - genetic operators) generally leads to improved solutions, often the best found to date by any method [5, 8]. For further details on the genetic operators, GA codification and GA implementation, the reader should report to Sethi [13]. Usually, each individual (chromosome; pseudo-solution) in the population (say 50 individuals) is represented by a binary string of 0's and 1's, coding the problem which we aim to solve.

In our case study, the aim is to analyse the combinatorial feature space impact on the recognition rate of nearest neighbour classifiers. In other words, the GA will explore the $N < 117$ features of the training set and their combinations. An efficient genetic coding for a feature sub-space is then represented by each GA individual, *i.e.* an hypothetical classification solution, via a reduced group of features. The GA fitness function is then given by the 1-NNR classifier recognition rate and by the number of features used on that specific NNR. In that way, for each GA generation and for each individual (pseudo-solution) is then performed the nearest neighbour classification. The results are then used on the GA again. The algorithm proceeds their search until a stopping criterion is achieved. After convergence, the final solution points out which features among 117 can maximise the fitness function. The overall GA search space is given by the set of all combinations of 116 features in 117, plus the 115 features in 117, plus all combinations of 2 features in 117. For the i individual the fitness function can be expressed as:

$$fit(i) = \alpha hits(i) - \beta nf(i) \qquad (5)$$

where α and β are real valued constants ($\alpha + \beta = 1$), and $hits(i)$ represents the number of images well recognised among the testing set, and $nf(i)$ the number of features used in the NNR classification. The representation (GA coding) for each solution is then achieved by means of a 117 bit long binary string, *i.e.*, if the nth bit is 1, then the nth feature is used on the NNR classification; if not (= 0), that specific feature is not used on the classification.

The results obtained are excellent. In fact, not only the overall recognition rate was improved, but also the number of features necessary for increasing that value was substantially reduced. It was possible to improve the recognition rate by 2%, achieving this way the maximum of 100% (all the samples of all types of granites were correctly classified) and simultaneously reduce the number of important features from 117 to 3 (representing a reduction of 97%). The three features selected by the GA to reach the maximum rate correspond to feature #70 (6th octil of the opening of size 5), feature #101 (1st octil of the opening of size 9) and feature #112 (3rd octil of the opening of size 10). The corresponding pairs of scatterplots resuming the 3D mapping are presented in figure 5.

In addition, we have tested our approach by studying two types of non-Portuguese granites. The additional samples (4 of each type) were submitted to the same previous procedure and the same features were used to plot the samples in the same feature space. Like is graphically demonstrated (see figure 6) the location of all of those samples of non-Portuguese granites is far enough from any of the previous clusters which prevents any misclassification.

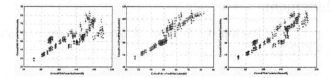

Figure 5. 2D scatterplots resuming the 3D mapping by pairs of features: #101 vs. #70 (left), #112 vs. #70 (centre) and #112 vs. #101 (right).

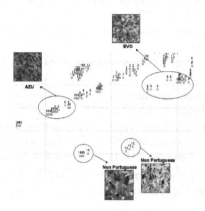

Figure 6. Zoom of feature space (features #70 vs. #101) with the location of the non-Portuguese granites in relation to the closest portuguese types.

5. Conclusions

Size-intensity diagram revealed to be appropriate to describe granite textures due its isotropic textural characteristics. On this context, this methodology points out the sub-features that are really important for successful discrimination. This way, it is possible to understand rigorously and to improve the knowledge on the proper nature of the structures we are working with.

Results show that this hybrid strategy is highly promising. Moreover, computation can still be reduced using new types of algorithms, strictly involved on the computation of nearest neighbours (as said before, typically achieved by a preordering of samples in the input space). However, this time is spent only once, *i.e.*, on the training phase. After that, its possible to classify images based on a reduced number of features, improving real time computer calculus on nearest neighbour relations, since the dimension of the feature space was significantly reduced.

Finally, the study of the visual aspect of the samples may also have a practical and direct industrial application for controlling counterfacting actions as it may be a useful tool to prevent products of different origins to be confounded.

Acknowledgments

This paper is developed in the frame of the project POC-TI/ECM/37998/2001.

References

[1] F.Z. Brill, D.E Brown, W.N. Martin. Fast genetic selection of features for neural network classifiers. *IEEE Transactions on Neural Networks*, 3(2):324–328, 1992.

[2] B.V. Dasarathy (ed.). *Nearest neighbour (NN) norms: NN pattern classification techniques.* IEEE Computer Society Press, 1973.

[3] E.R. Dougherty, R. Lotufo. *Hands-on morphological image processing.* SPIE Press, Bellingham, Washington, 2003.

[4] R.O. Duda, P.E. Hart. *Pattern classification and scene analysis.* J. Wiley & Sons, New York, 1973.

[5] D.E. Goldberg. *Genetic algorithms in search, optimization and machine learning.* Addison-Wesley, Reading, Massachusetts, 1989.

[6] R.A. Lotufo, E. Trettel. Integrating size information into intensity histogram. P. Maragos, R.W. Schafer, M.A. Butt, editors, *Mathematical Morphology and its Applications to Image and Signal Processing*, pages 281–288, Boston, 1996. Kluwer Academic Publishers.

[7] G. Matheron. *Random sets and integral geometry.* J. Wiley & Sons, New York, 1975.

[8] Z. Michalewicz. *Genetic algorithms + data structures = evolution programs.* 3rd edition Springer, Berlin, 1998.

[9] P. Pina, T. Barata. Petrographic classification at the macroscopic scale using a mathematical morphology based methodology. In F.J. Perales, A. Campilho, N. Pérez de la Blanca and A. Sanfeliu, editors, *LNCS 2652*, pages 758–765, Berlin, 2003. Springer.

[10] V. Ramos. Evolution and cognition in image analysis, (in portuguese). MSc thesis, Technical University of Lisboa, 1997.

[11] V. Ramos, F. Muge, P. Pina. Self-organized data and image retrieval as a consequence of inter-dynamic synergistic relationships in artificial ant colonies. J. Ruiz-del-Solar, A. Abraham, M. Köppen, editors, *Soft Computing Systems - Design, Management and Applications*, pages 500–509, Amsterdam, 2002. IOS Press.

[12] J. Serra. *Image analysis and mathematical morphology.* Academic Press, London, 1982.

[13] I.K. Sethi. A fast algorithm for recognizing nearest neighbors. *IEEE Transactions on Systems, Man and Cybernetics*, 11(3):245–248, 1981.

[14] W. Siedlecki, J. Sklansky. On automatic feature selection. *Int. Journal of Pattern Recognition and Artificial Intelligence*, 2(2):197–200, 1988.

[15] W. Siedlecki, J. Sklansky. A note on genetic algorithms for large-scale feature-selection. *Pattern Recognition Letters*, 10(5):335–347, 1989.

[16] P. Soille. *Morphological image analysis.* 2nd edition, Springer, Berlin, 2003.

[17] H. Yan. Building a robust nearest neighbour classifier containing only a small number of prototypes. *Int. J. Neural Systems*, 3:361–369, 1992.

[18] H. Yan. Prototype optimization of nearest neighbor classifiers using a 2-layer perceptron. *Pattern Recognition*, 26(2):317–324, 1993.

MORPHOLOGICAL SEGMENTATION APPLIED TO 3D SEISMIC DATA

Timothée Faucon[1,2], Etienne Decencière[1] and Cédric Magneron[2]

[1]*Centre de Morphologie Mathématique, Ecole des Mines de Paris*
35, rue Saint Honoré, 77305 Fontainebleau, France

[2]*Earth Resource Management Services*
16, rue du Chateau, 77300 Fontainebleau, France
timothee.faucon@cmm.ensmp.fr

Abstract   Mathematical morphology tools have already been applied to a large range of application domains: from 2d grey-level image processing to colour movies and 3D medical image processing. However, they seem to have been seldom used to process 3D seismic images. The specific layer structure of these data makes them very interesting to study. This paper presents the first results we have obtained by carrying out two kinds of hierarchal segmentation tests of 3D seismic data. First, we have performed a marker based segmentation of a seismic amplitude cube constrained by a picked surface called seismic horizon. The second test has consisted in applying a hierarchical segmentation to the same seismic amplitude cube, but this time with no *a priori* information about the image structure.

Keywords: 3D seismic data, mathematical morphology, hierarchical segmentation

Introduction

The application of image processing techniques to seismic 3D data has always been impaired by the huge volume of these data. However, recent algorithms and hardware performance improvements can partly solve this problem and seismic imaging is a domain that is expanding at present.

To our knowledge, morphological tools have been very rarely applied to 3D seismic data. This paper describes the first promising results of the application of morphological segmentation to seismic data.

The first section of the paper is devoted to the presentation of some generic information about seismic acquisition and processing techniques. In the second

475

C. Ronse et al. (eds.), Mathematical Morphology: 40 Years On, 475–484.
©2005 *Springer. Printed in the Netherlands.*

section, we briefly describe the morphological tools we have used to achieve our tests. The applications of these tools to seismic data and their results are also presented in this section.

1.   Seismic imaging

Seismic reflection

Seismic images result from acoustic propagation techniques. A surface source generates an acoustic signal into the subsurface of the earth. The signal propagates in the subsurface and is partly reflected by interfaces between rock layers of different acoustic impedance. The intensity of the reflected signal is proportional to the intensity of the impedance contrast. The residual part of the signal that is not reflected is transmitted and continues to propagate in the subsurface until it is reflected and transmitted by a new impedance contrast interface, and so on. The reflected part of the signal is recorded by several sensors located on the surface. This recorded information is processed, leading to an intensity - called amplitude - image of the subsurface. An interesting "echography" is then available thanks to this acquisition and processing technique. More information on seismic reflection techniques can be found in [8].

For years seismic data processing has been connected with image and signal processing. Among the techniques applied to seismic images, we find, for example, region growing and wavelet analysis [2, 7] which have been tested only on 2d data sets. On the contrary, mathematical morphology seems to have been rarely used to process seismic images.

3D seismic images

Common seismic 3D images are 16 bits amplitude images. They often contain several hundred millions of samples. A seismic image is defined by three main axes. The two horizontal ones are called inline and crossline. They correspond respectively to the acquisition direction and to the direction perpendicular to the acquisition direction. The vertical axis represents time (more or less equivalent to the recorded time of the reflected signal) or depth (after time to depth conversion).

As these images typically correspond to sedimentary underground, they are composed by a sequence of high intensity and low intensity surfaces roughly horizontal, called seismic horizons, which can be cut by faults (see *figure1(a)*).

Seismic Horizon

For the first test, we have chosen to work more specifically on a seismic interpretation context: horizon picking. Considering seismic amplitude images as grey level images, horizon picking consists in extracting continuous extreme grey surfaces corresponding to impedance contrast interfaces. These continuous surfaces are called seismic horizons.

Data

The data we have used for the tests are on the one hand a 500 megabytes seismic amplitude cube (16 bits coding) *(figure 1(a))*, on the other hand a seismic horizon *(figure 1(b))*. We will see in the next section how it has been generated. We must add that, as the seismic amplitude cube is very noisy, we expected some doubtful results from the first segmentation test.

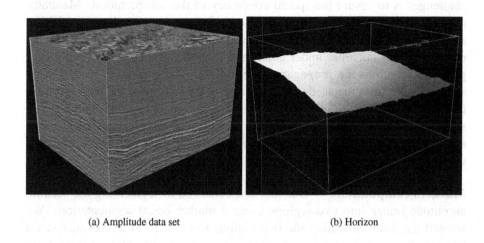

(a) Amplitude data set   (b) Horizon

Figure 1.   Data

2.   Segmentation with markers applied to seismic images

Markers based segmentation

Normally, the flooding used to produce a watershed starts from the image minima. The use of markers allows to impose the flooding sources. The resulting watershed separates the markers. This segmentation method was introduced by Beucher [1].

Tests and results

This first test consists in segmenting the seismic data cube into two regions using the seismic horizon segmented by an expert, called the reference horizon. We compare the limit separating the two regions of the segmentation with the reference horizon.

Justification. The watershed algorithm can provide the brighter and more continuous surfaces of an image. As previously mentioned, energy and continuity are the criteria chosen by interpreters to pick horizons from a seismic amplitude image. The objective of this test is to compare the precision of the horizon picked by a specialist with the result of the watershed. It justifies the use of the watershed algorithm for a horizon picking context.

Horizon picking technique. Main horizon picking techniques are semi-automatic. Interpreters control their interpretation by picking by hand some points of a targeted horizon. Being performed in 3D, one of the interpreter challenges is to ensure the spatial coherency of the interpretation. Manually picked points are considered as input seeds for an automatic horizon picking algorithm that fills the areas between points according to energetic and continuity criteria, sometimes under geometrical constraints. Generally, automatic picking algorithms are propagation algorithms. The horizon is built step by step, growing from the seeds, by searching in the image for the path of maximum energy and continuity - more or less the best spatial correlation way. The seismic horizon we used for our test is the result of the application of such a semi-automatic picking algorithm applied to our seismic amplitude cube. We call it the reference horizon.

Markers computation. Our first test consisted in segmenting the seismic amplitude image into two regions using a marker based segmentation. We wanted the limit separating the two regions to correspond to the interpreted horizon. For that purpose, we needed two markers. We obtained them from the reference horizon. First, we dilated the reference horizon to define the work area. Then we inversed the image to create the complementary region of the work area, only the minima being labelled. *Figure 2* explains the different steps of this markers computation.

Results and comments. The marker based segmentation is performed using the watershed algorithm. Results are very encouraging. In *figure 3* we compare the reference horizon and what we call the watershed horizon. The coordinates of the pictures are the inline and the crossline and the grey level gives the altitude of the horizon at the considered point. A first quick look does not reveal major differences between the reference horizon and the watershed horizon (*figure 3(a) and (b)*). But the difference map (*figure 3(c)*) is very instructive. It shows two main areas. The first one is a triangle observed in the northern part of the map. The second one is located in the south-western part of the map with differences higher than 20 mstwt (milliseconds two ways time). Several other small size differences are observed over the map.

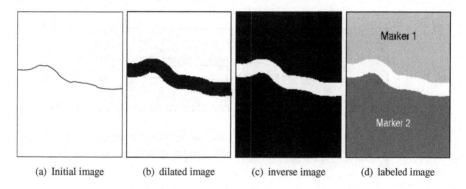

(a) Initial image   (b) dilated image   (c) inverse image   (d) labeled image

Figure 2.   Marker obtention (illustrated in 2d; performed in 3D)

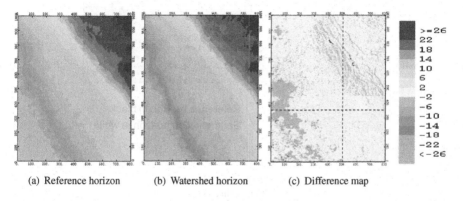

(a) Reference horizon   (b) Watershed horizon   (c) Difference map

Figure 3.   Horizons comparison

In *figure 4(a)* a section (inline 500) from the triangle region is shown. The watershed horizon appears in white and the reference horizon in black. It shows that the picking strategy is different in the triangle area from the rest of the map. The reference picking is focused on the maximum of amplitude everywhere except in the triangle area where it seems to be 4 mstwt above that maximum.

Figure 4(b) shows a section (crossline 360) which crosses the main horizon difference area in the south-western part of the field. It is obvious that these differences are due to a phase shift of the reference horizon. While the watershed horizon follows the continuous maximum of the amplitudes, the reference horizon jumps to an above maximum phase.

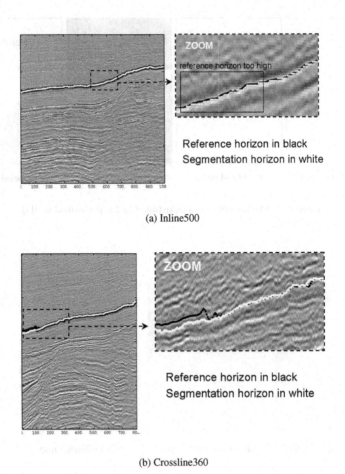

(a) Inline500

(b) Crossline360

Figure 4.   Example of horizon differences

Given the vertical resolution of the seismic amplitudes, a small horizon pick-ing difference can have a non negligible impact on the extracted amplitudes map. The extracted amplitude maps corresponding to the two horizons seem to be the same from a simple eye analysis. The difference map (*figure 5(b)*) is more instructive. It reveals high amplitudes differences in the triangle area. The normalized map (*figure 5(c)*), corresponding to the ratio difference abso-lute value / reference amplitude absolute value, shows mean differences higher than 20% and locally higher than 50%. Amplitude differences in the south-western zone are not relevant as they do not correspond to the same seismic event for the reference horizon and the watershed horizon.

This first test has allowed to highlight some interesting differences between a picked reference horizon and the segmentation horizon associated with it.

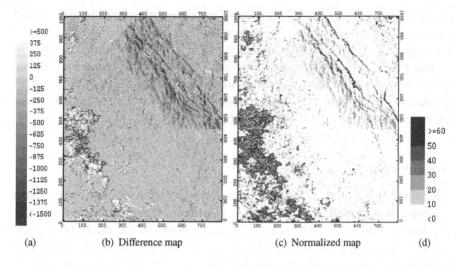

(a)   (b) Difference map   (c) Normalized map   (d)

Figure 5. Analysis

Phase jump and small deviation of the reference horizon are observed. The segmentation approach seems to be very powerful to localise and quantify these kind of anomalies.

3. Hierarchical segmentation applied to seismic images

For the second test we have applied the hierarchical segmentation using the watershed algorithm to the seismic amplitude cube. And we have played with the different levels of the hierarchy.

Hierarchical segmentation enables the control of the segmentation process. Regions defined by the hierarchical segmentation are classified according to a given criterion. The number of displayed regions is freely chosen. A series of tests has shown that the volume of catchment basins criterion gives the best results, therefore we have adopted it.

Morphological hierarchical segmentation

The hierarchical segmentation we have adopted was proposed in [3], [6], [9]. Its principle is to provide, in addition to the segmentation itself, a hierarchy between the catchment basins. To perform a hierarchical segmentation, we start with the minima of our grey-level image. These minima are considered as the flooding sources of the image. Each minimum represents one leaf of a tree and all minima are labelled with a different value. Then the image is flooded from the minima and each time two basins merge, the corresponding nodes are linked by a valued edge. The value of the edge corresponds to the merging criterion chosen (the smallest depth, surface or volume of the merging basins, for example). At the end of the flooding, we have a complete tree and we can choose the level of segmentation we want (the number of regions we search for). The resulting image comes from the progressive cutting out of the edges of the tree which have the highest value. It results in a number of subtrees equivalent to the aimed number of regions.

Application

Global segmentation .   We have applied the hierarchical segmentation using the watershed algorithm to the seismic amplitudes image with several hierarchy levels. With a low hierarchy level, the majority of the bright horizons is detected. With a higher hierarchy level, only the brightest horizons appear (see *figure 6(a)*). We can notice that the interface between layers is sometimes not continuous. This is due to the presence of "leaks" during the segmentation flooding step. Most of the time, these "leaks" are due to the presence of faults which introduce discontinuity into seismic layers. Then, the catchment basins related to the layers merge at a lower level in the hierarchy. The watershed between the two areas appears only at a lower hierarchy level. However these results may be difficult to interpret because no operational context has been defined for that test.

Cylinder segmentation .   To avoid "leak" problems during the watershed flooding in a horizon extraction context in a seismic cube image, we have decided to divide the initial image into vertical square based cylinders and segment each cylinder separately. Since these cylinders are perpendicular to the seismic layers, most of these layers are continuous across each cylinder except in the ones containing "leaks". The watershed flooding is then not problematic anymore in the "leak" free cylinders and the layers are well segmented. As a consequence, the resulting well defined horizons are well detected and the problem due to the "leaks" are restricted to certain cylinders. When all the cylinders are segmented, we obtain many horizon pieces which are not necessarily connected. It remains to connect them.

The natural solution to connect the cylinders is the use of overlapping cylinders: instead of computing a watershed on adjacent cylinders, we compute it on cylinders shifted at each step of less than one cylinder size. The resulting horizon pieces increment an accumulator. At the end, a simple threshold on this accumulator gives the connected horizons. An example of cylinder overlapping result is presented in *figure 6(b)*.

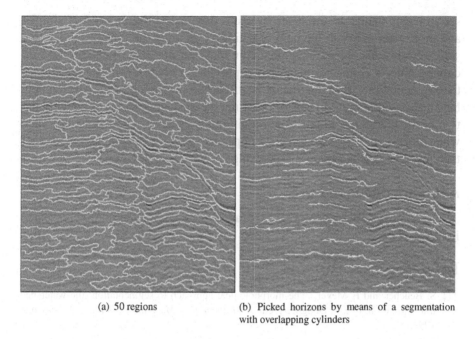

(a) 50 regions   (b) Picked horizons by means of a segmentation with overlapping cylinders

Figure 6. Segmentation with cylinders (image size 400 * 500)

Algorithm Performance

The algorithm we use for the watershed is based on hierarchical queues [4]. This fast algorithm makes the segmentation of several hundred megabytes seismic cubes operational. The hierarchy is described with a minimum spanning tree [5], which allows real time interaction with the hierarchy. The cylinder strategy improves the computation times and reduces memory management difficulties induced by one block seismic cube processing. We give two examples of hierarchical segmentation computation time of a 500 megabytes seismic cube (coding 16 bits) on a 3GHz PC :

1 divided in 4 cylinders : 45 minutes;

2 divided in 12 cylinders with an overlapping rate of half cylinder size : 1 hour.

4.   Conclusion

The application of mathematical morphology tools to seismic images leads to very encouraging results. These results show that morphological tools are relevant for seismic image processing. Segmentation appears to work well on such images thanks to the layer structure that makes the processing simpler and more effective with the help of cylinders dividing the image. On another side, not only hierarchical segmentation makes it possible to extract the well defined objects from the image but it provides a control on the precision of the segmentation. This interaction is still at the beginning of its development but further work should lead to a real interactivity concerning the manipulation of the hierarchy. Finally, the running time criterion – which is important when processing large size images – appears to be good in comparison to other techniques of seismic images processing.

Acknowledgements

The authors wish to thank Gaz De France for providing the images presented in this paper.

References

[1] S. Beucher and F. Meyer. The morphological approach to segmentation: the watershed transformation. In E.R. Dougherty, editor, *Mathematical Morphology in Image Processing*. 1993.

[2] I. Bournay Bouchereau. *Analyse d'images par transformée en ondelettes; application aux images sismiques*. PhD thesis, Université Joseph Fourier Grenoble, 1997.

[3] M. Grimaud. New measure of contrast : dynamics. *Image Algebra and Morphological Processing III, San Diego CA, Proc. SPIE*, 1992.

[4] F. Meyer. Un algorithme optimal pour la ligne de partage des eaux. In *8ème congrès de reconnaissance des formes et intelligence artificielle*, volume 2, pages 847–857, Lyon, France, November 1991.

[5] F. Meyer. Minimal spanning forests for morphological segmentation. In J. Serra and P. Soille, editors, *Mathematical Morphology and its applications to signal processing (Proceedings ISMM'94)*, Fontainebleau, France, September 1994. Kluwer Academic Publishers.

[6] L. Najman and M. Schmitt. Geodesic saliency of watershed contours and hierarchical segmentation. *IEEE Transactions on Pattern Analysis and Machine Intelligence*, pages 18 (12) : 1163–1173, December 1996.

[7] M.Q. N'Guyen. *Analyse multi-dimensionnelle et analyse par les ondelettes des signaux sismiques*. PhD thesis, institut National Polytechnique de Grenoble, 2000.

[8] O.Yilmaz. *Investigation in Geophysics*, volume vol. 2. Society of Exploration Geophysicist, 1994.

[9] C. Vachier and F. Meyer. Extinction values: A new measurement of persistence. *IEEE Workshop on Non Linear Signal/Image Processing*, pages 254–257, June 1995.

Index

Author Index